Life in the
World's Oceans

Alasdair D. McIntyre

The many Census scientists who contributed to this volume dedicate it to the memory of Alasdair McIntyre, who passed away near its completion. Some of us knew him for many years, while others for only a few, but we all greatly appreciated his wisdom, wit, and perpetual curiosity. His contributions to marine science have been many and he was one of the first to propose an international study of marine biodiversity. Alasdair will be greatly missed.

Life in the World's Oceans

Diversity, Distribution, and Abundance

Edited by

Alasdair D. McIntyre
The University of Aberdeen, Scotland, UK

WILEY-BLACKWELL

A John Wiley & Sons, Ltd., Publication

CENSUS
OF MARINE LIFE

Library of Congress Cataloging-in-Publication Data
Life in the world's oceans : diversity, distribution, and abundance / edited by Alasdair
D. McIntyre.
      p. cm.
   Includes bibliographical references and index.
   ISBN 978-1-4051-9297-2 (hardback : alk. paper)   1. Marine animals.   2. Marine
ecology.   3. Biodiversity.   4. Census of Marine Life (Program)   5. Zoological surveys.
6. Oceanography.   I. McIntyre, A. D.
   QH541.5.S3L54 2010
   577.7–dc22

                                                                            2010005814

A catalogue record for this book is available from the British Library.

Set in 9.5 on 12pt Classical Garamond BT by Toppan Best-set Premedia Limited
Printed and bound in Singapore by Fabulous Printers Pte Ltd
1   2010

Contents

CONTENTS

Foreword

FOREWORD

The Census of Marine Life is about the total richness of the sea:

The Census of Marine Life is the book of oceans' nature. This book reports total richness.

It reports richness of diversity, the richness of what.
It is about kinorhynchs, tardigrades, rotifers, gastrotrichs, and tantulocarids housed in Arctic polynyas.
It is about Antarctic actiniarians, pycnogonids, tunicates, and holothurians.
It is about the golden V kelp in the Aleutian Islands.
It is about polychaetes, bivalves, and isopods of the continental margins.
It is about sturgeon and salmon, sea turtles and pinnipeds, otters and sirenia.
It is about filter feeders.
It is about radiolaria and hydrozoa.
It is about lanternfishes and pearlfishes and roundnose grenadiers.
It is about a black, benthopelagic lobate ctenophore and a large pelagic worm with ten long cephalic tentacles.
It is about 10,000 crabs.
It is about 5,000 to 19,000 unique types of bacteria in each gram of sand.
It is about Upper Turonian diatoms.

This is a book about vastness and deepness.
It reports richness of distributions, the richness of where.
It is about the Western and Eastern Pacific, and about South American seas.
It is about Caribbean, European, and Polar seas and Indian and Atlantic oceans.
It is about the abyssal plains and basins beneath half of Earth's surface.
It is about the Porcupine Abyssal Plain.
It is about the canyons of the margins.
It is about large shallow banks and gravelly shorelines.
It is about Lizard Island and Ningaloo Reef.
It is about the architecture of seamounts.

This is a book of journeys.
It is about leatherback turtles tagged on their nesting beaches in Indonesia crossing Pacific longitudes to feed off central California.

It is about sooty shearwaters flying Pacific latitudes from New Zealand to the Bering Sea.
It is about water columns 5,000 meters high.
It is about the planet's busiest commute, the nightly rise of life from hundreds of meters deep to feed nearer the surface in the safety of darkness.
It is about circumpolar currents.

This is a book that explores abundance, the richness of how much.
It is about the northerly flowing Kuroshio Current along the southern Japanese coast characterized by high biodiversity but low biomass.
It is about shimmering shoals of herring swirling in numbers beyond counting.

This is a book about past richness.
It is about Greek merchants trading fish from the Black Sea and the Russian rivers to the Greek and later the Roman market.
It is about the decline of marbled rock cod and mackerel ice fish west of the Antarctic Peninsula.
This is a book of lost reefs.

This is a book about life and death.
It is about juvenile salmon and adult sturgeon.
It is about immature specimens carrying sperm packages.
It is about the loneliness of reproductive isolation.
It is about mass mortality.
It is about small dead coral heads.
It is about prey fields patrolled by marine hunters.
It is about fidelity to birthplace.

This is a book of paradoxes, where extreme is normal and rare is common.

This is a book of contrasts.
It is about the cosmopolitan and the local.
It is about glaciation and boiling seafloor geysers where metal would melt yet animals live.
It is about ancient assemblages and modern benthos.
It is about swimmers and drifters and sitters.

This is a book of mysteries.
It is about oceanic barriers to gene flow.

It is about trophic subsidies to carnivores.

It is about the immense volume of ocean still unexplored.

It is about 20 million marine microbes that might remain to be described.

It is about cryptic species.

It is a book of powerful prostheses.

It is a book of ships and sledges and gliders and pyrotags.

It is a book of attached identity cards and different mesh sizes.

It is a book that filters a million cubic meters of seawater.

It is a book of blue-water divers.

It is a book where yellow dots are actual observations of lionfish.

This book reports the known, unknown, and unknowable of the first Census of Marine Life.

This book is about the richness of 3.5 billion years.

Jesse H. Ausubel
Alfred P. Sloan Foundation

Introduction

Reflecting upon the successes of the Census of Marine Life over the past decade, I am recalling how my contemporaries and I first became interested in marine biodiversity. As an undergraduate, I had a small National Science Foundation grant to study the macrofaunal invertebrates in samples taken to examine sediment transport processes either side of a sand–mud transition at the edge of the Labrador Current off what is now the Cape Cod National Seashore. The leader of the expedition, John Zeigler, wanted to know how wave activity controlled sediment transport and the location of the transition from sandy to muddy sediments on the seafloor, and Howard Sanders wanted to know what lived either side of this transition. Except for the specimens from Vineyard Sound described by the great naturalist A.E. Verrill, most of the organisms belonged to undescribed species. Subsequently, in graduate school at Duke University, this experience led me to study bottom life either side of another sand–mud transition on the continental slope off North Carolina and, subsequently, to a career identifying and describing the diversity of life on continental shelves and coral reefs and in the deep sea.

In the 1980s, concern about loss of species diversity in all environments greatly increased with the realization that diversity of life in rain forests and coral reefs must be protected and studied. E.O. Wilson and Peter Raven were and still are articulate advocates for maintaining the diversity of life on the planet. In parallel with the efforts to protect terrestrial biodiversity, marine scientists met under the auspices of the International Association for Biological Oceanography and the UNESCO Working Group on High Diversity Marine Ecosystems. Participants at the meetings included leading marine biologists, such as Bruno Battaglia (Italy), Pierre Lasserre (France), Ramon Margalef (Spain), Alasdair McIntyre (UK), Tim Parsons (Canada), and Howard Sanders (USA). In 1990 Pierre Lasserre, Alasdair McIntyre, Carleton Ray (USA), and I wrote "A Proposal for an International Programme of Research: Marine Biodiversity and Ecosystem Function" (Grassle et al. 1991). When Diversitas, the international program of biodiversity science, was established in 1991, this marine program was incorporated (Ray & Grassle 1991). US support for marine biodiversity began with the establishment of the National Research Council's Committee on Biological Diversity in Marine Systems, chaired by Cheryl Ann Butman (now Zimmer) and James Carlton. Following a workshop attended by 54 leading US marine scientists, Butman and Carlton wrote *Understanding Marine Biodiversity: A Research Agenda for the Nation* (Committee on Biological Diversity in Marine Systems 1995), one of the most widely read reports published by the National Academy Press. However, this did not lead immediately to a program of research. Colleagues in Woods Hole urged me to talk with Jesse Ausubel at the Rockefeller University, who, unknown to me at the time, was also a program manager at the Alfred P. Sloan Foundation. Starting with initial discussions of the feasibility of a Census of the Fishes, Jesse built strong support within the marine biology community for a broader research approach, and I became the first chair of the Census of Marine Life Steering Committee.

With strong initial commitment, and sustaining support from the Alfred P. Sloan Foundation, this initiative will achieve its goal of a comprehensive Census of Marine Life by the end of 2010. Research on species diversity started with Evelyn Hutchinson's question in 1959: "Why are there so many kinds of animals?" (Hutchinson 1959). Hutchinson estimated there might be about 1 million species globally and, of these, three-quarters were insects. At that time, ocean life was very poorly known and only a very small proportion of species were thought to live in the ocean. Now, with nearly a decade of support from the Census, a rich diversity of previously unknown marine species has been discovered and previously unknown habitats are being described. The deep-sea floor is no longer considered a desert, characterized by a paltry diversity of species.

Marine scientists are at present unable to provide good estimates of the total number of species in any of the three domains of life that flourish in the ocean (Archaea, Bacteria, and Eukarya). From their molecular signatures, animals are in a relatively well-defined supergroup that also includes the fungi. There are at least five and probably more supergroups. If we consider only the kingdom Animalia, the number of species may be knowable, but it will probably take at least another decade of the Census before we can defensibly estimate the total number of marine species.

Before the Census, most marine biologists studying life in the ocean worked chiefly in shallow water or on continental shelves, where the prime scientific interest was food chains leading to harvestable populations of fish or shellfish. To achieve an estimate of the diversity, distribution, and abundance of life in the ocean in 10 years, the Census has endeavored to sample the full range of marine taxa from pole to pole and surface to abyssal depths. The products of the Census in 2010 constitute a quantum leap toward a full assessment of life in the oceans, and others are already planning for the second Census in the next decade. It is hard to recall how little we knew just 10 years ago and to predict how much we will learn in the next decade.

To provide the context for studying present-day life in the ocean, the History of Marine Animal Populations (HMAP) project of the Census has asked what lived in the ocean? Changes in the abundance and size of harvested marine populations are being documented from many sources, including records of fish catches, sales and shipping records, writings, photographs, and even restaurant menus. The Future of Marine Animal Populations (FMAP) project recognizes that predictions about future marine life depend on knowing what is being lost from unprotected marine habitats and the rates of recovery following their greater protection.

To learn about the many species in the present ocean, the Census drew together 14 teams of scientists specializing in diverse geographic environments or subject areas:

Coastal areas: Natural Geography in Shore Areas (NaGISA), Census of Coral Reef Ecosystems (CReefs), Gulf of Maine Area (GoMA), and Pacific Ocean Shelf Tracking (POST).

Deep-sea floor: Continental Margin Ecosystems on a Worldwide Scale (COMARGE) and Census of Diversity of Abyssal Marine Life (CeDAMar).

Central waters: Census of Marine Zooplankton (CMarZ) and Tagging of Pacific Predators (TOPP).

Deep-sea floor and Central waters: Patterns and Processes of the Ecosystems of the Northern Mid-Atlantic (MAR-ECO), Global Census of Marine Life on Seamounts (CenSeam), and Biogeography of Deep-Water Chemosynthetic Ecosystems (ChEss).

Polar regions: Arctic Ocean Diversity (ArcOD) and Census of Antarctic Marine Life (CAML).

Microbial life: International Census of Marine Microbes (ICoMM).

The Census has exceeded expectations and lived up to the goals set in the 1995 report, *Understanding Marine Biodiversity: A Research Agenda for the Nation*. The Census has enlisted oceanographers, ecologists, statisticians, and marine biologists to conduct research on a global array of topics:

- Ocean-scale distribution and abundance of marine species using the latest in oceanographic technologies and taxonomic expertise.
- Causes and consequences of changes in marine biological diversity.
- Tracks of individual marine species in estuaries, coastwide, and oceanwide settings.
- Effects of human activities on life in the ocean.
- Previously intractable, oceanwide biodiversity patterns, processes, and consequences.
- Predictions regarding future effects of human activities on marine biodiversity to facilitate use of the sea for societal needs while minimizing impacts on nature.
- Development of partnerships between ecology and taxonomy.
- Reinvigoration of the field of marine taxonomy and systematics, developing the Ocean Biogeographic Information System (www.iobis.org), and collaborating on the World Register of Marine Species (www.marinespecies.org).

The Census has discovered many new species and previously unknown habitats, especially in the deep sea and on coral reefs. Many of the species are rare and most are represented only by single individuals in samples. The new Census datasets are, or will soon be, maintained in the Ocean Biogeographic Information System (OBIS). At the time of this writing, OBIS contains more than 22 million distribution records representing more than 100,000 species.

The Census brings together many things I have wanted to see happen for marine biodiversity throughout my career in marine science, involving more focused scientific effort and better communication to the world of why marine biodiversity matters. This book represents the distillation of the labors of many people who have fostered my original idea to put marine biodiversity in the foreground of the scientific landscape. The Census has been developed and nurtured by fellow architects Jesse Ausubel and the other inspired members of the Scientific Steering Committee now chaired by Ian Poiner (see list on page xix), who guided the project leaders and were the builders who so aptly constructed all of the projects, through the many skilled scientific workers who actually did all of the complex work. The International Secretariat at the Consortium for Ocean Leadership, including senior scientists Ron O'Dor and Patricia Miloslavich and program managers Cynthia Decker, in the early years, and Kristen Yarincik, have been instrumental for the Census. We have had spectacular success with public outreach through the Education and Outreach Team, led by Sara Hickox, and the many contributions of Darlene Trew Crist are especially noted. The Mapping and Visualization Team, led by Pat Halpin, has created a wide range of wonderful and insightful illustrations for many Census projects. And the Synthesis Group, led by Paul Snelgrove and managed by Michele DuRand, has helped to bring together the many activities in the Census so that the sum is even greater than its many wonderful parts. My longtime friend and colleague Alasdair McIntyre, who passed away as this book was nearing completion, worked with the project leaders and authors of these chapters to distill the efforts of thousands of superb scientists into a single volume that will provide an excellent resource for scientists interested in marine biodiversity.

This volume is one of a suite of products of the first decadal Census of Marine Life. Census researchers have documented their new vision of life in the ocean in more than 2,500 scientific papers and about 30 books so far. There are many products for a variety of audiences. For those interested in what was learned about life in the global ocean from a national and regional perspective, I direct you to the online *PLoS ONE* collection of papers "Marine Biodiversity and Biogeography – Regional Comparisons of Global Issues". A book by my former student Paul Snelgrove, *Discoveries of the Census of Marine Life: Making Ocean Life Count*, details the Census findings and explains the implications of what has been learned for both a scientific and interested public audience. A richly illustrated narrative, *World Ocean Census*, written by Census colleagues Darlene Trew Crist, Gail Scowcroft, and James M. Harding, Jr, introduces the work of the Census to the public. Lastly, there is a delightful photographic guide to marine life written by Census colleague Nancy Knowlton, *Citizens of the Sea: Wondrous Creatures from the Census of Marine Life*. Separately and collectively, these documents serve as a tribute to the hard work, dedication, and true scientific achievements of the more than 2,600 scientists from more than 80 nations who accomplished this novel and important scientific endeavor known as the Census of Marine Life.

Fred Grassle
April 2010

References

Committee on Biological Diversity in Marine Systems (1995) *Understanding Marine Biodiversity: A Research Agenda for the Nation*. Washington, DC: National Academy Press. 114 pp.

Grassle, J.F., Laserre, P., McIntyre, A.D. & Ray, G.C. (1991) Marine Biodiversity and Ecosystem Function: A Proposal for an International Programme of Research. *Biology International*, Special Issue 23.

Hutchinson, G.E. (1959) Homage to Santa Rosalia, or why are there so many kinds of animals? *American Naturalist* 93, 145–159.

Ray, G.C. & Grassle, J.F. (1991) Marine biological diversity. *BioScience* 41, 453–457.

Acknowledgments

ACKNOWLEDGMENTS

Many individuals and organizations have contributed to the production of this book, more than could be fully named. First and foremost must be Jesse Ausubel and the Alfred P. Sloan Foundation, whose vision and support made the Census of Marine Life possible. The Census Steering Committee was chaired initially by Fred Grassle and later by Ian Poiner, and within the organization the International Secretariat at the Consortium for Ocean Leadership played a major coordinating role. The Synthesis Group, managed by Michele DuRand, was central to ensuring success of the book. The contribution of the Education and Outreach Team was much appreciated, while thanks are due to Jesse Cleary from the Mapping and Visualization Team for guaranteeing the quality of the illustrations, and to Catherine McIntyre for, among other things, copy-editing. The input of reviewers drawn from the Steering and Synthesis Groups was invaluable. Last but not least, the dedication of the authors and their colleagues involved in the Census projects is gratefully acknowledged.

Alasdair D. McIntyre

The Publishers would like to record our sincere thanks to Michele DuRand, who most kindly stepped in to finalize and carry through editorial aspects of the book in Alasdair McIntyre's stead.

Contributors

CONTRIBUTORS

Linda Amaral-Zettler
Marine Biological Laboratory, 7 MBL Street, Woods Hole, Massachusetts 02543, USA

Kelly Andrews
Northwest Fisheries Science Center, National Marine Fisheries Service, National Oceanic and Atmospheric Administration, 2725 Montlake Boulevard, Seattle, Washington 98112, USA

Luis Felipe Artigas
(1) Université Lille Nord de France, F-59000 Lille, France
(2) Université du Littoral, LOG, F-62930 Wimereux, France
(3) Unité Mixte de Recherche, CNRS 8187, F-62930 Wimereux, France

Maria C. Baker
School of Ocean and Earth Science, University of Southampton, National Oceanography Centre Southampton, Empress Dock, European Way, Southampton SO14 3ZH, UK

John Baross
University of Washington, 260 MSB, Box 357940, Seattle, Washington 98195, USA

Lisandro Benedetti-Cecchi
Department of Biology, University of Pisa, Via Derna 1, 1-56126 Pisa, Italy

Odd Aksel Bergstad
Institute of Marine Research, Flødevigen, NO-4817 His, Norway

Loka Bharathi P.A.
Microbiology Laboratory, National Institute of Oceanography, Dona Paula – 403 004, Goa, India

David S. M. Billett
National Oceanography Centre, Southampton, Empress Dock, European Way, Southampton SO14 3ZH, UK

Barbara A. Block
Stanford University, Department of Biology, Hopkins Marine Station, Pacific Grove, California 93950-3094, USA

Bodil A. Bluhm
School of Fisheries and Ocean Sciences, University of Alaska Fairbanks, Fairbanks, Alaska 99775-7220, USA

Antje Boetius
Max Planck Institute for Microbiology, Celsiusstrasse 1. D-28359 Bremen, Germany

Steven J. Bograd
Environmental Research Division, National Oceanic and Atmospheric Administration, Southwest Fisheries Science Center, 1352 Lighthouse Avenue, Pacific Grove, California 93950, USA

Russell E. Brainard
Coral Reef Ecosystem Division, Pacific Islands Fisheries Science Center, National Oceanic and Atmospheric Administration, 1601 Kapiolani Boulevard, Ste. 1110, Honolulu, Hawaii 96814, USA

Angelika Brandt
Zoologisches Museum und Biozentrum Grindel, Martin-Luther-King-Platz 3, 20641 Hamburg, Germany

Ann Bucklin
Department of Marine Sciences, University of Connecticut – Avery Point, 1080 Shennecossett Road, Groton, Connecticut 06340, USA

Ingvar Byrkjedal
University of Bergen, Bergen Museum, Department of Natural History, P.O. Box 7800, NO-5020 Bergen, Norway

M. Julian Caley
Australian Institute of Marine Science, PMB No. 3, Townsville, Queensland 4810, Australia

Robert S. Carney
Louisiana State University, Baton Rouge, Louisiana 70808-4600, USA

Dorairajasingam Chandramohan
19/3 3rd Street, Ratnapuri Colony, J.N. Salai, Koyambedu, Chennai-600107 (T.N.), India

Cedar Chittenden
Faculty of Biosciences, Fisheries and Economics, University of Tromsø, N-9037 Tromsø, Norway

Malcolm R. Clark
National Institute of Water and Atmospheric Research, P.B. 14-901, Wellington, New Zealand

Mireille Consalvey
National Institute of Water and Atmospheric Research, P.B. 14-901, Wellington, New Zealand

Nancy J. Copley
Biology Department, Woods Hole Oceanographic Institution, Woods Hole, Massachusetts 02543, USA

Erik E. Cordes
Department of Biology, Temple University BL248J, 1900 N 12th Street, Philadelphia, Pennsylvania 19122, USA

Daniel P. Costa
Department of Ecology and Evolutionary Biology, University of California, Santa Cruz, Long Marine Laboratory, 100 Shaffer Road, Santa Cruz, California 95060-5730, USA

Glenn Crossin
Centre for Applied Conservation Research, University of British Columbia, Vancouver, British Columbia, V6T 1Z4, Canada

Juan Jose Cruz-Motta
Departamento de Estudios Ambientales, Universidad Simon Bolivar, Sartenejas, Caracas 1080, Venezuela

Jan de Leeuw
The Royal Netherlands Institute for Sea Research, P.O. Box 59, 1790 AB Den Burg, Texel, The Netherlands

Nicole Dubilier
Max Planck Institute for Microbiology, Celsiusstrasse 1. D-28359 Bremen, Germany

Brigitte Ebbe
Forschungsinstitut Senckenberg, Deutsches Zentrum für Marine Biodiversitätsforschung, Sudstrand 44, D-26382 Wilhelmshaven, Germany

Kari Ellingsen
Norwegian Institute for Nature Research, Polar Environmental Centre, 9296 Tromsø, Norway

Sara L. Ellis
Aquatic Systems Group, University of Southern Maine, Portland, Maine 04101, USA

Tone Falkenhaug
Institute of Marine Research, Flødevigen, NO-4817 His, Norway

Charles R. Fisher
Pennsylvania State University, 208 Muller Laboratory, University Park, Pennsylvania 16802, USA

Rebecca Fisher
Australian Institute of Marine Science, The University of Western Australia Oceans Institute (M096) 35 Stirling Highway, Crawley, Western Australia 6009, Australia

Joëlle Galéron
Ifremer, BP 70, 29280 Plouzane, France

Andrey V. Gebruk
P.P. Shirshov Institute of Oceanology, Russian Academy of Sciences, Moscow 117997, Russia

Christopher R. German
Woods Hole Oceanographic Institution, Woods Hole, Massachusetts 02543, USA

Astthor Gislason
Marine Institute of Iceland, Skúlagötu 4, 121 Reykjavik, Iceland

Adrian Glover
Natural History Museum, Cromwell Road, London SW7 5BD, UK

Olav Rune Godø
Institute of Marine Research, P.O. Box 1870 Nordnes, NO-5817 Bergen, Norway

Fred Goetz
School of Aquatic and Fishery Sciences, University of Washington, Seattle, Washington 98195, USA

Rolf Gradinger
School of Fisheries and Ocean Sciences, University of Alaska Fairbanks, Fairbanks, Alaska 99775-7220, USA

J. Frederick (Fred) Grassle
Institute of Marine and Coastal Sciences, Rutgers University, 71 Dudley Road, New Brunswick, New Jersey, 08901-8521, USA

Julian Gutt
Alfred Wegener Institute, Columbusstrasse, D-27568 Bremerhaven, Germany

Mikko Heino
University of Bergen, Department of Biology, P.O. Box 7803, NO-5020 Bergen, Norway

Gerhard Herndl
The Royal Netherlands Institute for Sea Research, Den Burg, Texel, The Netherlands

Scott Hinch
Department of Forest Sciences, Centre for Applied Conservation Research, University of British Columbia, Vancouver, British Columbia, V6T 1Z4, Canada

Åge S. Høines
Institute of Marine Research, P.O. Box 1870 Nordnes, NO-5817 Bergen, Norway

Poul Holm
Trinity Long Room Hub, Trinity College Dublin, College Green, Dublin 2, Ireland

Russell R. Hopcroft
School of Fisheries and Ocean Sciences, University of Alaska Fairbanks, Fairbanks, Alaska 99775-7220, USA

Graham Hosie
Department of the Environment, Water, Heritage and the Arts, Australian Antarctic Division, Hobart, Australia

Katrin Iken
Institute of Marine Research, School of Fisheries and Ocean Science, University of Alaska Fairbanks, Fairbanks, Alaska 99775-7220, USA

Lewis S. Incze
Aquatic Systems Group, University of Southern Maine, Portland, Maine 04101, USA

Baban Ingole
National Institute of Oceanography, Dona Paula – 403 004, Goa, India

Tohru Iseto
Seto Marine Biological Laboratory, Field Station Education and Research Centre, Kyoto University, Shirahama-cho, Wakayama Prefecture 649-2211, Japan

Ian Jonsen
Biology Department, Dalhousie University, Halifax, Nova Scotia, B3H 4J1, Canada

Stefanie Keller
Forschungsinstitut Senckenberg, Deutsches Zentrum für Marine Biodiversitätsforschung, Sudstrand 44, D-26382 Wilhelmshaven, Germany

Edward Kimani
Kenya Marine and Fisheries Research Institute, P.O. Box 81651, Mombasa 80100, Kenya

Hiroshi Kitazato
Institute of Biogeosciences, Japan Agency for Marine-Earth Science and Technology, 2-15 Natsushimacho, Yokosuka 237-0061, Japan

Ann Knowlton
School of Fisheries and Ocean Sciences, University of Alaska Fairbanks, P.O. Box 757220, Fairbanks, Alaska 99775-7220, USA

Nancy Knowlton
(1) Center for Marine Biodiversity and Conservation, Scripps Institution of Oceanography, University of California San Diego, La Jolla, California 92093-0202, USA
(2) Department of Invertebrate Zoology, National Museum of Natural History, Smithsonian Institution, MRC 163, P.O. Box 37012, Washington, DC 20013-7012, USA

Kazuhiro Kogure
Ocean Research Institute, University of Tokyo, 1-15-1 Minamidai, Nakano, Tokyo 164, Japan

Brenda Konar
School of Fisheries and Ocean Sciences, University of Alaska Fairbanks, P.O. Box 757220, Fairbanks, Alaska 99775-7220, USA

Ksenia Kosobokova
P.P. Shirshov Institute of Oceanology, Russian Academy of Sciences, Moscow 117997, Russia

Elena Krylova
P.P. Shirshov Institute of Oceanology, Russian Academy of Sciences, 117997 Moscow, Russia

Helena P. Lavrado
Marine Biology Department, UFRJ Federal University of Rio de Janeiro, Brazil

Peter Lawton
Fisheries and Oceans Canada, St. Andrews Biological Station, St. Andrews, New Brunswick, E5B 2L9, Canada

Lisa A. Levin
Scripps Institution of Oceanography, Integrative Oceanography Division, 9500 Gilman Drive, La Jolla, California 92093-0218, USA

Phil Levin
Northwest Fisheries Science Center, National Marine Fisheries Service, National Oceanic and Atmospheric Administration, 2725 Montlake Boulevard, Seattle, Washington 98112, USA

Steve Lindley
Southwest Fisheries Science Center, National Marine Fisheries Service, National Oceanic and Atmospheric Administration, Santa Cruz, California 94930-1299, USA

Dhugal Lindsay
Japan Agency for Marine-Earth Science and Technology, 2-15 Natsushima-cho, Yokosuka City, 237-0021, Japan

Heike K. Lotze
Biology Department, Dalhousie University, Halifax, Nova Scotia, B3H 4J1, Canada

Ryuji J. Machida
Ocean Research Institute, University of Tokyo 164, Japan

Brian R. MacKenzie
National Institute for Aquatic Resources (DTU-Aqua), Technical University of Denmark, Kavalergaarden 6, DK-2920 Charlottenlund, Denmark

Marina Malyutina
A.V. Zhirmunsky Institute of Marine Biology, FEB Russian Academy of Sciences, Vladivostok, Russia

Anne Husum Marboe
Department of Environmental, Social and Spatial Change, Roskilde University, Building 3.2.2, DK-4000 Roskilde, Denmark

Pedro Martínez Arbizu
Forschungsinstitut Senckenberg, Deutsches Zentrum für Marine Biodiversitätsforschung, Sudstrand 44, D-26382 Wilhelmshaven, Germany

Scott McKinley
The University of British Columbia, West Vancouver Laboratory – Animal Science, 4160 Marine Drive, West Vancouver, British Columbia, V7V 1N6, Canada

Michael Melnychuk
Department of Zoology and Fisheries Centre, University of British Columbia, 2202 Main Hall, Vancouver, British Columbia, V6T 1Z4, Canada

Gui M. M. Menezes
Departamento de Oceanografia e Pescas, Universidade dos Açores, PT-9901-862, Horta, Portugal

Lenaick Menot
(1) DEEP/LEP, Ifremer, BP 70, 29280 Plouzane, France
(2) Institut Océanographique, Paris, France

Anna Metaxas
Department of Oceanography, Dalhousie University, Halifax, Nova Scotia, B3H 4J1, Canada

Patricia Miloslavich
Departamento de Estudios Ambientales, Universidad Simon Bolivar, Sartenejas, Caracas 1080, Venezuela

Megan Moews
UH Joint Institute for Marine and Atmospheric Research, Coral Reef Ecosystem Division, Pacific Islands Fisheries Science Center, National Oceanic and Atmospheric Administration, 1601 Kapiolani Blvd., Suite 1110, Honolulu, Hawaii 96814, USA

Tina Molodtsova
P.P. Shirshov Institute of Oceanology, Russian Academy of Sciences, 36 Nakhimovskii Prospect, Moscow 117218, Russia

Catherine Muir
Biology Department, Dalhousie University, Halifax, Nova Scotia, B3H 4J1, Canada

Phillip Neal
Marine Biological Laboratory, 7 MBL Street, Woods Hole, Massachusetts 02543, USA

Troy Nelson
Fraser River Sturgeon Conservation Society, Vancouver, British Columbia, V7B 0A2, Canada

Shuhei Nishida
Ocean Research Institute, University of Tokyo, 1-15-1 Minamidai, Nakano, Tokyo 164, Japan

John Payne
Pacific Ocean Shelf Tracking Project, Vancouver Aquarium, P.O. Box 3232, Vancouver, British Columbia, V6B 3X8, Canada

Carlos Pedrós-Alió
Institut de Ciències del Mar, Passeig Maritim de la Barceloneta 37-49, E-08003 Barcelona, Spain

Uwe Piatkowski
Leibniz-Institut für Meereswissenschaften, IFM-GEOMAR, Forschungsbereich Marine Ökologie, Düsternbrooker Weg 20, D-24105 Kiel, Germany

Laetitia Plaisance
(1) Center for Marine Biodiversity and Conservation, Scripps Institution of Oceanography, University of California San Diego, La Jolla, California 92093-0202, USA

(2) Department of Invertebrate Zoology, National Museum of Natural History, Smithsonian Institution, MRC 163, P.O. Box 37012, Washington, DC 20013-7012, USA

Gerhard Pohle
Huntsman Marine Science Centre, 1 Lower Campus Road, St. Andrews, New Brunswick, E5B 2L7, Canada

Gary Poore
Museum Victoria, GPO Box 666E, Melbourne, Australia

Bo Poulsen
Department of Environmental, Social and Spatial Change, Roskilde University, Building 3.2.1, DK-4000 Roskilde, Denmark

Imantes G. (Monty) Priede
University of Aberdeen, Oceanlab, Newburgh, Aberdeen AB41 6AA, Scotland, UK

Alban Ramette
Max Planck Institute for Marine Microbiology, Celsiusstrasse 1, D-28359 Bremen, Germany

Eva Z. Ramirez-Llodra
Institut de Ciències del Mar, Passeig Maritim de la Barceloneta 37-49, E-08003 Barcelona, Spain

Erin Rechisky
Department of Zoology and Fisheries Centre, University of British Columbia, 2202 Main Hall, Vancouver, British Columbia, V6T 1Z4, Canada

Michael Rex
University of Massachusetts, Department of Biology, 100 Morrissey Boulevard, Boston, Massachusetts 02125, USA

Ashley A. Rowden
National Institute of Water and Atmospheric Research, P.B. 14-901, Wellington, New Zealand

Gilbert T. Rowe
Texas A&M University, Galveston, Texas 77553, USA

Ricardo S. Santos
Department of Oceanography and Fisheries, University of the Azores, PT-9901-862, Horta, Portugal

Sigrid Schnack-Schiel
Alfred Wegener Institute for Polar and Marine Research, D-27568 Bremerhaven, Germany

Stefan Schouten
The Royal Netherlands Institute for Sea Research, P.O. Box 59, 1790 AB Den Burg, Texel, The Netherlands

Javier Sellanes
Facultad de Ciencias del Mar, Universidad Católica del Norte, Larrondo 1281, Coquimbo, Chile

Tim M. Shank
Biology Department, Woods Hole Oceanographic Institution, Woods Hole, Massachusetts 02543, USA

Yoshihisa Shirayama
Seto Marine Biological Laboratory, Field Science Education and Research Center, Kyoto University, Shirahama-cho, Wakayama Prefecture 649-2211, Japan

Myriam Sibuet
Institut Océanographique, 195 rue Saint Jacques, 75005 Paris, France

Boris Sirenko
Zoological Institute, Russian Academy of Sciences, St. Petersburg 199034, Russia

Henrik Skov
DHI, Agern Allé 5, DK-2970 Hørsholm, Denmark

Craig Smith
University of Hawaii at Manoa, Marine Sciences Building, 1000 Pope Road, Honolulu, Hawaii 96822, USA

Mitchell Sogin
Marine Biological Laboratory, 7 MBL Street, Woods Hole, Massachusetts 02543, USA

Henrik Søiland
Institute of Marine Research, P.O. Box 1870 Nordnes, NO-5817 Bergen, Norway

Lucas Stal
Netherlands Institute of Ecology, P.O. Box 140, NL-4400 AC Yerseke, The Netherlands

Karen I. Stocks
San Diego Supercomputer Center, University of California San Diego, 9500 Gilman Drive, La Jolla, California 92093, USA

Michael Stoddart
Institute for Marine and Antarctic Studies, University of Tasmania, Hobart, Australia

Tracey Sutton
Virginia Institute for Marine Science, College of William and Mary, P.O. Box 1346, Gloucester Point, Virginia 23062, USA

Anne Thessen
Marine Biological Laboratory, 7 MBL Street, Woods Hole, Massachusetts 02543, USA

Thomas Trott
Suffolk University, Boston, Massachusetts, USA

Anastasios Tselepides
Thalassocosmos, P.O. Box 2214, 71003 Heraklion, Crete, Greece

Paul A. Tyler
School of Ocean and Earth Science, University of Southampton, National Oceanography Centre Southampton, European Way, Southampton SO14 3ZH, UK

Cindy L. Van Dover
Duke University Marine Laboratory, Nicholas School of the Environment and Earth Sciences, 135 Marine Lab Road, Beaufort, North Carolina 28516, USA

Edward Vanden Berghe
Institute of Marine and Coastal Sciences, Rutgers University, 71 Dudley Road, New Brunswick, New Jersey 08901-8521, USA

Ann Vanreusel
Marine Biology Section, Ghent University, Ledeganckstraat 35, B-9000 Ghent, Belgium

Michael Vecchione
NMFS National Systematics Laboratory, National Museum of Natural History, MRC-153 Smithsonian Institution, P.O. Box 37012 Washington, DC 20013-7012, USA

Anders Warén
Swedish Museum of Natural History, Department of Invertebrate Zoology, Box 50007 (Frescativän 44) SE-10405, Stockholm, Sweden

David Welch
Kintama Research Corporation, 10-1850 Northfield Road, Nanaimo, British Columbia, V9S 3B3, Canada

Thomas de Lange Wenneck
Institute of Marine Research, P.O. Box 1870 Nordnes, NO-5817 Bergen, Norway

Jan Marcin Węsławski
Institute of Oceanology, Polish Academy of Sciences, Sopot 81-712, Powstancow Warszawy 55, Poland

Peter H. Wiebe
Biology Department, Woods Hole Oceanographic Institution, Woods Hole, Massachusetts 02543, USA

Nicholas H. Wolff
Aquatic Systems Group, University of Southern Maine, Portland, Maine 04101, USA

Boris Worm
Biology Department, Dalhousie University, Halifax, Nova Scotia, B3H 4J1, Canada

Craig M. Young
Oregon Institute of Marine Biology, P.O. Box 5389, Charleston, Oregon 97420, USA

The Scientific Steering Committee of the Census of Marine Life

Current Members

Dr. Ian Poiner (Chair), CEO, Australian Institute of Marine Science, Australia

Dr. Victor Ariel Gallardo (Vice Chair), Professor, Universidad de Concepción, Chile

Dr. Myriam Sibuet (Vice Chair), Senior scientist emerita in residence, Institut Océanographique de Paris, France

Dr. J. Frederick Grassle (Founder and Past Chair), Former Director, Institute of Marine & Coastal Sciences, Rutgers University, USA

Dr. Vera Alexander, Professor Emeritus, University of Alaska Fairbanks, USA

Dr. D. James Baker, Director, Global Carbon Measurement Program, William J. Clinton Foundation, USA

Dr. Patricio Bernal, Executive Secretary, Intergovernmental Oceanographic Commission, France/Chile

Dr. Dorairajasingam Chandramohan, Former Leader, Marine Microbiology, National Institute of Oceanography, India

Dr. David Farmer, Dean, Graduate School of Oceanography, University of Rhode Island, USA

Dr. Serge M. Garcia, Former Director, UN FAO Fisheries and Aquaculture Management Division, Italy/France

Dr. Carlo Heip, General Director of the Royal Netherlands Institute for Sea Research (NIOZ) and the Director of the Centre for Estuarine and Marine Ecology of the NIOO-KNAW Netherlands Institute of Ecology, Netherlands/Belgium

Dr. Poul Holm, Professor of Environmental History and Academic Director of the Trinity Long Room Hub, Trinity College Dublin, Ireland

Dr. Yoshihisa Shirayama, Director, Field Science Education and Research Center, Kyoto University, Japan

Dr. Michael Sinclair, Bedford Institute of Oceanography, Fisheries and Oceans, Canada

Dr. Song Sun, Director, Institute of Oceanology, Chinese Academy of Sciences, China

Dr. Meryl J. Williams, Honorary Life Member, Asian Fisheries Society, Australia/Malaysia

Ex-Officio Members

Mr. Jesse H. Ausubel, Vice President, Alfred P. Sloan Foundation, New York, USA

Dr. Daniel P. Costa, Professor of Ecology and Evolutionary Biology, University of California Santa Cruz, USA

Dr. Patrick Halpin, Associate Professor of Marine Geospatial Ecology, Nicholas School of the Environment, Duke University Marine Laboratory, USA

Ms. Sara Hickox, Director, Office of Marine Programs, University of Rhode Island, USA

Dr. Enric Sala, Fellow/Emerging Explorer, National Geographic Society, USA/Spain

Dr. Paul Snelgrove, Professor and Canada Research Chair in Boreal and Cold Ocean Systems, Ocean Sciences Centre, Memorial University of Newfoundland, Canada

Dr. Edward Vanden Berghe, OBIS Executive Director, Rutgers University, USA/Belgium

Former Members

Dr. Donald Boesch, Professor of Marine Science and President, Center for Environmental Science, University of Maryland, USA

Dr. Olav Rune Godø, Head of Research Group on Observation Methodology, Institute of Marine Research, Norway

Dr. Andrew Solow, Senior Scientist and Director, Marine Policy Center, Woods Hole Oceanographic Institution, USA

PART I

Oceans Past

Chapter 1

Marine Animal Populations: A New Look Back in Time

Poul Holm[1], Anne Husum Marboe[2], Bo Poulsen[2], Brian R. MacKenzie[3]

[1]*Trinity College Dublin, Ireland*
[2]*Department of Environmental, Social and Spatial Change, Roskilde University, Roskilde, Denmark*
[3]*National Institute for Aquatic Resources, Technical University of Denmark, Charlottenlund, Denmark*

1.1 Introduction

Since around 1980, marine-capture fisheries have stagnated at around 90 million tonnes per year, despite massive technological investments and the opening up of distant and deep waters in the Southern hemisphere. The oceans will simply not yield more. In fact catches are of increasingly smaller fish of less economic value and total returns on investments are dwindling. On a global scale, capture fisheries are doomed to be of less importance as a source of protein to a growing human population, while the fishing pressure remains extremely high. There is no sign that the rise of aquaculture in recent decades has eased the pressure on wild resources. The fisheries crisis is part of a general health alert for the oceans. Marine habitats are under severe pressure as a side effect of trawling and directly by dredging, harbor development, the concretization of large stretches of coastline, and especially from eutrophication caused by both agriculture and aquaculture (Lotze & Worm 2009).

But what is the scale of change? What used to be in the sea before humans began impacting marine ecosystems and habitats? What are the major long-term effects of human extractions of marine life? Are the impacts of recent or ancient origin? In other words what are the baselines against which we may evaluate some of the findings of the

Census of Marine Life field projects by 2010? Can we talk with confidence about the history of the sea, can we gauge how much has changed – and with what consequences to us humans? This was the grand challenge that was put to the scientific community some ten years ago when the Census endorsed the History of Marine Animal Populations (HMAP) Project to assess and explain the history of diversity, distribution, and abundance of marine life (Box 1.1).

Although the history of marine animal populations has long been one of the great unknowns, recent advances in scientific and historical methodology and new applications of existing methodology have enabled the HMAP teams to expand the realm of the known and the knowable (Holm 2002).

The analytical framework of HMAP embraces two basic premises, one concerning data, one concerning methodology. First, much of what we can know about the history of the oceans will be in the "human edges" of the ocean, those in the near shore and coastal zone. This is where humans most directly interacted with the sea in the past and therefore most historical records relate to these activities. However, in both the human edges and in the central oceanic waters there have been extensive fisheries for larger organisms, and the value of the organisms encouraged the creation and maintenance of archival material. As HMAP has evolved, new and unexpected data sources have been discovered, and we know now that vast repositories are still untapped.

Second, historical analysis must combine with ecological analysis in a truly interdisciplinary way. New insights are

Life in the World's Oceans, edited by Alasdair D. McIntyre
© 2010 by Blackwell Publishing Ltd.

Box 1.1

Regional and Species Focus of HMAP

HMAP is a collaborative effort by some 100 researchers around the globe participating in several region- or species-specific research teams. Twelve are based on marine areas, as follows: southeast Australian Shelf; New Zealand Shelf; Caribbean Sea; Gulf of Maine; Newfoundland and Grand Banks; Baltic Sea; North Sea; Mediterranean Sea; Black Sea; White and Barents Seas; southwest African Shelf; and the biodiversity of nearshore waters. Three case studies focus on the following species: whales, cod, and mollusks and one on Northern European fish bone assemblages. In addition, several smaller case studies have been undertaken in areas such as the Philippine Seas, the Wadden Sea, and the seas of Indonesia and northern Australia.

due to the introduction of established marine science methodology to historical data, notably standardizing fishing effort (catch per unit effort) (see, for example, Poulsen & Holm 2007), biodiversity counts of historical fisheries (Lotze *et al.* 2005), statistical modeling of historical data (Klaer 2005; Rosenberg *et al.* 2005), etc. Perhaps the most surprising results have come simply from the data-mining effort in itself, which has revealed a wealth of documentation for historical fisheries previously neglected by historians. Examples of this are of catch records spanning two to four centuries (Holm & Bager 2001; Starkey & Haines 2001; Lajus *et al.* 2005; B. Poulsen 2010). HMAP has provided inspiration to glean important information from surprising and sometimes unlikely sources such as restaurant menus (Jones 2008) and snapshots of sports fishermen's catches (McClenachan 2009). Archaeological techniques have been deployed in conjunction with historical methods and stable isotope analysis to explore the character and composition of fish catches during early medieval times (Barrett *et al.* 2008), and many more unconventional approaches could be cited.

In many ways the complicated interplay between man and nature calls for a new type of historical research. Science is a challenge to historians who have had little statistics, not to speak of modeling, as part of their training. Historical source-criticism is a challenge to scientists who are used to hard data. Although academic history through the 1990s concentrated on narrative and deconstructing skills, environmental history also demands command of both statistical and scientific methods.

The need for historians and scientists to work together is not uncontroversial. Some historians assert that history would carry no lessons for the future as events are never repeated in exactly the same form. Some scientists doubt the validity of historical data that are by definition "dirty data" in the sense that they are relics of events, not signals of a recurrent phenomenon, or experiment, established in a controlled environment such as a laboratory. In the early part of the Census some skeptics doubted the role of environmental history in this mega-science program. Would such a program not by default perpetuate the divide between science and the humanities? Indeed, as one critic put it, would the marriage of history and science not lead to scientists simply appropriating data for their own use (Van Sittert 2005)?

HMAP is founded on the belief that the divide between history and science needs to be bridged. History will never repeat itself but like the child learns to walk based on experience so does society base decisions and preferences on past experience. The historian may indeed detect trends and patterns of behavior behind diverse and unique events. Emphatic statements on the validity of and need for the HMAP approach have been made by some historians (Anderson 2006; Bolster 2006, 2008). Conversely, if we reduce science to controlled experiments we would never understand the fundamental principles of natural selection. More urgently, contemporary concerns about global climate change, biodiversity, and scarcity of resources are based on perceived changes of nature and availability of natural resources. Therefore, the history of nature itself – and the dependency and impact of human society on nature – has become a prime social, economic, and political concern, and scientists and historians need to address these very real issues, or decisions will be based on assumptions.

Environmental historians do not have to become biologists, nor do biologists need to become historians. However, we do need to understand enough of each other's language to exchange information and insight. Our experience of dialogue across the current divide of humanities and science has led to the emergence of the new scientific community of marine environmental history and historical marine ecology (Box 1.2).

A total of 205 books and papers have been published up to September 2009 and the HMAP database (www.hull.ac.uk/hmap) holds approximately 350,000 records, with some 80% available through OBIS (see Chapter 17). By late 2010, it is anticipated that up to 1,000,000 records will be available on the HMAP website. With such a massive output it is obvious that any overview of major findings will be highly selective. In the following, we shall establish first the state of knowledge before the beginning of the Census in 2000, then focus on some of the highlights from the HMAP case studies. By way of conclusion, the chapter closes with observations on what we do not know, how we may get to know it, and why some questions will remain unanswerable.

1.2 The Background

Marine ecology was born as a scientific discipline by the late nineteenth century and derived often from a strong interest in the fisheries (Smith 1994). The question of human impact on marine life was central not only from the perspective of economic interest (for example where are the fish and how do we catch them?) but from the perspective of human impact (for example what is the effect of extracting thousands of tonnes of fish and what damage to the seabed may be caused by certain fishing technologies?).

The central question of the possibility of overfishing was raised at the World Fisheries Exhibition in London in 1884 and drew two opposing answers. One came from one of the leading scientific figures of the day, Thomas Henry Huxley, who concluded that "... probably all the great fisheries are inexhaustible; that is to say that nothing we do seriously affects the number of fish" (Huxley 1883). A more conservative note was struck by Ray Lankester, a young professor of zoology, that "the thousands of *apparently* superfluous young produced by fishes are not really

superfluous, but have a perfectly definite place in the complex interactions of the living beings within their area" (Lankester 1890). To the credit of both men and to the academic community at the time the question of the possibility of harmful overfishing was put to the test. A rigorous series of trawls were undertaken in Scottish waters and were at first understood to support Huxley's view. In 1900, however, the tests were reanalyzed and further data from observations of commercial operations out of Grimsby were scrutinized. The conclusion by Walter Garstang was clear and had far-reaching implications: "... the rate at which sea fishes reproduce and grow is no longer sufficient to enable them to keep pace with the increasing rate of capture. In other words, the bottom fisheries are undergoing a process of exhaustion" (Garstang 1900; cf. Smith 1994, pp. 106–108).

This fundamental observation is at the heart of the question of human interaction with the oceans. Garstang established beyond scientific doubt that extractions might have an impact. Through the twentieth century, fisheries science concentrated on identifying optimal sustainable yields that would not extract more from the sea than marine life would be able to replenish. By the second half of the century, fisheries science had become highly sophisticated, equipped with research ships and advanced computer models. Scientific organizations like ICES, the International Council for the Exploration of the Sea, established in 1902 for the North Atlantic (Rozwadowski 2002), and a plethora of similar organizations for other ocean realms and migratory species, struggled to get both the science right and deliver management advice. Characteristically, fisheries studies were often based on very short time-series, although scientists were aware of long-term changes. The centennial variability of the Swedish Bohuslen herring fisheries provided a textbook example that fisheries may change dramatically over the long term. Nevertheless, perhaps because of the strong link with policy advice, the focus of cutting-edge science tended to be on recent data often obtained with new equipment, which by the very fact obliterated longer-term perceptions. Data observations over the long term were often discontinued for financial reasons. Few observation series are maintained today that span more than a few decades, the best-known of which is the Continuous Plankton Data Recorder survey, which has been maintained for the North Atlantic and North Sea since 1931 (Continuous Plankton Recorder 2009).

Marine science separated from fisheries science through the twentieth century as scientists developed the concept of ecology as a study of biodiversity, food webs, and biological processes and functions as a separate line of inquiry. To ecologists the ultimate question is not what is in nature for us, the humans, but how do we understand nature on its own, with the humans left out. Interest focused on biodiversity, the awesome richness of nature, and the exhilaration of understanding intricate and ingenious life-forms. By the 1960s ecologists did realize that ecosystems rarely remain steady for long, and "fluctuations lie in the very essence of the ecosystems and of every one of the … populations" (Margalef 1960 cited in Smith 1994, p. 33). Marine ecologists, however, perceived little or no need for history, with the exception of a few studies of correlations between contemporary and historical observations of animal populations and key environmental variables (Cushing 1982; Alheit & Hagen 1997; Southward 1995). Things were about to change, however, as demonstrated in a programmatic statement on the need to determine the historic structure of exploited ecosystems (Pitcher & Pauly 1998).

In a seminal study of the Caribbean ecosystem, Jeremy Jackson criticized ecologists for assuming that the natural or original condition is equal to the first scientific description of a phenomenon (Jackson 1997). Jackson turned to a concept developed a few years earlier by a fisheries scientist, Daniel Pauly, for a diagnosis of the problem, which was termed the shifting baseline syndrome (Pauly 1995). Pauly observed that equilibrium or steady-state models are based on a given dataset, often established by scientists within the past generation. However, what happens to equilibrium if older data are introduced? We cannot know from recent information the extent of the losses that have happened.

Jeremy Jackson, himself an American ecologist, son of a historian, used the British Empire trade statistics of the eighteenth century to learn of the trade in turtles from the Caribbean. When working out the numbers – hundreds of thousands of turtles killed in a single year – he realized that the ecosystem of the Caribbean would have looked very different to what conservation biologists supposed based on information from the past couple of decades (Jackson 1997). The lesson to ecologists of Jackson's historical analysis of Caribbean coral reefs was that textbook descriptions of reef ecosystems were limited by the fact that the systematic description by modern biology only began in the 1950s.

Jackson put the case squarely to the ecologists: they needed to turn to historical sources and rediscover the world.

Another influential development in reinstating the historical dimension in science was the development of paleo-ecology and archaeoichthyology in the past 30–40 years. The preservation of fish scales in anoxic bottom sediments off the coast of California provided scientists the opportunity to reconstruct 1,600 years of pelagic abundances (Soutar 1967; Baumgartner et al. 1992; Francis et al. 2001). The field of paleozoology provided one of the first clear examples of scientists working across the cultural divides of historical and ecological analysis. Analysis of fish remains from archaeological sites provided a possible avenue to understanding biodiversity distribution and abundance. In the 1960s the Swedish scientist Höglund analyzed fish bones excavated from eighteenth century production sites for train oil and found that the Bohuslen herring was spent (namely post-spawning) herring from the sub-population of the North Sea Buchan herring (Höglund 1972). In the 1990s studies clearly demonstrated the potential of bringing the different lines of inquiry together (Muniz 1996; Enghoff 1999).

What about the historians? Environmental history has been a growth field in the USA since the 1970s and a little later in Europe, Asia, and Australia, and indeed, despite institutional problems, also in South America and Africa. However, the focus by leading American environmental historians was strongly on human agency and perception whereas ecological factors were rarely allowed to play an explanatory role. On top of that, the discipline developed out of a strongly narrative and qualitative approach to history that had little rapport with the quantitative approach of ecologists. The focus was very much on frontier cultures of the prairies, bushlands, savannahs, and steppes, whereas the oceans were strangely disregarded. Maritime historians on the other hand were firmly embedded in economic and social history with a preoccupation for naval and shipping matters and had little regard for environmental issues. The few fisheries historians often found their subject of marginal interest to mainstream historians and a bit fuzzy as the ecological context of fishing could not be disregarded but on the other hand was little understood. The few substantial overviews of fisheries published generally adopted a national, regional, or port perspective whereas environmental considerations were accidental at best. It was only as late as 1995 that the North Atlantic Fisheries History Association was established, but even then few papers dealt with the impact of harvesting on the seas (Holm & Starkey 1995–99).

Signs were in the air, however, that things were about to change. In the North Atlantic, Holm & Starkey (1998) reported the results of a workshop titled "Fishing Matters" that brought together historians, social scientists, biologists, oceanographers, and fisheries managers to examine multidisciplinary approaches to understanding the past and current scale and character of the fisheries. In the North Pacific, Pauly *et al.* (1998b) similarly documented the results of a workshop aimed at mathematically reconstructing the state of the Strait of Georgia, off Vancouver Island. Participants were even more varied, and the focus was broader in attempting "to provide a vision for rebuilding the Strait's once abundant resources."

1.3 The HMAP Projects

Such was the state of play when a preparatory workshop of the Census in 1998 called attention to the need of a historical backdrop – a baseline – to observations of ocean life (Anon 1998). The challenges were apparent: there was no shared or agreed set of methodologies and not even agreement as to which research questions needed to be raised. Before the project started the first step was therefore to bring together a workshop in February 2000 to identify the hypotheses that could be tested against historical data, to identify the various sources of data, and the methodologies that might yield plausible answers. The workshop agreed that historical data were only sporadically available and that there was an urgent need to build consistent time-series of extractions and fishing effort for at least the best documented operations such as whaling and large commercial fisheries. Participants identified 10 hypotheses to direct work in the early years of the project. The focus was first of all to investigate the proposition that validated historical, archaeological, and paleoecological records can be used to gauge long-term change in the abundance, spatial distribution, and/or diversity of marine animal populations. Secondly we wanted to identify the environmental and human forces that might condition fish mortality. Thirdly, we wanted to understand better the drivers of these forces themselves, be they related to geophysical or human activity.

In May 2000 a Steering Group of historians and marine scientists was charged by the Census' Scientific Steering Committee to lead a global inquiry into the history of marine animal populations. A series of regional projects was proposed while we set up annual training workshops through the summers of 2001–2003, well knowing that as nobody had ever received academic training as marine historical ecologists or marine environmental historians, there was a need to train a new generation of two dozen young researchers to understand enough of several disciplines. As the project grew, the Oceans Past conferences of 2005 and 2009 attracted more than 100 researchers while many more worked in the field.

The identification of a viable project was not just a question of a top-down process. Although the Steering Group wanted to get projects started in Japan and in the American Pacific, we were confronted with the reality of needing to find like-minded people who would undertake not only

individual work but also lead a team for several years guided by an overarching research program. We were not always successful, but all projects that were begun proved viable. Although some were discontinued as the research was completed, other projects developed new agendas. A renewed focus on the evidence of archaeology brought new people and projects forward. By 2007 the focus shifted from data collecting to synthesis, both within projects and across projects. New collaboration with other Census projects emerged, in particular with Natural Geography in Shore Areas (NaGISA) for the History of the Near Shore project (see Chapter 2), which focused on providing historical data as baseline studies for ongoing fieldwork. In the following, we highlight selected research findings addressing two of the initial simple questions: what is the scale of change, and are changes of recent or ancient origin?

1.3.1 Mediterranean Sea and Black Sea

The Mediterranean and Black Seas are among the earliest heavily fished marine ecosystems in the world. Fish as a source of food was more important than meat in the ancient Mediterranean cultures (Fig. 1.1). Along the Nile, settlements with huge amounts of fish bones have been identified. Hundreds of full-time fishers were employed by the Lagash temple in Sumer around 2400 B.C. The fish was dried, salted, and stored. Babylonian sources from around 1750 B.C. show the importance of fishing. Greek merchants conducted an extensive fish trade from the Black Sea and the Russian rivers to the Greek and later the Roman market (Holm 2004).

However, the problem with assessing the impact of fisheries on ecosystems is that ancient records are rarely quantifiable and often we are not able to identify the fish species mentioned. Even worse, until quite recently, historians have assumed that the ancient fisheries were of minimal importance, technology was simple, and nets were cast close to the shoreline. A full reversal of this perception was only achieved as a result of an analysis of the evidence matched by an understanding of modern impact studies of pre-industrial fisheries technology. The Graeco-Roman world had seagoing vessels for hook-and-line as well as net fisheries. Ancient technology was neither ineffective nor unproductive, and indeed produced such large catches that the limiting factor was preservation and storage (Bekker-Nielsen 2005). The main fisheries for bluefin tuna (*Thunnus thynnus*), mackerel (*Scomber scombrus*), and other pelagic species took place in narrow straits such as the Strait of Gibraltar, Sardinia, Sicily, and Crimea in the Black Sea (Curtis 2005; Gertwagen 2008).

One solution to the problem of conserving the fish was to dry and salt the fish, which was done extensively and accounted for much of the Greek imports from the Black

Fig. 1.1
Polychrome mosaic ("Catalogo di pesci") found in
Pompeii, house VIII.2.16 and now in the National
Archaeological Museum, Naples. Last century B.C.
Size *ca.* 0.9 m × 0.9 m. Photograph courtesy of
Professor Dario Bernal Casasola, University of
Cádiz.

Sea. The most spectacular solution was, however, the reduction of fish to fish sauce, garum, essentially by throwing the catch into large containers to allow a fermenting process to result in a liquid that was then traded all around the Roman world to add flavor to the Roman cuisine. The large installations are especially found by the shores of the western Mediterranean and the Black Sea. They were probably privately owned commercial operations for export, and regularly had containers of several hundred cubic meters. The largest installation in present-day Mauretania had a capacity of over 1,000 cubic meters (Curtis 2005; Trakadas 2005).

As yet, there is no way to establish the quantities of catch, although evidently they will have been significant. One assessment of the distinctive amphora vessels for the oil, wine, and garum trades established that wine accounted for about 62% of relative volumes, whereas oil made up about 28%, and 10% contained garum. Fish sauce was sold all over the Roman Empire and was an essential part of the Roman dish, part of what made up Roman culture (Ejstrud 2005). There is no doubt that extractions will have been huge, and much will be learnt in coming years as this research continues.

Documentary records are especially rich for the Venetian lagoon and the Northern Adriatic Sea. Preliminary studies show that the marine system has been modified dramatically by human interventions since the medieval period. An ongoing project aims to reconstruct the dynamics of marine animal population in the Venetian Lagoon and in the Northern Adriatic Sea from the twelfth century up to the twenty-first century from historical and scientific sources (Gertwagen *et al.* 2008). Finally, the Catalan Sea has been studied carefully and data for twentieth-century fisheries have been made available for further study.

1.3.2 North Sea and Wadden Sea

The North Sea is another heavily fished and depleted marine system. The Mesolithic period about 6,000 years ago experienced a warm climate, which seems to have been conducive to extensive fisheries all over the Northern hemisphere. Many basic technologies for the fisheries were already developed by this time such as trap gear and fishing by hook-and-line from a boat. With domestication of animals and development of agriculture in the Neolithic period about 5,000 B.P., hunting and fishing became less important and settlements were no longer related to the seashore, and fishing seems to have been of minor importance through the Bronze and Iron Ages of Northern Europe. Rivers will have brought nutrition to the North Sea from the rich agricultural lands of Northern Europe already by the Bronze Age when major deforestation took place and increased the productivity of the sea (Enghoff 2000; Beusekom 2005).

Our knowledge of ancient fisheries is still deficient due to the lack of sieving of archaeological finds for small and easily overlooked fish bones. However, thanks to a thorough review of archaeological reports of dozens of medieval settlements we now know that the period *ca.* 950–1050 saw a major rise in fish consumption around the North Sea (Fig. 1.2) (Barrett *et al.* 2004, 2008). Early medieval sites are dominated by freshwater and migratory species such as eel and salmon, whereas later settlements reveal a widespread consumption of marine species such as herring (*Clupea harengus*), cod (*Gadus morhua*), hake (*Merluccius*), saithe (*Pollachius virens*), and ling (*Molva molva*). The "fish event" of the eleventh century reflected major economic and technological changes in coastal settlements and technologies, and formed the basis of dietary preferences that

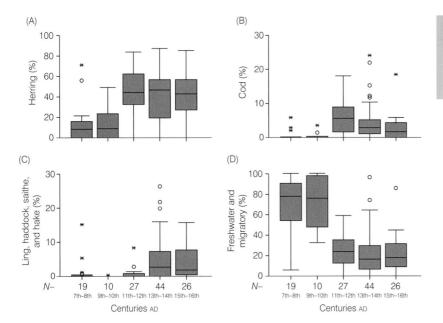

Fig. 1.2

Fish bones project: "Pristine" North Sea impacted *ca.* 950–1050, freshwater to marine species. Source: Barrett *et al.* (2004). Reproduced with permission.

were to last into the seventeenth century. In particular, the evidence of traded cod, "stock fish", which begins to show up in Northern European towns by the middle of the eleventh century, is clear evidence of the rise of commercial fisheries (Fig. 1.3). Barrett's group combines an osteological study of fish bones with analysis of their stable isotope signatures. The project has now identified traded cod in medieval settlements from Norway, England, Belgium, Germany, Denmark, Sweden, Poland, and Estonia. The evidence also supports a hypothesis that seagoing vessels were in wide use by the thirteenth century catching fishes at depths of 100–400 m such as ling. Commercial fisheries were well established by the high middle ages to feed a European population that had developed religious practices of fasting and abstinence of red meat in favor of fish on certain weekdays and through the 40 weekdays of Lent (Hoffmann 2004).

The first estimate of total removals of one species from the North and Baltic Seas comes from the sixteenth-century Danish inshore fisheries for herring in Scania and Bohuslen. Annual catches regularly reached a level of 35,000 tonnes (Holm 1999, 2003). By the late sixteenth century, the Dutch had taken the lead in Northern European herring fisheries with seagoing *buysen*, which harvested the rich schools off the coasts of Scotland and the Orkneys. They landed catches of 60,000–75,000 tonnes every year in the first quarter of the seventeenth century, and total removals with English, Scottish, and Norwegian landings amounted to upwards of 100,000 tonnes. Catches declined to about half by 1700, and only increased to about 200,000 tonnes in the late eighteenth century owing to Swedish and Scottish progress (B. Poulsen 2008).

By 1870 total removals reached a level of 300,000 tonnes, which equals the recommended Total Allowable Catch for 2007 for herring in the North Sea (ICES 2006). In the twentieth century, total catches repeatedly amounted to well over a million tonnes annually, causing collapses of herring stocks and the closure of fisheries for one or two decades to allow populations to rebuild.

This evidence demonstrates how fishermen in the age before steam and trawl were able to remove large quantities of biomass from the sea. The technologies of wind power and driftnets were practically unchanged in the Dutch fisheries from the seventeenth to the nineteenth centuries. There are indications that removals even at the much lower level than that recommended by modern standards had an effect on abundance. One study standardized the fishing power of North Sea herring fishing vessels across the technological divide from sail to motor-powered vessels from the sixteenth to the twentieth centuries. Even by a conservative estimate the returns of catch per unit effort indicated that stock abundance was ten times higher in the 1600s than in the 1950s, and already by the 1800s, well before the big technological change, it had dropped to 50–60% of the level of the 1600s (B. Poulsen 2008). The effects of early removals may therefore have been larger than we would have assumed.

The catches of two other commercially important species, ling (*Molva molva*) and cod (*Gadus morhua*), were abundant in the nineteenth century whereas the stocks showed signs of depletion by World War I. Detailed historical data are available from the Swedish fishery in the northeastern North Sea and Skagerrak, which make up about one-sixth of the entire North Sea. Minimum total biomass of cod in 1872 was estimated at about 47,000 tonnes for

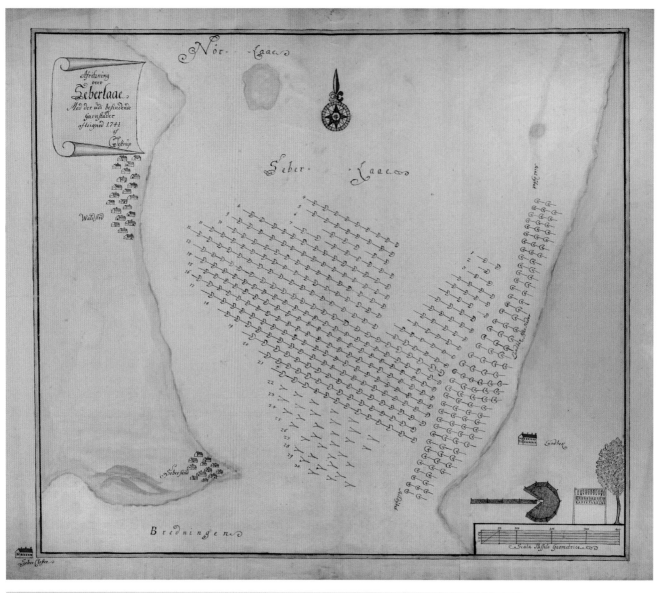

Fig. 1.3
Map of the pound nets in Sebberlaa area of the Limfjord. The district bailiff, Thestrup, drew the map in 1741, where hundreds of pound nets, each 70 meters long, were in use in this very narrow stretch of water. In the right hand bottom of the map, Thestrup has drawn a pound net scaled next to a tree and row of dried fish. Source: Royal Library, Copenhagen, Ny Kgl. Saml. 409d, fol.

this portion of the North Sea, but it may have been much higher, whereas the total biomass of ling was estimated at a total of 48,000 tonnes. These were very healthy stocks if the levels are compared with the modern biomass estimate for cod of 46,000 tonnes for the entire North Sea, Skagerrak, and Eastern Channel, whereas for ling no modern biomass estimate is available as the species is caught too infrequently. The cod population is today considered severely depleted throughout the North Sea, and the ling population may be considered commercially extinct from the region that once produced the major catches (R.T. Poulsen *et al.* 2007).

Ecosystem theory emphasizes the importance of top predators for the entire food web. Top predators play a controlling and balancing role for the abundance of other species further down the food chain, and an abundance of top predators is a sure sign of biodiversity (Baum & Worm 2008; Heithaus *et al.* 2008). Human hunting tends to focus on top predators as the big fish are of highest commercial value. When we take out the largest specimens, we remove one of the controls on the ecosystem. The mature fish are also highly important for the reproduction of the population as their eggs have been shown to be healthier and more plentiful than the spawn of younger and smaller specimens

(O'Brien 1999). Because the fish continues to grow through its entire life, a decline in the length of specimen caught is a clear indication that the fishery is changing the age structure and viability of the stock. Analysis has shown that whereas the average length of northeastern North Sea ling in the mid- to late nineteenth century was about 1.5 m, it had decreased to about 1.2 m by World War I, and ling caught today is less than 1 m on average (R.T. Poulsen *et al.* 2007). A century ago, cod landed from the North Sea was usually 1–1.5 m long whereas today it is only about 50 cm. This means that although cod used to live to an age of 8 or 10 years, today it is caught at less than three years of age; for example in 2007, 87% of the catch in numbers were aged two years or younger. As cod only spawns at the age of three years, this fishing pattern is inhibiting the population from maintaining itself and delaying recovery (ICES 2008).

The bluefin tuna (*Thunnus thynnus*) generally escaped human hunting activity until the twentieth century owing to its rapidity and superior strength, which made its capture difficult. By the 1920s superior hook-and-line technology was available and brought tuna within the reach of fishermen. Even more importantly, harpoon guns and purse-seining methods, eventually implemented with hydraulic winches, were developed in the 1930s and rapidly increased catches to thousands of individuals per year. By 1960, however, tuna catches were falling and ceased to be of commercial importance after the mid-1960s. Climate change and prey abundance seem unlikely causes for the sudden decline, and it seems now likely that the commercial extinction of bluefin tuna from the North Sea was caused by the heavy onslaught by humans in the mid-twentieth century (MacKenzie & Myers 2007).

In the southern North Sea, the haddock (*Melanogrammus aeglefinus*) fishery was of substantial size in the sixteenth and first half of the seventeenth centuries. The fishery declined in the later seventeenth into the eighteenth century, but by the 1770s the fishery was on the increase again. We have evidence of an abundant haddock fishery by German and Danish hand liners in the German Bight and along the Jutland coast in the late eighteenth century and first half of the nineteenth century. Statistics show substantial catches by 1875 declining rapidly in the last quarter of the century to nil around 1910. It would seem that the southern North Sea haddock stocks were rendered commercially extinct by the intensive German and Fanø-Hjerting fisheries of the late nineteenth century. Today, haddock is prevalent mainly in the northernmost part of the North Sea and in the Skagerrak (Holm 2005), whereas its former widespread presence in the southern part of the North Sea was not generally recognized by marine science until its regional history was revealed.

Major changes to the inshore habitats of the North Sea and thus to marine wildlife occurred in the Middle Ages. Hunting and fishing took its toll on the rich wildlife of the inshore areas of the Wadden Sea, a large intertidal zone off the coasts of the Netherlands, Germany, and Denmark. Dikes, traps, and other inshore coastal uses changed the wide mud flats. By the late nineteenth century industrial and chemical pollution began to build up in the sea. However, the major change to the ecosystem is likely to have come from direct effects of removals of animals by fishing and hunting (Beusekom 2005). Some marine species have been extirpated from the Wadden Sea such as pelicans (*Pelicanus crispus*), which disappeared about 2,000 years ago (Prummel & Heinrich 2005), the Atlantic gray whale (*Escherichtius robustus*), which went extinct not only from the nearshore habitats of the North Sea but as a species sometime in the late medieval period (Mead & Mitchell 1984), and the great auk (*Pinguinus impennis*), which disappeared from the North Sea by the medieval period before extinction from the North Atlantic by the nineteenth century (Meldgaard 1988). Several species have been so much reduced in numbers that they are considered regionally extinct or at least so rare that they have lost their ecosystem importance, and their previous commercial importance to the human economy. Sturgeon (*Acipenser sturio*) was previously caught in vast quantities and marketed in the hundreds, for instance at the Hamburg fish auction. By 1900, however, the fishery declined rapidly both because of river and inshore pollution and fisheries. As late as the 1930s sturgeon was still caught regularly in the northern Danish part of the Wadden Sea but is now extremely rare (Holm 2005).

A general survey of extirpations in the Wadden Sea concluded that major impacts occurred by the turn of the twentieth century, well before the introduction of modern industrial fishing technologies to this region. The major causes for species decline and indeed extirpations were associated with removals and habitat destruction whereas factors such as pollution, eutrophication, and climate change have been late and minor factors so far (Lotze *et al.* 2005).

1.3.3 Baltic Sea

One of the early research questions of HMAP was posed by fisheries scientists about Baltic cod (MacKenzie *et al.* 2002). In the absence of historical records before 1966, they wondered if the record high cod stock in the Baltic Sea in the late 1970s to early 1980s was a unique occurrence or likely to occur at regular intervals. The question was unequivocally answered by the work of the Baltic team.

Through the recovery of historical data back to 1925 we know now that abundant cod stocks corresponded to a favorable combination of four key drivers in the late 1970s: incursions of saline water to the brackish Baltic and hydrographic conditions allowing successful reproduction, low marine mammal predation, high productivity environment fuelled by nutrient loading, and reduced fishing pressure.

A similar situation did not occur at any other time in the twentieth century. The cod biomass in the 1920s–1940s was likely restricted by high abundance of marine mammals and low ecosystem productivity; and in the 1950s–1960s by high fishing pressure. Periods of deteriorated hydrographic conditions occurred throughout the twentieth century and were most pronounced in the past 20 years, thereby restricting cod recruitment (Eero *et al.* 2008).

Today, cod rarely ventures into the very brackish northern Baltic waters between Stockholm and the Gulf of Riga. In the late sixteenth and early seventeenth centuries the presence of a large cod fishery off southwest Finland indicates that cod abundance must have been very large. The abundance is all the more remarkable because the population of top predators such as seals would have been much larger than today (MacKenzie & Myers 2007). Climate clearly impacted fish distribution but there are some surprises which underline that some fish are quite resilient to change. Archaeological evidence of fish fauna in the Atlantic warm period (*ca.* 7000–3900 B.C.) shows many fish species in waters around Denmark that we would today expect to find in warmer waters. Indeed, comparison with contemporary data from surveys and commercial landings shows that many of these species are now re-appearing as temperatures rise. However, cod was very abundant in the Stone Age, even though temperatures were 2–4 °C warmer than late twentieth century temperatures. This finding suggests that commercially important cod populations can be maintained in the North Sea–Baltic region, even as temperatures rise due to global warming, provided that fishing mortalities are reduced (Enghoff *et al.* 2007).

During the Little Ice Age of the late seventeenth century, coldwater marine fish (herring, flounder (*Platichthys flesus*), and eelpout (*Zoarces viviparous*)) were of major importance in the Baltic Sea fisheries and the fishing season for the major pelagic fish was substantially later in the year compared with the present, much warmer conditions (Gaumiga *et al.* 2007). Similarly, catches of herring and other coastal fish (for example perch (*Perca fluviatilis*) and ide (*Leuciscus idus*)) near Estonia in the mid- and late nineteenth century varied, probably owing to climatic fluctuations, when fishing effort and methods were stable (Kraikovski *et al.* 2008). A major hydrographic event increased the salinity of the Limfjord in 1825; the saltwater intrusion destroyed the habitat for the freshwater whitefish (*Coregonus lavaretus*), but created conditions for saltwater species such as plaice (*Pleuronectes platessa*) (B. Poulsen *et al.* 2007).

Overall, fishing pressure was quite low in the inner parts of the Baltic. During the late seventeenth century, removals of fish biomass from the Gulf of Riga were at least 200 times less than at the end of the twentieth century, and most fisheries concentrated on the rivers. Migratory fish species, such as sturgeon, Atlantic salmon, brown trout, whitefish, vimba bream, smelt, eel, and lamprey were the most important commercial fish in the area, because they were abundant, had high commercial value, and were easily available. Over time, however, the main fishing areas moved downstream and to the sea. Owing to intensive fishing, populations of many migratory species, first of all sturgeon and Atlantic salmon, considerably declined and lost their commercial significance. Marine fish, especially Baltic herring, gained increased importance in the nineteenth century (Kraikovski *et al.* 2008).

1.3.4 Grand Banks, Gulf of Maine, and Scotian Shelf

Although the fisheries in the Northeast Atlantic developed later than in Northern Europe, they were no less intense in the past few centuries. The Grand Banks fishery for northern cod is a well-known example of the effects of sustained high fishing pressure ending in a sudden collapse of the stock (Myers *et al.* 1997). The HMAP research focused on correcting the historical landings statistics and showed that the combined efforts of British and French fishermen on the Grand Banks off Newfoundland yielded between 204,000 and 275,000 metric tonnes of cod in the years 1769–1774 (Starkey & Haines 2001), or two to three times higher than previous estimates, and at a level that was only eclipsed by the late nineteenth century when catches were on the order of 300,000 tonnes (Cadigan & Hutchings 2001). This level proved unsustainable and catches were only half as much in the 1940s. This finding underscores that – as happened in the North Sea herring fishery – extractions using pre-industrial technology could be similar to or indeed above modern levels. When the fishery finally collapsed in 1992, landings had reached a decadal peak of 268,000 tonnes only four years earlier.

Because of the open nature of the Grand Banks fishery, the data will always be incomplete. To understand the dynamics of the fisheries fully, we need to know how many people and boats participated in a particular fishery. The focus in the later stages of HMAP has therefore been on the Gulf of Maine and Scotian Shelf fisheries closer to the American mainland that were largely conducted by local vessels through the nineteenth and twentieth centuries. Luckily, these fisheries are exceptionally well documented thanks to a bounty that required fishing captains to keep and hand in their logbooks through the period 1852–1866. Thousands of these logbooks have been digitized and analyzed for content by a team at the University of New Hampshire.

The fishermen consistently removed 200,000 tonnes of live fish per year through the 1850s. For example, in 8.5 months during 1855, the hand-lining fishermen in 43 schooners from Beverly, Massachusetts, caught a little over 8,000 tonnes of cod on the Scotian Shelf, whereas in 15 months during 1999–2000 a total of just 7,200 tonnes of cod was extracted from the same waters by the entire

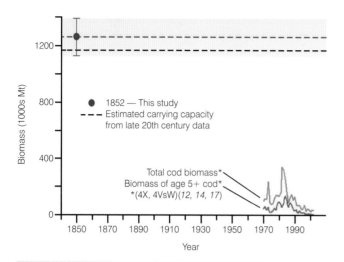

Fig. 1.4
Reduction of cod biomass, Scotian Shelf – estimated and historical.
Source: Rosenberg *et al.* (2005). Reproduced with permission of ESA.

Canadian mechanized fishing fleet and fell short of the full Total Allowable Catch by 11%, a comparison that points to a profound change in cod abundance on the Scotian Shelf over the past 150 years (Fig. 1.4) (Rosenberg *et al.* 2005).

Abundant as fish were, the fishermen perceived reductions in stock sizes sufficient to change to fishing grounds further at sea. By the end of the 1850s catches had declined sufficiently for many ships to undertake the longer voyage to the Gulf of St. Lawrence and the Grand Banks. Similar processes of moving from fishing ground to fishing ground in a relentless effort to earn marginal benefit are well-known for modern fisheries and well-documented for many historical fisheries, perhaps best of all for the mid-nineteenth century fishery off the Labrador coast (Myers 2001). This is a fishing strategy that is known as serial depletion and may be recognized again and again in historical records from all over the world.

Declining catches were offset by new technology. French fishermen introduced tub trawls to the Scotian Shelf fishery, and soon the Americans no longer used the traditional hand lines with two to four hooks per man but upwards of 400–500 hooks per crewman. Thus the catchment area of one boat increased immensely. Unfortunately, although catches went up in the short run, in a matter of a few years the fish stock was showing clear depletion signals, being caught at a smaller size and catch per unit effort of the fishermen declining. In the 1850s, based on the fishing effort, the adult cod biomass may be estimated to have been of the order of 1.26 million tonnes. The comparable estimate was of 50,000 tonnes in the 1990s (Rosenberg *et al.* 2005). The reduction in abundance is obvious and even starker than the decline of the cod and ling in the North Sea.

The American waters of the nineteenth century were incredibly rich and are today impoverished to a degree that present-day managers would not realize without historical research. It would be naive to suggest that restoration targets may simply be based on historical values. If an ecosystem regime shift has occurred, the ecosystem may never be able to rebuild to past abundance levels. However, analysis of the age structure of modern cod populations indicates that conservation measures in recent years have helped to rebuild a stock of older and better spawners, resembling the stock of the 1860s (Alexander *et al.* 2009).

An even more short-lived success than cod was the Atlantic halibut fishery, which became severely depleted owing to a rapidly developing taste for the halibut fins among American consumers from the 1840 to 1880s. This fishery has never regained its former strengths (Grasso 2008).

1.3.5 Southeast Australia

The Australian southeast shelf region was the first HMAP case study to be completed and the first case study to apply catch rate standardization methods rigorously, single species population models, and the Ecopath ecosystem modeling approach to historical data. Compared with other HMAP case studies, the Australian southeast shelf data set is of particularly high quality. It is comparatively short in duration, beginning only in 1915 with some years missing, but it was collected in a systematic manner since the beginning of the fishery and has data for a considerable number of species. The fishery was initially set up by the government and records kept to convince private enterprise of the profitability of the industry.

What the evidence allows us to see are the effects of a trawl fishery on a pristine marine ecosystem, or as untouched by humans as was ever documented (Klaer 2005). Indigenous fishing in the Sydney region was mainly concentrated on the snapper, which lives in nearshore waters, and the indigenous fishery may have impacted the population. European settlers in Sydney added problems of pollution and disturbance. However, the southeast Australian shelf and slope marine animal populations may largely be considered to have been in a pristine state until the Australian government began fisheries experiments with a single trawler at the turn of the century. The main impact and the start of historical documentation came with the arrival of three British trawlers, purchased to begin commercial fisheries for a state company by May 1915. The operation was privatized in 1923 and peaked with 17 steam trawlers in 1929. Danish seine vessels were brought in through the 1930s but during World War II activities almost came to a halt. Catches were resumed after the war. The fishery was primarily in shelf waters between 50 and 200 m depth. It targeted tiger flathead (*Neoplatycephalus richardsoni*), jackass morwong (*Nemadactylus macropterus*), and redfish (*Centroberyx affinis*) until the 1970s.

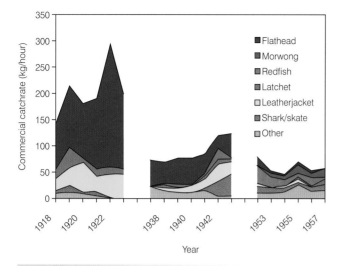

Fig. 1.5
Ecosystem effects of early trawling. Commercial catch rates by species on the southeast Australian continental shelf. Contribution per species to the total commercial catch per unit effort by year. Source: Klaer (2001), Fig. 10. Reproduced with permission.

The unique collection of 65,000 individual haul records, vessel logbooks, and landings data were used by Neil Klaer to develop relative indices of abundance for the major commercial fish species, to estimate the biomass of those species, and to examine ecosystem changes in the southeast Australian shelf over the period since the start of commercial fishing (Fig. 1.5). The results showed an overall decline in yields per haul over the history of steam trawling. Although initially the ships experienced larger catches as the men got acquainted with the new fishing grounds, the catch per hour trawled during 1937–43 was much lower than that of 1920–23. The fishing fleet moved further afield and into deeper waters as catch rates declined. In the early years, Botany Bay off Sydney yielded excellent catches of fish "very large and bursting with roe" and fishermen even talked of the "Botany Glut" from September to early December. However, by 1926 the glut failed to occur and by the 1930s the Botany Bay ground was no longer visited for commercial operations. As described for the Labrador coast, the Sydney fishermen began a mining operation that took them further up and down the coast and to deeper grounds. As the flathead was fished out, new, previously discarded fish began to be landed for the market. World War II gave a temporary reprieve for the stocks and the available biomass increased slightly, but catches quickly reduced the available biomass further.

Despite the substantial changes in the relative biomass of the main commercial species from 1915 to 1961, there were no great changes to relative biomass at lower trophic levels (megabenthos and lower). The biomass density of all large fish (flathead, latchet, other large fish, leatherjacket,

redfish, and morwong) decreased by 73% from 1915 to 1961, whereas the biomass density of invertebrates at the same time decreased by only 6% (Klaer 2004). In other words, the effect of the southeast Australian trawl fishery was a fishing down the food web, as described by Pauly *et al.* (1998a), resulting in fewer of the large fish, while the small fish, plankton, and crustaceans remained.

Klaer's analysis is unique both for the pristine nature of the ecosystem before documented trawling and because of the amount of data available throughout the fishery. By the time that conservation measures and restrictions on fishing effort were taken, the ecosystem had ceased to look anything like it had before, and a fishery management system that was informed only by recent data would have no knowledge of what had been lost, nor indeed of what might be if the system were allowed to rebuild by holding back on fishing effort. Until this study was made, no such information was available to managers.

1.3.6 Southwest Africa

Human activities were studied in the Benguela Current ecosystem (Griffiths *et al.* 2005). Like the southeast Australian fisheries, the main human impact occurred relatively recently, but is unfortunately less well documented. The aboriginal epoch until 1652 was characterized by low levels of mainly intertidal exploitation, whereas the pre-industrial epoch to about 1910 saw intense exploitation of few large, accessible species. The introduction of mechanized technology marked the beginning of the industrial epoch which had a huge increase in landings. Catches were stabilized in the post-industrial period after 1975, although there have been increasing impacts of non-fisheries on the system. Total extractions in the past 200 years were calculated at more than 50 million tonnes of biomass, with annual removals above one million tonnes in the 1960s. Subsequently landings have declined by over 50%. The short, sharp impact of fisheries in the twentieth century led to severe reduction of populations of whales, seals, and pelagic and demersal fish, which are now all showing signs of recovery thanks to declining fishing pressure and implementation of new management schemes. Inshore stocks, particularly abalone, rock lobster, and inshore linefish, remain severely depressed and are exposed to intense fishery and gathering for human subsistence.

1.3.7 Caribbean

Far from being a pristine ecosystem, the Caribbean was intensively fished already before the arrival of the Europeans, and subsequent removals happened on a massive scale through the seventeenth and eighteenth centuries. Twentieth-century fishing pressure has continued a trajectory of fishing down the food web.

The Caribbean HMAP team has continued Jeremy Jackson's pioneering work by documenting historical distributions of large marine vertebrates. In the absence of quantitative evidence for many species, the team analyzed a total of 271 descriptions from 1492 to the present, to demonstrate consistent patterns of decreased abundance and increased rarity through time. By assigning quantitative values to qualitative sightings of species, the team was able to build a comprehensive and statistically significant picture of decrease in abundance and increase in rarity in megafauna since the European settlement. The green turtle population, which had an abundance of 15.5 million to 116 million at the time of Columbus, is considered highly endangered today as only two nesting beaches with more than 100 nesting females now remain, whereas at least 23 nesting beaches have been eliminated. Similarly, 32 hawksbill turtle nesting beaches have been lost (McClenachan *et al.* 2006). The decline of the monk seal followed a clear path: exploitation first reduced the range of the population, which was estimated at 0.23–0.68 million around 1500. At the beginning of the twentieth century, monk seals occupied only 30% of their former range. Hunting in the two remaining breeding areas finally killed off the monk seal from Caribbean waters by 1952. The range of the American crocodile and the West Indian manatee was reduced in the eastern Caribbean before European settlement. Manatees were largely eliminated from the Lesser Antilles by 1700, and all but eliminated from all island sites by 1900. The decline of crocodiles was similar, except that they seem never to have been present in the Lesser Antilles. Estimates of abundance show probable declines of at least 90% for each species. The removal of large animals will have had significant consequences for food webs and the resilience of the marine system. Because hawksbill turtles consume primarily sponge matter, Bjorndal & Jackson (2003) suggest that large numbers of hawksbill turtles could have maintained high coral cover and sponge species diversity, with a concurrent increase in total benthic invertebrate diversity.

Other scientific achievements so far include an assessment of the degree of population change, ecological consequences of population change, and the historical distribution of large marine animals – including green turtles, hawksbill turtles, and monk seals – across the entire Caribbean basin and Gulf of Mexico (McClenachan *et al.* 2006). The team has documented and quantified changed Caribbean food webs as a whole (Bascompte *et al.* 2005). The Caribbean studies have brought out the potential effect of early non-mechanized fishing technologies on coral reef animals. Early declines in Jamaican coral reef fauna were a function of less than half a million people using simple gear. The research makes evident that sustainable levels of fishing in Jamaica are no more than 10–20% of current catch, or equivalent of a level of extractions already reached 100 years ago (Jackson 1997; Hardt 2008).

1.3.8 White and Barents Seas

The question of the impact of climate variability on fish populations is in the foreground of the work of the HMAP team working on the White and Barents Seas led by Julia Lajus. The climate effects are especially pronounced in high latitudes because many fish species occur there at the border of their distribution range. Moreover, during several centuries these effects were not masked by the human impact on ecosystems, which in the Russian north was minimal up to mid-twentieth century owing to very slow growth of the human population in the area and the late start of industrial development. Although fishing effort did increase steadily during this period, extractions were too low to influence fish populations significantly. Therefore historical data on fisheries provided a convenient research tool to trace the natural dynamics of populations, allowing reconstruction of effects of climate going back several centuries.

The team has analyzed landings records from monastic and governmental sources from the seventeenth to the twentieth centuries. The records of the Solovetsky Monastery, the largest monastery in the White and Barents Sea area, which controlled sea and river fisheries of the high north, proved to be especially rich. These records are possibly unique because in many cases they contain not only the number but also the weight of caught fish: Atlantic salmon, Atlantic cod, and halibut. Atlantic salmon was one of the most valuable products of the local economy. They were fished mostly in the lower parts of rivers, using weirs that were not changed technologically over the centuries (Fig. 1.6). This makes fishing effort commensurable over time and allows comparison of historical catch data of the seventeenth and eighteenth centuries (published in D. Lajus *et al.* 2007a) with official statistical data available since the last quarter of the nineteenth century.

Analysis of historical and statistical data from four different localities around the White and Barents Seas for the seventeenth and eighteenth centuries shows a positive correlation of catches with ambient temperature (D. Lajus *et al.* 2005). The conclusion was drawn that before the middle of the twentieth century the population dynamics of salmon was mostly driven by natural factors (D. Lajus *et al.* 2008a).

Signs of climate-related dynamics were observed also for other fish, such as cod, halibut, and herring, although correlation did not approach statistical significance (D. Lajus *et al.* 2005, 2007b). In particular, the White Sea herring fishery, of economic importance since the eighteenth century, showed considerable short-term fluctuations of catches both because of social and natural factors and their interaction, which may confound climate effects (D. Lajus *et al.* 2007b). Climate effects were also pronounced on Arctic marine mammals such as white whales, Greenlandic seals, narwhals, and others, which considerably changed their distribution patterns migrating to more southern

Fig. 1.6
Weir for catching Atlantic salmon at Kitsa River (tributary of Varzuga River). From the album "Risunki k issledovaniiu rybnykh i zverinykh promyslov na Belom i Ledovitom moriakh", St. Petersburg, 1863.

regions than usual in cold periods 1800–1809 and 1877–1903, and again in 1970–80 (D. Lajus *et al.* 2008b).

For marine mammals, anthropogenic pressure became a significant factor earlier than for fish. Hunting impacted the general dynamics of the population of the eastern walrus from at least the seventeenth century and may explain changes in its distribution range over several centuries. However, the walrus population was able to sustain itself as long as remote islands such as Franz Josef Land were not discovered by humans. Improvements of navigation and hunting techniques in the late nineteenth century resulted in a considerable decrease in the walrus population by the middle of the twentieth century. For fish, particularly Atlantic salmon, clear stress signals related to human activities such as overfishing and development of forestry with timber-rafting became apparent only by the end of the nineteenth century (Alekseeva & Lajus 2009).

Conducting fishing operations in such remote areas with severe climate conditions was especially difficult for humans in the pre-industrial age, causing clear interaction between natural and human factors. Fisheries productivity varied because of climate conditions, and, in particular, the price of salmon was negatively correlated with the level of catches and population abundance (J. A. Lajus *et al.* 2001). For herring, the long-term trend was a positive relation between catch size and human population in the area, likely reflecting an increase of fishing effort, emphasizing the importance of detailed historical analysis when reconstructing long-term trends of population abundance (D. Lajus *et al.* 2007b).

1.3.9 World whaling

Whaling was one of the most profitable extractive industries ever undertaken, and it was likely the one activity that impacted life in the oceans more than any other single pre-industrial activity. Relative to the fisheries it is extremely well documented and well researched (Fig. 1.7). Yet we still do not know how many whales there used to be in the ocean and where. Whereas historical fisheries research has really only developed in the past decade, the ecological history of whaling has been pursued for management purposes for many years. Since its origin in 1946 the International Whaling Commission has had a keen interest in estimating historical population sizes based on catch records in order to identify a conservation target for the rebuilding of whale populations. The approach taken by the HMAP team is to estimate historical abundance using population models based on the evidence of historical logbooks and catch records, and using present-day abundance estimates.

A global overview of the history of whaling identified 120 whaling operations grouped into 14 methodology-defined eras (Reeves & Smith 2006). Maps of the spatial and temporal extent of whaling in the nineteenth century allow resource managers to identify areas where populations have and have not recovered to their pre-whaling distribution, and to identify formerly occupied areas where whales are now essentially absent. Where recovery has been less than complete, human activities may need to be better managed to allow further recovery. Spatial distribution should become a more important element in assessing population recovery, in addition to the more usual measure based on current population size as a fraction of historical, or pre-whaling, population size.

The catch history of North Atlantic humpback (*Megaptera novaeangliae*) whaling was estimated by the HMAP team (Smith & Reeves 2006), and used in a stock assessment sponsored by the International Whaling Commission to estimate that current abundance is 37% to 70% of the historical abundance, which itself was 22,000–26,000. A previously unknown humpback whale feeding ground was

Fig. 1.7
Logbook for the ship *Abigail* of New Bedford, Benjamin Clark, master. The logbook was kept by Holder Wilcox during the November 19, 1831–June 12, 1835 whaling voyage to the North and South Atlantic and South Pacific Oceans. The logbook begins five days after the start of the voyage. Courtesy of New Bedford Whaling Museum.

identified based on nineteenth century whaling logbooks (Reeves *et al.* 2004), and the logbook data were used to help direct the Census Patterns and Processes of the Ecosystems of the Northern Mid-Atlantic (MAR-ECO) project (see Chapter 6). In the mid-nineteenth century some humpback whales migrating from breeding to feeding areas remained at mid-summer in oceanic habitats near the mid-Atlantic Ridge. Today humpbacks have only been known in summer months on coastal feeding grounds around the North Atlantic. Similarly, textbook assumptions on the distribution and abundance of North Pacific right whales (*Eubalaena japonica*) have been corrected (Josephson *et al.* 2008). The causes of failure of the North Pacific right whale to recover both numerically and spatially after the severe depletion of the 1840s continue to be a mystery.

Working as part of the HMAP New Zealand project, the team described the historical distribution and landings of southern right whales (*Eubalaena australis*) through analysis of over 150 whaling logbooks and other landings records. With 95% statistical confidence, population modeling shows that southern right whales numbered between 22,000 and 32,000 in the early 1800s, declining rapidly once whaling began. By 1925, perhaps as few as 25 reproductive females survived. Today the population has recovered to some 1,000 animals around sub-Antarctic islands south of New Zealand (Jackson *et al.* 2009).

Because of the strong need to establish past population sizes to inform conservation policy, the HMAP team is working closely with scientists who have proposed another possible modeling approach to working on historical landings data. This is the Whales Before Whaling project headed by Steven Palumbi at Stanford University. Palumbi's project

aims to measure the amount of genetic diversity of current populations and use knowledge of DNA mutation rates to estimate how many individuals a population must sustain over time to accumulate the measured diversity. Based on this method, Palumbi estimates that the pre-contact population size of the eastern Pacific gray whales (*Eschrichtius robustus*) was three to five times larger than the population size calculated by historical data. The HMAP team has therefore scrutinized the available landings and total removals and, in cooperation with the US National Marine Fisheries Service and the International Whaling Commission's Scientific Committee, is working to address this apparent inconsistency between historical whale removals, apparent population increases measured over the latter half of the twentieth century, and the genetic variability model (T. Smith, personal communication).

We now know more about the human drivers behind the whaling operations. In particular the project focused on how the enormously profitable so-called Yankee whaling changed from 1780 to 1924. The team has documented the nature of the changes in vessels, rigging, destinations, and catches over the lifespan of this fishery. They suggest that questions of the effect of whaling on the whale populations must be asked at regional rather than global levels, and that indeed regional depletion, even extirpation, was a frequent occurrence (see, for example, Josephson *et al.* 2008; Jackson *et al.* 2009). For example, contrary to Whitehead (2002), they show strong depletion of sperm whale abundance in the Pacific and raise the question why such depletion apparently did not occur in the Atlantic (Smith *et al.* in press). They also suggest that global economic analyses that do not account for these regional changes (see, for example, Davis *et al.* 1997) greatly oversimplify the dynamics of this fishery and are misleading about the causes of its decline.

1.3.10 Megamollusks

Shellfish have been heavily collected and used for meat and ornaments through history. Although some shells will be traded, most will be discarded. Shell middens have been known since the nineteenth century as excellent archaeological sources of information on coastal-dwelling peoples. Through the nineteenth and twentieth centuries, the oyster reefs of the US Atlantic and Pacific coasts were severely impacted by fishing (Kirby 2004) as were North European oyster banks (Holm 2005).

Thanks to the initiative of Andrzej Antczak of Venezuela, we now have a global series of studies of human–megamollusk interactions. Generally, mollusk populations are quite exposed to human impact as they may be collected close to the shoreline. The southwest African HMAP project showed that human gathering of inshore shellfish may reach a level where it threatens certain inshore species. In Papua New Guinea the exploitation of the giant clam (family Tridacnidae), which seems to have been at sustainable levels through a long period of history, has in recent decades necessitated a ban on collecting for export (Kinch 2008). Similarly, although ecological impacts such as declining size may be detected for the pre-Hispanic exploitation of queen conch (*Strombus gigas*) beds off Venezuela (Fig. 1.8), the exploitation was much less harmful to the mollusk population than the short-term modern fishery between 1950 and the 1980s (Antczak *et al.* 2008).

However, few human populations have been so dependent on mollusks for food to cause local or species extinctions (Bailey & Milner 2008).

The study of megamollusks is particularly rewarding for our understanding of human values and trade. The queen conch was heavily targeted between about 1100 and 1500 at the offshore islands of Los Roques, Venezuela, and both the meat and shells were brought to the mainland for consumption and redistribution. Ceremonial activity on the islands and the use of the queen conch as a symbol on the mainland indicate that the mollusk had achieved a central importance to north-central pre-Hispanic peoples in Amerindian Venezuela (Antczak and Antczak 2008).

1.3.11 Emerging projects

Two HMAP projects have not yet arrived at publication stage because they were only begun fairly recently: the New Zealand and the Southeast Asia projects. The Maori were experienced sailors and hunters, and on their arrival to New Zealand the Europeans encountered nothing like a pristine ecosystem. The Taking Stock project is therefore confronted with understanding fully the impact of pre-European, pre-industrial technologies on what was, until the arrival of the Maoris around 1300, a pristine marine and terrestrial ecosystem. The project will conclude by the end of 2010, but it is clear that the distributions and population sizes of seabirds, fur seals, and sea-lions were considerably impacted relative to pre-Maori conditions already by 1800. Fur seals, for instance, had been extirpated from North Island and only colonies at the southern tip of South Island awaited the arrival of European hunters to be ren-

Fig. 1.8
Pre-Hispanic mega-middens of queen conch (*Strombus gigas*) on La Pelona Island, Los Roques Archipelago, Venezuela, A.D. 1200–1500. Copyright Magdalena and Andrzej Antczak.

dered extinct. The Southeast Asia project covers a vast area and focuses on indigenous and American historical whaling in the Philippines, Taiwanese offshore tuna fishery, and shark fishing in Indonesia. All HMAP Asia research projects are now at an advanced stage, and a monograph (representing the main output of the project) is being prepared for publication by the end of 2010.

1.4 Conclusions

What is the big picture emerging from these regional and species projects? What is the scale of change between now at the completion of the Census and, say, 100 years, or between now and the origin of large-scale pre-industrial fisheries? When were the decisive moments? What were the main drivers? These are questions that we are grappling with now as the Census is coming to an end. Already we know some of the answers but many more will emerge as we have an overview of the vast amount of information that has been uncovered.

- The HMAP project has resolved the problem of the baseline. We now know that everywhere we look there is potential to know much more about the past and that we need to inform ourselves of the past both to enrich our understanding of the present and to inform our future preferences and decisions. The HMAP project is the beginning of the historical discovery of ocean and human interaction. Even after 10 years we have far from exhausted the archives and archaeology of the sea. We have made significant discoveries both of the importance of the sea to human life and of the impact of humans on the sea. Historical baselines should be an important element of future conservation plans. In some ecosystems, stocks will rebuild if given a chance. In other systems, regime shifts may have forever changed the food web so that past abundances of top predators will have a slim chance of rebuilding. Yet, environmental history has a very real role to play for future ocean policy by preserving the memory of what once lived in our seas. New management policies can be developed to promote recovery and prevent further declines of species and ecosystems.
- The distribution and abundance of marine animal populations change dramatically over time. The effects of climate variability during the Little Ice Age on marine mammals as well as fish stocks are clearly documented by the White and Barents Seas project and the Baltic Sea project, whereas the North Sea documents the effects of the past 20 years of warmer surface water for the introduction of southern species. Historical data will inform us of past patterns of distribution of species such as demonstrated for North

Atlantic humpback whales and North Pacific right whales, and indeed for the southern North Sea haddock.

- We now know that major extractions occurred more than 2,000 years ago in the Mediterranean and Black Seas, we know the basic outline of the origins of commercial fisheries in Northern Europe, and we have a good sense of developments in many regions around the globe during the past 500 years ranging from the Caribbean to the White Sea, from southeast Australia to South Africa. Pre-industrial technologies were sufficient to put marine animal populations under severe stress, and indeed by the late nineteenth century extractions in Europe, North America, and the Caribbean had reached levels that would be equivalent to today's Total Allowable Catches. The effects of large-scale removals in the seventeenth century North Sea herring fishery and the eighteenth century Grand Banks cod fishery may have been significant.
- Regime shifts may have occurred as a result of some pre-industrial fisheries such as the Caribbean, whereas the effects of industrial gear were striking in the southeast Australian case when a pristine ecosystem changed dramatically after 30 years of trawling. Collapses of stocks and serial depletion are widespread phenomena, even before the industrial era, but in most cases populations have been able to rebuild.
- Overall it seems that the removals of large marine animals have reduced abundance by an order of magnitude; a recent review concluded that 256 exploited populations declined 89% from historical abundance levels on average (range: 11–100%) (Lotze & Worm 2009). The detailed historical evidence for cod, ling, and bluefin tuna corroborate this general picture. Smaller animals have been less impacted and indeed may have replenished as larger predators have been removed.
- Human impacts on coastal environments have been similar across the globe, even in quite different ecosystems (Lotze et al. 2006). Although few exploitable marine species have gone extinct, there is concern that entire marine ecosystems have been depleted beyond recovery. Major impact on sensitive ecosystems such as the Wadden Sea may have happened before 1900, and there is therefore a need for deep historical assessment of ecosystem change.

We know now that we can push back the chronological limits of our knowledge.

More importantly perhaps, we now have the basis from which to start raising new questions: what more can we know about the drivers of change from the human perspective, can we extrapolate from the local or regional to the global, what about the continents or large countries that did not have an HMAP team such as much of South America and Africa, what about India and China? Can we unlock the sources to some of the large industrial fisheries in the deep seas that have become such important fisheries areas in recent decades for already endangered species such as orange roughy and Patagonian toothfish?

All these are challenging but certainly not impossible questions. They are questions that will not only be raised but answered in coming years as research continues beyond the HMAP project. Data rescue and digitization will provide vastly increased libraries of the kind we already know and that have served us well. We are beginning to understand the main drivers of change from the human perspective such as changing patterns of consumption, technology, price differentials, politics, and cultural preferences. We now know enough to begin to understand the importance of marine products for human consumption, and we have a much better basis from which to assess the main drivers of human marine exploitation. In the academic realm the historical turn of marine ecology is now a given, and the ecological challenge to traditional historical models cannot be neglected. In the future, new techniques and methodologies may be used to move what may now seem the unknowable into the realm of the knowable. If knowing the basics of marine ecological history seemed impossible some 10–15 years ago, we stand a good chance that in the next 10–15 years there will be several major breakthroughs. Scientific advances in fields such as genetics and stable isotope analysis have already impacted what we know and much more is to come. Advanced computer animations and geographic information systems (GIS) will be fully used to show changes in abundances, distributions over time, and how they could look in the future (under recovery situations), and we shall see new quantitative approaches for modeling changes in biodiversity, species' abundance, and distribution.

Perhaps new methodologies will enable us to lift the veil on what the pristine sea looked like before human contact. So far nearly all the information accessible to us relates to early human records of contact. Certainly we shall know much more about the implications of the ice ages for "trapped" species. Advances of molecular biology and ocean biogeography will tell us of the separation of species and subsequent development. Sediment cores will be unlocked as a library of past DNA of what used to be swimming in the water column above. All of this will underline what was one of the first steps of the HMAP project, the need to train the next generation of researchers in interdisciplinary skills.

Acknowledgments

We acknowledge the members of the HMAP Steering Committee for their help and assistance: Andrzej Antczak, Michaela Barnard, James Barrett, Ruthy Gertwagen, Jeremy Jackson, Julia Lajus, Alison Mcdiarmid, Henn Ojaveer, Andrew A. Rosenberg, Tim Smith, David J. Starkey, and Malcolm Tull.

References

Alekseeva, Ya.I. & Lajus, D.L. (2009) Interactive effect of fisheries and forestry on Atlantic salmon population abundance in the Russian North in the end of the 19th – beginning of the 20th C. In: *First World Congress on Environmental History (WCEH), Copenhagen.* Abstract, p. 97.

Alexander, K.E., Leavenworth, W.B., Cournane, J., *et al.* (2009) Gulf of Maine cod in 1861: historical analysis of fishery logbooks, with ecosystem implications. *Fish and Fisheries* 10, 428–449.

Alheit, J. & Hagen, E. (1997) Long-term climate forcing of European herring and sardine populations. *Fisheries Oceanography* 6, 130–139.

Anderson, K. (2006) Does history count? *Endeavour* 30, 150–155.

Anon. (1998) On the history of marine fisheries: a workshop sponsored by the Alfred P. Sloan Foundation. Woods Hole Oceanographic Institution (unpublished).

Antczak, A., Posada, J.M., Schapira, D., *et al.* (2008) A history of human impact on the queen conch (*Strombus gigas*) in Venezuela. In *Early Human Impact on Megamolluscs* (eds. A. Antczak & R. Cipriani), *British Archaeological Reports* S1865.

Antczak, Ma.M. & Antczak, A. (2008) Between food and symbol: the role of marine molluscs in the late pre-Hispanic north–central Venezuela. In *Early Human Impact on Megamolluscs* (eds. A. Antczak & R. Cipriani), *British Archaeological Reports* S1865.

Bailey, G. & Milner, N. (2008) Molluscan archives from European prehistory. In *Early Human Impact on Megamolluscs* (eds. A. Antczak & R. Cipriani), *British Archaeological Reports* S1865.

Barrett, J.H., Locker, A.M. & Roberts, C.M. (2004) The origins of intensive marine fishing in medieval Europe: the English evidence. *Proceedings of the Royal Society of London B* 271: 2417–2421.

Barrett, J.C., Johnstone, J., Harland, W., *et al.* (2008) Detecting the medieval cod trade: a new method and first results. *Journal of Archaeological Science* 35, 850–861.

Bascompte, J., Melián, C.J. & Sala, E. (2005) Interaction strength combinations and the overfishing of a marine food web. *Proceedings of the National Academy of Sciences of the USA*, **102**, 5443–5447.

Baum J.K. & Worm B. (2008) Cascading top-down effects of changing oceanic predator abundances. *Journal of Animal Ecology* 78, 699–714.

Baumgartner, T.R., Soutar, A. & Ferreira-Bartrina, V. (1992) Reconstruction of the history of Pacific Sardine and Northern Anchovy populations over the past two millennia from sediments of the Santa Barbara Basin. *California Cooperative Fishery Investigations Reports* 33, 24–40.

Bekker-Nielsen, T. (2005) The technology and productivity of ancient sea fishing. In: *Ancient Fishing and Fish Processing in the Black Sea Region* (ed. T. Bekker-Nielsen), pp. 83–95. Aarhus: Aarhus Universitetsforlag.

Beusekom, J.E.E. v. (2005) A historic perspective on Wadden Sea eutrophication. *Helgoland Marine Research* 59, 45–54.

Bjorndal, K.A. & Jackson, J.B.C. (2003) Roles of sea turtles in marine ecosystems: reconstructing the past. In: *The Biology of Sea Turtles*, vol. II (eds. P.L. Lutz, J.A. Musick & J. Wyneken), pp. 259–273. Boca Raton, Florida: CRC Press.

Bolster, J. (2006) Opportunities in marine environmental history. *Environmental History* 11, 567–597.

Bolster, J. (2008) Putting the Ocean in Atlantic history: maritime communities and marine ecology in the Northwest Atlantic, 1500–1800. *American Historical Review* 113, 19–47.

Cadigan, S.T. & Hutchings, J.A. (2001) Nineteenth-century expansion of the Newfoundland fishery for Atlantic cod: an exploration of underlying causes. In: *The Exploited Seas: New Directions for Marine Environmental History, Research in Maritime History*, vol. 21 (eds. P. Holm, T.D. Smith & D.J. Starkey), pp. 31–65. St. John's, Newfundland: International Maritime Economic History Association.

Continuous Plankton Recorder (2009) Continuous plankton recorder (CPR) data from the Sir Alister Hardy Foundation for Ocean Science (SAHFOS). Available at http://iobis.org.

Curtis, R. (2005) Sources for production and trade of Greek and Roman processed fish. *Ancient Fishing and Fish Processing in the Black Sea Region* (ed. T. Bekker-Nielsen), pp. 31–46. Aarhus: Aarhus Universitetsforlag.

Cushing, D.H. (1982) *Climate and Fisheries*. Bury St Edmunds, UK: Academic Press.

Davis, L.E., Gallman, R.E. & Gleiter, K. (1997) In *Pursuit of Leviathan: Technology, Institutions, Productivity and Profits in American Whaling, 1816–1906*. Chicago: University of Chicago Press.

Eero, M., Köster, F.W. & MacKenzie, B.R. (2008) Reconstructing historical stock development of the eastern Baltic cod (*Gadus morhua*) before the beginning of intensive exploitation. *Canadian Journal of Fish Aquaculture and Science* 65, 2728–2741.

Ejstrud, B. (2005) Size matters: estimating trade of wine, oil and fish-sauce from amphorae in the first century AD. In: *Ancient Fishing and Fish Processing in the Black Sea Region* (ed. T. Bekker-Nielsen), pp. 171–181. Aarhus: Aarhus Universitetsforlag.

Enghoff, I.B. (1999) Fishing in the Baltic region from the 5th century BC to the 16th century AD: evidence from fish bones. *Archaeofauna* 8, 41–85.

Enghoff, I.B. (2000) Fishing in the southern North Sea region from the 1st to the 16th century A.D.: evidence from fish bones. *Archaeofauna* 9, 59–132.

Enghoff, I.B., MacKenzie, B.R. & Nielsen, E.E. (2007) The Danish fish fauna during the warm Atlantic period (ca. 7,000–3,900 BC): forerunner of future changes? *Fisheries Research* 87, 285–298.

Francis, R.C., Field, J., Holmgren, D. & Strom, A. (2001) HMAP and the Northern California Current ecosystem. In: *The Exploited Seas: New Directions for Marine Environmental History* (eds. P. Holm, T.D. Smith & D.J. Starkey). *Research in Maritime History* 21, 123–140.

Garstang, W. (1900) The impoverishment of the sea: a critical summary of the experimental and statistical evidence bearing upon the alleged depletion of the trawling grounds. *Journal of the Marine Biological Association of the United Kingdom* 6, 1–69.

Gaumiga, R., Karlsons, G., Uzars, D. & Ojaveer, H. (2007) Gulf of Riga (Baltic Sea) fisheries in the late 17th century. *Fisheries Research* 87, 120–125.

Gertwagen, R. (2008) Approccio multidisciplinare allo studio dell'ambiente marino e della pesca nel Medio Evo nel Mediterraneo orientale. In: *Il Mare Come Era* (eds. R. Gertwagen, S. Raicevich, T. Fortibuoni & O. Giovanardi), pp 144–182. Proceedings of the II HMAP Mediterranean and the Black Sea Project. Chioggia (Italy), 27th–29th September 2006.

Gertwagen, R., Raicevich, S., Fortibuoni, T. & Giovanardi, O. (eds.) *Il Mare Come Era*. Proceedings of the II HMAP Mediterranean and the Black Sea Project. Chioggia (Italy), 27th–29th September 2006.

Grasso, G.M. (2008) What appeared limitless plenty: the rise and fall of the 19th-century Atlantic halibut fishery. *Environmental History* 13, 1.

Griffiths, C., van Sittert, L., Best, P.B., *et al.* (2005) Impacts of human activities on marine animal life in the Benguela: a historical overview. *Oceanography and Marine Biology* 42, 303–392.

Hardt, M.J. (2008) Lessons from the past: the collapse of Jamaican coral reefs. *Fish and Fisheries* 10, 143–158.

Heithaus, M.R., Frid, A., Wirsing, A.J. & Worm, B. (2008) Predicting ecological consequences of marine top predator declines. *Trends in Ecology and Evolution* 23, 202–210.

Hoffmann, R.C. (2004) A brief history of aquatic resource use in medieval Europe. *Helgoland Marine Research* 59, 22–30.

Holm, P. (1999) Fiskeriets økonomiske betydning i Danmark 1350–1650. Sjæk'len 1998 (Esbjerg), 8–42.

Holm, P. (2002) History of marine animal populations: a global research program of the Census of marine life. *Oceanology Acta* 25, 207–211.

Holm, P. (2003) The Bohuslen herring: interlude to Dutch supremacy in the European fish market, 1556–1589. In *het kielzog. Maritiem-historische studies aangeboden aan Jaap R.Bruijn, bij zijn vertrek als hoogleraar zeegeschiedenis aan de Universiteit Leiden.* (eds. L.M. Akveld, F. Broeze, F.S. Gaastra, *et al.*), pp. 282–288. Amsterdam.

Holm, P. (2004) *Fishing. Encyclopedia of World Environmental History* (eds. S. Krech III, J.R. McNeill & C. Merchant), pp. 529–534.

Holm, P. (2005) Human impacts on fisheries resources and abundance in the Danish Wadden Sea, c. 1520 to the present. *Helgoland Marine Research* 59, 39–44.

Holm, P. & Bager, M. (2001) The Danish Fisheries, c. 1450–1800: Medieval and Early Modern Sources and their Potential for Marine Environmental History. In: *The Exploited Seas: New Directions for Marine Environmental History* (eds. P. Holm, T.D. Smith & D.J. Starkey) *Research in Maritime History* 21, 97–122.

Holm, P. & Starkey, D.J. (eds.) (1995–1999) *Studia Atlantica* 1–4. Esbjerg.

Holm, P. & Starkey, D.J. (1998) Fishing matters: interdisciplinary approaches to the long-term development of fisheries (unpublished).

Höglund, H. (1972) On the Bohuslän herring during the great herring fishery. Institute of Marine Research, Lysekil. Series Biology Report, No. 20.

Huxley, T.H.H. (1883) Opening address. Fisheries exhibition, London (1883). Available at http://math.clarku.edu/huxley/SM5/fish.html.

ICES (2006) North Sea Herring (Sub-area IV, Division VIId-e and Div IIIa (autumn spawners)). Available at http://www.marine.ie/NR/rdonlyres/5A2F51D5-72C2-4BFB-805B-F503907CD52E/0/HerringinIV_NorthSea06.pdf.

ICES (2008) Report of the working group on demersal stocks in North Sea and Skagerrak – combined spring and autumn. ICES CM 2008/ACOM: 09745-834.

Jackson, J., Baker, S., Carroll, E., *et al.* (2009) Taking stock of the New Zealand southern right whale (tohora): a revised history of the impact of early whaling. Unpublished conference paper, Oceans Past II, Vancouver 2009.

Jackson, J.C.B. (1997) Reefs since Columbus. *Coral Reefs* 16, 23–32.

Jones, G.A. (2008) Quite the choicest protein dish: the costs of consuming seafood in American restaurants, 1850–2006. In: *Oceans Past: Management Insights from the History of Marine Animal Populations* (eds. D.J. Starkey, P. Holm & M. Barnard), pp. 47–76. London: Earthscan.

Josephson, E.A., Smith, T.D. & Reeves, R.R. (2008) Depletion within a decade: the American 19th century North Pacific right whale fishery. In: *Oceans Past: Management Insights from the History of Marine Animal Populations* (eds. D.J. Starkey, P. Holm & M. Barnard), pp. 133–149. London: Earthscan.

Kirby, M. X. (2004) Fishing down the coast: historical expansion and collapse of oyster fisheries along continental margins. *Proceedings of the National Academy of Sciences of the USA* 101, 13096–13099.

Kinch, J. (2008) From prehistoric to present: giant clam (Tridacnidae) use in Papua New Guinea. In *Early Human Impact on Megamolluscs* (eds. A. Antczak & R. Cipriani), *British Archaeological Reports* S1865.

Klaer, N.L. (2001) Steam trawl catches from south-eastern Australia from 1918 to 1957: trends in catch rates and species composition. *Marine and Freshwater Research* 52, 399–410.

Klaer, N.L. (2004) Abundance indices for main commercial fish species caught by trawl from the south-eastern Australian continental shelf from 1918 to 1957. *Marine and Freshwater Research* 55, 561–571.

Klaer, N.L. (2005) Changes in the structure of demersal fish communities of the south east Australian continental shelf from 1915 to 1961. PhD thesis, University of Canberra.

Kraikovski, A., Lajus, J. & Lajus D. (2008) Fisheries on the south-eastern coast of the Gulf of Finland and the adjoining river basins, 15th–18th centuries. In: *Leva vid Östersjöns kust. En antologi om resursutnyttjande på båda sidor av Östersjön ca 800–1800* (ed. S. Lilja), pp. 195–211. Stockholm: Södertörn.

Lajus D.L., Lajus J.A., Dmitrieva Z.V., *et al.* (2005) Use of historical catch data to trace the influence of climate on fish populations: examples from the White and Barents Sea fisheries in 17th–18th centuries. *ICES Journal of Marine Sciences* 62, 1426–1435.

Lajus, D., Dmitrieva, Z., Kraikovski, A., *et al.* (2007a) Historical records of the 17th – 18th century fisheries for Atlantic salmon in northern Russia: methodology and case studies of population dynamics. *Fisheries Research* 87, 240–254.

Lajus, D., Alekseeva, Y. & Lajus J. (2007b) Herring fisheries in the White Sea in the 18th – beginning of the 20th centuries: spatial and temporal patterns and factors affecting the catch fluctuations. *Fisheries Research* 87, 255–259.

Lajus D.L., Alekseeva, Ya.I., *et al.* (2008a) Climate effect on abundance of Atlantic salmon in Barents and White Sea Basins (17–20th cc.). In: *SCAR/IASC IPY Open Science Conference "Polar Research – Arctic and Antarctic perspectives in the International Polar Year"*. Abstract volume. St. Petersburg, Russia, July 8–11, 2008, p. 253.

Lajus D.L., Alekseeva Ya.I. & Lajus J.A. (2008b) Climate influence on the Barents and White Sea ecosystems. In: *World Conference on Marine Biodiversity*. Book of Abstracts, 11–15 November, Valencia, Spain, p. 2–3.

Lajus J.A., Alexandrov D.A., Alekseeva Y.A., *et al.* (2001) Status and potential of historical and ecological studies of Russian fisheries in the White and Barents Seas: the case of the Atlantic salmon (*Salmo salar*). In: *The Exploited Seas: New Directions for Marine Environmental History* (eds. P. Holm, T.D. Smith & D.J. Starkey). *Research in Maritime History* 21, 67–96.

Lankester, E.R. (1890) The scientific results of the International Fisheries Exhibition, London 1883. In: *The Advancement of Science. Occasional Essays and Addresses* (ed. E.R. Lankester), pp. 193–223. London.

Lotze, H., Reise, K., Worm, B., *et al.* (2005) Human transformations of the Wadden Sea ecosystem through time: a synthesis. *Helgoland Marine Research* 59, 84–95.

Lotze, H.K., Lenihan, H.S., Bourque, B.J., *et al.* (2006) Depletion, degradation, and recovery potential of estuaries and coastal seas. *Science* 312, 1806.

Lotze, H.K. & Worm, B. (2009) Historical baselines for large marine animals. *Trends in Ecology and Evolution* 24, 254–262.

MacKenzie, B.R., Alheit, J., Conley, D.J., *et al.* (2002) Ecological hypothesis for a historical reconstruction of upper trophic level biomass in the Baltic Sea and Skagerrak. *Canadian Journal of Fisheries and Aquatic Sciences* 59, 173–190.

MacKenzie, B.R. & Myers, R.A. (2007) The development of the northern European fishery for north Atlantic bluefin tuna, *Thunnus thynnus*, during 1900–1950. *Fisheries Research* 87, 2–3.

McClenachan, L., Jackson, J.B.C. & Newman, M.J.H. (2006) Conservation implications of historic sea turtle nesting beach loss. *Frontiers in Ecology and the Environment* 4, 290–296.

McClenachan, L. (2009) Documenting loss of large trophy fish from the Florida Keys with historical photographs. *Conservation Biology* 23, 636–643.

Mead, J.G. & Mitchell, E.D. (1984) Atlantic gray whales. In *The Gray Whale*, Eschrichtius robustus (eds. M.L. Jones, S.L. Swatz & S. Leatherwood), pp. 225–278. New York: Academic.

Meldgaard, M. (1988) The great auk, *Pinguinus impennis* (L) in Greenland. *Historical Biology* I, 145–178.

Muniz, A.M. (1996) The evolution of the I.C.A.Z. Fish remains working group from 1981 to 1995. *Archaeofauna* 5, 13–20.

Myers, R.A. (2001) Testing ecological models: the influence of catch rates on settlement of fishermen in Newfoundland, 1710–1833. *Research in Maritime History* 21, 13–29.

Myers, R.A., Hutchings, J.A. & Barrowman, N.J. (1997) Why do fish stocks collapse? The example of cod in Atlantic Canada. *Ecological Applications* 7, 91–106.

O'Brien, L. (1999) Factors influencing the rate of sexual maturity and the effect on spawning stock for Georges Bank and Gulf of Maine Atlantic cod *Gadus morhua* stocks. *Journal of Northwest Atlantic Fisheries Science* 25, 179–203.

Pauly, D. (1995) Anecdotes and the shifting baseline syndrome of fisheries. *Trends in Ecology and Evolution* 10, 430.

Pauly, D., Christensen, V., Dalsgaard, J., *et al.* (1998a) Fishing down marine food webs. *Science* 279, 860–863.

Pauly, D., Pitcher, T., Preikshot, D. & Hearne, J. (eds) (1998b). Back to the future: reconstructing the Strait of Georgia ecosystem. Fisheries Centre Research Reports, University of British Columbia, 6.

Pitcher, T.J. & Pauly, D. (1998) Rebuilding ecosystems, not sustainability, as the proper goal of fishery management. In: *Reinventing Fisheries Management* (eds. T.J. Pitcher, P.J.B. Hart & D. Pauly), pp. 311–329. London.

Poulsen, B., Holm, P. & MacKenzie, B.R. (2007) A long-term (1667–1860) perspective on impacts of fishing and environmental variability on fisheries for herring, eel and whitefish in the Limfjord, Denmark. *Fisheries Research* 87, 181–195.

Poulsen, B. (2008) *Dutch Herring: An Environmental History, c. 1600–1860*. Amsterdam: Aksant.

Poulsen, B. (2010) The variability of fisheries and fish populations prior to industrialized fishing: an appraisal of the historical evidence. *Journal of Marine Systems* 79, 327–332.

Poulsen, R.T., Cooper, A.B., Holm, P. & MacKenzie, B.R. (2007). An abundance estimate of ling (*Molva molva*) and cod (*Gadus morhua*) in the Skagerrak and the northeastern North Sea, 1872. *Fisheries Research* 87, 196–207.

Poulsen, R.T. & Holm, P. (2007) What can fisheries historians learn from marine science? The concept of catch per unit effort (CPUE). *International Journal of Maritime History* XIX, 89–112.

Prummel, W. & Heinrich, D. (2005) Archaeological evidence of former occurrence and changes in fishes, amphibians, birds, mammals and molluscs in the Wadden Sea area. *Helgoland Marine Research* 59, 55–70.

Reeves, R.R., Smith, T.D., Woolmer, G., *et al.* (2004) Seasonal distribution of blue, right and humpback whales in the North Atlantic Ocean based on analysis of logbooks of American whalers in the 19th century. *Marine Mammal Science* 20, 774–786.

Reeves, R.R. & Smith, T.D. (2006) A taxonomy of world whaling: operations and eras. In: *Whales, Whaling and Ocean Ecosystems* (eds. J.A. Estes *et al.*), pp. 82–101. Berkeley: University of California Press.

Rosenberg, A.A., Bolster, J., Cooper, A., *et al.* (2005) The history of ocean resources: modeling cod biomass using historical records. *Frontiers in Ecology and the Environment* 3, 84–90.

Rozwadowski, H.M. (2002) *The Sea Knows No Boundaries: A Century of Marine Science Under ICES*. Copenhagen: The International Council for the Exploration of the Sea.

Smith, T.D. (1994) *Scaling Fisheries: The Science of Measuring the Effects of Fishing, 1855–1955*. Cambridge, UK: Cambridge University Press.

Smith, T.D. & Reeves, R.R. (2006) Pre-20th century whaling: implications for management in the 21st century. In: *Whaling and History II* (ed. J.E. Ringstad), pp. 119–134. Sandefjord, Norway: publication 331, Kommandør Chr. Christensens Hvalfangstmuseum.

Smith, T.D., Lund J.N., Josephson, E.A. & Reeves, R.R. (in press). Spatial dynamics of American offshore whaling in the 19th century: were sperm whales depleted? In: *Whaling and History III* (ed. J.E. Ringstad). Sandefjord, Norway: Sandefjordmuseene.

Soutar A. (1967) The accumulation of fish debris in certain Californian coastal sediments. *California Cooperative Fishery Investigations Reports* **11**, 136–139.

Southward, A.J. (1995) The importance of long time-series in understanding the variability of natural systems. *Helgoland Marine Research* **49**, 329–333.

Starkey, D.J. & Haines, M. (2001) The Newfoundland fisheries, c. 1500–1900: a British perspective. In: *The Exploited Seas: New Directions for Marine Environmental History* (eds. P. Holm *et al.*), pp. 1–11.

Trakadas, A. (2005) The archaeological evidence for fish processing in the Western Mediterranean. Ancient Fishing and Fish Processing in the Black Sea Region. In: *Ancient Fishing and Fish Processing in the Black Sea Region* (ed. T. Bekker-Nielsen), pp. 47–82. Aarhus: Aarhus Universitetsforlag.

Van Sittert, L. (2005) The other seven tenths. *Environmental History* **10**, 106–109.

Whitehead, H. (2002) Estimates of the current global population size and historical trajectory for sperm whales. *Marine Ecology Progress Series* **242**, 295–304.

PART II
Oceans Present – Geographic Realms

Chapter 2

Surveying Nearshore Biodiversity

Brenda Konar[1], Katrin Iken[1], Gerhard Pohle[2], Patricia Miloslavich[3], Juan Jose Cruz-Motta[3], Lisandro Benedetti-Cecchi[4], Edward Kimani[5], Ann Knowlton[1], Thomas Trott[6], Tohru Iseto[7], Yoshihisa Shirayama[7]

[1]*School of Fisheries and Ocean Sciences, University of Alaska Fairbanks, Alaska, USA*
[2]*Huntsman Marine Science Centre, St. Andrews, New Brunswick, Canada*
[3]*Departamento de Estudios Ambientales, Universidad Simon Bolivar, Caracas, Venezuela*
[4]*Department of Biology, University of Pisa, Italy*
[5]*Kenya Marine and Fisheries Research Institute, Mombasa, Kenya*
[6]*Suffolk University, Boston, Massachusetts, USA*
[7]*Seto Marine Biological Laboratory, Kyoto University, Japan*

2.1 Introduction

The nearshore region is defined here as the area from the high intertidal down to 20 m water depth, which is the focus of the Census of Marine Life Natural Geography in Shore Areas (NaGISA) project (www.nagisa.coml.org, Fig. 2.1 and Box 2.1). The overarching goal of NaGISA is to produce nearshore biodiversity baselines with global distribution from which new scientific questions and hypothesis testing can arise, long-term monitoring can be designed, and management plans can be implemented. One of NaGISA's goals is to create accurate biodiversity estimates by producing species lists for nearshore sites around the world. Previous overall marine biodiversity estimates, which include the nearshore, range from 178,000 to more than 10 million species (Sala & Knowlton 2006). To narrow this large range and obtain specific assessments for the nearshore, more species lists from more nearshore regions of the world are needed such as those produced during the NaGISA project.

One example of the use of NaGISA baseline data is to examine latitudinal trends in biodiversity. Thus far, there have been few truly global nearshore biodiversity comparisons attempted because of the lack of comparable data (e.g. Witman *et al.* 2004; Kerswell 2006). NaGISA contributes to our ability to make latitudinal and other spatial comparisons by establishing a standardized sampling protocol ensuring comparability of datasets and by greatly increasing the data coverage over a large latitudinal and longitudinal range. NaGISA also has initiated a growing network of scientists that will continue to accumulate data in the years to come. This project and its goals are particularly timely because of the changes in nearshore biodiversity that are resulting from increasing anthropogenic impacts and the changing climate.

NaGISA is a Japanese word that translates into the "area where the sea meets the land". Specifically, the goal of NaGISA is to assess nearshore biodiversity in rocky macroalgal and soft-bottom seagrass areas from the high intertidal to a water depth of 20 m. Within NaGISA these nearshore habitat types were chosen for two reasons. First, these habitats are known to have high biodiversity because of the three-dimensional structure provided by the macrophytes. Even in nearshore areas

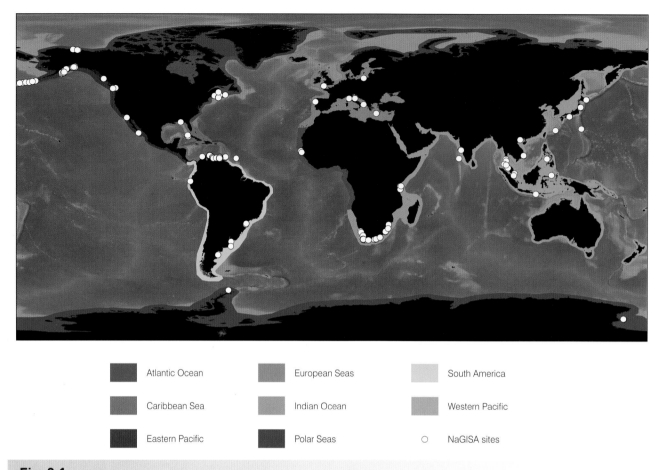

	Atlantic Ocean		European Seas		South America
	Caribbean Sea		Indian Ocean		Western Pacific
	Eastern Pacific		Polar Seas	○	NaGISA sites

Fig. 2.1

Global map showing the NaGISA regions with associated sites that have been sampled.

Box 2.1

NaGISA Genesis

NaGISA began from the coastal component of Diversitas International of the Western Pacific Asia (DIWPA; diwpa. ecology.kyoto-u.ac.jp). DIWPA is an international program that aims to promote and facilitate biodiversity research in the Western Pacific region. This program, supported by UNESCO, the International Union of Biological Sciences (www.iubs.org), and other international organizations, aimed to increase international biodiversity studies and thus created the International Biodiversity Observation Year (IBOY; www.nrel.colostate.edu/projects/iboy/index2. html). The target of the IBOY program was a matrix of selected taxa in major coastal ecosystems including temperate, subtropical, and tropical regions. The Census of Marine Life selected this program as one of its field projects under the name NaGISA, and extended spatial and taxonomic coverage so that spatial patterns of coastal marine biodiversity in all global coastal regions could be analyzed.

where soft sediments dominate, small macrophyte oases have a higher biodiversity than the surrounding soft sediments (Dunton & Schonberg 2000). Second, these habitats are fairly globally distributed, in contrast to other habitats like nearshore coral reefs that are typically restricted to warmer waters.

One of NaGISA's largest legacies is the development of a standardized sampling protocol for nearshore rocky macroalgal and seagrass habitats (Rigby *et al.* 2007). This protocol ensures comparability among all NaGISA data to make an evaluation of large-scale to global nearshore biodiversity patterns possible. In addition to data comparability, a major hurdle for many nearshore biodiversity surveys is a lack of taxonomic information for many groups beyond conspicuous macrofauna and flora, especially for many smaller organisms that make up much of the existing biodiversity. NaGISA's network of scientists includes local taxonomists as well as taxonomic training to ensure accurate and reliable identifications for all major taxonomic groups. However, given the comprehensive coverage resulting from NaGISA collections, a lack of taxonomic expertise still exists for many of the smaller and less charismatic organisms and in many regions of the world.

For organizational purposes, NaGISA divided the world's shorelines into eight regions: Western Pacific, Eastern Pacific, South American Seas, Caribbean Seas, Indian Ocean, Atlantic Ocean, European Seas, and Polar Seas (Fig. 2.1). As of May 2010, the NaGISA project has sampled 253 sites, of which 179 were macroalgal sites, 71 were seagrass sites, and one each was a rhodolith site, a sandy beach, and a mudflat (Table 2.1). NaGISA also organizes the world's coastline into 20-degree bins and has data coverage (at least one sampling site per bin) in about 45% of these nearshore bins so far. Also, of the 253 sites, 64 sampled so far have been sampled more than once and many are on their way to becoming long-term monitoring sites. This initial census (2000–2010) provided a baseline dataset for long-term monitoring and the information needed to answer fundamental ecological questions about spatial patterns in nearshore biodiversity. Building on this growing baseline, NaGISA data will eventually help identify the drivers that structure these nearshore communities on local, regional to global scales. Apart from its scientific value, the strength of NaGISA is that it involves local interests and stakeholders, from local community groups to elementary, high school, and university students. This allows stakeholders to become vested in the nearshore and build an on-the-ground force that uses NaGISA data to solve local management problems. NaGISA data are part of the OBIS database (Ocean Biogeographic Information System; www.iobis.org) and are thus publicly available. As of May 2010, NaGISA contributed over 47,700 records towards OBIS distributional maps with a total of over 3,100 taxa.

Table 2.1

Sites sampled by NaGISA by region and habitat type. A total of 253 sites have been sampled during NaGISA activities in macroalgal and seagrass habitats. Other habitat types include rhodolith beds, mudflats, and sandy beaches. Data as of May 25, 2010.

Region	Sampling effort		Site habitat types		
	Point data sites (single sampling)	Monitoring sites (multiple samplings)	Macroalgal	Seagrass	Other
Atlantic Ocean (AO)	8	5	5	7	1
Caribbean Sea (CS)	59	22	45	35	1
Eastern Pacific (EPAC)	45	13	51	6	1
European Seas (ES)	5	4	8	1	0
Indian Ocean (IO)	32	7	34	5	0
Polar Seas (PS)	8	11	19	0	0
South American Seas (SAS)	4	2	4	2	0
Western Pacific (WPAC)	28	0	13	15	0
Total	189	64	179	71	3

2.2 The Status of Regional Nearshore Biodiversity Knowledge

The nearshore region is highly accessible and, as such, has historically received much taxonomic and ecological attention from scientists and naturalists. As with other ocean biomes, taxonomic and biodiversity knowledge differs depending on geographic region and taxonomic group. Even with this varying knowledge base, nearshore field guides and scientific publications exist for most regions of the world. It is therefore surprising that before NaGISA, very few regional estimates for nearshore biodiversity existed and information regarding biodiversity patterns on the regional scale was scarce. The following is a brief highlight that describes the status of nearshore biodiversity knowledge in each of the eight NaGISA regions (Fig. 2.1).

2.2.1 Eastern Pacific (EPAC)

NaGISA sites sampled in the Eastern Pacific region span from approximately 61°N (south–central Alaska) to 24°N (Baja Mexico). Fifty-eight sites have been established in various locations along the coasts of the United States (Alaska and California), Canada (British Columbia), and Mexico (Baja) (Table 2.1). Of these sites, 13 have been sampled more than once and are becoming established monitoring sites. Some sites in Alaska were established with the assistance of local native communities, and some sites in both Alaska and California are being maintained with the assistance of various high school and university classes.

Although much research has been done in this relatively well-known region, there are no estimates for overall nearshore biodiversity. Nonetheless, some latitudinal descriptions of this region do exist. Early work demonstrated that benthic processes, such as competition and predation, caused a north–south gradient of decreasing recruitment of intertidal sessile invertebrates from Oregon to California (Connolly & Roughgarden 1998). Along the Pacific coast of North America biogeographical and oceanographic discontinuities separate rocky intertidal communities into 13 distinct spatial groups (Blanchette *et al*. 2008). In general, they found strong correlations between species similarity and both geographical position and sea surface temperature. Supporting this view is the observed latitudinal gradient in the recruitment of intertidal invertebrates for this region (Connolly *et al*. 2001). Interestingly, in this same region, Schoch *et al*. (2006) suggested that wave run-up was the most significant physical parameter that affected community structure. NaGISA has added much knowledge to this region by starting the first extensive nearshore monitoring in Alaska and by adding to existing datasets, which will allow for a more complete longitudinal comparison along the Northwestern American coast.

2.2.2 Western Pacific (WPAC)

NaGISA sites in the Western Pacific region span from approximately 43°N (Eastern Hokkaido, Japan) to 8°S (Indonesia). Twenty-eight sites have been established in various locations in Japan, Vietnam, Philippines, Thailand, Malaysia, and Indonesia (Table 2.1). Although so far none of these sites has been sampled more than once, current WPAC efforts are trying to establish several monitoring sites.

Although much research has been done in this region, particularly in Japan, there are no nearshore biodiversity estimates. Nonetheless, some latitudinal descriptions do exist along some major ocean current regimes. Along the northern Japanese coast, the subarctic, southerly flowing Oyashio current is characterized by high biomass, large individuals, and low biodiversity. In contrast, the warm, northerly flowing Kuroshio current along the southern Japanese coast is characterized by high biodiversity but low biomass (Nishimura 1974). The high biodiversity in the Kuroshio region occurs because this current transports species living in the high diversity Coral Triangle around the Philippines, Indonesia, and Malaysia to the northern subtropical and temperate regions of the western Pacific. The high biodiversity in the south Asian coastal area has sparked much research, including important taxonomic work. NaGISA has contributed to some of these publications, such as field guides on echinoderms (Yasin *et al*. 2008), hermit crabs (Rahayu & Wahyudi 2008), and seagrasses (Susetiono 2007).

2.2.3 European Seas (ES)

NaGISA sites sampled within the European Seas region span from approximately 55°N (Poland) to 35°N (Crete). Sampling sites have been established in the North Sea, the Baltic Sea, the East Atlantic Ocean, the Northwest Mediterranean, the Northern and Southern Adriatic Sea, and the Aegean Sea, with collaborators from Italy, the United Kingdom, Portugal, Greece, and Poland. A total of nine sites have been sampled, four of which have been sampled more than once (Table 2.1).

Although the biodiversity of individual regions within the European Seas has been the focus of intense research (Frid *et al*. 2003), an exhaustive analysis of biodiversity estimates, patterns, and trends is lacking. One pattern that has been noted is the replacement of large canopy algae that dominate at higher latitudes with seagrasses that become dominant in the Mediterranean, where relict kelp populations persist only in the Strait of Messina and in the Sicily Channel (Lüning 1990). NaGISA information in the ES is allowing researchers to explore nearshore processes more thoroughly than before. For example, NaGISA data have helped to show that rare species may become more abundant when the environment is variable (Benedetti-Cecchi *et al.* 2008).

2.2.4 Indian Ocean (IO)

The Indian Ocean NaGISA sites range latitudinally from 28°N (Egypt) to 34°S (South Africa) and are found in Kenya, Tanzania, Mozambique, India, Egypt, and South Africa. Of the 39 sites that have been sampled, seven have been sampled more than once and are on their way to becoming monitoring sites. Two of the sites in Tanzania were established and are being monitored with the assistance of high school students, both local and from the United States.

As a result of several landmark expeditions (see, for example, Ekman 1953) and later research, taxonomic knowledge of the Indian Ocean region has been expanding. However, although biodiversity estimates do exist for certain groups in particular areas, latitudinal biodiversity descriptions for this region are lacking. The southern region of the African continent is particularly high in coastal biodiversity, with estimates of over 12,000 species from southern Mozambique in the Indian Ocean to northern Namibia in the east Atlantic, representing 6% of all coastal marine species known worldwide (Branch *et al.* 1994; Gibbons *et al.* 1999; Adnan Awad *et al.* 2002; Griffiths 2005). Other coastal regions of the IO are largely unknown, such as the island marine fauna in India, which have been estimated to be approximately 75% unknown (Venkataraman & Wafar 2005). In the IO region, NaGISA efforts are focusing to contribute specifically to areas of currently little existing information such as India.

2.2.5 Atlantic Ocean (AO)

The Atlantic Ocean region was sampled at 13 sites ranging from approximately 47°N (Canada) to 13°N (Senegal). These sites have been located along the coasts of Canada, the United States (Maine to Connecticut), and Senegal. Sites in Canada and the United States have largely involved elementary, high school, and university students for their sampling. Of the AO sites, five have been sampled multiple times and are considered monitoring sites (Table 2.1). In 2010, at least 12 additional sites will be established and monitored in collaboration with summer science camps from Connecticut to Maine in the Unites States.

In the AO region, it is generally recognized that biodiversity increases with decreasing latitude when comparing boreal with tropical regions (Udvardy 1969). Various environmental factors, such as local habitat heterogeneity can complicate this trend at the local scale. For example, NaGISA sampling has helped to show that Cobscook Bay at the US/Canada border, contrary to the general trend, has substantially higher macroinvertebrate species diversity than areas

further south (Trott 2009). In addition, there are distinct biogeographic regions in the Northwest Atlantic, including the Polar, Acadian, Virginian, and Carolinian Provinces, with distinct regional diversity patterns (Pollock 1998).

2.2.6 South American Seas (SAS)

The South American Seas sites extend from a latitude of 2°S (Ecuador) to 42°S (Argentina) and include the countries of Argentina, Ecuador, and Brazil. A total of six sites have been sampled, with both Argentinean sites being sampled twice (Table 2.1). All sites in the SAS region were sampled with the assistance of local university students.

Although much local knowledge exists within various countries in this region, good nearshore biodiversity estimates and discussions of latitudinal trends are scarce. In Brazil, 540 taxa were described associated with seagrass beds, mostly polychaetes, fish, amphipods, decapods, mollusks, foraminiferans, macroalgae, and diatoms (Couto *et al.* 2003). Other areas, such as the fjords in southern Chile, have received little attention so far, and recently explorations have discovered 50 new species associated with them (Haussermann & Forsterra 2009). In Chile, several marine invertebrate taxa were found to decrease in biodiversity with increasing latitude between 18° and 40–45°S, and then increase further south, probably because of the presence of sub-Antarctic fauna (Gallardo 1987; Clarke & Crame 1997; Fernandez *et al.* 2000). NaGISA is contributing to the overall biodiversity effort in the SAS region by attempting to establish well-distributed NaGISA sites that will greatly enhance communication among countries so that larger-scale comparisons can be made.

2.2.7 Caribbean Sea (CS)

The Caribbean Sea sites span from approximately 10°N (Venezuela) to 30°N (Florida). Although latitudinally this is the shortest NaGISA region, it has an impressive total of 81 sites from the countries of Cuba, Trinidad and Tobago, Venezuela, Colombia, and the United States (Florida). Of the 81 sites, 22 have been sampled more than once (Table 2.1). Many of the sites in Venezuela have involved university students in their sampling, and the Florida site was initiated by a high school group, which has also gone on to help other high school groups with NaGISA sampling around the world, including Greece, Zanzibar, and Egypt.

It should be noted that for the Caribbean Seas, NaGISA is the first attempt to establish a monitoring program that does not target coral systems. This is particularly important for this region because the massive changes that have occurred in coral reefs over the past several decades (Gardner *et al.* 2003), including an 80% drop in live coral cover in 25 years (Wilkinson 2004), have prompted an increase in hard substrate availability, which in turn might

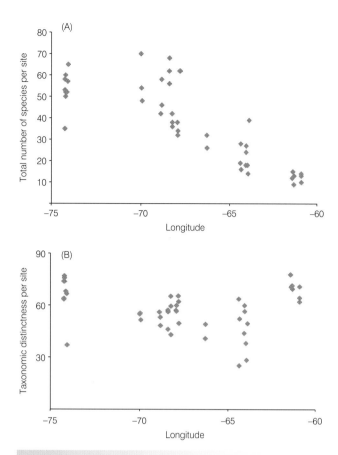

Fig. 2.2

(A) Total number of species per sampling site (48 sites) along the Southern Caribbean Coast (Colombia–Venezuela–Trinidad & Tobago) declines with decreasing longitude. **(B)** Index of taxonomic diversity (Clarke & Warwick 2001) for the same 48 sites does not differ with longitude.

were off of the United States McMurdo Station and one was off of the Uruguayan Artigas Research Base at the Antarctic Peninsula. Of the sites around McMurdo Station, three have been sampled more than once (Table 2.1).

Biodiversity estimates are scarce for both polar regions for most taxa (see also Chapters 10 and 11). However, for macroalgae it is estimated that there are as many as 120 macroalgal species in the Antarctic (Wiencke & Clayton 2002) and slightly more in the Arctic (Wilce 1997) but with a much higher percentage of endemic species in the Antarctic. The polar regions also have little information available regarding latitudinal trends. Typically, Arctic nearshore systems are thought to be less diverse than northern temperate systems (see, for example, Kuklinski & Barnes 2008; Wlodarska-Kowalczuk *et al.* 2009). In the Arctic nearshore, it seems that higher diversity is typically found at more southern locations compared with northern locations (see, for example, Kedra & Wlodarska-Kowalczuk 2008). In the Antarctic, the Peninsula, which spans approximately six degrees of latitude from 62° to 68° S, shows a latitudinal macroalgal decline (Moe & DeLaca 1976). Extending this gradient further south to the Ross Sea (77° S), the southernmost location of open water, only two species of fleshy macroalgae occur (Miller & Pearse 1991). This latitudinal decline is mainly driven by reduced light availability with increasing latitude due to strong seasonality, low solar angle, and extended periods of ice cover.

2.3 Historical Knowledge of Global Nearshore Biodiversity

2.3.1 *Biodiversity gradients*

Latitudinal gradients of increasing species diversity from the poles to the tropics have often been touted as a fundamental concept in terrestrial ecology (Willig *et al.* 2003). Many mechanisms have been proposed to explain this latitudinal gradient, but changes in temperature have been targeted as the most plausible factor in terrestrial systems. The variation in ocean temperatures over the same distance, however, is significantly smaller and the overall importance of temperature versus other physical factors has only begun to be discussed (Blanchette *et al.* 2008). Other mechanisms driving latitudinal trends of rocky nearshore biodiversity are primarily large-scale oceanographic conditions and local biological interactions, which can include nutrient content and, thus, primary productivity, local assemblages of herbivores and predators, the prevalence of larval stages with differing dispersal ranges, speciation rates, and so forth (Connolly & Roughgarden 1998; Roy *et al.* 2000; Broitman *et al.* 2001; Connolly *et al.* 2001;

result in a phase shift from coral-dominated communities to hard-bottom macroalgal communities.

With the exception of general field guides and some specific scientific publications, no nearshore biodiversity estimates or biodiversity trends are known to exist. However, NaGISA is contributing to this knowledge, by producing the first longitudinal comparison in the CS region, which has shown that diversity decreases from west to east (J.J. Cruz-Motta, personal communication; Fig. 2.2).

2.2.8 *Polar Seas (PS)*

The Polar Seas region includes both the Arctic and the Antarctic. There are 13 Arctic NaGISA sites that were sampled around 70° N, off the United States coast of Alaska. Eight of these sites have been sampled multiple times and are monitoring sites (Table 2.1). In the Antarctic, six sites have been sampled at 62° S and 78° S. Five of these sites

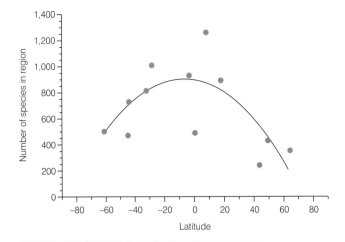

Fig. 2.3

Regional species richness as a function of latitude. Reproduced with permission from Witman *et al.* (2004). Copyright 2004 National Academy of Sciences, USA.

Rivadeneira *et al.* 2002; Okuda *et al.* 2004; Kelly & Eernisse 2007).

Debate still surrounds the existence of nearshore latitudinal biodiversity trends, especially on the global scale. The reason for this is the lack of studies actually completed at the global scale. It is time intensive and costly to sample sites globally and literature reviews are difficult to compare owing to the various biases associated with using different sampling protocols. Even with these constraints, there are two excellent examples of global studies. In one study, field sampling found that shallow subtidal boulder communities tended to have higher species numbers at equatorial sites compared with sites closer to the poles (Fig. 2.3) (Witman *et al.* 2004). In contrast, a study based on a literature search of nearshore algal genera found that more biodiversity hot spots occurred in temperate regions compared with tropical or polar (Kerswell 2006). Although both studies are ground-breaking as they were the first to attempt global comparisons, it should be noted that they are limited in that one was completed on a specific habitat (subtidal rock walls in 12 biogeographic regions, totaling 49 local sites) and the other focused on one taxonomic group (macroalgae). NaGISA is assisting to broaden the knowledge of global biodiversity by increasing the number and distribution of sites, increasing the range of habitats (including intertidal and subtidal rocky shores and seagrass beds), and increasing the number of taxa examined. Based on NaGISA's main target taxa, global latitudinal comparisons will be possible for macroalgae, seagrasses, mollusks, echinoderms, polychaetes, and decapods, in addition to comparisons of overall community composition in rocky shores and seagrass systems.

2.3.2 Biogeographic breaks

We cannot discuss biodiversity gradients without mentioning biogeographic breaks. Biogeographic breaks are important because biodiversity gradients do not always change continuously but sometimes are abrupt owing to these breaks. Breaks can be driven by the dynamic interaction of two or more distinct water masses. This creates active transition zones where species mingle across their respective boundaries, for example, the biogeographic provinces associated with cold- and warm-water masses. These transition zones include species pools from both systems, often resulting in a high level of biodiversity at the breaks.

Biogeographic breaks are worldwide. For example, in the east Pacific, a well-studied biogeographic break is Point Conception in California. Offshore of Point Conception, the continental shelf is broad and the south-flowing California Current is deflected offshore (Brink & Muench 1986; Browne 1994). Point Conception is a "transition zone" between the warm Californian Province and the cooler water regime of the Oregonian Province, resulting in different fish, invertebrate, and algal communities on either side of this break (Horn & Allen 1978, Murray & Littler 1981; Murray & Bray 1993). Similarly, in the eastern Atlantic along the western African coast, the coastal waters of Mauritania and Senegal and adjacent areas form a transition zone between a more temperate northern zone and a warmer tropical zone farther south. Despite variations in local conditions, biodiversity patterns of fishes, invertebrates, and particularly macroalgae reflect this change within a relatively narrow 400–500 km band (Lawson & John 1987). For Eastern South African macroalgae, a biogeographic break occurs at St. Lucia, 135 km south of the Mozambique border. Here, there is a transition from a tropical Indian Ocean flora to a temperate South African flora. As another example, a biogeographic break is found in the Gulf of Maine at Penobscot Bay, Maine, where the Maine coastal current splits to flow southwest from eastern Maine. One of the resulting branches travels east and the other continues in a southwestern direction. The communities above and below this break are statistically distinct, but not within either of the two regions (Trott 2007; see also Chapter 3). The already mentioned boundary of the subtropical, warm Kuroshio current and the subpolar, cold Oyashio current forms an important biogeographic break along the eastern coast of Japan, influencing patterns of diversity and biomass. There are other biogeographic breaks around the world; these are just a few to highlight their importance to biodiversity.

Some biogeographic breaks are still under investigation and highlight the need for more biodiversity studies. For example, in the Aleutian Archipelago in Alaska, a biogeographic break may exist that drives the presence of the canopy-forming kelp from only *Eualaria fistulosa* to the west to primarily *Nereocystis luetkeana* to the east (Miller

& Estes 1989). However, more oceanographic and biological data are needed to identify the exact location and drivers of this possible break (Ladd *et al.* 2005). NaGISA is assisting in this discussion by establishing sites along the Aleutian Archipelago.

2.3.3 Nearshore biodiversity hot spots

A biodiversity hot spot is a biogeographic location that contains an unusually high number of species. Hot spots may occur along a coastline where habitats are homogeneous but for some reason a particular location has high biodiversity. A hot spot also may occur at a site where habitat type is different than the surrounding environment, as commonly seen in deeper waters at seamounts surrounded by soft sediment. There are many reasons why species diversity may be higher in certain locations and these reasons are often site specific. Reasons may include change in substrate, water mass, topography, nutrient intrusions, or geologic history.

An example of a NaGISA site that is a hot spot because of a substrate change is in the Arctic Beaufort Sea (in the PS region). Here, the typically soft-bottom seafloor contains a low-diversity fauna, with only about 30 infaunal species, mainly polychaetes and amphipods (Feder & Schamel 1976; Carey & Ruff 1977; Carey *et al.* 1984). In this region, local biodiversity hot spots occur where boulders provide colonizable hard substrate for macroalgae and sessile epibenthic macrofauna, which attract other organisms including more than 150 species of macroalgae, invertebrates, and fish (Dunton *et al.* 1982).

Hot spots also can be created by oceanographic conditions, such as in the Gulf of Maine (Buzeta *et al.* 2003; Trott & Larsen 2003). The NaGISA site in Cobscook Bay has the highest species richness of macroinvertebrates of any bay similar in size and habitat characteristics in the Gulf of Maine, with approximately 800 known species representing all major phyla (Trott 2004). The high biodiversity of Cobscook Bay appears to result from wave exposure and the extraordinary tides this system experiences (Campbell 2004). Additional hot spots were also identified in the Bay of Fundy where NaGISA assisted the Department of Fisheries and Oceans Canada in an effort to determine Ecologically and Biologically Significant Areas (EBSA), which resulted in the identification of five EBSA's in the Quoddy Region (Buzeta & Singh 2008).

2.4 Closing Information Gaps

The field of taxonomy, traditionally based primarily on morphology, has expanded in recent years to include molecular information (Blaxter 2003; Hebert *et al.* 2003). This has not only enhanced our understanding of evolutionary relationships but also our knowledge of biodiversity and species distributions ranging from algae to fishes (Saunders 2005, 2008; Blum *et al.* 2008; Pfeiler *et al.* 2008; Thacker 2009). Nonetheless, our ability to identify organisms in some areas and for some taxa is still limited, leaving gaps in taxonomic knowledge as well as for particular regions of the world's coasts (see Box 2.2). Many developing coun-

---

Box 2.2

NaGISA Contributions to the Effort of Closing Taxonomic Gaps

NaGISA conducts workshops to train new taxonomists.

- This assists in the taxonomy of lesser known groups or in areas where taxonomists are rare.
- Workshops have included the taxonomy of macroalgae, polychaetes, amphipods, ascideans, decapods, gastropods, echinoderms, harpacticoid copepods, stomatopods, tanaids, and nematodes.

NaGISA creates public ownership for coastal marine diversity.

- NaGISA researchers give public lectures and involve children's camps, school groups, university classes, local and native communities, and interested local

naturalists in site selection processes and sampling activities.

- The most prominent example of this is the NaGISA High School Initiative established at Niceville High School in northwestern Florida. They have been sampling annually since 2003 and have visited other schools and countries to encourage the involvement of other schools. Niceville students helped Kizimkazi High School in Zanzibar start their NaGISA efforts in 2007, as well as the Heraklion School of the Arts in Crete, Greece in 2009. The Niceville team visited Sharm el Sheikh College in Egypt in early 2010 for sampling activities in the Red Sea.

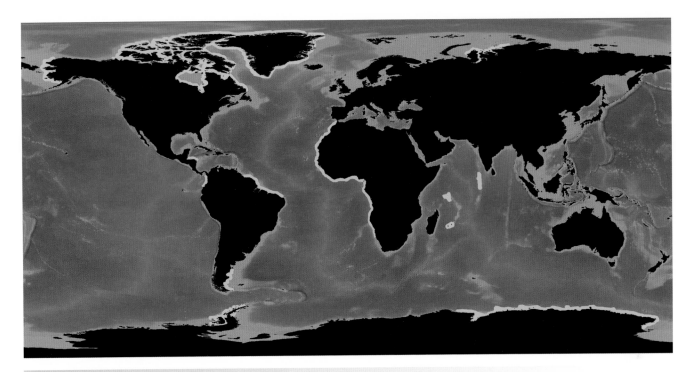

Fig. 2.4
Global map showing current major biodiversity gaps (defined as missing information for most taxonomic groups) in NaGISA-focused habitats. These gaps are based on estimates by NaGISA researchers.

tries and remote regions lack financial support, technology, and taxonomic information for their fauna and flora. This is particularly true for smaller, less charismatic organisms of no economic importance. Access to these remote regions is difficult due to logistical and financial constraints, leaving gaps in data coverage. Each of the eight NaGISA regions contains areas that have not been sufficiently explored (Fig. 2.4). The western coast of Alaska in the Eastern Pacific region, the western African coast in the Atlantic region, all of the Arctic coastline except where research stations allow access, the eastern Antarctic coast, and remote islands in the Indian Ocean are just a few examples.

2.5 NaGISA's Major Findings

Although the nearshore region is probably among the most-studied parts of the ocean because of its accessibility and obvious interest to humans as a resource, the lack of information on biodiversity and its large-scale and long-term patterns in more than a handful of locations is particularly surprising. Also surprising is the lack of integrated information so that regional and global trends and patterns can be discussed. NaGISA is the first project to undertake the

ambitious step to create such large-scale baselines with the establishment of standardized protocols and a growing global network of nearshore researchers. With over 250 sites located around the world, and still growing, and 28 different countries involved, NaGISA is the largest-ever attempt to address truly global-scale biodiversity issues.

The central idea of the NaGISA standardized sampling protocol is a fully nested design. Replicate samples along various tidal heights are collected at each site, and multiple sites are sampled within regions of specific latitude and longitude. This hierarchical design of the protocol with replicate samples within a site, which is then nested within latitude or longitude, allows a statistically appropriate and powerful method to analyze biodiversity patterns across several spatial and temporal scales (Benedetti-Cecchi 2007). Not only can biodiversity patterns be analyzed on local, regional, and up to global scales, but it can also be determined at which of these scales most variability occurs.

NaGISA's protocols include various independent sampling levels, from cover estimates to actual collections and detailed taxonomic identification of all organisms. This design allows flexibility in sampling effort, so where the full sampling effort is not possible due to logistical or financial constraints, parts of the protocols can be used to create important local information that can be compared with large-scale NaGISA data. For example, cover estimates can be done relatively quickly, and students, agencies, or local people can be trained to do so with high scientific accuracy. This opens opportunities to perform long-term monitoring at specific sites and/or the expansion of quantitative near-shore coverage with the inclusion of added manpower from local stakeholders. NaGISA's regionally organized network of nearshore researchers allows local scientists across the world to participate in this effort and thus make the final product larger than the sum of its individual parts.

NaGISA's specific scientific findings from its field surveys include inventories of marine flora and fauna, data on their abundance and biomass, new species records, species range extensions, habitat range extensions, biodiversity hot spots, and explanations of nearshore ecological processes and biodiversity drivers. All findings can now be analyzed on regional as well as global scales.

Several new species were found and subsequently described during the NaGISA inventories. These new species discoveries included some small and inconspicuous species, like two cumaceans from the Gulf of Alaska (*Cumella oculatus* and *C. alaskensis*; Gerken 2009). These cumaceans are not only new species but their discovery was surprising as the genus *Cumella* is typically tropical rather than boreal–Arctic. Cumaceans, as filter feeders and surface deposit feeders, are ecologically important in energy transfer within the benthic food web and on the Alaskan shelf as they are important food for grey whales. Other, more conspicuous new species discovered were the golden V kelp in the Aleutian Islands, Alaska (*Aureophycus aleuticus*; Kawai *et al.* 2008). This kelp grows up to 3 m in length, and histological and genetic analyses show that it may not be closely related to other kelp species in the region. This opens interesting evolutionary and distributional questions about kelps in the North Pacific, where they form important habitats for associated biodiversity.

Another significant NaGISA accomplishment has been the discovery of the anomalodesmatan bivalve *Pholadomya candida* living in a *Thalassia testudinum* seagrass bed at Santa Marta, Colombia. This bivalve species belongs to the ancient family Pholadomyidae, a group of burrowing bivalves living on Earth since at least the Early Carboniferous (330 million years bp), which reached a high degree of diversification in Jurassic to Cretaceous times. *Pholadomya candida* had been collected alive only twice, with the last record in 1842, and, because living specimens had not been recorded for nearly 140 years, some authors considered the species extinct. The evolutionary implications of this re-discovery are remarkable. Comparative molecular sequencing of *P. candida* with other anomalodesmatan species and with representatives of other presumably related groups may provide clues of the evolution of the Anomalodesmata, as well as indications on the origin of the Myoida (Díaz *et al.* 2009).

Several new species distributional records and range extensions have been found during NaGISA sampling efforts. In the western Pacific, the solitary entoproct *Loxosomella* sp. was found in a seagrass sample from Akajima, Okinawa Prefecture, Japan. This is the first record of this animal group from sandy seagrass habitats in this region. Another interesting discovery was made in the Eastern Pacific with the coralline alga *Phymatolithon calcareum*. During NaGISA sampling, this species was found in its gametangial reproductive state (Konar *et al.* 2006). Although this species is relatively common and globally distributed, it was previously found only once in this reproductive state and that record was off the Atlantic coast of France (Mendoza & Cabioc'h 1998). In the well-studied Cobscook Bay of the Gulf of Maine in the Atlantic Ocean region, NaGISA surveys found tens of benthic faunal taxa previously unreported from the area, from such diverse groups as hydrozoans (for example *Clytia gracilis*), mollusks (for example *Spisula solidissima*, *Astarte portlandica*), crustaceans (for example *Nebalia bipes*, *Metopella carinata*), polychaete worms (for example *Aricidea albatrossae*, *Euchone papillosa*), and bryozoans (for example *Haplota clavata*, *Cribrilina punctata*). Similarly, new species records for five macroalgal species were found at NaGISA sites in the Arctic Beaufort Sea, including the brown algae *Sphacelaria plumosa* and *S. arctica*, and the red algae *Rhodomela tenuissima* and *Scagelia* cf *americana*. Also at these sites, the common red alga *Phyllophora truncata* was often infested with what has been tentatively identified as an endophytic alga *Chlorochytrium*.

In addition to species-level discoveries, NaGISA also had some significant discoveries of habitat extensions. A major range extension was the discovery of a rhodolith habitat in the Eastern Pacific region (Konar *et al.* 2006). Rhodoliths are unattached calcareous red algae that form extensive beds, which provide habitat for many associated, sometimes commercially important species. Although rhodolith beds are widely distributed in temperate and tropical areas, the rhodolith bed discovery in Alaska's Prince William Sound in the North Pacific Ocean represents a significant northward extension of known rhodolith distribution. Also, in the Arctic Beaufort Sea, a new boulder field providing substrate for a diverse community of macroalgae and invertebrates was mapped in Camden Bay through NaGISA efforts (Iken & Konar 2007).

Regional comparisons have already yielded new insights into biodiversity patterns. Longitudinal comparisons in the Caribbean Seas region have shown that there is a decrease in species numbers from west to east. At the same time, this gradient in species numbers is not similarly reflected in the taxonomic structure of the communities (based on the index of taxonomic distinctiveness) as this is the same along that longitudinal gradient (Fig. 2.2). The nested design of the NaGISA sampling protocol was used in the Gulf of Alaska in the EPAC region to analyze the contributions of local versus regional scales of variability in nearshore communities (Konar *et al.* 2009). Interestingly, most variability was associated with the local scale and very little with regional scales. On the local scale, the depth gradient was the most important factor contributing to variability, which was also found when only echinoderm distribution was analyzed over the same spatial scales (Chenelot *et al.* 2007). The number of species generally increases from the high intertidal to a depth of 1 m and then decreases with increasing subtidal depths. The large tidal range in the region effectively renders the 1 m depth stratum low intertidal and thus a suitable interface for a large variety of intertidal and subtidal organisms (Konar *et al.* 2009). Seasonal comparisons in the South American region (Puerto Madryn, Argentina) found that local biodiversity varies throughout an annual cycle in close relation to the presence of an invasive brown algal species (*Undaria pinnatifida*), which is sensitive to warm temperatures. During the austral winter, *U. pinnatifida* invades the rocky substrates replacing the natural community, but also attracts another community of gastropods, polychaetes, sea urchins, and other invertebrates that feed on the algae. As the water temperature increases in the austral summer, *U. pinnatifida* dies, and the natural community returns.

Along with reporting community patterns and biodiversity trends, it also is important to explain why and how these trends and patterns exist. Some research has already examined drivers of community patterns and biodiversity trends at various spatial scales (Coleman *et al.* 2006; Kuklinski *et al.* 2006; Scrosati & Heaven 2007; Wulff *et al.* 2009). NaGISA in the European Seas conducted an experimental study using a combination of long-term observations and field manipulations to show that rare species take advantage of environmental variability, becoming less rare in fluctuating environments (Benedetti-Cecchi *et al.* 2008). Hence, an increase in environmental variability, such as that expected under climate change models, may lead to major shifts in species composition within assemblages, with the prediction that currently rare species may become more dominant with increasing levels of environmental heterogeneity.

2.6 Remaining Questions

Although NaGISA efforts are greatly contributing to the field of nearshore biodiversity, many issues and questions remain. First and foremost, true estimates for global nearshore biodiversity do not exist. The million dollar question of how many organisms live in nearshore waters, still cannot be answered. It appears that the more regions that are sampled and the more taxonomists that are involved, the more new species and range extensions are found. We may never know exactly how many species live in the nearshore, but we can and should continue working towards increasingly accurate estimates.

Another question that still remains open is why certain areas have higher biodiversity (or abundances or biomass) than other areas. The more we learn about biodiversity trends and the physical and biological attributes that contribute to biodiversity hot spots, the easier it will be to answer the question of why these hot spots exist. From the data currently available, it appears that many of the hot spots occur because of various site-specific parameters (that is, hard substrate in an otherwise soft substrate environment, local oceanographic conditions). However, more information is needed to determine if and what biological and physical parameters will result in the existence of a hot spot and if large-scale generalizations of such relationships can be made.

Along with the questions, some problems also remain. One such problem that still exists in many regions is the surveying of remote and isolated areas. With the advances in remote sensing, these areas are becoming more accessible. The intertidal zone can be surveyed with remote sensing, using Ikonos satellite imaging (www.satimagingcorp.com/gallery-ikonos.html), followed by hyperspectral imaging and ground-truthing (Larsen *et al.* 2009). In some areas, like the northern part of the Eastern Pacific region, programs exist that have already mapped nearshore coasts, such as the ShoreZone project (alaskafisheries.noaa.gov/habitat/shorezone/szintro.htm), and these images are available online. In many areas, the subtidal areas can be mapped and information such as bottom type and depth can be acquired from multibeam sonar acoustic mapping. This can then be ground-truthed with benthic sampling. This type of information will make discovering new biodiversity hot spots and describing patterns and processes in the nearshore much easier. Nevertheless, although such mapping efforts can supply guidance and large-scale coverage, the need for local ground-truthing and traditional establishment of biodiversity remains.

Although there has been much advancement in the knowledge of nearshore biodiversity, education in developing countries must continue. It has become evident that there is a need for expert services and facilities to process field samples efficiently and completely. Some regions have this service, such as the Atlantic Ocean region through the Atlantic Reference Centre (Huntsman Marine Science Centre, New Brunswick, Canada), which is in charge of processing, quality control/assurance, and archiving all Atlantic Ocean regional samples.

Current NaGISA data will culminate in 2010 with assessing spatial (for example latitudinal or longitudinal) trends of overall community patterns in rocky macroalgal systems and seagrass beds, as well as of selected taxonomic groups. Because of the relatively good taxonomic expertise available in most regions of the world, NaGISA is focusing in this first phase on patterns in macroalgae, polychaetes, gastropods, echinoderms, and decapods. However, there are many other taxonomic groups that are yet unexplored, but are no less ecologically important. Not only do we know that biodiversity trends vary depending on the taxonomic group examined, but these patterns may be quite different for the rarer groups than the more common taxa. Similarly we have learned that biodiversity trends often depend on the depth strata examined, but without better knowledge of small-scale biodiversity patterns, overall trends will be difficult to determine. There are many questions that remain unanswered here, such as what is the global latitudinal trend for cnidarians, sponges, or bryozoans, and do these trends vary with depth.

2.7 Conclusions

NaGISA's major legacies thus far can be summarized as the following.

1) The creation of the first standardized global baseline of coastal biodiversity in rocky shores and seagrass beds from the intertidal zone to water depths of up to 20 m.

2) The establishment of a standardized sampling protocol that is suitable to analyze biodiversity trends on multiple spatial and temporal scales.

3) The improvement of benthic taxonomy.

4) The network of scientists and new scientific capacity-building around the world, a network that is now working together to address major questions in nearshore biodiversity.

5) The elucidation of the scales of temporal and spatial variability in nearshore habitats.

6) The addition of knowledge on the interactive effects of multiple drivers, including human activities, on spatial patterns of marine coastal biodiversity at the global scale.

7) The identification of hot spots of marine coastal biodiversity that can be suggested for new Marine Protected Areas.

The NaGISA project has sampled many sites throughout the world, but the efforts are still dwarfed compared with the vastness of the world's nearshore region. Some of the sampled sites have now been established for long-term monitoring. In all regions, there will and should be continued monitoring of selected NaGISA sites. This monitoring will be done by a combination of researchers, elementary, high school, and university students, local communities, and other stakeholders. NaGISA has particularly enhanced stakeholder "ownership" at many sites, similar to sponsorship of roadside clean-up programs. Although information for truly global comparisons is still lacking in many areas and for certain taxonomic groups, patterns in biodiversity are beginning to emerge. More sites are continually being

added and more taxonomists are being engaged. The momentum that NaGISA has started must continue if we are to get an increasingly accurate description of global diversity.

The NaGISA monitoring sites will assist with the identification of inter-annual variability. This is crucial to be able to distinguish short-term variability from longer-term changes that may be driven by climatic changes or anthropogenic pressures. Such long-term changes will become measurable over time from NaGISA sites that are part of the long-term monitoring. In addition, the NaGISA–History of Marine Animal Populations collaboration, the History of the Nearshore (HNS) project, is identifying changes in nearshore communities that have occurred over decadal scales. By comparing historical baselines with present-day data, regional changes within the various HNS studies may be detected. Changes revealed by comparisons of several Atlantic HNS regions could, for example, produce a Pan-Atlantic pattern and identify driving factors.

NaGISA has done much to not only advance the knowledge and appreciation of nearshore biodiversity, but it has started a momentum through its outreach, networking, and capacity building. We may never be able to answer how many species live in the nearshore, but we will continue to produce a more accurate estimation and to explain why there are so many nearshore species and why they are distributed as they are.

Acknowledgments

We thank the many unnamed scientists, students, and interested people who have helped to sample the nearshore environment all over the world to contribute to the NaGISA effort. Specifically for the EPAC region, we thank Rafael Riosmena-Rodriguez (Universidad Autónoma de Baja California Sur) and Matthew Edwards (San Diego State University) for data and input into various sections of this chapter. In the SAS region, Gabriela Palomo (Museo Argentino de Ciencias Naturales MACN, Argentina), Manuel Ortiz (Universidad de La Habana, Cuba), Gregorio Bigatti (Centro Nacional Patagonico CENPAT, Argentina), Manuel Cruz (INOCAR and Facultad Ciencias Naturales, Universidad de Guayaquil, Ecuador), and Paulo Lana (Universidade Federal do Paraná, Brazil) provided data summarized in Table 2.1. In the Caribbean Sea region, the following individuals provided information and feedback: Diana Isabel Gómez (INVEMAR, Colombia), Judith Gobin (University of West Indies, Trinidad and Tobago), Manuel Ortiz (Universidad de La Habana, Cuba), and Andrea Bueno (Universidad Simon Bolivar, Venezuela). Lastly, we remember P. Robin Rigby (1977–2007) whose leadership has been essential in creating the NaGISA network.

References

Adnan Awad, A., Griffiths, C.L. & Turpie, J. (2002) Distribution of South African marine benthic invertebrates applied to the selection of priority conservation areas. *Diversity and Distributions* 8, 129–145.

Benedetti-Cecchi, L. (2007) Biodiversity and sampling design. In: *Sampling Biodiversity in Coastal Communities*. (eds. P. R. Rigby, K. Iken & Y. Shirayama), pp. 7–10. Kyoto University Press.

Benedetti-Cecchi, L., Bertocci, I., Vaselli, S., *et al.* (2008) Neutrality and the response of rare species to environmental variance. *PLoS ONE*, 3 (7), e2777.

Blanchette, C.A., Miner, C.M., Raimondi, P.T., *et al.* (2008) Biogeographical patterns of rocky intertidal communities along the Pacific coast of North America. *Journal of Biogeography* 35, 1593–1607.

Blaxter, M. (2003) Counting angels with DNA. *Nature* 421, 122–124.

Blum, M.J., Neely, D.A., Harris, M.H., *et al.* (2008) Molecular systematics of the cyprinid genus *Campostoma* (Actinopterygii: Cypriniformes): disassociation between morphological and mitochondrial differentiation. *Copeia* 2008, 360–369.

Branch, G.M., Griffiths, C.L., Branch, M.L., *et al.* (1994) *Two Oceans: A Guide to the Marine Life of Southern Africa*. Cape Town and Johannesburg: D. Phillip.

Brink, K.H. & Muench, R.D. (1986) Circulation in the Point Conception-Santa Barbara Channel region. *Journal of Geophysical Research C* 91, 877–895.

Broitman, B.R., Navarette, S.A., Smith, F., *et al.* (2001) Geographic variation of southeastern Pacific intertidal communities. *Marine Ecology Progress Series* 224, 21–34.

Browne, D.R. (1994) Understanding oceanic circulation in and around the Santa Barbara Channel. In: *The Fourth California Islands Symposium: Update on the Status of Resources* (eds. W.L. Halvorson & G.J. Maender), pp. 27–34. Santa Barbara Natural History Museum.

Buzeta, M.-I., Singh, R. & Young-Lai, S. (2003) Identification of significant marine and coastal areas in the Bay of Fundy. Canadian Manuscript Report of Fisheries and Aquatic Sciences 2635, xii + 177 pp.

Buzeta, M.-I. & Singh, R. (2008) Identification of ecologically and biologically significant areas in the Bay of Fundy, Gulf of Maine. Vol. 1. Areas identified for review and assessment of the Quoddy region. Canadian Technical Report of Fisheries and Aquatic Sciences 2788: vii + 80 p.

Campbell, D.E. (2004) Evaluation and energy analysis of the Cobscook Bay Ecosystem. *Northeastern Naturalist* 11 (Special Issue 2), 355–424.

Carey, A.G. Jr., & Ruff, R.E. (1977) Ecological studies of the benthos in the western Beaufort Sea with special reference to bivalve molluscs. In: *Polar Oceans*. (ed. M. J. Dunbar), pp. 505–530. Calgary: Arctic Institute of North America.

Carey, A.G. Jr., Scott, P.H. & Walters, K.R. (1984) Distributional ecology of shallow southwestern Beaufort Sea (Arctic Ocean) bivalve Mollusca. *Marine Ecology Progress Series* 17, 125–134.

Chenelot, H.A., Iken, K., Konar, B., *et al.* (2007) Spatial and Temporal Distribution of Echinoderms in Rocky Nearshore Areas of Alaska. In: *Selected Papers of the NaGISA World Congress 2006* (eds. P.R. Rigby & Y. Shirayama), pp. 11–28. Publications of the Seto Marine Biological Laboratory, Special Publication Series, Vol. VIII.

Clarke, A. & Crame, J.A. (1997) Diversity, latitude and time: patterns in the shallow sea. In: *Marine Biodiversity*. (eds. R.F.G. Ormond, J.D. Gage & M.V. Angel) pp. 122–145. Cambridge University Press.

Clarke, K.R., & Warwick, R.M. (2001) *Change in marine communities. An approach to statistical analysis and interpretation*. 2nd edition. Plymouth, UK: PRIMER-E.

Coleman, A.R., Underwood, A.J., Benedetti-Cecchi, L., *et al.* (2006) A continental scale evaluation of the role of limpet grazing on rocky shores. *Oecologia* 147, 556–564.

Connolly, S.R. & Roughgarden, J. (1998). A latitudinal gradient in Northeast Pacific intertidal community structure: evidence for an oceanographically based synthesis of marine community theory. *American Naturalist* 151, 311–326.

Connolly, S.R., Menge, B.A. & Roughgarden, J. (2001) A latitudinal gradient in recruitment of intertidal invertebrates in the Northeast Pacific Ocean. *Ecology* 82, 1799–1813.

Couto, E., Lang Da Silveira, F. & Rocha, G. (2003) Marine biodiversity in Brazil: the current status. *Gayana*, 67, 327–340.

Díaz, J.M., Gast, F. & Torres, D.C. (2009) Rediscovery of a Caribbean living fossil: *Pholadomya candida* G.B. Sowerby I, 1823 (Bivalvia: Anomalodesmata: Pholadomyoidea). *Nautilus* 123, 19–20.

Dunton, K.H., Reimnitz E. & Schonberg, S. (1982). An Arctic kelp community in the Alaskan Beaufort Sea. *Arctic* 35, 465–484.

Dunton, K.H. & Schonberg, S. (2000) The benthic faunal assemblage of the Boulder Patch kelp community. In: *The Natural History of an Arctic Oil Field*. (eds. J. C. Truett & S. R. Johnson) pp. 371–398. San Diego, California: Academic Press.

Ekman, S. (1953) *Zoogeography of the Sea*. Sidgwick & Jackson.

Feder, H.M. & Schamel, D. (1976) Shallow water benthic fauna of Prudhoe Bay. In: *Assessment of the Arctic Marine Environment: Selected Topics* (eds. D. W. Hood & D. Burrell), pp. 329–359. Institute of Marine Science Occasional Publications No. 4, University of Alaska, Fairbanks.

Fernandez, M., Jaramillo, E., Marquet, P.A., *et al.* (2000) Diversity, dynamics and biogeography of Chilean benthic nearshore ecosystems: an overview and guidelines for conservation. *Revista Chilena de Historia Natural* 73, 797–830.

Frid, C., Hammer, C., Law, R., *et al.* (2003) *Environmental Status of the European Seas*. Copenhagen: ICES.

Gallardo, V.A. (1987) The sublittoral macrofaunal benthos of the Antarctic shelf. *Environment International* 13, 71–81.

Gardner, T., Cote, I., Gill, J., *et al.* (2003). Long-term region-wide declines in Caribbean Corals. *Science* 301, 958–960.

Gerken, S. (2009) Two new *Cumella* (Crustacea: Cumacea: Nannastacidae) from the North Pacific, with a key to the North Pacific *Cumella*. *Zootaxa* 2149, 50–61.

Gibbons, M.J., Abiahy, B.B., Angel, M., *et al.* (1999) The taxonomic richness of South Africa's marine fauna: a crisis at hand. *South African Journal of Science* 95, 8–12.

Griffiths, C.L. (2005) Coastal marine biodiversity in East Africa. *Indian Journal of Marine Sciences* 33, 35–41.

Haussermann, V. & Forsterra, G. (2009) *Fauna Marina Bentónica de la Patagonia Chilena*. Nature in Focus, Santiago, Chile.

Hebert, P.D.N., Cywinska, A., Ball, S.L., *et al.* (2003) Biological identifications through DNA barcodes. *Proceedings of the Royal Society of London B* 270, 313–321.

Horn, M.H. & Allen, L.G. (1978) A distributional analysis of California coastal marine fishes. *Journal of Biogeography* 5, 23–42.

Iken, K. & Konar, B. (2007) Essential habitats in our Arctic front yard: nearshore benthic community structure. Alaska Sea Grant, Final Report.

Kawai, H., Hanyuda, T., Lindeberg, M., *et al.* (2008) Morphology and molecular phylogeny of *Aureophycus aleuticus* gen. et sp. nov. (Laminariales, Phaeophyceae) from the Aleutian Islands. *Journal of Phycology* 44, 1013–1021.

Kedra, M. & Wlodarska-Kowalczuk, M. (2008) Distribution and diversity of sipunculan fauna in high Arctic fjords (west Svalbard). *Polar Biology* 31, 1181–1190.

Kelly, R.P. & Eernisse, D.J. (2007) Southern hospitality: a latitudinal gradient in gene flow in the marine environment. *Evolution* 61, 700–707.

Kerswell, A.P. (2006) Global biodiversity patterns of benthic marine algae. *Ecology* 87, 2479–2488.

Konar, B., Riosmena-Rodriquez, R. & Iken, K. (2006) Rhodolith bed: a newly discovered habitat in the North Pacific Ocean. *Botanica Marina* 49, 355–359.

Konar, B., Iken, K. & Edwards, M. (2009) Depth-stratified community zonation patterns on Gulf of Alaska rocky shores. *Marine Ecology* 30, 63–73.

Kuklinski, P. & Barnes, D.K.A. (2008). Structure of intertidal and subtidal assemblages in Arctic vs temperate boulder shores. *Polish Polar Research* 29, 203–218.

Kuklinski, P., Barnes, D.K. A. & Taylor, P.D. (2006) Latitudinal patterns of diversity and abundance in North Atlantic intertidal boulder-fields. *Marine Biology* 149, 1577–1583.

Ladd, C., Hunt, G.L., Mordy, C.W., *et al.* (2005) Marine environment of the eastern and central Aleutian Islands. *Fisheries Oceanography* 14, 22–38.

Larsen, P.F., Phinney, D.A., Rubin, F., *et al.* (2009). Classification of boreal macrotidal littoral zone habitats in the Gulf of Maine: comparisons of IKONOS and CASI multispectral imagery. *Geocarto International* 24, 457–472.

Lawson, G.W. & John D.M. (1987) The marine algae and coastal environment of tropical West Africa. *Beihefte Nova Hedwigia* 93, 1–415.

Lüning, K. (1990) *Seaweeds: Their Environment, Biogeography, and Ecophysiology*. New York: Wiley.

Mendoza, M.L. & Cabioc'h, J. (1998) Study compares reproduction of *Phymatolithon calcareum* (Pallas) Adey & Mc-Kibbin and *Lithothamnion corallioides* (P. & H. Crouan) P. and H. Crouan (Corallinales, Rhodophyta), and reconsiderations of the definition of the genera. *Canadian Journal of Botany* 76, 1433–1445.

Miller, K.A. & Estes, J.A. (1989) Western range extension for *Nereocystis luetkeana* in the North Pacific Ocean. *Botanica Marina* 32, 535–538.

Miller, K.A. & Pearse, J.S. (1991) Ecological studies of seaweeds in McMurdo Sound, Antarctica. *American Zoologist* 31, 35–48.

Moe, R.L. & DeLaca, T.E. (1976) Occurrence of macroscopic algae along the Antarctic Peninsula. *Antarctic Journal of the United States* 11, 20–24.

Murray, S. N. & Littler, M.M. (1981) Biogeographical analysis of intertidal macrophyte floras of southern California. *Journal of Biogeography* 8, 339–351.

Murray, S.N. & Bray, R.N. (1993) Benthic macrophytes. In: *Ecology of the Southern California Bight: A Synthesis and Interpretation*. (eds. M. D. Dailey, D. J. Reish, & J. W. Anderson) pp. 304–368. Berkeley: University of California Press.

Nishimura, S. (1974) *History of Japan Sea: Approach from Biogeography*. Tokyo: Tsukiji-shokan. [In Japanese.]

Okuda, T., Noda, T., Yamamoto, T., *et al.* (2004) Latitudinal gradient of species diversity: multi-scale variability in rocky intertidal sessile assemblages along the Northwestern Pacific coast. *Population Ecology* 46, 159–170.

Pfeiler, E., Bitler, B.G., Ulloa, R., *et al.* (2008) Molecular identification of the bonefish *Albula escuncula* (Albiformes: Albulidae) from the tropical eastern Pacific, with comments on distribution and morphology. *Copeia* 2008, 763–770.

Pollock, L.W. (1998) *A Practical Guide to the Marine Animals of Northeastern North America*. New Jersey: Rutgers University Press.

Rahayu, D.L & Wahyudi, A.J. (2008) *Common littoral hermit crabs of Indonesia* (eds. Y. Susetiono, Y. Shirayama & A. Asakura). Kyoto University Press.

Rigby, P.R., Iken, K. & Shirayama, Y. (2007) *Sampling Biodiversity in Coastal Communities: NaGISA Protocols for Seagrass and Macroalgal Habitats*. Kyoto University Press.

Rivadeneira, M.M., Fernández, M. & Navarrete, S.A. (2002) Latitudinal trends of species diversity in rocky intertidal herbivore assemblages: spatial-scale and the relationship between local and regional species richness. *Marine Ecology Progress Series* 245, 123–131.

Roy, K., Jablonski D. & Valentine, J.W. (2000) Dissecting latitudinal diversity gradients: functional groups and clades of marine bivalves. *Proceedings of the Royal Society of London B* 267, 293–299.

Sala, E. & Knowlton, N. (2006) Global marine biodiversity. *Annual Review of Environment and Resources* 31, 93–122.

Saunders, G.W. (2005). Applying DNA barcoding to red macroalgae: a preliminary appraisal holds promise for future applications. *Philosophical Transactions of the Royal Society of London B* 360, 1879–1888.

Saunders, G.W. (2008) A DNA barcode examination of the red algal family Dumontiaceae in Canadian waters reveals substantial cryptic species diversity. 1. The foliose *Dilsea–Neodilsea* complex and *Weeksia*. *Botany* 86, 773–789.

Schoch, G.C., Menge, B.S., Allison, G., *et al.* (2006) Fifteen degrees of separation: latitudinal gradients of rocky intertidal biota along the California current. *Limnology and Oceanography* 51, 2564–2585.

Scrosati, R. & Heaven, C. (2007) Spatial trends in community richness, diversity, and evenness across rocky intertidal environmental stress gradients in eastern Canada. *Marine Ecology Progress Series* 342, 1–14.

Susetiono (2007). Lamun dan Fauna. Teluk Kuta, Pulau Lombok. [Marine fauna and flora in Kuta Bay, Lombok Island.] [In Indonesian.] Jakarta: Pusat Penelitian Oseanografi, LIPI Press. 155 pages.

Thacker, C.E. (2009) Phylogeny of Gobioidei and placement within Acanthomorpha with a new classification and investigation of diversification and character evolution. *Copeia* 2009, 93–104.

Trott, T.J. (2004) Cobscook inventory: a historical checklist of marine invertebrates spanning 162 years. *Northeastern Naturalist* 11 (Special Issue 2), 261–324.

Trott, T.J. (2009). Location of biological hotspot in the Gulf of Maine. In: *Proceedings of the Gulf of Maine Symposium*, St. Andrews, New Brunswick, 4–9 October 2009.

Trott, T.J. (2007) Zoogeography and changes in macroinvertebrate community diversity of rocky intertidal habitats on the Maine coast. In: *Challenges in Environmental Management in the Bay of Fundy-Gulf of Maine*. (eds. G.W. Pohle, P.G. Wells & S.J. Rolston) pp. 54–73. Proceedings of the 7th Bay of Fundy Science Workshop, St. Andrews, NB, Oct. 24–27, 2006. Bay of Fundy Ecosystem Partnership, Technical Report No. 3. Wolfville, Nova Scotia: Acadia University.

Trott, T.J. & Larsen, P.F. (2003) Cobscook Bay, Maine: the crown jewel of regional biodiversity in the Gulf of Maine. In: *Proceedings of the 32nd Annual Benthic Ecology Meeting*, Mystic, Connecticut.

Udvardy, M. (1969). *Dynamic Zoogeography*. New York: Van Nostrand Reinhold.

Venkataraman, K. & Wafar M. (2005) Coastal and marine biodiversity of India. *Indian Journal of Marine Sciences* 34, 57–75.

Wiencke, C. & Clayton, M.N. (2002) Antarctic seaweeds. In: *Synopses of the Antarctic Benthos*. (eds. J. W. Wägele & J. Sieg), vol. 9., 239 pp. Ruggell: ARG Gantner.

Wilce, R.T. (1997) The Arctic subtidal as a habitat for macrophytes. In: *Seaweed Ecology and Physiology*. (eds. C.S. Lobban & P.J. Harrison) pp. 89–92. Cambridge University Press.

Willig, M.R., Kaufman, M. & Stevens, R.D. (2003) Latitudinal gradients of biodiversity: pattern, process, scale, and synthesis. *Annual Review of Ecology and Evolutionary Systematics* 34, 273–309.

Wilkinson, C. (2004). *Status of Coral Reefs of the World: 2004*. Townsville, Queensland, Australia: Australian Institute of Marine Science.

Witman, J.D., Etter, R.J. & Smith, F., *et al.* (2004) The relationship between regional and local species diversity in marine benthic communities: a global perspective. *Proceedings of the National Academy of Sciences of the USA* **101**, 15664–15669.

Wlodarska-Kowalczuk, M., Kuklinski, P., Ronowicz M., *et al.* (2009). Assessing species richness of macrofauna associated with macroalgae in Arctic kelp forests (Hornsund, Svalbard). *Polar Biology* **32**, 897–905.

Wulff, A., Iken K., Quartino, M.L., *et al.* (2009). Biodiversity, biogeography and zonation of benthic micro- and macroalgae in the Arctic and Antarctic. *Botanica Marina* **52**, 491–507.

Yasin, Z., Kwang, S.Y., Shau-Hwai, A.T., *et al.* (2008) *Field Guide to the Echinoderms (Sea Cucumbers and Sea Stars) of Malaysia*. Kyoto University Press.

Chapter 3

Biodiversity Knowledge and its Application in the Gulf of Maine Area

Lewis S. Incze[1], Peter Lawton[2], Sara L. Ellis[1], Nicholas H. Wolff[1]

[1]*Aquatic Systems Group, University of Southern Maine, Portland, Maine, USA*
[2]*Fisheries and Oceans Canada, St. Andrews Biological Station, St. Andrews, New Brunswick, Canada*

3.1 Introduction

The diversity of life at all levels, from ecosystems to genes, is part of our natural heritage, an inheritance molded by more than three billion years of evolutionary innovation, adaptation, and chance (Raup 1976; Knoll 2003; Falkowski *et al.* 2008). By comparison with Earth's long and complex history of biological, chemical, and geophysical change, modern humans are relative newcomers (Liu *et al.* 2006), albeit with enormous capacity to alter the environment, its species composition, and functioning (Millennium Ecosystem Assessment 2005). Despite our technological prowess, we depend on natural ecosystems for life support, economic activity, and pleasure. What will happen as human populations occupy, use, and transform ever-increasing portions of the environment (Rockström *et al.* 2009)? The question has practical, as well as ethical and aesthetic, dimensions. Managing human activities in ways that preserve the ability of ecosystems to provide goods, critical services, natural beauty, and wonder into the future is one of the great challenges we face as a society.

Many have advocated a comprehensive approach to the sustainable use of the marine environment, including the supporting role of the ecosystem in general, and the con-servation of biodiversity specifically (Grumbine 1994; Pew Oceans Commission 2003; Ragnarsson *et al.* 2003; Sinclair & Valdimarsson 2003; US Commission on Ocean Policy 2004; McLeod *et al.* 2005; Rosenberg & McLeod 2005; Palumbi *et al.* 2009). Ecosystem-based management (EBM) is an integrated approach that considers the entire ecosystem, including humans, and circumscribes a broad set of objectives and principles designed to guide decision making whenever the environment might be impacted (Murawski 2007; McLeod & Leslie 2009). EBM is an evolving practice, and explicit incorporation of ecosystem considerations into management of human interactions has recently increased dramatically (McLeod & Leslie 2009; Rosenberg *et al.* 2009). Conserving biodiversity as a cornerstone of EBM, however, is challenging because most biodiversity is still unknown, most species are comparatively rare, and the "importance" (function) of many non-dominant species is difficult to quantify and impossible to predict. Even if it can be shown that a species plays no significant role today, its contribution to the future remains unknowable. This need not require evolutionary time scales for expression, because systems experiencing rapid change – whether by climate, major natural disturbance, or human disturbance – may suddenly favor a different set of genes or species (Yachi & Loreau 1999; Bellwood *et al.* 2006). Biodiversity is the reservoir of options that enables species (whose populations contain genetic diversity) and systems at all higher levels of organization to respond to changes over time, and

Life in the World's Oceans, edited by Alasdair D. McIntyre
© 2010 by Blackwell Publishing Ltd.

biodiversity is the encyclopedia of information about life itself. Thus, there are many reasons, practical and otherwise, to document, understand, and conserve it.

This chapter describes recent efforts by the Gulf of Maine Area (GoMA) project of the Census of Marine Life to improve our understanding of biodiversity in the Gulf of Maine Area (Fig. 3.1) and suggests ways this information can be used to support EBM in the marine environment. Most projects within the Census were focused on species discovery in remote and under-explored areas of the ocean (O'Dor & Gallardo 2005). Early on, however, the Census recognized the need for an integrative study of biodiversity on an eco-system-wide scale, covering a range of trophic levels (from microbes to mammals) and habitats (from shallow intertidal to deep offshore). The Gulf of Maine was selected as the ecosystem project because it is a well-studied, comparatively data-rich body of water with a long history of commercial exploitation and associated management needs. Its moderate size and intermediate levels of biodiversity were other potential advantages in terms of tractability. Although there was a large body of knowledge about the region, there had not yet been any coordinated effort to summarize the Gulf's biodiversity in an accessible format (Foote 2003), or to consider how biodiversity information could be used to improve management of a system of this size.

3.2   Environmental and Biogeographic Setting and History of Human Use

Biodiversity of the Gulf of Maine Area has been shaped over geologic time by geophysical and evolutionary processes, and, more recently, by anthropogenic pressures. During the Last Glacial Maximum (*ca.* 20,000 years before present (B.P.)), ice sheets extended onto the eastern North American continental shelf south of 41°N latitude, scouring the bedrock and depositing moraines that shape the present-day submarine topography of the Gulf of Maine and the Scotian Shelf (Knott & Hoskins 1968). Maximum present-day depths exceed 250 m in Georges and Emerald Basins (Figs. 3.1A and B), and the interior of the Gulf and the Scotian Shelf are generally deep except for a few large offshore banks and a narrow coastal fringe. The shoreline is diverse, consisting of extensive regions of tectonically deformed metamorphic rock, granites and other igneous intrusions, as well as sandy and gravelly shorelines of varying lengths. Salt marshes are mostly small and comparatively infrequent in rock-dominated sections of the coast, but are substantial in the aggregate and extensive along some sections of coast in the Bay of Fundy and in the southern Gulf (Gordon *et al.* 1985; Jacobson *et al.* 1987). Rocky sections are typically highly indented, with

numerous bays, peninsulas, and islands providing a wide variety of habitat types.

The dominant circulation in the upper 100 m is southward over the Scotian Shelf and counterclockwise around the Gulf of Maine, with most water exiting around the northern end of Georges Bank (Xue *et al.* 2000; Smith *et al.* 2001; Townsend *et al.* 2006). The banks and shoals along the outer periphery of the Gulf of Maine restrict exchanges between the Gulf and the open Atlantic and lengthen the path and increase the residency time of water as it travels along the southern flank of Georges Bank, thus contributing to the temperature contrast between the interior of the Gulf and the more temperate region to the south (Fig. 3.1C). Deeper water enters the Gulf from the upper slope through the Northeast Channel (sill depth approximately 190 m) and may be of northern (Labrador Sea) or southern (Mid-Atlantic) origin (Greene & Pershing 2003). Sources of slope water influence the temperature, salinity, and nutrient ratios of water and are themselves under the influence of larger-scale climate forcing (Greene & Pershing 2003, 2007; Townsend *et al.* 2010).

Tidal ranges vary from less than 2 m along the Nova Scotia Atlantic coast and approximately 3 m in the southern Gulf of Maine to 16 m in the northeastern Bay of Fundy (Minas Basin), reputedly the largest tidal range in the world (Archer & Hubbard 2003; O'Reilly *et al.* 2005). Where the tidal range is large, the difference between neap and spring tides exceeds the entire tidal range of locations in the southern Gulf (Dohler 1970). In the northern Gulf and over the crest of many of the offshore banks and shoals, turbulence created by strong tidal bottom friction contributes to unstratified or only weakly stratified conditions even during warm months of the year (Garrett *et al.* 1978), whereas elsewhere there is strong seasonal stratification induced by salinity and temperature (Fig. 3.1C).

From a global perspective, the Gulf of Maine Area has relatively low diversity (Witman *et al.* 2004), and is generally less diverse than waters farther south along the US east coast (Fautin *et al.*, unpublished observations) and in the northeast Atlantic (Vermeij *et al.* 2008). The intertidal and subtidal zone of the Cobscook/Passamaquoddy Bay region (US–Canadian border) may prove to be an exception (Larsen 2004; Trott 2004; Buzeta & Singh 2008). Cape Cod, which partly defines the western boundary of our study area (Fig. 3.1A), is generally recognized as the transition between the southern Virginian and the northern Acadian biogeographic provinces (Engle & Summers 1999; Wares & Cunningham 2001; Wares 2002). Some argue that the transition may be focused slightly south of the Cape in association with changes in water mass properties (Wares 2002; Jennings *et al.* 2009), but many Virginian and Acadian species occur well north and south, respectively, of this transition (Fautin *et al.*, unpublished observations). The modern biogeographic provinces are aligned with a steep latitudinal gradient in surface water

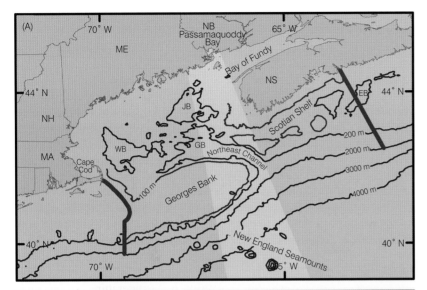

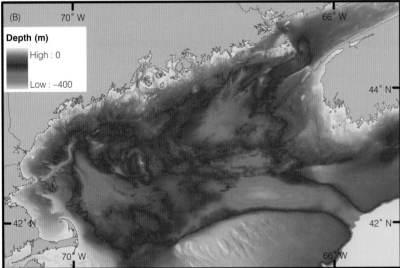

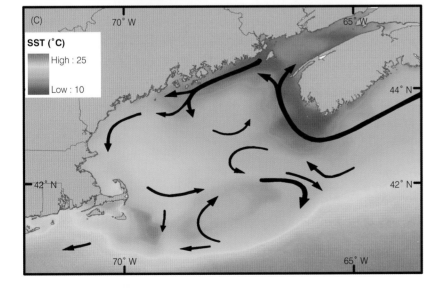

Fig. 3.1

Gulf of Maine study area.
(A) Major physiographic features and names. Isobaths in dark grey (200–4,000 m) show the continental slope, Northeast Channel, and major basins. EB, GB, JB, and WB are Emerald, Georges, Jordan, and Wilkinson basins, respectively. A portion of the 100 m isobath is shown in dark blue to illustrate the major banks and the inner Scotian Shelf (see next panel for details of inner Gulf). Canadian provinces (Nova Scotia, New Brunswick) and US states (Maine, New Hampshire, and Massachusetts) are abbreviated (NS, NB, ME, NH, and MA, respectively). The highlighted sector across the northern Gulf is the "Discovery Corridor", which roughly straddles the Canada–US border. The GoMA study area is bounded by the two red lines and the 2,000 m isobath (later extended to 3,500 m), plus Bear Seamount, the most western of the New England Seamount chain and located between 2,000 and 3,000 m. **(B)** Bottom topography of the Gulf of Maine showing the complex structure and generally deep bathymetry of the interior, as well as the principal channels into the system (data from US Geological Survey). Complex structures pose extra challenges to assessing and describing benthic diversity patterns and ecological functioning. **(C)** Climatological (1997–2008) satellite-derived (NOAA-AVHRR) sea surface temperatures (SST) for August, with schematic of the major surface circulation features. (SST data from Andrew Thomas, University of Maine, Orono, Maine, USA; circulation based on Beardsley *et al.* (1997)).

temperatures, with lower annual means and smaller annual ranges in the north. The transition has undergone large changes during the Holocene (a significant northward expansion and retraction of warm-water biota; Pielou 1991) and is likely to be affected by expected global warming (Hayhoe *et al.* 2007). The current regional warming trend of more than a decade is probably already affecting the distributions of some organisms (Fogarty *et al.* 2008), although the trajectory of future temperature changes may be affected by accelerated melting of Arctic ice (Häkkinen 2002; Smedsrud *et al.* 2008) and variations in ocean circulation (Greene & Pershing 2003, 2007; Fogarty *et al.* 2008; Townsend *et al.* 2010).

Humans have affected biodiversity of coastal systems around the world, and the Gulf of Maine is no exception (Jackson *et al.* 2001; Lotze *et al.* 2006). There is evidence of human habitation along coastal Gulf of Maine as early as 8,500 to 6,000 years B.P. (Bourque 2001; Bourque *et al.* 2008). Although some evidence suggests that prehistoric hunter-gatherers had negligible impacts on the coastal marine environment (Lotze & Milewski 2004), archaeological studies of faunal remains in middens have shown changes in the relative abundance of available prey species by 3,500 years B.P., indicating a decline in local cod (*Gadus morhua*) and changes in the food web (Bourque *et al.* 2008). Europeans started coming to the Gulf of Maine regularly in the mid-1500s to take advantage of rich natural resources, and colonized the area in the 1700s (Bourque *et al.* 2008). They rapidly transformed the coastal environment by multiple "top-down" (exploitation), "bottom-up" (nutrient loading), and "side-in" (habitat destruction, pollution) impacts, causing widespread changes in abundance and diversity at all trophic levels, from primary producers to top predators (Lotze & Milewski 2004). On the Scotian Shelf, regional cod stocks were severely reduced by 1859 (Rosenberg *et al.* 2005), and by 1900 most large vertebrates in the productive southwestern region of the Bay of Fundy were severely overexploited, leading to the extinction of three species of mammals and six bird species (Lotze & Milewski 2004).

In the early twentieth century, human pressures on the Gulf became more intense and far-reaching. Mechanized fishing technologies beginning in the 1920s led to a rapid decline in numbers and body size of many species, especially coastal cod in the Gulf of Maine (Steneck *et al.* 2004) and on Georges Bank (Sherman 1991). Starting in the middle of the twentieth century, commercial fish stocks experienced significant reductions (Cohen & Langton 1992; Sinclair 1996) and many important stocks remain at low levels. In 2007, cod landings in the entire Gulf of Maine were only 5–6% of those in 1861 (Alexander *et al.* 2009), and many historical fishing grounds along the coast from Massachusetts to Maine and Nova Scotia are no longer very productive (Ames 2004; Frank *et al.* 2005). The decline of large predatory fish has been used to explain cascading effects

at lower trophic levels involving various combinations of macroinvertevbrates and their invertebrate and algal prey (Steneck *et al.* 2004; Frank *et al.* 2005). Fluctuating abundances of sea urchins (caused by trophic cascades, direct fishing on urchins, and disease) and kelp (caused by predation by urchins and other factors (see, for example, Schmidt & Scheibling 2006)) have attracted particular attention because of the structuring role of kelp in shallow subtidal communities (Scheibling *et al.* 2009). The naturally low diversity of the Gulf of Maine kelp ecosystem may have facilitated the rapidity of these changes (Steneck *et al.* 2004).

Today, fishing remains the anthropogenic activity with the greatest impact on the Gulf of Maine system through removals and trophic effects (Steneck *et al.* 2004; Frank *et al.* 2005; Lotze *et al.* 2006), impacts on bottom biota and habitats (Auster *et al.* 1996; Collie *et al.* 1997, 2000; Watling & Norse 1998; Norse & Watling 1999; Myers and Worm 2003; Simpson & Watling 2006), and possible genetic effects. Modern means of harvesting as well as expanding human development along shorelines can be significantly disruptive or destructive of habitat, and virtually all areas of the Gulf from the intertidal to deep basins have been affected to some extent by human activities. Over the past three decades such impacts have generated growing concern, and a long series of restrictions on participation, gear, season and areas fished have been implemented, with historical emphasis on "catch" management and an emerging consideration of habitats, species of special concern, and biodiversity (Auster & Shackell 2000; Murawski *et al.* 2000; Lindholm *et al.* 2004; Buzeta & Singh 2008; Gavaris 2009).

3.3 Objective, Approaches, and Progress

The Gulf of Maine is an international body of water shared by Canada and the United States. The GoMA Project involved scientists from both countries and the area of study was defined as the Gulf of Maine proper (waters between Cape Cod, Massachusetts, and Cape Sable in southeastern Nova Scotia, and inside Georges Bank), Georges Bank, the Great South Channel, the western Scotian Shelf, the neighboring continental slope down to 3,500 m, and Bear Seamount (Fig. 3.1A). It is difficult to know how to conceptualize biodiversity and its functioning in a physically and oceanographically complex ecosystem of this size, and when GoMA was initiated in 2003 there was little regional consensus on how to integrate biodiversity information into management decision making. More fundamentally, what is the biodiversity of the Gulf of Maine Area? At an early meeting organized by the Census in Woods Hole, Massachusetts, in 1999, one of the region's taxonomic experts asked a much simpler question: "How many *named* species are there in the Gulf of Maine?" No one knew.

GoMA played a convening role in the region to consolidate and summarize existing data, identify gaps in knowledge, and stimulate new research. In addition, the project is developing a framework that can be shared by managers and scientists, of how knowledge of regional marine biodiversity could be used in management. The purpose is not to make recommendations on how to manage, but to encourage thinking about how biodiversity information could be used outside its purely scientific realm.

GoMA's objectives were the following:

- Synthesize current knowledge of biodiversity, including patterns of distribution, drivers of biodiversity patterns and change, and how biodiversity patterns affect function of the Gulf of Maine ecosystem.
- Assess the extent of unknown biodiversity.
- Lead and support development of information systems to increase access to data.
- Support selected field projects and emerging research technologies.
- Work with the scientific community and federal agencies in the US and Canada to help develop a framework for incorporating biodiversity information into EBM.
- Make recommendations for future research and monitoring.
- Educate the public on the role and importance of marine biodiversity.

In examining progress made toward these objectives during the first Census, we cover different aspects of how biodiversity is organized within the Gulf of Maine system, at diverse levels from the ecoregion to genes. We start with basic compositional features, proceed through considerations of how structure and function must be understood at multiple scales, and conclude with some perspectives on generating and using biodiversity knowledge.

3.3.1 The known regional biodiversity

One of our responses to the unanswered question of how many named species there are in our region was to assemble a Gulf of Maine Register of Marine Species (GoMRMS) based on species either known to exist here (using a variety of sources) or expected in the region based on a larger Northwest Atlantic register. The goal of GoMRMS (not yet complete) is to provide references and electronic links to taxonomic histories, descriptions, ecological and distributional information, museum holdings, and relevant databases, such as the Encyclopedia of Life (EOL; www.eol.org)

and the Ocean Biogeographic Information System (OBIS, see Chapter 17; www.iobis.org). In addition to being a resource for researchers interested in particular species, a well-developed and maintained list enables biogeographic comparisons (see, for example, Brunel *et al.* (1998) for the Gulf of St. Lawrence; the European Register of Marine Species for the North Sea), and can help answer the question "What kind of system is this?" The answer to this question helps to identify the extent to which systems may be similar and can be compared, which is one way of gaining insights into natural processes and responses to management actions (Murawski *et al.* 2010).

Currently, regional and global species registers are still works in progress that must be maintained with updated species entries, changing taxonomies, and documentation of sources, and they require a rigorous process of validation. As of November 2009, GoMRMS listed 3,141 species in the Gulf of Maine Area, with just under a third of the entries validated. To continue to build the register we have searched several databases to identify potential additions to the species already named in GoMRMS. Databases came from both countries and covered the shelf, interior basins, Northeast Channel, and the upper slope to 2,000 m. Data were from demersal trawl assessment surveys used for fisheries management, benthic surveys of infauna and epifauna, and planktonic collections from research and monitoring programs. In total, these data came from more than 11,000 trawls, 4,000 benthic samples, and 39,000 plankton samples collected since 1961. Most of the demersal trawl and benthic data were from depths shallower than 400 m, whereas plankton samples included the slope sea. Macrofaunal diversity of the slope and seamounts and microbial communities were evaluated by Expert Groups assembled for the purpose, and results are discussed later.

The database searches revealed location, date, and count data for 1,828 species: 1,403 from benthic/demersal samples (245 from near shore) and 559 from the net plankton (almost all metazoan, with some redundancies due to species with biphasic life histories). Of these, 821 were not listed in GoMRMS, bringing the provisional new total to 3,962 species. Significantly, nearly half of the species in GoMRMS now have spatial information, and the provisional additions provide guidance for prioritizing further work on the register. Other sources of information are being analyzed to assemble a better description of the system from work that has already been done, and new sampling programs for biodiversity studies are underway. In terms of species, large gains can be expected with increased effort directed at smaller organisms, and on all organisms in deep water environments. At all depths, however, closer looks reveal more species.

Recent subtidal sampling in Cobscook Bay, Maine, which has been studied for more than 160 years, produced 13 species not previously on the historical checklist (Trott 2004) of this well-studied bay (amphipods, polychaetes, a mysid, a mollusk, and a cumacean; P.F. Larsen, unpublished observations). These are species that occur widely throughout the Gulf of Maine and were therefore not a surprise, but this example poses a challenge: when is a system adequately described, and what are pragmatic standards and approaches for doing this? In somewhat deeper (50–56 m) water and within 20 km of the coast in the southwestern Gulf of Maine, a study of a small sample area found 70 genera of nematodes in 27 families from a total of 1,072 individuals (Abebe *et al.* 2004); eight of the genera had no previous representatives in GoMRMS. The nematode

diversity was considered to be quite high (Abebe *et al.* 2004), and the number of local additions at the level of genus reflects the scant number of previous investigations of small infaunal organisms.

Farther from the coast, researchers from the Canadian Department of Fisheries and Oceans, Canadian Atlantic region universities, and the Centre for Marine Biodiversity have been documenting new species records within the offshore portion of the Gulf of Maine Discovery Corridor (Figs. 3.1A and 3.2A; see also Section 3.3.3). A current student thesis project (A.E. Holmes, unpublished

Fig. 3.2

Examples of Gulf of Maine fauna.
(A) Rich suspension-feeding community dominated by sponges and sea anemones, discovered on a deep (188 m) bedrock ridge (dubbed "The Rock Garden") in Jordan Basin in 2005 by Canadian researchers working in the Discovery Corridor. Subsequent cruises in 2006 and 2009 have provided additional information on the overall extent of these hard substratum features within the otherwise sediment-dominated basin. Most species have not yet been identified below family and/or genus level owing to the predominant use of video- and still-imagery survey approaches (photograph: Department of Fisheries and Oceans, Bedford Institute of Oceanography, Dartmouth, Nova Scotia, Canada). **(B)** Winter skate (*Leucoraja ocellata*) cruising past deep-sea corals, *Primnoa resedaeformis* (sea corn) and *Paragorgia arborea* (bubble gum coral), in Northeast Channel (668 m) (photograph: ROPOS deep submergence vehicle, Canadian Scientific Submersible Facility, Sidney, British Columbia, Canada).
(C) Humpback whale (*Megaptera novaeangliae*) feeding on a surface patch of krill (*Meganyctiphanes norvegica*) formed by interactions of krill with internal waves over a small offshore bank (photograph: H. McRae, New England Aquarium, Boston, Massachusetts, USA).

observations) sampled three soft sediment sites at 200–220 m depth in Jordan Basin during the first Discovery Corridor mission in 2005, with three 0.5 m² replicates per site sieved through 0.5 mm mesh screens. Thirty-two of the 183 species in the samples were not in GoMRMS, including several in minor phyla. Some represent northerly or southerly range extensions, but others may be new observations for the region.

During the 2005 mission, and again in 2006, dense stands of large, habitat-forming corals were surveyed within the Northeast Channel Coral Conservation Area, which lies within the corridor. Although the diversity of coral species may be higher elsewhere (Cogswell *et al.* 2009), this conservation area is the heart of the greatest known abundance of deep-sea corals in the region, particularly of *Primnoa resedaeformis* (sea corn) and *Paragorgia arborea* (bubble gum coral) (Fig. 3.2B). Abundance and colony height of these two corals were greater at depths more than 500 m than had been reported from previous surveys in shallower waters (Watanabe *et al.* 2009). Relationships between the size of a colony and the size of its attachment stone were typically stronger and less variable for *P. resedaeformis* than for *P. arborea*, suggesting that factors such as topographic relief may play an additional role in regulating distributions of *P. arborea* (Watanabe *et al.* 2009).

In deeper waters outside the Coral Conservation Area, but still within the corridor, two species of black corals, *Stauropathes arctica* and *Bathypathes patula*, were recorded for the first time in regional and Canadian waters, respectively (K. MacIsaac, unpublished observations). Using the remotely operated vehicle ROPOS, small samples were collected from coral colonies for genetic analyses to help future definition of coral populations and connectivity between corals in the corridor and elsewhere. Additional species that are potentially new to regional or Canadian waters include the amphipod crustaceans *Eusirus abyssi* and *Leucothoe*

spinicarpa, the holothurians *Psychropotes depressa* and *Benthodytes* cf. *sordida*, the carnivorous chiton *Placiphorella atlantica*, and the bone-devouring pogonophoran worm *Osedax* (K. MacIsaac, unpublished observations). More new species may emerge as samples continue to be processed.

These closer looks at the environment reveal not only new additions to knowledge of what lives in the Gulf of Maine, but also habitat features that previous ocean-sounding data had overlooked, and organism densities that were sometimes surprising. None of these were extensive efforts. Thus, the nature, extent, and patchiness of biological communities in the Gulf of Maine are all significantly under-characterized. Indeed, even within this comparatively well-studied environment, the question "What lives here?" remains only partly answered, and an understanding of abundance and patterns of distribution much less so. With such a large heterogeneous area to examine more closely, and interest not only in composition but also structure and function, a strategy is needed to make the discovery process efficient. More is said on this topic later.

The best example of a well-documented pattern of distribution and abundance at Gulf-wide scale is for the fishes (Fig. 3.3), which have been sampled by fishery-independent assessment surveys for more than 40 years. The average number of species per tow (sample diversity), averaged over all tows, is highest around the periphery of the Gulf and lowest in the deep basins, the Northeast Channel, and parts of the slope and Scotian Shelf. This is slightly affected by dominance patterns, as rarefaction curves show the highest *total* fish diversity on the upper slope and Georges Bank, followed by the coastal shelf between Cape Cod and Maine, and then other regions (L.S. Incze & N.H. Wolff, unpublished observations). The basins, Northeast Channel, and shelf regions south and east of Nova Scotia group together and have much lower total diversity. Fishes have habitat preferences such that certain species and communities can

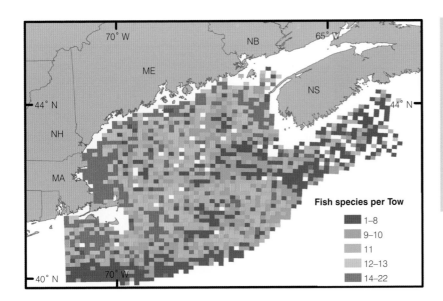

Fig. 3.3

Species diversity of fish in the Gulf of Maine (average number of species per tow per 10 km × 10 km cell), based on fall trawl surveys of the Northeast Fisheries Science Center (Woods Hole, Massachusetts), 1963–2008. Fall surveys took place between September and December (92% in October and November) and include 8,717 tows. Samples included 197 species of fish, with 15 elasmobranchs. Species richness groupings are quintiles of the frequency distribution of the samples. There is no correlation between species richness and the number of tows per cell.

Fish species per Tow

1–8
9–10
11
12–13
14–22

serve as proxies for seafloor habitat distributions (Auster *et al.* 2001; Auster & Lindholm 2005). The extensive fish data then become an information resource that can be linked with other biological and physical data to help characterize diversity of the Gulf of Maine system at subregional scales. Watling & Skinder (2007) showed this with invertebrate assemblages. The above patterns resulted from analysis of abundance/tow data that are now available from OBIS, and Ricard *et al.* (2010) have shown that OBIS data provide a very similar view to that obtained by more detailed analysis using comprehensive source databases from the surveys.

The continental slope and seamounts have not been studied as much or in the same way as the shelf, and so the status of biodiversity knowledge for this sub-region has been assessed separately by an ongoing Expert Group contributing to the Gulf of Maine Census (N.E. Kelly *et al.*, unpublished observations). Information has been assembled for benthic (infauna and epibenthic macro- and megafauna), demersal, mesopelagic, and bathypelagic taxa, comprising mostly adult stages, although a few larval fish were included. Sources of information include peer-reviewed literature, US and Canadian technical reports, OBIS, online museum collections and databases, and data provided by group members. Data extend west of GoMA, to 71.3°W, and from 150 to 3,500 m depth (Fig. 3.4). Although there have been studies on several of the western seamounts, only Bear Seamount was included in these analyses. So far, 899 species have been identified from the slope (mostly above 2,000 m) and 633 are associated with Bear Seamount; 240 were found in both locations. Bray-Curtis similarity (Clarke & Warwick 2001) between the slope and the GoMRMS species list is a little over 30%, and between Bear Seamount and GoMRMS is approximately 10%. A map of species numbers (Fig. 3.4) illustrates that many of the high values are associated with the seamount and major canyons. These values are

not corrected for effort or sampling method, so at this time they reflect the pattern of biodiversity knowledge, rather than intrinsic diversity patterns.

The smallest but most numerous and diverse organisms in the Gulf of Maine, as elsewhere, belong to a group of unicellular, prokaryotic, and eukaryotic organisms known collectively as marine microbes. The group includes viruses, bacteria, archaea, phytoplankton (for example diatoms), flagellates, ciliates, and other protists. We know most about the eukaryotic microalgae ("phytoplankton": 696 names in 193 genera), and much less about the other groups. Heterotrophic and mixotrophic protists include some familiar groups (the Dinophyceae) as well as others that are rarely identified below the level of genus (amoeboid organisms and ciliates). For the bacteria and viruses, the basic unit of diversity, the species, is probably inadequate and several approaches have been considered to express diversity in these groups (see Cohan 2002; Pedrós-Alió 2006). A Microbial Expert Group assembled for GoMA (W.K.W. Li *et al.*, unpublished observations) estimated the diversity of prokaryotes and phytoplankton in operational taxonomic units (OTUs) for the purpose of placing GoMA in a global context. The calculation is based on scaling arguments using the total number of individuals in the community (for instance, the bacterioplankton) and the number of individuals comprising the most abundant members of the community (the corresponding group for GoMA is the SAR11 cluster *Candidatus* Pelagibacter; for methods see Curtis *et al.* (2002); Morris *et al.* (2002)). Population sizes were estimated from the depth-dependent average of cell densities from a time series on the Scotian Shelf and neighboring slope (an extension of work published earlier by Li and Harrison (2001)) times the volume at depth in GoMA derived from a hypsometric analysis (L.S. Incze & N.H. Wolff, unpublished observations). The calculations indicate, as a very rough approximation, that GoMA could have between 10^5 and 10^6 taxa of prokaryotes and between

Fig. 3.4

Species diversity knowledge for the slope, canyons, and Bear Seamount, depicted as number of species per 0.2 degree square. Red lines mark eastern and western ends of the GoMA study area to 2,000 m. Species counts are divided into quintiles and have not been corrected for effort or sampling method. Black arrow points to the grid over Bear Seamount where the highest species count (494) was recorded. Data compiled by N.E. Kelly, Centre for Marine Biodiversity, Dartmouth, Nova Scotia, Canada.

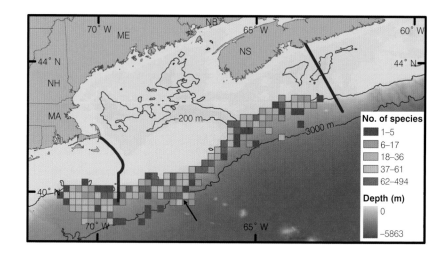

10^3 and 10^4 taxa of phytoplankton (because the assessment techniques used autofluorescence as a discriminator, the phytoplankton estimate includes the cyanobacteria). More specifically, the taxonomic richness of bacterioplankton in our study area is estimated to be 4×10^5 OTUs. This is 20% of the maximum global estimate of bacterioplankton diversity (2×10^6 OTUs; Curtis *et al.* 2002), which suggests a very diverse microbial community in Gulf of Maine Area.

The taxonomic distribution of biodiversity knowledge in the Gulf of Maine Area is summarized in Table 3.1, alongside a recent estimate of the global known marine biodiversity (Bouchet 2006). The table includes GoMRMS and the provisional additions from the survey databases, but does not include the above slope and seamount assessment because it has not been completed. The estimated diversity (OTUs) of the bacteria calculated above cannot be compared with the species estimate given by Bouchet (2006). How do the general patterns of named diversity in the Gulf of Maine Area compare with the global pattern, aside from the huge differences in numbers of species? Relatively speciose groups in both lists include the cnidarians, annelids, crustaceans, mollusks, bryozoans, and echinoderms, reflecting relatively high species richness in these groups in general, as well as conspicuousness, human interest, and relative ease of sampling and description by methods that have been established for many years. Among other speciose groups globally, the named marine algae and fish comprise a higher proportion of named species in the Gulf of Maine Area compared with the global list, and for the Gulf the proportion is lower for urochordates, Porifera, platyhelminthes, and nematodes. For the Porifera, the diversity has not been elucidated but may be comparatively low, whereas for the nematodes, a lack of significant effort on the group must be a major factor. These are general reflections on the state of knowledge for the Gulf as a whole. Valuable comparisons of species occurrence, distribution, and abundance across the Atlantic and north and south along the North American coast can be made within well-studied groups to study past and ongoing ecological changes (Vermeij *et al.* 2008).

To convey how much is known and unknown about diversity in the Gulf of Maine Area, we used a length-based approach for all adult stages of biota from viruses to the largest whales (Fig. 3.5). This is a coarse and subjective approximation because animal size (length) can vary greatly within a phylum and it was not practical to try to perfect this estimate by assigning "best approximate sizes" to all the named species! The smoothed line indicating the known (named) taxa approximates species numbers for groups of organisms contained within size groupings of $10^x \pm 10^{0.5x}$ m, where x is a whole number from −8 to +1. OTUs are used for viruses, bacteria, and archaea because there is no agreement on what constitutes species for these organisms. Trends and relative numbers are the important features being depicted (Fig. 3.5). "Monitored" species are those for

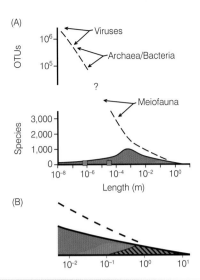

Fig. 3.5

Biodiversity size spectrum.
(A) Length-based schematic of Gulf of Maine biodiversity, showing the approximate size distribution of named species (solid line; blue shading is for emphasis), and a suggestion of the possible extent of the unknown biodiversity (broken line). For the prokarya and viruses, diversity is expressed as operational taxonomic units (OTUs), because there is no agreement on what makes a species in these groups. The shape of the curve of "unknowns" from meiofauna to viruses, and the maximum number of OTUs are unknown. The orange shape and orange squares are for monitored species, including harmful algae and coliform bacteria. Meiofauna is shown because it contains many unknown species, but there are other ecological and taxonomic groups that could be listed (see text). **(B)** Enlarged view of the lower right portion of the size-diversity curve, illustrating where most "monitored" (orange) and "managed" species (diagonal stripes) occur. Coliform bacteria, which are managed through effluent waste regulations, are not shown (see upper panel).

which we have some information on abundance over space and time (for example unmanaged species caught in fisheries assessment surveys, seabird abundances at long-term study sites); and "managed" species are those with management plans such as commercial fish, crustaceans and mollusks, cetaceans, and threatened or endangered species. At the far right end of the size spectrum, virtually all species are known, at least by name. "Unknowns" are dealt with in the next section. The schematic illustrates the point that the organisms of most concern to humans, whether for practical, aesthetic, ethical, or spiritual reasons, are a small fraction of the diversity in the system, and are supported by that diversity in ways that are only partly known.

3.3.2 Extent of unknown biodiversity

In general, we know less about the diversity of organisms as they get smaller, have softer bodies, inhabit more remote

Table 3.1

Comparison of the number of named species in the Gulf of Maine area with global estimates of marine species. Gulf of Maine totals are based on the Gulf of Maine Register of Marine Species and provisional additions from other sources (see section 3.3.1 for details on provisional additions to GoMRMS).

Taxon	GoMA species (this paper)[a]	Global species (Bouchet 2006)	Taxon	GoMA species (this paper)[a]	Global species (Bouchet 2006)
Bacteria	1[b]	4,800	Acanthocephala	27	600
Cyanophyta/Cyanobacteria	9	1,000	Entoprocta		165–170
Ciliophora	1	?	Gnathostomulida		97
Radiolaria		550	Priapulida		8
Foraminifera	2	10,000	Loricifera		18
Fungi		500	Cycliophora		1
Chlorophyta	98	2,500	Sipuncula	12	144
Bacillariophyta	224	5,000	Echiura	3	176
Phaeophyta	154	1,600	Annelida	489	12,000
Rhodophyta	148	6,200	Pogonophora		148
Dinomastigota	60	4,000	Tardigrada		212
Other protoctista	3	750	Crustacea	762	44,950
Plantae			Chelicerata (non-arachnid)	21	2,267
Porifera	31	5,500	Mollusca	504	52,525
Placozoa			Phoronida	1	10
Cnidaria	186	9,795	Bryozoa/Ectoprocta	119	5,700
Ctenophora	5	166	Brachiopoda	1	550
Platyhelminthes	72	15,000	Echinodermata	110	7,000
Dicyemida/Rhombozoa		82	Chaetognatha	12	121
Orthonectida		24	Hemichordata	5	106
Nemertea	35	1180–1230	Urochordata	44	4,900
Rotifera	4	50	Cephalochordata		32
Gastrotricha		390–400	Pisces	578	16,475
Kinorhyncha		130	Reptilia	2	–[c]
Nematoda	28	12,000	Aves	182	–[c]
Nematomorpha	2	5	Mammalia	27	110
			Total	3,962[a]	229,175[d]

[a] Total includes named species in GoMRMS plus provisional additions (see text).
[b] A new estimate for bacterioplankton OTUs in the Gulf of Maine is 4×10^5 (W.K.W. Li et al., unpublished observations), but this is not directly comparable with species (see text).
[c] These taxa are not included in Bouchet (2006).
[d] For taxa with a range of estimates, the average was used.

(deeper and offshore) places, and live within, rather than on, the bottom. Although the number of unknown species is impossible to estimate accurately, the more essential point is to illustrate where knowledge is most deficient. From this perspective, we can consider how these deficiencies affect our understanding of local communities and marine ecosystem processes, and what strategies might be used to understand better and conserve viable and functional populations of these poorly known and unknown parts of the ecosystem.

Recent studies have revealed a stunning level of diversity among marine prokaryotes (Sogin *et al.* 2006) and protists (Massana & Pedrós-Alió 2008), but many questions remain about how best to characterize it (Cohan 2002; Pedrós-Alió 2006; Not *et al.* 2009). The bacterial diversity was estimated in the section above with the "knowns" because there was a reasonable and interesting basis for making the calculation. We plot it as an "unknown" (Fig. 3.5), however, because OTUs do not necessarily correspond with phylogenetic relationships, and because the estimation is still very preliminary. The diversity of viruses was assumed to scale with abundance, and we have used a multiplier of ten (W.K.W. Li *et al.*, unpublished observations).

In the Gulf of Maine Area, we know that benthic and pelagic heterotrophic protists are seldom identified to species level despite the unquestioned importance of the "microbial loop" – the series of interactions among viruses, bacteria, archaea, and protists responsible for the large amount of cycling of organic matter and elements that occur in the water column (Sherr & Sherr 2000; Steele *et al.* 2007). A study by Savin *et al.* (2004) and subsequent work by the International Census of Marine Microbes comparing molecular methods with taxonomic assessments of microeukaryotic diversity in the Bay of Fundy show that the extent of diversity is not close to being understood. We know, too, that the soft-bottom infauna from all depths, and especially the meiofauna (Hicks 1985), are severely under-sampled and under-studied for their diversity, community composition, species–habitat relationships, and function. The nematodes provide an example within our area. One of the most diverse of marine animal phyla (*ca.* 12,000 named species estimated worldwide (Bouchet 2006)) and the most abundant of meiofaunal organisms (Chen *et al.* 1999), GoMRMS lists only 42 species. By contrast, the European Register of Marine Species (ERMS) lists over 1,800 (over a wider geographic and environmental range, but surely also incomplete). The addition of eight nematode genera to GoMRMS from a very small sample area (Abebe *et al.* 2004; Section 3.3.1) makes it reasonable to estimate that hundreds of nematode species, perhaps more than a thousand, have yet to be identified in the Gulf of Maine Area.

Small infaunal crustaceans (mostly harpacticoid copepods) and polychaetes are also abundant and diverse members of infaunal communities (Li *et al.* 1997;

Vanaverbeke *et al.* 1997). The polychaetes are better represented in the provisional list for the GoMA (411 of the 455 annelid species are polychaetes), but may be locally as diverse as the nematodes (De Bovée *et al.* 1996; Gobin & Warwick 2006). By comparison, only nine harpacticoid species are currently identified among the Crustacea listed in Table 3.1, and the diversity of this group in coastal, shelf, and slope waters from elsewhere (Baguley *et al.* 2006) suggests that a significant number of species living in the area have yet to be confirmed. It is likely that platyhelminth worms, another highly diverse marine phylum with only 73 named species in GoMRMS, are also significantly undercounted among the GoMA animal phyla. Recent field work demonstrated that it is still easy to add to the list of named species in our region (see earlier discussion), hinting at the size of the gap to be filled. With few exceptions, GoMRMS and the above discussion refer only to free-living forms (see discussion of symbionts by Bouchet (2006)).

Collectively, we estimate that the list of invertebrates yet to be identified must number in the thousands and extend from nearshore to the outer limits of our study area. Although these numbers are dominated by smaller organisms, the larger macrofauna and megafauna of the slope, canyons, and seamounts, including some fishes and cephalopods, are also incompletely known owing to their less accessible nature and the lack of widespread sampling across these regions so far. Research cruises conducted during the past decade have identified species new to science as well as additional specimens thought to be new species (Moore *et al.* 2003, 2004; Cairns 2007; Watling 2007; Hartel *et al.* 2008). Genetic studies will almost certainly add to the assessment of species composition. In addition to confirming suspected species splits (based on subtle characteristics, such as behavior, reproduction, habitat, morphotype, or physiology), genetics can also reveal cryptic species where single species were once thought to exist. The potential for cryptic species among "familiar" organisms is well illustrated by algae, which have simple morphologies, high rates of convergence, and phenotypic variation in varying environmental conditions, and for which there is often a lack of authoritative understanding of the complete life history (Saunders 2008). Increasingly, molecular tools are being used to resolve cryptic algal species (Saunders 2005, 2008; Kucera & Saunders 2008). Far from being mere taxonomic "splitting", such revelations are important to our understanding of the biological and ecological processes operating in the Gulf of Maine Area ecosystem.

3.3.3 Resolving structure and function

In the preceding sections we focused on which species are present in the Gulf of Maine Area. However, organisms exist in a context; they interact with each other and with their chemical and physical environment, and the resulting patterns of species distributions, abundances, vital rates, and behaviors affect the properties and functioning of an ecosystem. An over-arching series of questions, then, is (1) how are species distributed (including patterns of abundance and community composition), (2) what determines these patterns (including their temporal variations), and (3) how do the patterns affect the ecosystem? At small scales (meters to tens of meters) and in shallow water, patterns of biodiversity can be observed directly and intensively, but at larger scales (hundreds of meters to hundreds of kilometers) and in deeper water, observations are more difficult to make and there usually are trade-offs between intensive and extensive data collection. To understand biodiversity at the level of an ecosystem, new strategies and technologies are needed to obtain, analyze, and interpret data at multiple scales. We examine research at a few size scales in our area that bear on achieving ecosystem-level understanding.

Microbes – viruses, bacteria, archaea, and unicellular eukaryotic autotrophs, mixotrophs, and heterotrophs – are the most numerous and diverse organisms in the sea, and they constitute the largest reservoir of biomass (Kirchman 2008). Even among microbial groups with extensive morphology to support traditional methods of classification, molecular techniques have indicated far greater species diversity than previously thought (Savin *et al.* 2004). Because microbes play fundamental roles in primary fixation of carbon, recycling of elements, and gene transfer (some leading to disease), their dynamics are unquestionably relevant to the ecology of multicellular organisms and the Gulf of Maine ecosystem. However, ecosystem processes are usually considered at a high level of biological organization, often in the context of the major players, with a much reduced emphasis on diversity at the lower levels where speciation processes are more important. Thus a question arises about microbial diversity: to what extent is it linked to patterns and processes evident at the ecosystem level? At this early stage of discovery, microbial lessons may not be easily transferred to an understanding of diversity in multicellular organisms and the multifarious trophic dependencies. Nonetheless, the Gulf of Maine Area has a rich history of research on marine microbes, and it is impossible to predict what insights to local processes might come from continuing work. Given modern techniques for high throughput analysis (Stepanauskas & Sieracki 2007), one foreseeable application is that microbial diversity might prove to be a sensitive, integrative signal of environmental change, owing to the great dispersive capacity of microbial populations.

Biological activity is patchily distributed in space and time and is often driven by hydrodynamic effects on the distribution of small particles. These effects are often tied to interactions between water movement and bottom depth, thus, banks, ridges, and other areas of steep topographic

change are frequently biological "hot spots" that attract attention and study because of the concentration of biomass and interactions, and the putative importance of these centers of activity to biological production over a larger area (Yen *et al.* 2004; Cotté & Simard 2005; Stevick *et al.* 2008). Small and isolated features offer other advantages for study because the signals of interest can be distinguished from the surrounding background, one can examine the whole system and not just part of it, and the feature may be less frequently disturbed by humans than areas closer to the coast (although distance does not offer the protection it once did). In the Gulf of Maine Area, many such features have been implicated as hot spots, often because they are frequented by upper trophic level predators such as seabirds (Huettmann & Diamond 2006) and cetaceans (Kenney & Winn 1986). One challenge when examining structure and processes and calculating the ecological functioning of these features is to distinguish unique aspects from those that can be generalized. Studies of Cashes Ledge in the central Gulf of Maine illustrate this point: the top of the ledge system is shallow enough that it protrudes into the internal wave field, causing large fluctuations in energy and temperature, and mixing nutrients into the surface layer (Witman *et al.* 1993). Although the degree of mixing is unusual, the biological community includes mature kelp beds and large predatory fish reminiscent of what coastal hard substrate communities probably looked like many decades ago (Steneck *et al.* 2004). The vertical zonation, although made somewhat unique by the shallow top and unusually steep sides, provides a mesocosm for studying communities from a range of depths within a relatively small area. Studies at another site, a small offshore bank named Platts Bank in the southwestern Gulf, show that the depth of the bank interacts with the internal wave field to cause surface patches of krill and sometimes other plankton, attracting whales and birds to a small crest region where they feed intensively while the krill and other small prey are abundant (Stevick *et al.* 2008; Fig. 3.2C). A multi-year investigation of the bank shows that it is often inactive, however, the difference perhaps being in the vertical movements and abundance of krill in this portion of the Gulf of Maine. It appears that the bank is nonetheless frequently visited by whales, especially humpback whales (*Megaptera novaeangliae*) which may be moving among a network of potential feeding sites and switching between krill and fish as their primary prey. We do not know what this network looks like, which makes it difficult to speculate on the ecosystem dynamics supporting the summer population of this species. Significantly, little else on Platts Bank has been studied, although there is commercial and recreational fishing on it.

The Stellwagen Bank National Marine Sanctuary in the southwest corner of the Gulf of Maine is a little over 2,100 km^2 (6% the size of Georges Bank) and presents a very heterogeneous environment with well-mapped mud, sand, gravel, and boulder habitats in a bank and basin topography from 19 to more than 60 m depth. The sanctuary is heavily used by sea life and by people, and has been an important fishing area for more than 400 years (Claesson & Rosenberg 2009). It does not have a high level of protection by sanctuary standards, but its status as a sanctuary has attracted considerable research on fishes (Auster *et al.* 2001, 2006; Auster 2002), benthic communities (Blake *et al.* 1993; Cahoon *et al.* 1993), plankton and hydrography (Clark *et al.* 2006), seabirds (Pittmann & Huettmann 2006), and marine mammals (Pittmann *et al.* 2006). A sliver of the bank and a large section of the seafloor northeast of the sanctuary have been protected from bottom fishing gear for several years, and the combined areas offer opportunities for study of altered and, in some areas and to some extent, recovering communities. To do this requires expanded, non-intrusive sampling techniques that operate with high resolution and high location accuracy over significant sampling tracks. Although several vehicles have been under development, a towed habitat camera system (HabCam) developed at the Woods Hole Oceanographic Institution resolves over significant distances the patterns of organism–organism relationships (abundance, species, and distance), organism–substrate relationships, and other oceanographic parameters, and affords population assessments for resolved species (Fig. 3.6A; York 2009). A significant development is that much of the data processing is automated, including a growing proportion of the acquired images. These types of systems will make quantitative assessments of the bottom and epibenthic communities possible, a large step forward in sampling, understanding, and monitoring the seafloor. Analysis of demersal fish assemblages on Stellwagen Bank as sampled by the trawl surveys has shown surprisingly good concordance between predicted ecological associations for certain fish with the bottom type, and the mapped benthic substrates. The surprise is that the patterns would be resolved by trawl assessments within the mixture of habitats on Stellwagen Bank, and the promise is that we can start to draw the two types of datasets together.

Detailed sampling, which HabCam and similar developments make possible, must be nested within a larger geography of habitat space determined by oceanography and other broad-scale biological datasets (see, for example, Watling *et al.* 1978; Watling & Skinder 2007). One question is the degree to which factors such as location in the Gulf (for example, distance from the Northeast Channel and Scotian Shelf inflows), water mass and hydrographic characteristics, chlorophyll production, temperature properties, depth, substrate, and substrate spatial heterogeneity affect benthic community types and processes. As one of the synthesis projects of the Census, Canadian and US scientists are working with Australian researchers to apply a Random Forests statistical analysis (a bootstrapped randomized tree statistical method: Breiman (2001); see Peters *et al.* (2007) and Knudby *et al.* (2010) for recent application examples) to shelf-scale biological and physical datasets from the temperate Gulf of Maine and two tropical/subtropical

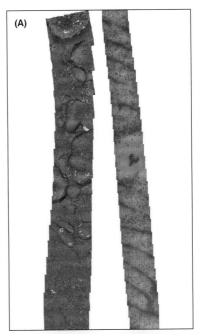

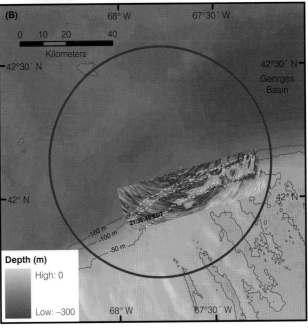

Fig. 3.6

Automated imaging of biological patterns, from sub-meter to tens of kilometers. **(A)** Mosaic imagery of benthic habitat from two towed transects taken by HabCam (towed habitat camera system) on Stellwagen Bank. Serrations mark the corners of individual 1 m² images that have been automatically adjusted for light, color, and elevation and then stitched together (photographs courtesy of S. Gallager, Woods Hole Oceanographic Institution, Woods Hole, Massachusetts, USA, and The HabCam Group®). **(B)** Synoptic view of schooling herring (*Clupea harengus*) on the northern edge of Georges Bank, sampled during a 75 s scan (full sweep of red circle) by OAWRS (Ocean Acoustic Waveguide Remote Sensing), 3.4 h after sunset on September 29, 2006. School densities are –45 (blue) to –33 dB (red) (data from N.C. Makris, Massachusetts Institute of Technology, Cambridge, Massachusetts, USA). This is one of a sequence of images showing the formation and movement of herring toward the bank (Makris *et al.* 2009).

systems, the Gulf of Mexico and the Great Barrier Reef. The statistical analysis involves a modification to Random Forests that collates numerous split values and change in deviance information for each physical variable and species (Pitcher *et al.* unpublished observations). The results are presented as cumulative distributions of splits, weighted by deviance, averaged over multiple species within selected levels of aggregation. Species come from the spatial datasets on fishes and benthic invertebrates that we presented in the section on the "known biodiversity". The results represent patterns of biological change along gradients for each physical variable. The outputs also summarize the overall prediction performance of physical surrogates and identify the physical variables that contribute most to the prediction. The statistical techniques being developed in this work will contribute to understanding the importance of physical drivers in the marine environment and should allow a first-order prediction of macrofaunal benthic and demersal fish biodiversity and community patterns based on seabed and environmental characterizations. This would provide an intermediate level of spatial resolution that could facilitate design of the next higher-resolution stage of sampling and evaluation, and eventual monitoring.

Although much of the above information has focused on resolving benthic and demersal community structure at sub-meter to 100-m scales, another technological development has provided unprecedented views of pelagic fish (Makris *et al.* 2006, 2009). In this case, low-frequency, long-range acoustic sampling was used to provide a synoptic assessment of the behavior, density distribution, and scale of herring schools as they emerge from depth off the northern flank of Georges Bank (Fig. 3.6B), and move onto the bank for spawning. The large-scale view (tens of kilometers) afforded by this technique was coupled to traditional transect acoustic sampling and biological (net) sampling to supply biological detail, but the large, synoptic dataset described, as no other method could, the magnitude, coordinated nature, and timing of the event. Further development and testing could make these approaches applicable to geophysically more complicated environments, opening up opportunities for researchers to measure and understand behavior of some sound-scattering pelagic organisms in other settings.

A recent Canadian research initiative aims to contribute to our understanding of intermediate scales of biodiversity structure within the northern Gulf of Maine by creating a focus area for long-term research. In 2004, the Canadian Department of Fisheries and Oceans, with several Canadian academic institutions, launched the Discovery Corridor Initiative. This "corridor" in the sea begins in the intertidal zone at the US–Canadian border (Passamaquoddy Bay) and extends across the banks and basins of the northern Gulf of Maine to the base of the continental slope (Fig. 3.1A). There have been three offshore research missions so far that have used surface-deployed video,

digital photography, and benthic grab-sampling tools, as well as a deep-submergence vehicle to sample benthic habitats in water depths from 60 to 2,500 m (Figs. 3.2A and B). The corridor concept has recently been embraced by a new national marine biodiversity research program (the Canadian Healthy Oceans Network) that will establish similar corridors in the Arctic and Pacific oceans, based in part on the model developed in the Gulf of Maine. Passamaquoddy Bay and the adjacent Cobscook Bay are also the site of joint US–Canadian studies for Natural Geography in Shore Areas (NaGISA) and History of the Near Shore (HNS). The corridor concept is strategically useful – it is large enough to enable a complementary range of spatially resolved sampling and experimental designs to be undertaken, taking advantage of both commercial fisheries management questions and conservation planning needs (for example, the North East Channel Coral Conservation Area), to inform EBM approaches. A corridor from the shore to the deep sea also captures public imagination and becomes a vehicle for education.

In this brief review of some of the ongoing research into the structure of biological communities in the Gulf of Maine Area, we should point out that some important ecological functions in our current ecosystem state are performed by species whose distributions we know quite well. Examples include the copepod *Calanus finmarchicus* as a food source for some planktivorous fish and whales; the important role that certain planktivorous fish, such as herring (*Clupea harengus*) and sand lance (*Ammodytes americanus*), play in the food web of large predators such as tunas and whales; the importance of mud-flat amphipods (*Corophium volutator*) to the diet of benthic fish (McCurdy *et al.* 2005) and migrating seabirds (Hamilton *et al.* 2006); and the effect of kelp in structuring nearshore benthic communities (Steneck *at al.* 2004). Although their distributions and ecological roles are generally well known, there are few ecosystem-level assessments of their impacts. There are other "well-known" and important species that we know have large trophic roles, but we understand relatively little about them in terms of population patterns. These include organisms such as krill (especially the abundant *Meganyctiphanes norvegica*), gelatinous zooplankton, and squids. These are difficult to sample, but as important consumers and prey, their distribution, abundance, behavior, and dynamics are also an important part of understanding regional biodiversity. Finally, interesting and important perspectives can be gained by examining historical data and considering shifting baselines in the Gulf (Lotze & Milewski 2004; Steneck *et al.* 2004; Rosenberg *et al.* 2005; Bourque *et al.* 2008; Alexander *et al.* 2009; Claesson & Rosenberg 2009). These provide insights into previous states of the system, the magnitude and nature of previous impacts, and factors determining the current state of the system. These can be a basis for discussing trade-offs and setting goals for the future.

3.3.4 A framework for representing biodiversity in EBM applications

In the preceding overview of sources of scientific information that contribute to our present-day knowledge of regional biodiversity, we presented examples that span a range of spatial and temporal scales, and also a wide array of species and environments. This reflects the bewildering complex of details and uncertainties that need to be incorporated into an understanding of biodiversity patterns and processes at a large (GoMA) scale, and incorporated into decision making on human usage of the oceans. To ensure development of a realistic and useful concept of how biodiversity knowledge can be used in public policy and management, GoMA has worked with other groups and projects that emphasized stakeholder involvement, planning and implementation. These included US and Canadian fisheries agencies, which are working on implementing ecosystem approaches to fisheries management (Ecosystem Assessment Program 2009; Gavaris 2009) or integrated management approaches (O'Boyle & Jamieson 2006; O'Boyle & Worcester 2009), as well as academic, industry, non-federal management, and conservation groups.

In the search for ways to capture the complexity of biodiversity organization within an ecoregion, and in relation to the development of indicators and monitoring programs to assess status and trends of regional biodiversity, various hierarchical frameworks have been proposed. One approach stems from an adaptation of earlier conceptual frameworks on ecosystem structure and function put forward by Noss (1990) and articulated around then-available techniques for monitoring terrestrial biodiversity. Recently, Cogan and Noji (2007) refined Noss's schema for application toward marine biodiversity research and monitoring. Cogan *et al.* (2009) also found use in this approach in helping to codify how marine habitat mapping could be logically connected to EBM principles and implementation.

As Noss (1990) did for terrestrial applications, these later marine frameworks (Cogan and Noji 2007; Cogan *et al.* 2009) deconstructed biodiversity organization into three principal elements, each of which is further represented through a hierarchical (nominally spatially referenced) structure ranging from the ecoregion to genetic levels.

Compositional diversity elements represent the identity and variety of biodiversity at different levels within the system from biogeographic provinces and ecoregions to genes. Commonly used biodiversity metrics for species composition are inventory diversities (alpha, gamma, epsilon; Whittaker 1977; Stoms & Estes 1993).

Structural diversity elements are concerned with the physical organization or pattern within the system, from ecoregion- to habitat-scales, including both biotic and abiotic variables that modulate patterns. Common biodiversity metrics that help define structure are differentiation diversities (beta, delta) that characterize the amount of change in species composition between different structural features in the environment (Whittaker 1977; Stoms & Estes 1993). At the ecoregion level, structural diversity may be thought of in relation to the arrangement of physiographic regions and landscapes that contribute to the internal makeup of the ecoregion. At finer scales, we must grapple with ways to understand and represent the dynamics of graininess of biotic and abiotic processes.

In these monitoring frameworks, *functional* diversity *elements* are those abiotic and biotic factors (or processes) that are influential in either maintaining characteristic biodiversity features within the system, or which contribute to changes. These range from genetic processes to regional natural and anthropogenic forcing variables that operate at various spatial, ecological and evolutionary scales. It is important to distinguish here that the term functional diversity *elements* relates to *processes* and is thus different from functional diversity as a *property* within a system, such as the depth of membership or number of feeding guilds, or an aggregate set of functions and services.

We have found this hierarchical view, organized around compositional, structural, and functional diversity elements (which we term a CSF template), to be useful in deriving overviews of the status of biodiversity knowledge, but alone it does not fully encompass features of biodiversity organization that need to be taken into account in developing a comprehensive view of the regional biodiversity science/management framework. As shown in Section 3.3.2 (Fig. 3.5), only a very minor component of the regional biodiversity is routinely monitored, and much of it is completely unknown. Indeed, organism size and the number of known and potentially unknown species (or OTUs) fundamentally influence how the science community approaches research on different ecosystem "compartments". Much of the existing work on the microbial system, for example, emphasizes functional diversity as a way of characterizing composition, especially at the prokaryote level.

Cogan *et al.* (2009) considered the CSF template to be a good fit into marine habitat mapping applications articulated within adaptive EBM approaches. We agree with this assertion, but advocate that an additional overarching conceptual model that integrates the complexity and spatial extent of different structural features in the system may help clarify the connections and transfer of knowledge between research and management applications. The system can be considered as a nested set of spatial domains that range from fine-scale micro-habitat features influencing occupancy, to seascapes, and at the broad-scale the ecoregion itself. Oceanographically modulated processes, such as bentho-pelagic coupling, and transport and mixing, promote connectivity across these spatial domains and represent an important consideration for spatial management under EBM. Layered on top of this spatial matrix of habitats, organisms, and linkages is temporal variability, including year-to-year variations (natural and human) and secular change (climate and community trajectories). Conceptually, it can be argued that although discovery-based research is interested in investigation across the full range of these spatial domains, monitoring and spatial planning largely remain focused on periodic assessments and decision making above a certain minimum domain size and at lower levels of complexity.

Marine conservation has traditionally focused on individual species or populations. More recently, there has been a shift of emphasis toward managing specific marine habitat spaces, species assemblages, and hot spots of biodiversity (Buzeta & Singh 2008). Marine protected areas, long-term ecological research sites, and other types of natural heritage sites have been established as a means for conserving biodiversity, and they can serve as important experimental and control areas for long-term studies (Lubchenco *et al.* 2003; Satie *et al.* 2003; Cook & Auster 2006, 2007; Palumbi *et al.* 2009) and as a means to educate the public. Within the Gulf of Maine there are more than 200 coastal and marine protected areas, comprising parks, sanctuaries, research reserves, critical habitat areas, and restricted fishing areas (Baumann *et al.* 1998; Recchia *et al.* 2001). These have been created for numerous purposes, by many types of organizations. Most allow considerable usage. Significantly, from the perspective of EBM, most of these were not developed with consideration as to how they provide for conservation of biodiversity and function at the higher level of the ecosystem.

3.4 Perspectives on Generating and Using Biodiversity Information

Large ecosystems consist of mixtures of habitats and communities that exist at various sizes and patterns of distribution. Organisms within these communities exhibit great differences in size, abundance, mobility, life expectancy, and recruitment patterns, and they participate in a wide variety of ecological interactions. A single species may act differently in different environments or over time. These variations tend to focus scientific investigations in specific ways, perhaps emphasizing structure, or composition, or function of a particular community or habitat. Often, and for pragmatic reasons, a study must focus on a small subset of species, conditions, and time. In science, many questions are of potential interest, but there will always be compartments of knowledge that are hard to connect to one another, or to connect directly to management needs. A challenge to establishing a regional scale of understanding and management is to integrate across the types of knowledge that tend to be generated for different groups of organisms and spatial and temporal scales of processes. It is difficult to do this well after the fact, suggesting that a conceptual framework is needed to help identify needs and opportunities. But beware: a framework should stimulate rather than dictate.

The challenge of understanding so much information at multiple scales in a spatially and temporally heterogeneous environment is daunting. The three principal components of the framework that we have discussed (CSF template) provide a means by which to gauge and communicate the relative completeness of our inventory and understanding of biodiversity organization within a regional ecosystem. Our references to knowledge along the size spectrum of organisms and to spatial hierarchies and function are meant to illustrate gaps in knowledge as well as ways that we might connect biodiversity information and processes to EBM at the scale of a regional ecosystem. A framework, discussed and improved over time, can help draw the long connections between scientific investigations, which typically focus on details, and EBM, which must operate at longer, larger, average, and less certain scales. In assembling the first list of species already known to exist in the Gulf of Maine, it became clear that even that basic information was not very accessible, and so much more is needed to advance research and support application. To make progress in this coupled research–application framework, biodiversity informatics must be developed as a component of ocean observing and analysis (O'Dor *et al.* 2010; Ricard *et al.* 2010).

3.5 Future Directions

Exploration of the oceans is essential to our understanding and conservation of biodiversity, but such an undertaking will take many years and is expensive. How do we conduct investigations and monitoring so that scientific and societal objectives can both be met, and the efforts and benefits sustained? The ecological questions are multi-scaled, multi-layered, and complex. The past focus on dominant organisms must somehow accommodate the larger and growing list of rarer species; the individual and collective role of rarer species must be incorporated into the immediate and longer-term perspective on ecosystem function and adaptation; community-wide patterns and dynamics need to be understood; and the relation between biodiversity, ecosystem functioning, and societal benefits must be elucidated.

Strategies might include a program to evaluate rigorously multi-scale relationships between community types, organism abundances, habitat types, and broader patterns of distribution. For example, how do patterns vary within and between basins around the Scotian Shelf and Gulf of Maine? In what ways are they the same or different, and how interdependent are the basins in terms of population dynamics? What methods would be needed to answer these questions? What about the same questions applied to banks, ridges, and outcroppings? What are the relationships between shallower and deeper parts of the region; among the coastal sections from Nova Scotia to Massachusetts; and between the coast and the interior of the Gulf? How do community types and functional groups relate to ecosystem function in these various environments, and how are these related to services that society depends on? Nested within this are some of the more small-scale questions about the specific biodiversity patterns and processes within communities, including their dynamics, responses to and recovery from disturbance, and what is needed to conserve them. Monitoring for function and identifying indicators must become part of integrated ocean observing, assessment, and education. We need the ability to detect change, distinguish between natural and anthropogenic forcing, and respond in an informed way, which includes precautionary steps, previous experience (from here and elsewhere), and assessed risk.

To gain a better understanding of how ecosystem function and adaptability may be linked to biodiversity, we need the means to conduct experimental investigations at a range of scales. Consolidating some regional research capacities within "defined ocean spaces" where ecological structure and function can be assessed across different temporal and spatial scales, along with evaluation of comprehensive data integration and modeling techniques, could represent a key step toward testing and implementing EBM approaches across the region.

Understanding the inextricable links between human interactions and the natural system is the basis of ecosystem management. Because EBM regulates human activities, public literacy at local, regional, national, and international levels is fundamental to its implementation (Novacek 2008). Biologist Rachel Carson's popular books of the mid-twentieth century, *The Sea Around Us* (Carson 1951) and *Silent Spring* (Carson 1962), helped create a societal shift toward support of the environmental policies of the 1960s and 1970s. Scientists must convey to the public and policy-makers the connection between biodiversity and the sustainability of goods and services provided by ecosystems. To build an ecosystem-literate public, one must first acknowledge that there is truly no "general public", but collections of individuals with varying backgrounds, interests, and values. Now, more than ever in human history, societies supported by marine ecosystems – in the Gulf of Maine region and around the world – are made up of direct and indirect stakeholders with different socio-cultural values, economic concerns, and perceived connections to the natural world. Recognizing this human diversity is essential to building public support for research and acceptance of indicated management actions.

Acknowledgments

The GoMA project was managed through the University of Southern Maine (USM) and the Centre for Marine Biodiversity (CMB). We acknowledge the input of colleagues and friends over the years as we endeavored to define what a Gulf of Maine Area Census should and could accomplish, and then as we sought to follow through on those ideas. Thankfully, there were many. Peter Auster, Michael Fogarty, Ron O'Dor, and Robert Stephenson deserve special thanks for their formative contributions throughout the project, and Fred Grassle and Michael Sinclair for providing focus at critical times. We thank the many participants in GoMA's six Expert Groups for discussion and insights; we especially thank the leaders: Catherine Johnson, Noreen Kelly, Scott Kraus, Peter Larsen, William Li, Jeffrey Runge, Michael Sieracki, and Kent Smedbol. The following colleagues generously shared data, ideas, and good times on and under the water: Scott Gallager, Stephen Hale, Ellen Kenchington, Nick Makris, Anna Metaxas, Tom Noji, Joan Palmer, Gerhard Pohle, Purnima Ratilal, Paul Snelgrove, Thomas Trott, Lou Van Guelpen, Michael Vecchione, and Les Watling. Kenneth Foote and Evan Richert deserve credit for organizing and leading GoMA during its initial and formative years. We thank Susan Ryan for her ongoing efforts at outreach and public education on behalf of the Census, and for keeping us mindful of their importance to achieving conservation goals. We also recognize the contributions of GoMA team members Jennifer Ecker and Adrienne Adamek (USM), and Michelle Greenlaw, Victoria Clayton, Ashley Holmes, and Chelsie Archibald (CMB). We are grateful for financial support from the Alfred P. Sloan Foundation, Fisheries and Oceans Canada, and the University of Southern Maine. Finally, we express our thanks to Jesse Ausubel for the vision, dedication, patience, energy, critical eye, and boundless enthusiasm it took to guide all of the Census projects, including this one, from conception through to the present benchmark – but a stepping stone in the process of describing Earth's marine biodiversity and its importance to humanity.

References

Abebe, E., Grizzle, R.E., Hope, D., *et al.* (2004) Nematode diversity in the Gulf of Maine, USA, and a Web-accessible, relational database. *Journal of the Marine Biological Association of the United Kingdom* **84**, 1159–1167.

Alexander, K.E., Leavenworth, W.B., Cournane, J., *et al.* (2009) Gulf of Maine cod in 1861: historical analysis of fishery logbooks, with ecosystem implications. *Fish and Fisheries* **10**, 428–449.

Ames, E.P. (2004) Atlantic cod stock structure in the Gulf of Maine. *Fisheries Research* **29**, 10–28.

Archer, A.W. & Hubbard, M.S. (2003) Highest tides in the world. *Geological Society of America Special Paper* **370**, 151–173.

Auster, P.J. & Lindholm, J. (2005) The ecology of fishes on deep boulder reefs in the western Gulf of Maine. In: *Diving for Science 2005, Proceedings of the American Academy of Underwater Science*, pp. 89–107. Groton, Connecticut: Connecticut Sea Grant.

Auster, P.J. & Shackell, N.L. (2000) Marine protected areas for the temperate and boreal Northwest Atlantic: the potential for sustainable fisheries and conservation of biodiversity. *Northeastern Naturalist* 7, 419–434.

Auster, P.J. (2002) Representation of biological diversity of the Gulf of Maine region at Stellwagen Bank National Marine Sanctuary (Northwest Atlantic): patterns of fish diversity and assemblage composition. In: *Managing Protected Areas in a Changing World* (eds. S. Bondrup-Nielson, T. Herman, N.W.P. Munro, *et al.*), pp. 1096–1125. Wolfville, Nova Scotia: Science and Management of Protected Areas Association.

Auster, P.J., Joy, K. & Valentine, P.C. (2001) Fish species and community distributions as proxies for seafloor habitat distributions: the Stellwagen Bank National Marine Sanctuary example (Northwest Atlantic, Gulf of Maine). *Environmental Biology of Fishes* 60, 331–346.

Auster, P.J., Malatesta, R.J., Langton, R.W., *et al.* (1996) The impacts of mobile fishing gear on the sea floor habitats in the Gulf of Maine (Northwest Atlantic): implications for conservation of fish populations. *Reviews in Fisheries Science* 4, 185–200.

Auster, P.J., Clark, R. & Reed, R.E.S. (2006) Chapter 3. Marine fishes. In: *An Ecological Characterization of the Stellwagen Bank National Marine Sanctuary Region: Oceanographic, Biogeographic, and Contaminants Assessment* (eds. T. Battista, R. Clark & S. Pittmann), pp. 89–229. Silver Spring, Maryland: National Center for Coastal Ocean Science, NOAA. NOAA Technical Memorandum NOS NCCOS 45.

Baguley, J.G., Montagna, P.A., Lee, W., *et al.* (2006) Spatial and bathymetric trends in Harpacticoida (Copepoda) community structure in the Northern Gulf of Mexico deep-sea. *Journal of Experimental Marine Biology and Ecology* 330, 327–341.

Baumann, C., Brody, S., Fenton, D., *et al.* (1998) *A GIS database of existing coastal and marine protected areas, conservation zones, and restricted fishing areas in the Gulf of Maine*. Gulf of Maine Council on the Marine Environment.

Beardsley, R.C., Butman, B., Geyer, W.R., *et al.* (1997) Physical oceanography of the Gulf of Maine: An update. In: *Proceedings of Gulf of Maine Ecosystem Dynamics: a Scientific Symposium and Workshop, 16–20 September 1996, St. Andrew's, N.B.* (eds. G. Braasch & G. Wallace), pp. 39–52. RARGOM Report 97-1.

Bellwood, D.R., Hughes, T.P. & Hoey, A.S. (2006) Sleeping functional group drives coral–reef recovery. *Current Biology* 16, 2434–2439.

Blake, J.A., Hilbig, B. & Rhoads, D.C. (1993) *Soft–bottom Benthic Biology and Sedimentology, 1992 Baseline Conditions in Massachusetts and Cape Cod Bays*. Report for Massachusetts Bay Outfall Monitoring Program. Prepared for the Massachusetts Water Resources Authority.

Bouchet, P. (2006) The magnitude of marine biodiversity. In: *The Exploration of Marine Biodiversity: Scientific and Technological Challenges*. (ed. C.M. Duarte), pp. 31–64. Bilbao, Spain: Fundación BBVA.

Bourque, B.J. (2001) *Twelve Thousand Years: American Indians in Maine*. Lincoln, Nebraska: University of Nebraska Press.

Bourque, B.J., Johnson, B.J. & Steneck, R.S. (2008) Possible prehistoric fishing effects on coastal marine food webs in the Gulf of Maine. In: *Human Impacts on Ancient Marine Ecosystems*. (eds. T.C. Rick & J.M. Erlandson), pp. 165–185. Berkeley, California: University of California Press.

Breiman, L. (2001) Random forests. *Machine Learning* 45, 5–32.

Brunel, P., Bossé, L. & Lamarche, G. (1998) Catalogue of the marine invertebrates of the estuary and Gulf of St. Lawrence. *Canadian Special Publication of Fisheries and Aquatic Sciences* 126.

Buzeta M.-I. & Singh, R. (2008) Identification of ecologically and biologically significant areas in the Bay of Fundy, Gulf of Maine. Volume 1: Areas identified for review and assessment of the Quoddy region. *Canadian Technical Report of Fisheries and Aquatic Sciences* 2788.

Cahoon, L.B., Beretich, G.R., Thomas, C.J., *et al.* (1993) Benthic microalgal production at Stellwagen Bank, Massachusetts Bay, USA. *Marine Ecology Progress Series* 102, 179–185.

Cairns, S.D. (2007) Studies on western Atlantic Octocorallia (Gorgonacea: Primnoidae). Part 8: New records of Primnoidae from the New England and Corner Rise Seamounts. *Proceedings of the Biological Society of Washington* 120, 243–263.

Carson, R. (1951) *The Sea Around Us*. New York: Oxford University Press.

Carson, R. (1962) *Silent Spring*. Boston, Massachusetts: Houghton Mifflin.

Chen, G., Herman, R. & Vincx, M. (1999) Meiofauna communities from the Straits of Magellan and the Beagle Channel. *Scientia Marina* 63, 123–132.

Claesson, S.H. & Rosenberg, A.A. (2009) *Stellwagen Bank Marine Historical Ecology: Final Report*. Gulf of Maine Cod Project, University of New Hampshire.

Clark, R., Manning, J., Costa, B., *et al.* (2006) Chapter 1. Physical and oceanographic setting. In: *An Ecological Characterization of the Stellwagen Bank National Marine Sanctuary Region: Oceanographic, Biogeographic, and Contaminants Assessment* (eds. T. Battista, R. Clark & S. Pittmann), pp. 1–58. Silver Spring, Maryland: National Center for Coastal Ocean Science, NOAA. NOAA Technical Memorandum NOS NCCOS 45.

Clarke, K.R. & Warwick, R.M. (2001) *Change in Marine Communities: An Approach to Statistical Analysis and Interpretation*. 2nd ed. Plymouth, UK: PRIMER-E.

Cogan, C.B. & Noji, T.T. (2007) Marine classification, mapping, and biodiversity analysis. In: *Mapping the Seafloor for Habitat Characterization* (eds. B. Todd and H.G. Greene), pp. 129–139. St. John's, Newfoundland: Geological Association of Canada.

Cogan, C.B., Todd, B.J., Lawton, P., *et al.* (2009) The role of marine habitat mapping in ecosystem-based management. *ICES Journal of Marine Science* 66, 2033–2042.

Cogswell, A.T., Kenchington, E.L.R., Lirette, C.G., *et al.* (2009) The current state of knowledge concerning the distribution of coral in the Maritime Provinces. *Canadian Technical Report of Fisheries and Aquatic Sciences*, 2855, v +66 pp.

Cohan, F.M. (2002) What are bacterial species? *Annual Review of Microbiology* 56, 457–487.

Cohen, E.B. & Langton, R.W. (1992) The ecological consequences of fishing in the Gulf of Maine. In: *The Gulf of Maine. NOAA Coastal Ocean Program, Regional Synthesis Series, No. 1* (eds. D.W. Townsend & P.F. Larsen), pp. 45–69. Washington, DC: NOAA Coastal Ocean Program.

Collie, J.S., Escanero, G.A. & Valentine, P.C. (1997) Effects of bottom fishing on benthic megafauna of Georges Bank. *Marine Ecology Progress Series* 155, 159–172.

Collie, J.S., Hall, S.J., Kaiser, M.J., *et al.* (2000) A quantitative analysis of fishing impacts on shelf-sea benthos. *Journal of Animal Ecology* 69, 785–798.

Cook, R.R. & Auster, P.J. (2006) *Developing alternatives for optimal representation of seafloor habitats and associated communities in Stellwagen Bank National Marine Sanctuary*. Silver Spring, Maryland: NOAA Office of Marine Sanctuaries.

Cook, R.R. & Auster, P.J. (2007) *A bioregional classification of the continental shelf of northeastern North America for conservation analysis and planning based on representation*. Marine Sanctuaries Conservation Series NMSP-07-03. Silver Spring, Maryland: US Department of Commerce, National Oceanic and Atmospheric Administration, National Marine Sanctuary Program.

Cotté, C. & Simard, Y. (2005) Formation of dense krill patches under tidal forcing at whale feeding hot spots in the St. Lawrence Estuary. *Marine Ecology Progress Series* 288, 199–210.

Curtis, T.P., Sloan, W.T. & Scannell, J.W. (2002) Estimating prokaryotic diversity and its limits. *Proceedings of the National Academy of Sciences USA* 99, 10494–10499.

De Bovée, F., Hall, P.O.J., Hulth, S., *et al.* (1996) Quantitative distribution of metazoan meiofauna in continental margin sediments of the Skagerrak (Northeastern North Sea). *Journal of Sea Research* 35, 189–197.

Dohler, G.C. (1970) *Tides in Canadian waters.* Ottawa, Ontario: Canadian Hydrographic Service, Department of Energy, Mines and Resources.

Ecosystem Assessment Program (2009) *Ecosystem status report for the northeast United States Continental Shelf Large Marine Ecosystem.* Woods Hole, Massachusetts: National Marine Fisheries Service Northeast Fisheries Science Center Reference Document 09–11.

Engle, V.D. & Summers, J.K. (1999) Latitudinal gradients in benthic community composition in Western Atlantic estuaries. *Journal of Biogeography* 26, 1007–1023.

Falkowski, P.G., Fenchel, T. & DeLong, E.F. (2008) The microbial engines that drive Earth's biogeochemical cycles. *Science* 320, 1034–1039.

Fogarty, M.J., Incze, L.S., Hayhoe, K., *et al.* (2008) Potential climate change impacts on Atlantic cod (*Gadus morhua*) off the northeastern USA. *Mitigation and Adaptation Strategies for Global Change* 13, 453–466.

Foote, K.G. (2003) Pilot census of marine life in the Gulf of Maine: contributions of technology. *Oceanologica Acta* 25, 213–218.

Frank, K.T., Petrie, B., Choi, J.S., *et al.* (2005) Trophic cascades in a formerly cod–dominated eosystem. *Science* 308, 1621–1622.

Garrett, C.J.R., Keeley, J.R. & Greenberg, D.A. (1978) Tidal mixing versus thermal stratification in the Bay of Fundy and Gulf of Maine. *Atmosphere-Ocean* 16, 403–423.

Gavaris, S. (2009) Fisheries management planning and support for strategic and tactical decisions in an ecosystem approach context. *Fisheries Research* 100, 6–14.

Gobin, J.F. & Warwick, R.M. (2006) Geographical variation in species diversity: a comparison of marine polychaetes and nematodes. *Journal of Experimental Marine Biology and Ecology* 330, 234–244.

Gordon, D.C., Jr., Cranford, P.J. & Desplanque, C. (1985) Observations on the ecological importance of salt marshes in the Cumberland Basin, a macrotidal estuary in the Bay of Fundy. *Estuarine, Coastal and Shelf Science* 20, 205–227.

Greene, C.H. & Pershing, A.J. (2003) The flip side of the North Atlantic Oscillation and modal shifts in slope–water circulation patterns. *Limnology and Oceanography* 48, 319–322.

Greene, C.H. & Pershing, A.J. (2007) Climate drives sea change. *Science* 315, 1084–1085.

Grumbine, R.E. (1994) What is ecosystem management? *Conservation Biology* 8, 27–38

Häkkinen, S. (2002) Freshening of the Labrador sea surface waters in the 1990s: another great salinity anomaly? *Geophysical Research Letters* 29, 2232.

Hamilton, D., Diamond, A. & Wells, P.G. (2006) Shorebirds, snails, and the amphipod (*Corophium volutator*) in the upper Bay of Fundy: top-down vs. bottom-up factors, and the influence of compensatory interactions on mudflat ecology. *Hydrobiologia* 567, 285–306.

Hartel, K.E., Kenaley, C.P., Galbraith, J.K., *et al.* (2008) Additional records of deep-sea fishes from off greater New England. *Northeastern Naturalist* 15, 317–334.

Hayhoe, K., Wake, C.P., Huntington, T.G., *et al.* (2007) Past and future changes in climate and hydrological indicators in the U.S. Northeast. *Climate Dynamics* 28, 381–407.

Hicks, G.R.F. (1985) Meiofauna associated with rocky shore algae. In: *The Ecology of Rocky Coasts* (eds. P.G. Moore & R. Seed), pp. 36–56. London: Hodder & Stoughton.

Huettmann, F. & Diamond, A. (2006) Large-scale effects on the spatial distribution of seabirds in the Northwest Atlantic. *Landscape Ecology* 21, 1089–1108.

Jackson, J.B.C., Kirby, M.X., Berger, W.H., *et al.* (2001) Historical overfishing and the recent collapse of coastal ecosystems. *Science* 293, 629–637.

Jacobson, H.A., Jacobson, G.L., Jr. & Kelley, J.T. (1987) Distribution and abundance of tidal marshes along the coast of Maine. *Estuaries* 10, 126–131.

Jennings, R.M., Shank, T.M., Mullineaux, L.S., *et al.* (2009) Assessment of the Cape Cod phylogeographic break using the bamboo worm *Clymanella torquata* reveals the role of regional water masses in dispersal. *Journal of Heredity* 100, 86–96.

Kenney, R.K. & Winn, H.E. (1986) Cetacean high–use habitats of the northeast United States continental shelf. *Fishery Bulletin* 84, 345–347.

Kirchman D.L. (2008) *Microbial Ecology of the Oceans.* 2nd edn. Wiley-Blackwell, New Jersey.

Knoll, A.H. (2003) *Life on a Young Planet: The First Three Billion Years of Evolution on Earth.* Princeton, New Jersey: Princeton University Press.

Knott, S.T. & Hoskins, H. (1968) Evidence of Pleistocene events in the structure of the continental shelf off the northeastern United States. *Marine Geology* 6, 5–43.

Knudby, A., Brenning, A. & LeDrew, E. (2010) New approaches to modeling fish–habitat relationships, *Ecological Modeling* 221, 503–511.

Kucera, H. & Saunders, G.W. (2008) Assigning morphological variants of *Fucus* (Fucales, Phaeophyceae) in Canadian waters to recognized species using DNA barcoding. *Botany* 86, 1065–1079.

Larsen, P.F. (2004) Notes on the environmental setting and biodiversity of Cobscook Bay, Maine: a boreal macrotidal estuary. *Northeastern Naturalist* 11 (Special Issue 2), 13–22.

Li, J., Vincx, M., Herman, P.M.J., *et al.* (1997) Monitoring meiobenthos using cm-, m- and km-scales in the Southern Bight of the North Sea. *Marine Environmental Research* 43, 265–278.

Li, W.K.W. & Harrison, W.G. (2001) Chlorophyll, bacteria and picophytoplankton in ecological provinces of the North Atlantic. *Deep-Sea Research II* 48, 2271–2293.

Lindholm, J., Auster, P. & Valentine P. (2004) Role of a large marine protected area for conserving landscape attributes of sand habitats on Georges Bank (Northwest Atlantic). *Marine Ecology Progress Series* 269, 61–68.

Liu, H., Prugnolle, F., Manica, A., *et al.* (2006) A geographically explicit genetic model of worldwide human-settlement history. *American Journal of Human Genetics* 79, 230–237.

Lotze, H.K. & Milewski, I. (2004) Two centuries of multiple human impacts and successive changes in a North Atlantic food web. *Ecological Applications* 14(5), 1428–1447.

Lotze, H.K., Lenihan, H.S., Bourque, B.J., *et al.* (2006) Depletion, degradation, and recovery potential of estuaries and coastal seas. *Science* 312, 1806–1809.

Lubchenco, J., Palumbi, S.R., Gaines, S.D., *et al.* (2003) Plugging a hole in the ocean: the emerging science of marine reserves. *Ecological Applications* 13(1) (Supplement: Marine Reserves), S3–S7.

Makris, N.C., Ratilal, P., Jagannathan, S., *et al.* (2009) Critical population density triggers rapid formation of vast oceanic fish shoals. *Science* 323, 1734–1737.

Makris, N.C., Ratilal, P., Symonds, D.T., *et al.* (2006) Fish population and behavior revealed by instantaneous continental shelf-scale imaging. *Science* **311**, 660–663.

Massana, R. & Pedrós-Alió, C. (2008) Unveiling new microbial eukaryotes in the surface ocean. *Current Opinion in Microbiology* **11**, 213–218.

McCurdy, D.G., Forbes, M.R., Logan, S.P., *et al.* (2005) Foraging and impacts by benthic fish on the intertidal amphipod *Corophium volutator. Journal of Crustacean Biology* **25**, 558–564.

McLeod K.L., Lubchenco, J., Palumbi, S.R., *et al.* (2005) Communication Partnership for Science and the Sea scientific consensus statement on marine ecosystem-based management. www.compassonline.org/pdf_files/EBM_Consensus_Statement_v12.pdf.

McLeod, K.L. & Leslie, H. (2009) *Ecosystem-based Management for the Oceans.* Washington, DC: Island Press.

Millennium Ecosystem Assessment (2005) *Ecosystems and Human Well-being: Biodiversity Synthesis.* Washington, DC: World Resources Institute.

Moore, J.A., Hartel, K.E., Craddock, J.E., *et al.* (2003) An annotated list of deepwater fishes from off New England, with new area records. *Northeastern Naturalist* **10**, 159–248.

Moore, J.A., Vecchione, M., Collette, B.B., *et al.* (2004) Selected fauna of Bear Seamount (New England Seamount chain), and the presence of "natural invader" species. *Archive of Fishery and Marine Research* **51**, 1–3

Morris, R.M., Rappé, M.S., Connon, S.A., *et al.* (2002) SAR11 clade dominates ocean surface bacterioplankton communities. *Nature* **420**, 806–810.

Murawski, S. (2007) Ten myths concerning ecosystem approaches to marine resource management. *Marine Policy* **31**, 681–690.

Murawski, S.A., Brown, R., Lai, H.-L., *et al.* (2000) Large-scale closed areas as a fishery-management tool in temperate marine systems: the Georges Bank experience. *Bulletin of Marine Science* **66**, 775–798.

Murawski, S.A., Steele, J.H., Taylor, P., *et al.* (2010) Why compare marine ecosystems? *ICES Journal of Marine Science* **67**, 1–9.

Myers, R.A. & Worm, B. (2003) Rapid worldwide depletion of predatory fish communities. *Nature* **423**, 280–283.

Norse, E. & Watling, L. (1999) Impacts of mobile fishing gear: the biodiversity perspective. In: *Fish habitat: essential fish habitat and rehabilitation.* (ed. L. Benaka), pp. 31–40, Bethesda, Maryland: American Fisheries Society Symposium 22.

Noss, R.F. (1990) Indicators for monitoring biodiversity: A hierarchical approach. *Conservation Biology* **4**, 355–364.

Not, F., del Campo, J., Balagué, V., *et al.* (2009) New insights into the diversity of marine picoeukaryotes. *PLoS ONE* **4**(9), e7143.

Novacek, M.J. (2008) Engaging the public in biodiversity issues. *Proceedings of the National Academy of Sciences of the USA* **105**, 11571–11578.

O'Boyle, R. & Jamieson, G. (2006) Observations on implementation of ecosystem–based management: Experiences on Canada's east and west coasts. *Fisheries Research* **79**, 1–12.

O'Boyle, R. & Worcester, T. (2009) Eastern Scotian Shelf. In: *Ecosystem-based Management for the Oceans* (eds. K. McLeod & H. Leslie), pp. 253–267. Washington, DC: Island Press.

O'Dor, R.K., Acosta, J., Bergstad, O.A., *et al.* (2010) Bringing life to ocean observation. In: *Proceedings of OceanObs'09: Sustained Ocean Observations and Information for Society (Vol. 2), Venice, Italy, 21–25 September 2009* (eds. J. Hall, D.E. Harrison & D. Stammer). European Space Agency Publication WP-306.

O'Dor, R.K. & Gallardo, V.A. (2005) How to census marine life: ocean realm field projects. *Scientia Marina* **69** (Supplement 1), 181–199.

O'Reilly, C.T., Solvason, R. & Solomon, C. (2005) Resolving the world's largest tides. In: *The Changing Bay of Fundy: Beyond 400 Years. Proceedings of the 6th Bay of Fundy Workshop* (eds. J.A.

Percy, A.J. Evans, P.G. Wells, *et al.*), pp. 153–157. Environment Canada – Atlantic Region, Occasional Report No. 23. Dartmouth, NS and Sackville.

Palumbi, S., Sandifer, P., Allan, J.D., *et al.* (2009) Managing for ocean biodiversity to sustain marine ecosystem services. *Frontiers in Ecology and the Environment* **7**, 204–211.

Pedrós-Alió, C. (2006) Marine microbial diversity: can it be determined? *Trends in Microbiology* **14**, 257–263.

Peters, J., De Baets, B., Verhoest, N.E.C., *et al.* (2007) Random forests as a tool for ecohydrological distribution modelling. *Ecological Modelling* **207**, 304–318.

Pew Oceans Commission (POC) (2003) *America's Living Oceans: Charting a Course for Sea Change. A Report to the Nation.* Arlington, Virginia: Pew Oceans Commission.

Pielou, E.C. (1991) *After the Ice Age: the Return of Life to Glaciated North America.* Chicago, Illinois: University of Chicago Press.

Pittmann, S. & Huetmann, F. (2006) Chapter 4. Seabird distribution and diversity. In: *An Ecological Characterization of the Stellwagen Bank National Marine Sanctuary Region: Oceanographic, Biogeographic, and Contaminants Assessment* (eds. T. Battista, R. Clark & S. Pittmann), pp. 230–263. Silver Spring, Maryland: National Center for Coastal Ocean Science, NOAA. NOAA Technical Memorandum NOS NCCOS 45.

Pittmann, S., Cosat, B., Kot, C., *et al.* (2006) Chapter 5. Cetacean distribution and diversity. In: *An Ecological Characterization of the Stellwagen Bank National Marine Sanctuary Region: Oceanographic, Biogeographic, and Contaminants Assessment* (eds. T. Battista, R. Clark & S. Pittmann), pp. 264–324. Silver Spring, Maryland: National Center for Coastal Ocean Science, NOAA. NOAA Technical Memorandum NOS NCCOS 45.

Ragnarsson, S.A., Jaworski, A., Scott, C., *et al.* (2003) *European Fisheries Ecosystem Plan: The North Sea Significant Web.* Report to the European Union.

Raup, D.M. (1976) Species diversity in the Phanerozoic: a tabulation. *Paleobiology* **2**, 279–288.

Recchia, C., Farady, S., Sobel, J., *et al.* (2001) *Marine and Coastal Protected Areas in the United States Gulf of Maine Region.* Washington, DC: The Ocean Conservancy.

Ricard, D., Branton, R.M., Clark, D.W., *et al.* (2010) Extracting groundfish survey indices from the Ocean Biogeographic Information System (OBIS): an example from Fisheries and Oceans Canada. *ICES Journal of Marine Science* **67**, 638–645.

Rockström, J., Steffen, W., Noone, K., *et al.* (2009) A safe operating space for humanity. *Nature* **461**, 472–475.

Rosenberg, A.A. & McLeod, K.L. (2005) Implementing ecosystem-based approaches to management for the conservation of ecosystem services. *Marine Ecology Progress Series* **300**, 270–274.

Rosenberg, A.A., Bolster, W.J., Alexander, K.E., *et al.* (2005) The history of ocean resources: modelling cod biomass using historical records. *Frontiers in Ecology and the Environment* **3**, 84–90.

Rosenberg, A.A., Mooney-Seus, M.L., Kiessling, I., *et al.* (2009) Lessons from national-level implementation across the world. In: *Ecosystem-Based Management for the Oceans* (eds. K. McLeod & H. Leslie), pp. 294–313. Washington, DC: Island Press.

Satie, A., Dugan, J.E., Lafferty, K.D., *et al.* (2003) Applying ecological criteria to marine reserve design: a case study from the California Channel Islands. *Ecological Applications* **3** (Supplement: Marine Reserves), 170–184.

Saunders, G.W. (2005) Applying DNA barcoding to red macroalgae: a preliminary appraisal holds promise for future applications. *Philosophical Transactions of the Royal Society of London B* **360**, 1879–1888.

Saunders, G.W. (2008) A DNA barcode examination of the red algal family Dumontiaceae in Canadian waters reveals substantial cryptic species diversity. 1. The foliose *Dilsea–Neodilsea* complex and *Weeksia*. *Botany* **86**, 773–789.

Savin, M.C., Martin, J.L., LeGresley, M., *et al.* (2004) Plankton diversity in the Bay of Fundy as measured by morphological and molecular methods. *Microbial Ecology* **48**, 51–65.

Scheibling, R.E., Kelly, N.E. & Raymond, B.G. (2009) Herbivory and community organization on a subtidal cobble bed. *Marine Ecology Progress Series* **382**, 113–128.

Schmidt, A.L. & Scheibling, R.E. (2006) A comparison of epifauna and epiphytes on native kelps (*Laminaria* species) and an invasive alga (*Codium fragile* ssp. *tomentosoides*) in Nova Scotia, Canada. *Botanica Marina* **49**, 315–330.

Sherman, K. (1991) The large marine ecosystem concept: research and management strategy for living marine resources. *Ecological Applications* **1**, 349–360.

Sherr, E. & Sherr, B. (2000) Marine microbes: an overview. In: *Microbial Ecology of the Oceans.* (ed. D.L. Kirchman), pp. 13–46. New York: Wiley-Liss.

Simpson, A.W. & Watling, L. (2006) An investigation of the cumulative impacts of shrimp trawling on mud-bottom fishing grounds in the Gulf of Maine: effects on habitat and macrofaunal community structure. *ICES Journal of Marine Science* **63**, 1616–1630.

Sinclair, M. & Valdimarsson, G. (eds.) (2003) *Responsible Fisheries in the Marine Ecosystem.* Rome, Italy; Food and Agriculture Organization.

Sinclair, M. (1996) Recent advances and challenges in fishery science. In: *Proceedings of Gulf of Maine Ecosystem Dynamics: a Scientific Symposium and Workshop, 16–20 Sept. 1996, St. Andrew's, N.B.* (eds. G. Braasch & G. Wallace), pp. 193–209, RARGOM Report 97-1.

Smedsrud, L.H., Sorteberg, A. & Kloster, K. (2008) Recent and future changes of the Arctic sea-ice cover. *Geophysical Research Letters* **35**, L20503.

Smith, P.C., Houghton, R.W., Fairbanks, R.C., *et al.* (2001) Interannual variability of boundary fluxes and water mass properties in the Gulf of Maine and Georges Bank: 1993–1997. *Deep-Sea Research II*, **48**, 37–70.

Sogin, M.L., Morrison, H.G., Huber, J.A., *et al.* (2006) Microbial diversity in the deep sea and the under-explored "rare biosphere". *Proceedings of the National Academy of Sciences of the USA* **103**, 12115–12120.

Steele, J.H., Collie, J.S., Bisagni, J.J., *et al.* (2007) Balancing end-to-end budgets of the Georges Bank ecosystem. *Progress in Oceanography* **74**, 423–448.

Steneck, R.S., Vavrinec, J. & Leland, A.V. (2004) Accelerating trophic-level dysfunction in kelp forest ecosystems of the western North Atlantic. *Ecosystems* **7**, 523–552.

Stepanauskas, R. & Sieracki, M.E. (2007) Matching phylogeny and metabolism in the uncultured marine bacteria, one cell at a time. *Proceedings of the National Academy of Sciences of the USA* **104**, 9052–9057.

Stevick, P.T., Incze, L.S., Kraus, S.D., *et al.* (2008) Trophic relationships and oceanography on and around a small offshore bank. *Marine Ecology Progress Series* **363**, 15–28.

Stoms, D.M. & Estes, J.E. (1993) A remote sensing research agenda for mapping and monitoring biodiversity. *International Journal of Remote Sensing* **14**, 1839–1860.

Townsend, D.W., Rebuck, N.D., Thomas, M.A., *et al.* (2010) A changing nutrient regime in the Gulf of Maine. *Continental Shelf Research* **30**, 820–832.

Townsend, D.W., Thomas, A.C., Mayer, L.M., *et al.* (2006) Oceanography of the Northwest Atlantic continental shelf. In: *The Sea*, Vol. 14 (eds. A.R. Robinson & K.H. Brink), pp. 119–168. Cambridge, Massachusetts: Harvard University Press,.

Trott, T.J. (2004) Cobscook Bay inventory: a historical checklist of marine invertebrates spanning 162 years. *Northeastern Naturalist* **11** (Special Issue 2), 261–324.

US Commission on Ocean Policy (USCOP) (2004) *An Ocean Blueprint for the 21st Century.* Final Report. Washington, DC.

Vanaverbeke, J., Soetaert, K., Heip, C., *et al.* (1997) The metazoan meiobenthos along the continental slope of the Goban Spur (NE Atlantic) *Journal of Sea Research* **38**, 93–107.

Vermeij, G.J., Dietl, G.P. & Reid, D.G. (2008) The trans-Atlantic history of diversity and body size in ecological guilds. *Ecology* **89** (11, Supplement), S39–S52.

Wares, J.P. & Cunningham, C.W. (2001) Phylogeography and historical ecology of the North Atlantic intertidal. *Evolution* **55**, 2455–2469.

Wares, J.P. (2002) Community genetics in the northwestern Atlantic intertidal. *Molecular Ecology* **11**, 1131–1144.

Watanabe, S., Metaxas, A., Sameoto, J., *et al.* (2009) Patterns in abundance and size of two deep-water gorgonian octocorals, in relation to depth and substrate features off Nova Scotia. *Deep-Sea Research I* **56**, 2235–2248.

Watling, L., Kinner, P. & Maurer, D. (1978) The use of species abundance estimates in marine benthic studies. *Journal of Experimental Marine Biology and Ecology* **35**, 109–118.

Watling, L. & Norse, E.A. (1998) Disturbance of the seabed by mobile fishing gear: A comparison to forest clearcutting. *Conservation Biology* **12**, 1180–1197.

Watling, L. & Skinder, C. (2007) Video analysis of megabenthos assemblages in the central Gulf of Maine. In: *Mapping of the Seafloor for Habitat Characterization* (eds. B.J. Todd & H.G. Greene), pp. 369–377. Geological Association of Canada Special Paper 47.

Watling, L. (2007) A review of the genus *Iridogorgia* (Octocorallia: Chrysogorgiidae) and its relatives, chiefly from the North Atlantic Ocean. *Journal of the Marine Biological Association of the United Kingdom* **87**, 393–402.

Whittaker, R.H. (1977) Evolution of species diversity in land communities. *Evolutionary Biology* **10**, 1–67.

Witman, J., Etter, R.J. & Smith, F. (2004) The relationship between regional and local species diversity in marine benthic communities: a global perspective. *Proceedings of the National Academy of Sciences of the USA* **101**, 15664–15669.

Witman, J.D., Leichter, J.J., Genovese, S.J., *et al.* (1993) Pulsed phytoplankton supply to the rocky subtidal zone: influence of internal waves. *Proceedings of the National Academy of Sciences of the USA* **90**, 1686–1690.

Xue, H., Chai, F. & Pettigrew, N.R. (2000) A model of the seasonal circulation in the Gulf of Maine. *Journal of Physical Oceanography* **10**, 1111–1135.

Yachi, S. & Loreau, M. (1999) Biodiversity and ecosystem functioning in a fluctuating environment: the insurance hypothesis. *Proceedings of the National Academy of Science of the USA* **96**, 1463–1468.

Yen, P.P.W., Sydeman, W.J. & Hyrenbach, K.D. (2004) Marine bird and cetacean associations with bathymetric habitats and shallow-water topographies: implications for trophic transfer and conservation. *Journal of Marine Systems* **50**, 79–99.

York, A.D. (2009) HabCam sheds light on invading tunicate: habitat mapping camera system monitors ecosystem change over time and space. *Sea Technology* **50**(8), 41–46.

Chapter 4

Coral Reef Biodiversity

Nancy Knowlton[1,2], Russell E. Brainard[3], Rebecca Fisher[4], Megan Moews[3], Laetitia Plaisance[1,2], M. Julian Caley[5]

[1]Center for Marine Biodiversity and Conservation, Scripps Institution of Oceanography, University of California San Diego, La Jolla, California, USA
[2]Department of Invertebrate Zoology, National Museum of Natural History, Smithsonian Institution, Washington, DC, USA
[3]Coral Reef Ecosystem Division, Pacific Islands Fisheries Science Center, National Oceanic and Atmospheric Administration, Honolulu, Hawaii, USA
[4]Australian Institute of Marine Science, The University of Western Australia Oceans Institute, Crawley, Western Australia, Australia
[5]Australian Institute of Marine Science, Townsville, Queensland, Australia

4.1 Introduction

Coral reefs are often called the rainforests of the sea, but not because of their vastness. Being largely limited to warm shallow waters, their extent is surprisingly small – in total only 260,000–600,000 km², or approximately 5% that of rainforests, less than 0.1% of the earth's surface, or 0.2% of the ocean's surface (Reaka-Kudla 1997) – and thus smaller in total land area than France. One might therefore think that assessing the diversity of coral reefs would be far easier than for some of the other realms, geographic regions, and taxonomic groups that make up the 14 field projects of the Census of Marine Life that concern ocean life today. Yet coral reefs are the most diverse marine habitat per unit area, and perhaps the most diverse marine habitat overall – the deep sea being the only other contender, in part because of its huge area. As with rainforests, most of this diversity is not found in the organisms that create the three-dimensional structure of reefs – there are, in fact, fewer than 1,000 species of stony corals (scleractinians) that build reefs (Cairns 1999). Rather, the multitude of small organisms living with corals – the equivalent of the insects in the forest – are responsible for the staggering numbers of species associated with reefs.

Coral reefs share another, more dubious, characteristic with tropical rainforests, namely their vulnerability to human impacts, both global (associated with CO_2 emissions) and local (poor water quality, destructive and overfishing, invasive species). Although today concerns about the future of coral reefs dominate the news and the literature (see, for example, Bellwood *et al.* 2004), alarm about the state of coral reefs was slow in coming (Knowlton 2006). The first modern concerns arose in the late 1960s because of mass mortalities associated with population explosions of the crown-of-thorns sea star, *Acanthaster planci*, in the Pacific (Chesher 1969). The collapse of coral reefs in the Caribbean, caused by diseases (both of branching corals and the keystone urchin herbivore *Diadema antillarum*) coupled with overfishing, further added to the alarm (Hughes 1994), with paleontological analyses indicating that these levels of mortality had no precedents in the past several thousand years (Pandolfi 2002). More recently, surprisingly high losses to Pacific reefs, where

Life in the World's Oceans, edited by Alasdair D. McIntyre
© 2010 by Blackwell Publishing Ltd.

most coral reef diversity is found, have been documented (Bruno & Selig 2007). The dangers associated with rising temperatures have been recognized since the El Niño warming event of 1983 bleached and killed many eastern Pacific corals (Glynn 1993), and more recently the perhaps even graver threat of ocean acidification has come to the fore (Kleypas *et al.* 2006; De'ath *et al.* 2009). In late 2007 and mid-2008, two sobering reports appeared: one declared that one-third of all corals were at risk of extinction (Carpenter *et al.* 2008), making them the most endangered group of animals on the planet, and the other predicted that based on current trends in greenhouse gas emissions, coral reefs would cease to exist meaningfully by 2050 (Hoegh-Guldberg *et al.* 2007). As recently as October 2009, the Center for Biological Diversity petitioned the United States Government to list 83 species of corals under the Endangered Species Act owing to ocean warming and acidification.

Given these dire prognoses, it is surprising that relatively little attention has been paid to threats to coral reef diversity itself. Terrestrial conservation biologists often invoke the specter of a sixth mass extinction, originally based on loss of tropical rain forest to agriculture and livestock, with the disruptive effects of warming gaining more attention of late. However, marine scientists in general, and coral reefs scientists specifically, have been curiously less vocal when it comes to overall biodiversity loss. With a few exceptions (for example, the analysis of reef hot spots by Roberts *et al.* 2002), most scientific studies have focused instead on the loss of fishes, corals, and the ecosystem services that they provide. Indeed, even Roberts *et al.* based their biodiversity analysis on just four groups: fishes, corals, snails, and lobsters. The lack of serious attention to overall biodiversity loss stems not only from the assumption that extinction is less common in the ocean (McKinney 1998), but also from the sheer magnitude of the unknown diversity associated with coral reefs, which makes it difficult to assess its loss. People in general, and conservation groups in particular, tend to focus on what can be more easily measured.

Thus a central purpose of the Census of Coral Reef Ecosystems (CReefs) project has been to make the unmeasured measurable, and thus to make the unknown if not known at least knowable. In this chapter we summarize what we knew about coral reef diversity when we started, what the Census has contributed to our understanding, and what the findings suggest for the future of coral reef diversity, both as a topic of scientific study and as a heritage that may or may not be with us when this century draws to a close. A second and equally important goal is to chart the path that will make assessing coral reef diversity, and marine diversity generally, both locally and globally, a realistic endeavor, one that will contribute both to basic diversity science and coral reef and ocean management.

4.2 Background: The "Known" before the Census

Although the Census began in 2000, CReefs was not launched until 2005. At that time, there were two key studies that attempted to estimate the global diversity of coral reefs. The first (Reaka-Kudla 1997) extrapolated from the diversity of tropical rainforests. The second (Small *et al.* 1998) extrapolated from the diversity of a large tropical aquarium. Extrapolation is of course the only way to make such an estimate: the key issues are the reliability of the assumptions underpinning the extrapolation and the scale of the extrapolation. Thus we explain the logic underlying these two analyses in some detail (Fig. 4.1), because it is important to understand their limits.

4.2.1 Estimates from rainforest diversity

To estimate the likely total number of coral reef species, Reaka-Kudla (1997) started with three estimates of rainforest diversity: 1,305,000, 2,000,000, and 20,000,000 species. Then, assuming that forests and reefs have the same species–area relations ($S = cA^{.25}$), she used the forest estimates to calculate the constant c for each of the three forest diversity estimates, and then calculated a total coral reef diversity assuming reefs occupy 5% of the area of tropical forests. From this she arrived at a reef estimate of approximately 618,000–9,477,000 species. Of course, the estimates of rainforest diversity are just that – estimates, and rough ones – and we know very little about how the diversity of a square kilometer of forest compares with that of a square kilometer of reef, or how heterogeneous that diversity is with distance.

4.2.2 Estimates from a mesocosm

Small *et al.* (1998) took an entirely different approach to estimate the likely number of coral reef organisms. They identified to species or morphospecies the nonbacterial/archaeal occupants of a coral reef mesocosm that was largely created by two collections totaling 5 m² of reef from a single locality in the Bahamas. The analysis occurred seven years after the last addition to the mesocosm, and the number obtained was 532 species. Using the same species–area relation as Reaka-Kudla and an estimate of Caribbean reef area of 23,000 km², they estimated a minimum total Caribbean reef diversity of approximately 138,000 species. Assuming that they missed 30% of the species in the mesocosm tank and that 20% did not survive, the figures increased to approximately 180,000 and 216,000 Caribbean reef species, respectively. Finally,

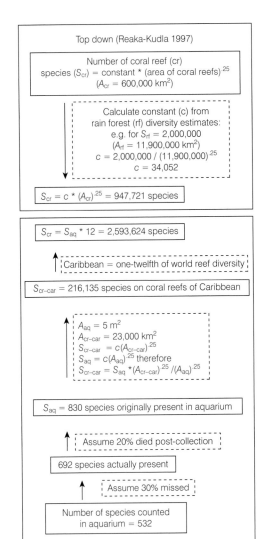

Fig. 4.1

Steps used for estimating coral reef biodiversity based on extrapolation from rainforests (Reaka-Kudla 1997) and a mesocosm (Small *et al.* 1998). *S*, number of species; *A*, area; *c*, constant in species–area equations; cr, coral reefs; rf, rain forests; aq, aquarium; cr-car, Caribbean coral reefs.

the estimates of total area occupied by reefs vary by a factor of two, the total number of species on reefs is not surprisingly a highly uncertain figure. For this reason, most analyses of reef diversity have focused on patterns (with respect to latitude, longitude, depth, etc.; see, for example, Mora *et al.* 2003) rather than actual numbers.

4.2.3 Estimates of the undescribed

Reaka-Kudla (1997) took a published estimate of the total number of described species of 1,868,000, calculated that approximately 15% of all described species are marine based on taxon-by-taxon reviews, and used another analysis indicating that approximately 80% of all marine species are coastal, to arrive at an estimate of 219,000 described coastal species. She then used the percentage of coasts that are tropical (24%) and a species–area calculation (estimated number of species $S = cA^z$; see Fig. 4.1 legend), with the assumption that tropical coasts are twice as diverse as other coasts) to get an estimate of 195,000 described tropical coastal species. Finally, she used the proportion of tropical coasts that are reefs (6%), and the assumption that reefs are twice as diverse per unit area as other tropical coastal habitats, to arrive at a figure of 93,000 described coral reef species. By these calculations the proportion of coral reef species that remains to be described ranges from 85% to 99%.

Such attempts to estimate reef diversity based on existing data have continued over the lifetime of the Census. In 2005, as part of the Census effort, Reaka-Kudla (2005) updated and refined her analysis using similar methods. She estimated that there were approximately 95,000 described coral reef species, representing 35% of all marine species. Her summary estimate of total coral reef species was 1 million to 3 million (based on global estimated totals of all species of 10 million, 14 million, and 20 million), with approximately 30,000 being found in the Caribbean, and only about 5% described. Combining Reaka-Kudla's approach with Chapman's (2009) recent estimate of 11 million species globally would suggest a figure of less than 2 million for all coral reef species. Bouchet (2006) reviewed a variety of methods for estimating global diversity: extrapolation from samples, from known faunas and regions, from ecological criteria (species–area relations), and from taxonomists' estimates of ratios of known species to unknown species. The estimate that he found most credible was based on brachyuran crabs. With currently 212 species from Europe representing 4% of the world's total of 5,200 crab species and a total of 29,713 European marine species in all taxa overall, a global estimate of 728,809 marine species results. If one accepts that there are really 250 European brachyuran crabs, 10,000 brachyuran crabs globally, and 35,000–40,000 marine species in all taxa in

assuming that Caribbean reefs have one-twelfth of all reef species, they calculated total reef diversity figures of approximately 1,656,000, 2,163,000, and 2,594,000 species, respectively. If the less conservative estimate of total reef area in the world that Reaka-Kudla (1997) used (600,000 km²) is applied to these calculations, the figure for reefs overall rises to approximately 3.2 million species.

Although it is gratifying, and indeed somewhat surprising, that these two different approaches yield estimates that overlap, the many untested assumptions that went into the calculations make them "guestimates" at best (though they remain extraordinarily valuable efforts). After all, if even

Europe, then one arrives at a global estimate of approximately 1.4 million to 1.6 million marine species. If coral reefs contain 35% of the marine total, then the total number of coral reef species is about 490,000–560,000.

We thus have estimates of coral reef diversity that range from approximately 500,000 to nearly 10 million. Yet even the smallest of these numbers greatly exceeds the numbers of species recorded for individual tropical locations. For example, Paulay (2003) reported 5,640 marine species from the Mariana Islands, Wehrtmann *et al.* (2009) reported 6,778 marine species from the Atlantic and Pacific coasts of Costa Rica, and Bouchet *et al.* (2002) reported 2,738 species of marine mollusks from the west coast of New Caledonia. There is essentially no bridge between coarse global analyses and intensive local biodiversity assessments for most coral reef organisms. Moreover, most of these geographically focused studies have used traditional (that is primarily morphological) criteria for recognizing species.

One way to examine the limits of such compilations is to look at a handful of species in detail. Just as CReefs was launched, Meyer *et al.* (2005) published such an analysis, and it clearly indicated that the geographic scale for endemism could be far finer for some marine organisms than traditionally assumed. In a study of the snail *Astralium* "*rhodostomum*" in the western Pacific and eastern Indian Ocean, they documented two deep clades (estimated age more than 30 million years), which together contained seven subclades (estimated ages 11 million to 20 million years). The subclades themselves comprised evolutionary significant units (with estimated ages of 2 million to 7 million years) that in total represented at least 30 divergent, geographically isolated units, some separated by as little as 180 km. Given that marine organisms separated by the Isthmus of Panama (at least 3 million years ago) are typically reproductively isolated (Knowlton *et al.* 1993; Lessios 2008) and that many of the evolutionary significant units of *Astralium* also have differences in color pattern, it could be argued (depending on the species concept used (Knowlton & Weigt 1997)) that all the clades/subclades, and perhaps most of the evolutionary significant units merit recognition at the level of species.

The question for global estimates of reef diversity, of course, is how typical is the pattern documented by Meyer *et al.* (2005)? As they note, many previous studies of phylogeography concern groups, like fishes and sea urchins, that have relatively widely dispersing larvae, and in these groups fine-scaled endemism is rarer, although hardly unknown. *Astralium* has limited dispersal and other species with similar larval dispersal are very likely to have similar patterns. Moreover, limited dispersal is associated with small size (Strathmann 1985; Knowlton & Jackson 1993), and much of the diversity on reefs, as in other biological communities (May 1988), comprises small organisms. For example, Bouchet *et al.* (2002) found that the most diverse size class of mollusks (25% of the species)

was the 1.9–4.1 mm size class, with about one-third of all diversity of that size or smaller. It is not inconceivable that at least 30% of all reef species have patterns of endemism like *Astralium*.

4.3 The Census of Coral Reef Ecosystems Approach

Because of the limits to previous analyses, a primary goal of the CReefs project has been to develop new methods needed to address the challenge of assessing the enormous diversity of coral reefs and begin to apply these methods. A comprehensive global assessment of the diversity of the world's coral reefs, both healthy and in distress (that is, at least one-quarter and perhaps over one-third of the diversity of the oceans overall) was clearly beyond the scope of a five-year project, but developing the methods and ground-truthing them was not. The two key methodological components upon which CReefs focused were molecular analyses and standardized sampling.

4.3.1 Molecular analyses

As can be seen from the above summary of previous efforts, the most significant limitation has been the sparseness of available diversity data. It is difficult to extrapolate reliably from a few tiny samples, fundamentally because we do not know the rules for doing so. We cannot develop the rules because it is too expensive and labor intensive to analyze a greater number of samples and we lack adequate scientific expertise for many taxonomic groups.

Molecular methods can reduce these constraints. Instead of depending on taxonomic expertise for species identification, or even sorting into operational taxonomic units of all collected material, one can identify organisms by their genes. This is the concept that underpins genetic barcoding (Hebert *et al.* 2003), and sequencing of the mitochondrial cytochrome oxidase (*COI*) gene works well for many marine organisms, although some technical problems remain. For some groups, such as corals, the *COI* gene is not variable enough to be used at the species level (Shearer & Coffroth 2008), and for others, such as certain crustaceans, there are problems with either amplifying the genes before sequencing (success rates are rarely above 80% overall (Plaisance *et al.* 2009)) or with pseudogenes (extra, non-functional and independently evolving copies in the nucleus; see, for example, Williams & Knowlton (2001); Plaisance *et al.* (2009)). Nevertheless, for taxonomically well-known groups it has great power and potential when applied carefully.

However, even when a gene like *COI* is effective and speeds up the process of sorting organisms to taxonomic groups considerably, it does not by itself provide a name.

For example, in CReefs analyses (described in more detail below) of crustaceans from the Northern Line Islands and Moorea (Plaisance *et al.* 2009), not a single barcode sequence obtained matched any genetic sequence in GenBank at the species level (match of at least 95%), and most matched by only 80–84%; Puillandre *et al.* (2009) similarly found that only one of 24 neogastropod egg cases in the Philippines could be tentatively identified to genus. Given the enormous scope of the unnamed, as reviewed above, this is likely to remain the case for the foreseeable future. Indeed, a mechanism for using standardized barcodes as names as they accumulate, though resisted in many traditional circles and certainly requiring planning for implementation, may well be the only viable solution when so much of biodiversity will remain unstudied by traditional approaches.

Moreover, even though barcoding represents a vast improvement in efficiency compared with visual sorting and traditional identification, it remains an intermediate step towards the goal of obtaining a truly efficient biodiversity assessment methodology. This is because barcoding still involves removing and sampling or sub-sampling individual organisms from collected material, a process that can be time-consuming and costly, especially for small organisms where most of the diversity lies. The next part of the journey towards having a truly efficient method for assessing diversity involves a new and still evolving technology – next generation sequencing – which can be used to obtain large numbers of very short sequences from a sample. This technology has already transformed our ability to study microbial communities (see Chapter 12), and is probably the most significant technological advance for DNA-based studies of diversity since the development of the polymerase chain reaction (PCR).

Although now routinely used in analyzing the genomes of single organisms and environmental samples ("environmental genomics") of microbes, using this method to study the diversity of multicellular life poses challenges. The essence of the problem is the following: if you throw a sample in a blender at one end and get out a list of DNA sequences at the other, to what extent does the list of sequences faithfully reflect what went into the blender? Although this problem of representativeness affects all environmental genomics, it is particularly severe for multicellular organisms for several reasons. First, many multicellular organisms, including common members of coral reefs such as sponges and tunicates, produce an assortment of substances that interfere with critical reactions needed for amplification of DNA. Second, even ignoring the very large organisms that can better be identified in reef transects, multicellular life varies enormously in size, from clumps of algae or sponge of several cubic centimeters to tiny amphipods and worms, so the amount of DNA from different organisms in a sample will vary correspondingly. Finally, some types of DNA amplify well with standard primers

used to start the amplification reaction, and others amplify poorly or not at all (for example, caridean shrimp in the studies of Plaisance *et al.* (2009)). Thus because of the general problem of inhibition and the fact that some organisms might be over- or under-represented because of differences in DNA amount or amplificability, there is no necessary relation between what goes into the blender and what comes out of the sequencer.

To tackle this problem, CReefs is engaged in experimental analyses to estimate the extent and patterns of bias associated with mass sequencing of coral reef community samples. For example, from a given sample one can remove all mobile organisms (which overall are less likely to produce inhibitory substances), remove one subsample from each for individual amplification (to test for any taxonomically based primer problems), remove another subsample to mix with the collection of other similar-sized subsamples before amplification (yielding a mixture where the amount of tissue is approximately equal for all individuals, to test for the effects of different amounts of DNA), and compare these two methods with results when the rest of the body parts, with their very different sizes, are mixed together before amplification (as would be the case in any large-scale sampling protocol). Results so far suggest that genes that are optimal for species-level diagnosis (for example, *COI*) may not be ideal for getting a representative assessment of the community composition, so some compromise between sensitivity and comprehensiveness may be needed. However, if the extent and pattern of bias are known and relatively constant, it should be possible to compare across space and time, which is what is most needed.

4.3.2 Systematic sampling

The second component of the CReefs strategy is systematic sampling, with samples analyzed using the molecular techniques described above, widely applied. Hand sampling by divers ranging over a reef remains the most efficient way to find species (both already known and new) when they are large enough to be seen (Fig. 4.2). However, the effort required to enumerate diversity properly can be daunting (May 2004), and divers vary enormously in their abilities in this regard, making it very difficult to compare work from different places and times involving different people. Two particular methods have been developed by CReefs: assessments of the organisms (especially crustaceans) living in heads of dead *Pocillopora* coral, and assessments of all marine organisms settling into autonomous reef monitoring structures (ARMS) placed on the reef for one to three years.

Using communities of invertebrates living in dead heads of the coral *Pocillopora* (Fig. 4.3) as proxies for reef diversity had the advantage that it could be implemented immediately, an important consideration given the short time-frame for CReefs from its founding to the close of the

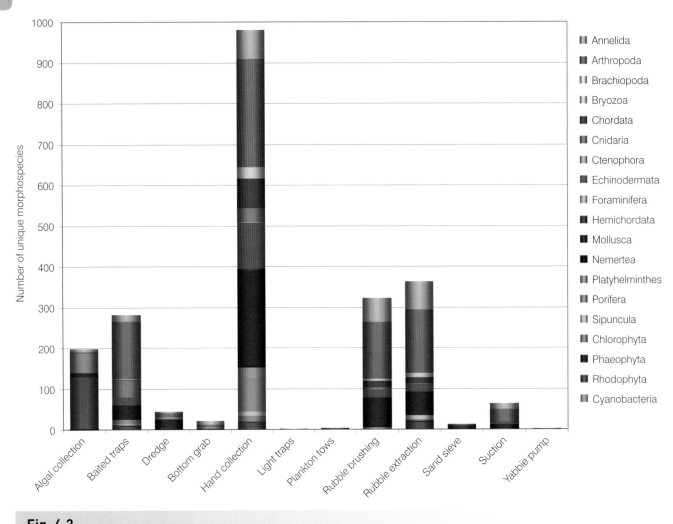

Fig. 4.2

Numbers of species (by higher-level taxon) obtained using different methods on the CReefs cruise to French Frigate Shoals, Northwestern Hawaiian Islands.

Fig. 4.3

Laetitia Plaisance at work on barcoding for CReefs project.
(A) Preparing to break open dead head of *Pocillopora* coral to extract resident invertebrates.
(B) Examining extracted DNA before sequencing. A, B, Juergen Freund © FreundFactory.

Census. In addition, because these are natural habitats, interpretation is not confounded by concerns associated with artificial substrates. Assessment of *Pocillopora* heads can be replicated from the Red Sea to the Eastern Pacific (and with some adjustments, reef rubble in the Caribbean can also be compared). On the other hand, the sizes of dead coral heads can be only roughly standardized (for example, fitting snuggly into standard buckets), their ages cannot be known with any precision (although one can collect heads that are old enough to be covered with fouling sessile organisms but young enough not to have been substantially bioeroded, to provide some standardization), and all collections involving removal of reef require permits (which, for example, were denied for the initial CReefs cruise to the Northwestern Hawaiian Islands).

The second systematic sampling method, ARMS, has been developed as a standard method to mimic the structural complexity of coral reef habitats and attract colonizing invertebrates and algae. ARMS evolved from "artificial reef matrix structures" (note: this is different from "autonomous reef monitoring structures"), originally designed and tested in the eastern Caribbean to collect as much diversity as possible (Zimmerman & Martin 2004). Their original design involved several layers of concrete with different sized openings and a variety of microhabitats, including a mesh basket containing coral rubble suspended from a PVC frame to allow different occupants to colonize the structure. It was determined that such structures were heavy, over-sampled sites, and that it was difficult, time-consuming, and costly to extract and process specimens.

Rather than attempting to collect and document all of the diversity of coral reefs, CReefs developed the current generation ARMS as a simple, cost-effective, standardized tool to assess spatial patterns and temporal trends of indices of cryptic diversity systematically on a global scale. After numerous design modifications and test deployments, CReefs settled on an ARMS design consisting of 23 cm × 23 cm gray, type 1 PVC plates stacked in alternating series of open and obstructed formats (created by x-shaped inserts dividing each space into four sectors), topped by plastic pond filter mesh and a final plate, and attached to a base plate of 35 cm × 45 cm, which is then affixed to the reef (Fig. 4.4A). In December 2008, using some experimental ARMS deployed off Oahu, CReefs partners conducted a workshop to develop protocols for retrieval, sampling, and processing, including sample preservation and molecular analyses.

DNA barcode analyses were also conducted to characterize crustacean biodiversity associated with ARMS in comparison to the dead *Pocillopora* heads from other sites in the Pacific. These results suggest that coupling ARMS with taxonomic and molecular analyses can be an effective method to assess and monitor understudied coral reef invertebrate biodiversity. In the long run, ARMS will be a much more powerful tool than assessment of dead *Pocillopora* heads, because they can be deployed nearly anywhere (including non-reef sites), do not involve destructive sampling of natural habitats, are much easier to remove organisms from (especially true for sessile organisms), and can be highly standardized. Permits for deploying and subsequently collecting ARMS are also in general easier to obtain than those for collection of live rubble. ARMS have the disadvantage of not being natural habitats (being made of PVC and lacking many small nooks and crannies), but early assessments suggest that the diversity captured is representative of the communities in which they are placed (Gustav Paulay, personal communication; see also Fig. 4.4B).

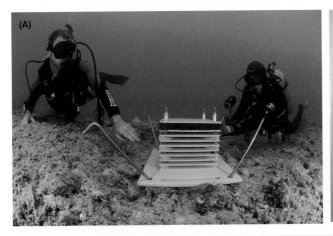

Fig. 4.4

Autonomous reef monitoring structures (ARMS) in Australia. **(A)** ARMS being installed on reef (Juergen Freund © FreundFactory). **(B)** ARMS layer after removal from reef one year after deployment (Gustav Paulay, Florida Museum of Natural History).

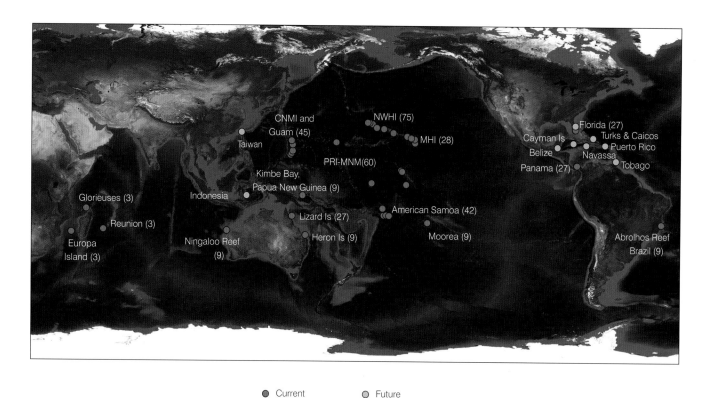

Fig. 4.5

Map of current and planned deployment sites for ARMS. See text for full listing of abbreviated names.

Over 400 ARMS have been widely deployed throughout the Pacific between 2006 and 2009, with smaller, yet increasing efforts in the Indian Ocean and the Caribbean (Fig. 4.5). They were successfully deployed throughout the Papahānaumokuākea Marine National Monument (MNM) and Main Hawaiian Islands, the recently established Pacific Remote Islands MNM (Line Islands, Phoenix Islands, and Wake Atoll), American Samoa (including Rose Atoll MNM), Australia (the Great Barrier Reef's Lizard and Heron Islands, Ningaloo Reef), Brazil (Abrolhos Reef), Guam, Northern Mariana Islands (including Marianas Trench MNM), French Polynesia (Moorea), western Indian Ocean (Reunion, Europa, and Glorieuses Islands), Panama, Papua New Guinea (Kimbe Bay), and Florida. Additional deployments are planned for 2010 in Puerto Rico, the Cayman Islands, Belize, Taiwan, Indonesia, and other locations within the Coral Triangle. The approach has been adopted as a key biodiversity assessment tool by the National Oceanic and Atmospheric Administration (NOAA)'s Pacific Reef Assessment and Monitoring Program and as a central component of the Smithsonian's Marine Initiative and NOAA's Biodiversity Alternative. To ensure consistency and comparability, and to reduce costs, efforts so far have been led by CReefs' Hawaii Node (NOAA's Pacific Islands Fisheries Science Center, Coral Reef Ecosystem Division), with ARMS being centrally produced.

Finally, it should be noted that dead *Pocillopora* heads and ARMS are not the only solution to standardized sampling. Analyses of fixed amounts of sediment or of fixed amounts of material vacuumed from a reef (for example, see methods of Bouchet *et al.* 2002) are complementary approaches that target different components of coral reef communities. The key features that all of them share are (1) they can be at least in some sense standardized, so that results from different studies can be compared, and (2) they lend themselves to molecular analyses using either barcoding or environmental genomics.

4.4   CReefs Results

Analyses of dead *Pocillopora* heads from five islands in the Central Pacific (Kirimati, Tabuaeran, Palmyra, Kingman, and Moorea) have now been published (Plaisance *et al.* 2009), and several surprising results from these analyses are already apparent. First, the total number of crustaceans recorded was exceptionally high for such a small sample. A total of 22 small dead coral heads (combined length + width + height dimensions of each approximately 90 cm, total basal area less than 2 m²) yielded 789 individual crustaceans. Of these, 500 were sequenced (all rare plus representatives of abundant morphospecies were selected), which yielded 403 usable sequences, from which 135 operational taxonomic units were distinguished. Of these, 65 were brachyuran crab species, a number equivalent to approximately 30% of the entire described brachyuran fauna of European seas and approximately 1% of the global total! Second, most species were rare and locally distributed: 44% of all species were sampled just once, and another 33% were only found on one of five islands. Even more surprisingly, 48 of the 70 decapod species found in the Northern Line Islands were not only not found in the Moorea *Pocillopora* head survey, but were not recorded in the extensive cross-habitat collections associated with the Moorea Biocode Project (www.mooreabiocode.org). Third, despite the marked anthropogenic impacts on the abundance of corals and fishes on the two inhabited islands in the Northern Line Islands (Sandin *et al.* 2008), there does not appear to be a comparable negative impact on the diversity of small crustaceans. Therefore, our reliance so far on corals and fishes as surrogates for coral reef biodiversity may need to be re-examined.

Data from other expeditions are still being analyzed, but the patterns remain consistent with these results. For example, the numbers of species/percentage singleton figures from Australian dead coral samples were 58/43% (Ningaloo Reef, 7 heads), 113/47% (Heron Island, 16 coral heads), and 48/60% (Lizard Island, 11 heads). Likewise, at French Frigate Shoals in the Northwestern Hawaiian Islands, one-third of all invertebrate morphospecies collected were singletons or found at only one site, and one-third of the crustaceans from ARMS were singletons.

Finally, from the beginning, one goal of CReefs has been to build taxonomic expertise and information for those groups of coral reef organisms that are poorly known, to complement the molecular approaches. As noted above, most ecological studies focus on corals or fishes; mollusks (and to a lesser extent crustaceans) are better known than most other groups, but even for these groups many gaps remain. Much progress was made possible by several CReefs cruises and expeditions, beginning with the cruise to French Frigate Shoals in the Northwestern Hawaiian Islands. By 2009, scientists from the French Frigate Shoals efforts had already found that of the nearly 400 algal specimens (approximately 160 morphospecies) catalogued, many were not on the list of 179 described taxa previously reported (Vroom *et al.* 2006). Also, at least 50 new invertebrate species and over 100 new records were identified for the region, including probable new species among sponges, corals, anemones, flatworms, segmented worms, crabs, bivalves, gastropods, octopuses, sea cucumbers, sea stars, and sea squirts (six octopuses were collected repre-

senting six different species, three may be new). As a result of the repeated expeditions to Heron Island, Lizard Island, and Ningaloo Reef in Australia, hundreds more new species and records are being identified. The taxonomic papers published under the aegis of CReefs are now appearing, but initial estimates suggest that there are about 100 new species among the 4,150 sample lots and approximately 2,100 morphospecies in Hawaii (PIFSC 2007), and more than 1,000 new species from over 15,000 sample lots from Australia; one new family has already been described.

Initial results from the original French Frigate Shoals ARMS showed that prototype ARMS were most productive in sampling mollusks (28%), ascidians (24%), crustaceans (19%), and bryozoans (11%) in fore reef and lagoonal patch reef habitats. Of the 12 prototype ARMS recovered from French Frigate Shoals, new records for the Northwestern Hawaiian Islands were recorded for two non-native (alien) species of solitary tunicates, *Cnemidocarpa irene* and *Polycarpa aurita* (Godwin *et al.* 2008). The results from the standardized, globally distributed ARMS await 2010–2012 retrieval and analyses. Further analyses will take place beyond 2012 as the ARMS are used for continued monitoring to assess the biodiversity impacts of climate change and ocean acidification.

Identification and analysis of specimens can be a time-consuming process, thus there are likely to be many more discoveries as the specimens from these efforts are further analyzed. Such discoveries will be documented in multiple joint publications and the data placed in the global Ocean Biogeographic Information System (Chapter 17) (more than 400,000 records submitted so far). By providing scientists and managers with a more complete picture of what exists in coral reef ecosystems, they will be better equipped to manage them and in particular watch for and manage changes over time. Furthermore, with the integration of future investigations, there can be a greater understanding of biodiversity over gradients of human disturbance.

4.5 Gaps in Knowledge

There are unlimited questions at various scales that could be asked about coral reef diversity, but CReefs has focused on developing the methods needed to answer these four:

1) How many species occur on coral reefs and what are the patterns of species diversity for all reef species across gradients of human disturbance?

2) What kinds/percentages of species are obligately associated with healthy coral reefs and how widely are they distributed?

3) What are the prospects for maintenance of species diversity on reefs suffering various levels of human impacts?

4) How much and what kinds of taxonomic and ecological information are required to manage reef biodiversity effectively, and are cost-effective proxies possible?

Ironically, it is the last of these that we have answered first: we do now have a method that is cost effective for assessing diversity, and it has the potential to work far more effectively than using individual taxa as proxies. Moreover, in a DNA analogue to Moore's Law, sequencing costs have dropped substantially since the start of CReefs, and will continue to drop over the coming decade, further increasing the use of these approaches.

Although we do not have a firm answer to question (3), results so far suggest that moderately impacted reefs continue to support large amounts of diversity but seriously degraded reefs do not. This pattern is easiest to understand as a nonlinear relation between diversity and disturbance such as predicted by the intermediate disturbance hypothesis. For example, human disturbance may initially increase the types of habitat available to small invertebrates living in the reef, by causing corals and algae to coexist more equally (unimpacted reefs have very little macroalgae; see, for example, Sandin *et al.* (2008)). Thus diversity appears to be relatively unaffected by human disturbance in the Northern Line Islands, probably because even the most degraded of the Northern Line Islands are comparatively pristine (Knowlton & Jackson 2008). In contrast, Caribbean reefs show a clear pattern of decreased invertebrate diversity with lower coral cover and three-dimensionality (Idjadi & Edmunds 2006), probably because almost all Caribbean reefs are seriously degraded (Gardner *et al.* 2003; Pandolfi *et al.* 2003).

The first two questions are the hardest to answer – we cannot reliably estimate global reef diversity from $2\,m^2$ of dead coral heads from the Central Pacific, any more than Small *et al.* (1998) could from $5\,m^2$ of Caribbean reef placed and later sampled in a mesocosm. Just comparing these two analyses points to the many possibilities for error in the assumptions. For example, Small *et al.* (1998) found eight species of decapods and assumed that a comparable sample in the Pacific would be 12 times more diverse, yet in a sample less than half that size from a part of the Pacific not renowned for its diversity, Plaisance *et al.* (2009) found 108 decapod species. The extremely high prevalence of singletons at any site and the related absence of overlap between sites, a pattern characteristic of all the CReefs results, clearly imply that much more geographically dense sampling is needed to determine the level of endemism, which in turn profoundly affects extrapolations from single sites to the world at large. However, despite the challenges now, much better answers to these questions will be possible with the analysis over the next few years of the hundreds of ARMS that are currently deployed (Fig. 4.5).

4.6 Advancing Knowledge

4.6.1 Current limits to our knowledge

The greatest limit to knowledge has been the lack of a biodiversity assessment protocol that can be implemented globally. Both lack of money and lack of agreement on an appropriate method have played a role, but providing the latter would go a long way towards acquiring the necessary financial commitments. This is why CReefs has focused on methodological advancements.

With the establishment of an agreed approach (widely accepted by scientists and managers globally), and an appropriate financial commitment for both the regular analysis of samples and the maintenance of databases, coral reef biodiversity studies could be standardized at a global scale, with adequately dense sampling. Given estimates that coral reefs represent perhaps over one-third of all the diversity of marine life, it is ludicrous to assume that this diversity can be understood by tiny and unsystematic assessments – 22 heads of dead corals from the central Pacific, $5\,m^2$ of reef from the Bahamas, species lists or numbers from a handful of well-studied locations, or even a many-month expedition in the heart of reef diversity, the Coral Triangle. Alone, these do not begin to provide enough information, even to know what the appropriate geographic scale of sampling is.

4.6.2 Why the knowledge is needed

The scientific justifications for estimating coral reef diversity go well beyond simple curiosity about the total number of organisms living on reefs – given the threats that reefs face, better knowledge of how reef diversity is likely to be impacted by loss of living coral is clearly essential for conservation and management. Although diversity may remain poorly understood, the nature of the threats is far clearer. Three are globally pervasive – overfishing and destructive fishing, poor water quality, and the effects of carbon dioxide in the atmosphere. A fourth threat – invasive species – also represents a serious problem in an increasing number of locations (for example, seaweeds in the Pacific and lionfish in the Caribbean). This combination of human-induced impacts has resulted in a situation where globally about 60% of coral reefs have been degraded or lost (Jackson 2008).

Coral reefs have sometimes been referred to as the equivalent of a canary in a coal mine, an unmistakable warning that humankind is in the process of doing irreparable damage to the planet. Although it is certainly true that coral reefs are among the first victims of the combined onslaughts of local impacts and global change, it is worth remembering that coral reefs represent far more than a "canary". Whether the number is 0.5 million or 2 million or 10 million species, accepting the potential demise of the ecosystems that support perhaps 35% of the global marine diversity of the planet seems ill-advised at best. Even in strictly economic terms, the value of coral reefs has been assessed at approximately 30 billion US dollars annually (Conservation International 2008). This does not include the potential medical and industrial benefits associated with compounds yet to be discovered from reef organisms.

The loss of corals and the potential threat to biodiversity in general terms is well established, but we still remain largely ignorant of the details. The efforts of CReefs have provided a means to begin rapidly narrowing some of the key gaps in our knowledge. As noted above, conservation priorities are often based on what can be measured. Providing a reliable method that estimates biodiversity across space and through time is essential for motivating the public to protect biodiversity, for designing the specifics of marine protected areas, and for monitoring their effectiveness. In addition to assessing changes in diversity generally, these approaches can also be used to detect invasive species (see, for example, Godwin *et al.* 2008).

4.6.3 Ideas to move the goal posts

On land in general, and in particular for rainforests, two approaches have transformed our understanding of their diversity, both originating in the tropical forests of Panama. On the one hand, Erwin (1982) catapulted studies of rainforest diversity forward by counting the total diversity of insects associated with 19 trees, and then extrapolating from these results to the world at large. Although Erwin's estimates remain highly debated and the inspiration for continuing research, the basic principle is sound – an intensive survey of a limited area. The closest analogue to the Erwin approach, amplified by the use of molecular methods and a broader taxonomic and spatial base, is the Moorea Biocode Project. The second approach is the establishment of the Center for Tropical Forest Science, which involves the repeated census of trees greater than 5 cm diameter at breast height in 50-hectare forest plots now scattered around the world. This provides a different kind of data, more superficial in terms of diversity estimates in any one location, but with the potential to monitor differences and similarities across space and through time. In the sea, no comparable program exists; the establishment of such a program would be transformative.

4.6.4 Blueprint for the future

We will never know every life form that lives in the oceans, or on coral reefs. However, with standardized sampling at many locations and intensive sampling at a few, both made newly practical by molecular techniques, we can extrapolate from what we know to reefs at large. This approach involves far more than a static inventory of DNA sequences. By selecting sites strategically, we can address specific questions about factors responsible for biodiversity patterns that have challenged scientists for decades. Repeated samples will detect biodiversity loss and change and the arrival and impact of invasive species. Conservation biologists will know where biodiversity is concentrated, where it is disappearing, and where it is most resilient.

A blueprint for the future would thus consist of the following elements.

1) A global array of ARMS sites. At any site, these are currently deployed in three groups of three: three closely spaced ARMS constitute a replicate set, and three sets of three in one general location can be used to assess very fine-scale replicability of results. If comparisons of ARMS deployed for one versus two versus three years are desired, then three sets of the array of nine are needed. At broader scales, the results of Meyer *et al.* (2005) clearly indicate that every archipelago should have at least one set of nine (or 27 for multi-year analyses). As part of the international OceanObs '09 conference, systematic assessments and monitoring of biodiversity using standardized ARMS were included in Community White Papers on global ocean observing systems for coral reef ecosystems (Brainard *et al.* 2010) and ocean acidification (Feely *et al.* 2010).

2) Two or three all-taxa inventories, on the model of the Moorea Biocode Project but with the addition of bringing a sequencer on site to analyze everything as it comes ashore. These are essential to ground-truth the ARMS, that is, to determine the relation of diversity measured by ARMS versus diversity measured by much more intensive efforts (which cannot be practically implemented at many sites or many times).

3) An ARMS biodiversity database. Sequence data, including environmental genomics data, can be stored in GenBank, but many additional data are associated with ARMS. Moreover, if ARMS are placed in permanently established quadrats, essentially small versions of 50-hectare forest plots, the sequence data become connected with other ecological data, greatly enhancing the value of the diversity measures. Fifty-hectare plots have stimulated many other ecological observations and experiments, and ARMS themselves could be elaborated upon, and even be subjected to manipulations designed to test hypotheses.

4.7 Conclusions

Coral reefs are enormously diverse marine ecosystems, perhaps harboring one-quarter to one-third of all marine species. Most coral reef species are undescribed, and will remain so for the foreseeable future. Because coral reef diversity is so hard to measure in any comprehensive fashion, almost all previous attempts have focused on a very limited number of taxonomic groups, or have been very limited in geographic scope. This has resulted in our knowledge of coral reef diversity being based on groups that do not represent most of the diversity on reefs (for example, especially fishes and corals) or on a patchwork and idiosyncratic set of broader analyses that are almost impossible to compare with one another.

Molecular approaches, applied systematically on a global scale, offer reef scientists the ability for the first time to assess coral reef diversity. The key to success will be to use techniques that can quickly survey a representative fraction of biodiversity, and to use these techniques across a dense global sampling grid. CReefs has developed autonomous reef monitoring structures (ARMS) that can be deployed, recovered, and analyzed by next generation sequencing methods in a cost-effective way. When combined with more intensive ground-truthing surveys at a limited number of sites, these methods will allow us to better understand diversity patterns generally, and to evaluate, monitor, and enhance the effectiveness of management strategies for coral reef ecosystems.

Acknowledgments

We thank the scientists from around the world who took part in the numerous workshops and meetings held to determine priorities and implementation for the CReefs Project. We acknowledge those who volunteered their time and resources for the coral reef censuses in the Northwestern Hawaiian Islands Marine National Monument and Australia, joined forces to create globally standardized protocols for ARMS and DNA sequencing, and assisted with the processing and identification of specimens. Special thanks go to Gustav Paulay, Scott Godwin, Joel Martin, Philippe Bouchet, Christopher Meyer, Amy Driskell, Andrea Ormos, Tito Lotufo, and in particular, Ian Poiner and Jesse Ausubel, for their guidance and support throughout the CReefs mission. Our appreciation further goes out to Vaiarii Terorotua, Florent Angly, Mary Wakeford, Shawn Smith, Penny Dockry, the NOAA Pacific Islands Fisheries Science Center, University of Hawaii Joint Institute for Marine and Atmospheric Research, and the University of Hawaii Institute of Marine Biology for their extensive field, outreach, and administrative support. We gratefully acknowledge the Alfred P. Sloan Foundation, the Gordon and Betty Moore Foundation, the Moore Family Foundation, the National Geographic Society, the Australian Institute of Marine Science, the Great Barrier Reef Foundation, NOAA's Coral Reef Conservation Program and Pacific Region Integrated Data Enterprise, the Scripps Institution of Oceanography, and the National Museum of Natural History for their resource and monetary support of the CReefs efforts.

References

Bellwood, D.R., Hughes, T.P., Folke, C., *et al.* (2004) Confronting the coral reef crisis. *Nature* **429**, 827–833.

Bouchet, P. (2006) The magnitude of marine biodiversity. In: *The Exploration of Marine Biodiversity: Scientific and Technological Challenges* (ed. C.M. Duarte), pp. 32–64. Madrid, Spain: Fundación BBVA.

Bouchet, P., Lozouet, P., Maestrati, P., *et al.* (2002) Assessing the magnitude of species richness in tropical marine environments: exceptionally high numbers of molluscs at a New Caledonia site. *Biological Journal of the Linnean Society* **75**, 421–436.

Brainard, R., Bainbridge, S., Brinkman, R., *et al.* (2010) An international network of coral reef ecosystem observing systems (I-CREOS). In: *Proceedings of OceanObs'09: Sustained Ocean Observations and Information for Society (Vol. 2), Venice, Italy, 21–25 September 2009* (eds. J. Hall, D.E. Harrison, D. Stammer). ESA Publication WPP-306.

Bruno, J.F. & Selig, E.R. (2007) Regional decline of coral cover in the Indo-Pacific: timing, extent, and subregional comparisons. *PLoS ONE*, 8, e711.

Cairns, S.D. (1999) Species richness of recent Scleractinia. *Atoll Research Bulletin* **459**, 1–46.

Carpenter, K.E., Abrar, M., Aeby, G., *et al.* (2008) One-third of reef-building corals face elevated extinction risk from climate change and local impacts. *Science* **321**, 560–563.

Chapman, A.D. (2009) *Numbers of Living Species in Australia and the World*, 2nd edn. Australian Biological Resources Study, Department of the Environment, Water, Heritage and the Arts.

Chesher, R.H. (1969) Destruction of Pacific corals by the sea star *Acanthaster planci*. *Science* **165**, 280–283.

Conservation International (2008) *Economic Values of Coral Reefs, Mangroves, and Seagrasses: A Global Compilation*. Arlington, Virginia: Center for Applied Biodiversity Science, Conservation International.

De'ath G., Lough J.M. & Fabricius K.E. (2009) Declining coral calcification on the Great Barrier Reef. *Science* **323**, 116–119.

Erwin, T.L. (1982) Tropical forests: their richness in Coleoptera and other arthropod species. *The Coleopterists Bulletin* **36**, 74–75.

Feely, R.A., Fabry, V.J., Dickson, A., *et al.* (2010) An international observational network for ocean acidification. In: *Proceedings of OceanObs'09: Sustained Ocean Observations and Information for Society (Vol. 2), Venice, Italy, 21–25 September 2009* (eds. J. Hall, D.E. Harrison, D. Stammer). ESA Publication WPP-306.

Gardner, T.A., Côté, I.M., Gill, J.A., *et al.* (2003) Long-term region-wide declines in Caribbean corals. *Science* **301**, 958–960.

Glynn, P.W. (1993) Coral reef bleaching: ecological perspectives. *Coral Reefs* **12**, 1–17.

Godwin, S., Harris, L., Charette, A., *et al.* (2008) The marine invertebrate species associated with the biofouling of derelict fishing gear in the Pāpahanaumokuākea–Marine National Monument: A focus on marine non-native species transport. Hawaii Institute for Marine Biology, Kaneohe, HI, 26 pp.

Hebert, P.D.N., Cywinska, A., Ball, S.L., *et. al.* (2003) Biological identifications through DNA barcodes. *Proceedings of the Royal Society of London B* 270, 313–321.

Hoegh-Guldberg, O., Mumby, P.J., Hooten, A.J., *et al.* (2007) Coral reefs under rapid climate change and ocean acidification. *Science* 318, 1737–1742.

Hughes, T.P. (1994) Catastrophes, phase shifts, and large-scale degradation of a Caribbean coral reef. *Science* 265, 1547–1551.

Idjadi, J.A. & Edmunds, P.J. (2006) Scleractinian corals as facilitators for other invertebrates on a Caribbean reef. *Marine Ecology Progress Series* 319, 117–127.

Jackson, J.B.C. (2008) Evolution and extinction in the brave new ocean. *Proceedings of the National Academy of Sciences of the USA* 105 (Suppl. 1), 11458–11465.

Kleypas, J.A., Feely, R.A., Fabry, V.J., *et al.* (2006) *Impacts of Ocean Acidification on Coral Reefs and Other Marine Calcifiers: A Guide for Future Research*. Washington, DC: NSF, NOAA, and the US Geological Survey.

Knowlton, N. (2006) Coral reef coda: what can we hope for? In: *Coral Reef Conservation* (eds. I. Côté & J. Reynolds), pp. 538–549. Cambridge, UK: Cambridge University Press.

Knowlton, N. & Jackson, J.B.C. (1993) Inbreeding and outbreeding in marine invertebrates. In: *The Natural History of Inbreeding and Outbreeding* (ed. N.W. Thornhill), pp. 200–249. Chicago, Illinois: University of Chicago Press.

Knowlton, N. & Jackson, J.B.C. (2008) Shifting baselines, local impacts, and global change on coral reefs. *PLoS Biology*, 6, e54.

Knowlton, N. & Weigt, L.A. (1997) Species of marine invertebrates: a comparison of the biological and phylogenetic species concepts. In: *Species: the Units of Biodiversity* (eds. M.F. Claridge, H.A. Dawah, M.R. Wilson). *Systematics Association (UK) Special Volume Series* 54, pp. 199–219. London: Chapman and Hall.

Knowlton, N., Weigt, L.A. Solorzano, L.A., *et al.* (1993) Divergence in proteins, mitochondrial DNA, and reproductive compatibility across the Isthmus of Panama. *Science* 260, 1629–1632.

Lessios, H.A. (2008) The Great American Schism: Divergence of marine organisms after the rise of the Central American Isthmus. *Annual Review of Ecology, Evolution and Systematics* 36, 63–91.

May, R.M. (1988) How many species are there on Earth? *Science* 241, 1441–1449.

May, R.M. (2004) Tomorrow's taxonomy: collecting new species in the field will remain the rate limiting step. *Philosophical Transactions of the Royal Society of London B* 359, 733–734.

McKinney, M.L. (1998) Is marine biodiversity at less risk? Evidence and implications. *Diversity and Distributions* 4, 3–8.

Meyer, C.P., Geller, J.B. & Paulay, G. (2005) Fine scale endemism on coral reefs: archipelagic differentiation in turbinid gastropods. *Evolution* 59, 113–125.

Mora, C., Chittaro, P.M., Sale, P.F., *et al.* (2003) Patterns and processes in reef fish diversity. *Nature* 421, 933–936.

Pandolfi, J.M. (2002) Coral community dynamics at multiple scales. *Coral Reefs* 21, 13–23.

Pandolfi, J.M., Bradbury, R.H., Sala, E., *et al.* (2003) Global trajectories of the long-term decline of coral reef ecosystems. *Science* 301, 955–958.

Paulay, G. (2003) Marine biodiversity of Guam and the Marianas: overview. *Micronesica* 35–36, 3–25.

Pacific Island Fisheries Science Center (PIFSC) (2007) Oscar Elton Sette/Cruise OES-06-11 (OES-47). U.S. Department of Commerce, National Oceanic and Atmospheric Administration, National Marine Fisheries Service (NMFS), PIFSC Cruise Report CR-07-007, pp. 11–36. Honolulu: NOAA NMFS PIFSC.

Plaisance, L., Knowlton, N., Pauley, G., *et al.* (2009) Reef-associated crustacean fauna: biodiversity estimates using semi-quantitative sampling and DNA barcoding. *Coral Reefs* 28, 977–986.

Puillandre, N., Strong, E.E., Bouchet, P., *et al.* (2009) Identifying gastropod spawn from DNA barcodes: possible but not yet practicable. *Molecular Ecology Resources* 9, 1311–1321.

Reaka-Kudla, M.L. (1997) The global biodiversity of coral reefs: a comparison with rain forests. In: *Biodiversity II: Understanding and Protecting Our Biological Resources* (eds. M.L. Reaka-Kudla, D.E. Wilson & E.O. Wilson), pp. 83–108. Washington, DC: Joseph Henry Press.

Reaka-Kudla, M.L. (2005) Biodiversity of Caribbean coral reefs. In: *Caribbean Marine Biodiversity: The Known and the Unknown* (eds. P. Miloslavich & E. Klein), pp. 259–276. Lancaster, Pennsylvania: DEStech Publications.

Roberts, C.M., McClean, C.J., Veron, J.E.N., *et al.* (2002) Marine biodiversity hotspots and conservation priorities for tropical reefs. *Science* 295, 1280–1284.

Sandin, S.A., Smith, J.E., DeMartini, E.E., *et al.* (2008) Baselines and degradation of coral reefs in the northern Line Islands. *PLoS ONE* 3, e1548.

Shearer, T.L. & Coffroth, M.A. (2008) Barcoding corals: limited by interspecific divergence, not intraspecific variation. *Molecular Ecology Resources* 8, 247–255.

Small, A.M., Adey, W.H. & Spoon, D. (1998) Are current estimates of coral reef biodiversity too low? The view through the window of a microcosm. *Atoll Research Bulletin* 458, 1–20.

Strathmann, R.R. (1985) Feeding and nonfeeding larval development and life-history evolution in marine invertebrates. *Annual Review of Ecology and Systematics* 16, 339–361.

Vroom, P.S., Page, K.N., Peyton, K.A., *et al.* (2006) Marine algae of French Frigate Shoals, northwestern Hawaiian islands: species list and biogeographic comparisons. *Pacific Science* 60, 81–95.

Wehrtmann, I.S., Cortés, J. & Echeverría-Sáenz, S. (2009) Marine biodiversity of Costa Rica: perspectives and conclusions. In: *Marine Biodiversity of Costa Rica, Central America* (eds. I.S. Wehrtmann & J. Cortés), pp. 521–533. Springer.

Williams, S.T. & Knowlton, N. (2001) Mitochondrial pseudo-genes are pervasive and often insidious in the snapping shrimp genus *Alpheus*. *Molecular Biology and Evolution* 18, 1484–1493.

Zimmerman, T.L. & Martin, J.W. (2004) Artificial Reef Matrix Structures (ARMS): an inexpensive and effective method for collecting coral reef-associated invertebrates. *Gulf and Caribbean Research* 16, 59–64.

Chapter 5

New Perceptions of Continental Margin Biodiversity

Lenaick Menot[1,2], Myriam Sibuet[2], Robert S. Carney[3], Lisa A. Levin[4], Gilbert T. Rowe[5], David S. M. Billett[6], Gary Poore[7], Hiroshi Kitazato[8], Ann Vanreusel[9], Joëlle Galéron[1], Helena P. Lavrado[10], Javier Sellanes[11], Baban Ingole[12], Elena Krylova[13]

[1]Ifremer, Plouzane, France
[2]Institut Océanographique, Paris, France
[3]Louisiana State University, Baton Rouge, Louisiana, USA
[4]Scripps Institution of Oceanography, La Jolla, California, USA
[5]Texas A&M University, Galveston, Texas, USA
[6]National Oceanography Centre, Southampton, UK
[7]Museum Victoria, Melbourne, Australia
[8]Institute of Biogeosciences, Japan Agency for Marine-Earth Science and Technology, Yokosuka, Japan
[9]Marine Biology Section, Ghent University, Ghent, Belgium
[10]Marine Biology Department, UFRJ Federal University of Rio de Janeiro, Brazil
[11]Facultad de Ciencias del Mar, Universidad Católica del Norte, Coquimbo, Chile
[12]National Institute of Oceanography, Goa, India
[13]P.P. Shirshov Institute of Oceanology, Russian Academy of Sciences, Moscow, Russia

5.1 Introduction: Diversity Re-examined as Slope Complexity is Disclosed

The Census of Marine Life has promoted synergetic approaches to assess and explain the diversity, distribution, and abundance of life in the ocean, focusing on domains where new approaches allowed discoveries and evident new steps in science. The field project "Continental Margin Ecosystems on a Worldwide Scale", COMARGE, is one of the five Census projects concerned with the deep ocean. It was launched in 2005 to focus on the complex and active continental margins (Box 5.1), where unique ecosystems including canyons, oxygen minimum zones, cold seeps, and reef-like coral mounds were only recently discovered and studied owing to the development of new oceanographic equipments.

The complexity of the slope seabed has until recently limited the exploration of continental margins to major marine laboratories in developed countries. Such studies shaped our original, sometimes naive, conceptions of what lives on these steep depth gradients. The first impression was that the deep ocean is azoic, owing to the rapid decline in abundances with depth in the Mediterranean (Forbes 1844). This was later disproved by the telegraph cable-laying industry and an ensuing international race to sample to the greatest depths on an ocean scale (reviewed in Mills 1983). Trawl records in the Atlantic off Western Europe revealed that depth ranges of many species were limited to sometimes little more than several hundred meters,

Life in the World's Oceans, edited by Alasdair D. McIntyre

Box 5.1

What are Continental Margins?

COMARGE focuses on the deep continental margins, excluding the continental shelf. The upper boundary is delineated by the shelf break at *ca.* 140 m depth over most of the margins except in Antarctica where it can be as deep as 1,000 m. It coincides with a sharp turn-over in species composition. From a geological point of view, the margin ends at the boundary between the continental and oceanic crusts, thus including trenches. From a biological point of view, however, the lower boundary of the margin is more elusive and usually located at the bottom of the continental slope or rise, between 2,000 m and 5,000 m depth. For the purpose of computations and mining in a georeferenced database, we set up the lower boundary at 3,500 m depth. Between these upper and lower boundaries, deep continental margins cover approximately 40 million km^2 or 11% of the ocean surface. Their width ranges from 10 to over 500 km and their slope from 6° to 1° in active and passive settings, respectively.

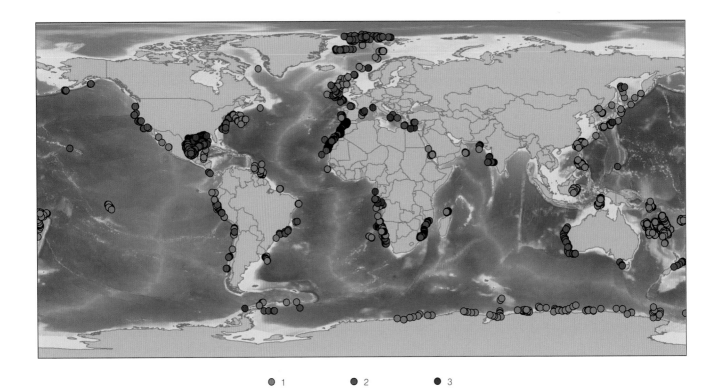

● 1   ● 2   ● 3

Fig. 5.1

Location map of sampling on continental margins showing the evolution of field strategies from description of faunal communities (biogeography, 1 in green) to the addition of environmental factors (ecology, 2 in purple) to the integration of energy fluxes (ecosystem functioning, 3 in red).

creating an intricate zonation of populations lining the slope (Le Danois 1948). The true extent of deep-sea biodiversity in the seemingly monotonous sediment environment became evident when Sanders *et al.* (1965) found that benthic communities along a transect between southern New England and the Bermuda islands were more diverse on the middle of the continental slope than on the shelf or the abyss (Sanders 1968; Rex 1981). Over the past 50 years, biological research on continental margins increased, our perception of deep habitats greatly improved, and descriptive exploration has given way to more functional studies (Fig. 5.1).

Of utmost importance in this evolution of deep-sea science have been the technological developments which disclosed the complexity of the slope environment. These include the use of trawled cameras, manned submersibles, remotely operated vehicles (ROVs), and autonomous underwater vehicles (AUVs), as well as high-resolution sidescan sonar, multibeam bathymetry mapping systems, high-resolution sub-seabed profilers, precision sampling qualitatively and quantitatively, and video and photographic imaging systems. Analyses of these data have been greatly enhanced by advances in digital processing, network databases, and visualization. Geophysical tools have been used to classify and map habitats over large areas because they can discriminate seabed type (mud, sand, rock). Higher-resolution tools have allowed the characterization of ecological features such as coral mounds, outcropping methane hydrate, mud volcanoes, and seabed roughness. Newly developed, near-bottom swathbathymetry operated from ROVs now resolves seafloor structures as small as 20 centimeters. The deep bottom is no longer as remote as it once was. Our perceptions are now of a much higher resolution and reveal that continental margins are both very complex and active regions ecologically, geologically, chemically, and hydrodynamically (Wefer *et al.* 2003). Collectively, these processes create unique ecosystems such as methane seeps, coral reefs, canyons, or oxygen minimum zones (OMZs). These hot spots are characterized by unusually high biomasses, productivity, physiological adaptations, and apparent high species endemicity (Fig. 5.2).

Fundamental patterns of species distribution first observed and explained in the context of monotonous slopes had thus to be re-evaluated in light of this newly recognized heterogeneity and its interplay with large-scale oceanographic features. The question was timely as the concurrent development of human activities already threatened margin hot spots and triggered urgent needs for sound scientific advice on the evaluation and conservation of continental margin biodiversity (Rogers *et al.* 2002). Large integrated projects had already begun to address these issues at a regional scale in the European North Atlantic (Weaver *et al.* 2004), the Gulf of Guinea (Sibuet & Vangriesheim 2009), and the Gulf of Mexico (Rowe & Kennicutt 2008). COMARGE benefited from these programs and expanded their scope to a global scale to address questions that had to be tackled through synergies within an international network of scientists (Box 5.2). This chapter summarizes the progress made so far and underscores in conclusion the major unknowns that may guide future research on continental margins during the next decade and beyond.

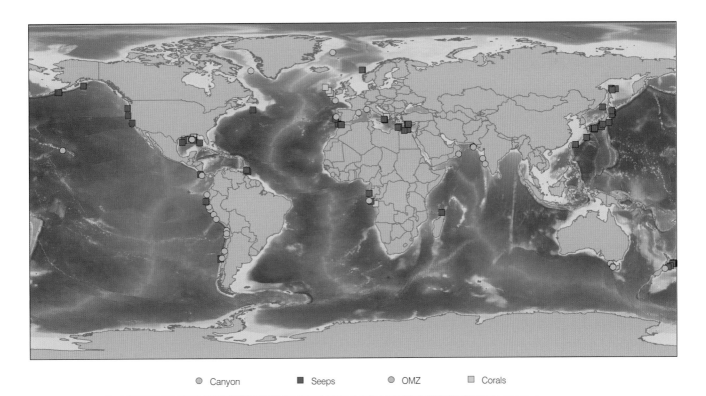

○ Canyon　■ Seeps　◎ OMZ　□ Corals

Fig. 5.2
Location map of known hot spots sampled for biological purposes and included in COMARGE syntheses.

Box 5.2

COMARGE Questions and Strategies

The COMARGE science plan was discussed and finalized during a community workshop held in 2006. Three main questions have been identified as major unknowns regarding continental margin ecology:

1] What are the margin habitats and what is the relation between diversity and habitat heterogeneity?

2] Are large-scale biodiversity patterns such as zonation or diversity-depth trends ubiquitous and what are their drivers?

3] Is there a specific response of continental margin biota to anthropogenic disturbances?

The strategy to tackle these issues was to create a network of scientists, promote discussions and syntheses through workshops, and foster data integration. The COMARGE network grew to bring together over a hundred researchers

and students. Four workshops were organized that addressed: (1) the classification of margin habitats globally, (2) the roles of habitat heterogeneity in generating and maintaining continental margin biodiversity (Levin et al. 2010b), (3) the effects of both large-scale oceanographic features and habitat heterogeneity on nematode diversity (Vanreusel et al. 2010), and (4) the biogeography of marine squat lobsters (Baba et al., 2008). Data integration has been achieved either through the Ocean Biogeographic Information System (OBIS; www.iobis.org) or via the COMARGE Information System (COMARGIS), connected to OBIS. The originality of COMARGIS resides in the fact that it is ecologically oriented. The database has been built on an existing system (Fabri et al. 2006) that allows archiving comprehensive sampling metadata for both biological and environmental data.

5.2 Roles of Habitat Heterogeneity in Generating and Maintaining Continental Margin Biodiversity

Following early exploration, ecological studies were mainly directed at understanding the mechanisms that promote high species richness, with greatest focus on processes that operate at small spatial scales (reviewed in Snelgrove & Smith 2002), or the influence of energy flux on the structure of benthic communities (see, for example, Laubier & Sibuet 1979). In recent years new scientific questions have emerged about the relation between diversity and various forms and scales of margin heterogeneity that are closely linked to growing environmental concern about the deep sea. In the past quarter of a century this interest has focused the study of margins on several key environments. Cold seep communities (see Chapter 9) have been discovered and investigated in conjunction with tectonic studies in active margins and with oil and gas development on passive margins (Sibuet & Olu 1998; Sibuet & Olu-Le Roy 2003). There has been considerable interest in determining the extent of deep coral reef habitats to minimize the impact of deep-sea fisheries (Freiwald 2002). Canyons that cut across margins

are now seen not only as novel, and somewhat specialized habitats, but also conduits for pollutant transport into the deeper abyss and sometimes sites of intensive fishing. Of the many environmental gradients that occur on the margins, oxygen minimum zones are seen as special habitats that mirror effects of coastal eutrophication and that may expand in response to climate change (Levin 2003).

COMARGE has brought together scientists working on these and other aspects of margins to evaluate and understand the relations between habitat heterogeneity and diversity.

5.2.1 Types of habitat heterogeneity that affect diversity: scales in space and time

Margin heterogeneity exists in many forms (Figs. 5.3 and 5.4) and on multiple space and time scales; it is also perceived differently depending on the size, mobility, and lifestyles of the species considered. The COMARGE focus on how different sources and scales of heterogeneity influence margin biodiversity has spanned a wide range of taxa, from Protozoa to megabenthos, in diverse settings across the globe. Workshop discussions, synthetic papers, and regional analyses published in a special volume of

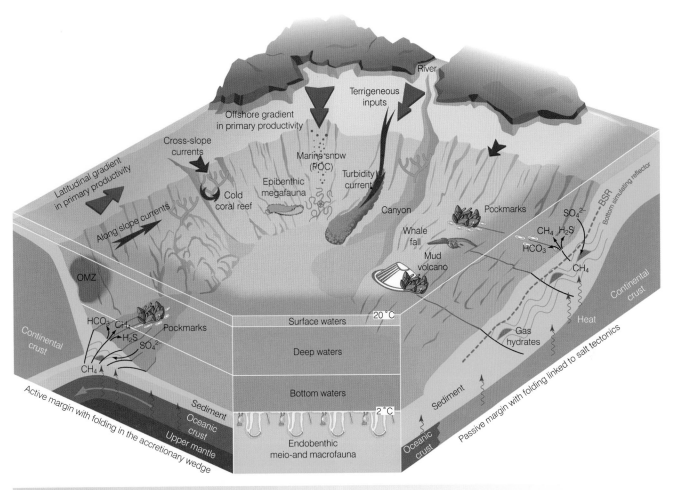

Fig. 5.3

Diagram summarizing the main geological, hydrological, and biological factors driving habitat heterogeneity on active and passive continental margins. The figure illustrates strong depth-related and geographic variations in water masses, productivity, and currents superimposed on typical margin habitats such as cold coral reefs, canyons, chemosynthetic communities linked to cold methane seep structures (pockmarks or mud volcanoes), whale falls, and oxygen minimum zones (OMZs). All these features create a complex mosaic of influences shaping margin biodiversity.

the journal *Marine Ecology* (Levin *et al.* 2010b) have generated several major results.

Perhaps the most universal finding is that heterogeneity acts in a hierarchical, scale-dependent manner to influence margin diversity. At the largest scale with strong effect is hydrography associated with water masses (in particular temperature and oxygen) and overlying productivity. The impingement of water masses on the slope interact with depth and latitude (productivity) to shape levels of diversity and community composition (De Mello E Sousa *et al.* 2006; Priede *et al.* 2010; Sellanes *et al.* 2010; Williams *et al.* 2010). Productivity influences the water masses and food supply to the sea floor; both positive and negative diversity influences may result (Levin *et al.* 2001; Corliss *et al.* 2009).

However, hydrographic influences on diversity are modulated by variations in substrate and flow regime (Williams *et al.* 2010). At meso-scales (tens of kilometers) there is topographic control in the form of canyons, banks, ridges, pinnacles, and sediment fans. Deposition regimes (canyon floor and deep-sea fan) and substrate vary within these (Baguley *et al.* 2008; Ramirez-Llodra *et al.* 2010). At smaller scales there are earth and tectonic processes that control fluid seepage and sediment disturbance forming seeps (Olu-Le Roy *et al.* 2007b; Cordes *et al.* 2010; Menot *et al.* 2010). And at the smallest scales there are habitats formed by ecosystem engineers that influence diversity through provision of substrate, food, refuge, and various biotic interactions. These habitats include coral and sponge reefs, mytilid, vesicomyid, and siboglinid beds (Cordes *et al.* 2009, 2010). In some cases the biotic influence arises from decay processes at whale, wood, and kelp falls (Fujiwara *et al.* 2007; Pailleret *et al.* 2007).

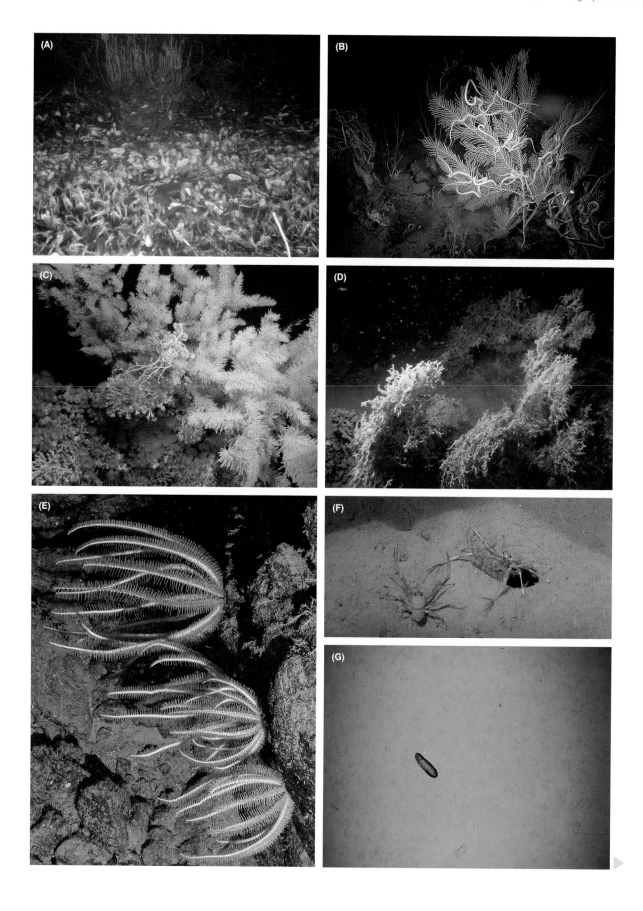

Our focus on specialized margin settings has revealed that stressed habitats associated with hypoxia or high sediment sulfide levels exhibit depressed alpha diversity relative to open slope systems. Importantly, these settings contribute significantly to regional diversity patterns and to beta diversity (species turnover) on margins, ultimately adding to the species richness. This is true for macro- and megabenthos on slopes with oxygen minimum zones (Gooday *et al.* 2010; Levin *et al.* 2010; Sellanes *et al.* 2010), for all taxon sizes on slopes with methane seeps (Cordes *et al.* 2010; Menot *et al.* 2010; Van Gaever *et al.* 2010), and for an even broader range of habitats occupied globally by nematodes (Vanreusel *et al.* 2010). To address the question of how diversity accumulates across habitats a new analytical approach was developed that examines the change in slopes of species and genus accumulation curves as habitats are included. Analysis of invertebrate diversity at methane seeps from four very different regions illustrates addition of species as hypoxic, microbial mat, vesicomyid clam, and tube worm habitats are added, but with different rates depending on taxon and location (Fig. 5.5). A strong diversity response to habitat heterogeneity was found in Gulf of Mexico habitats; there was a much slower increase in the rate of species accumulation with habitat heterogeneity for the nematode fauna of the Haakon Mosby mud volcano (Cordes *et al.* 2010).

Several global analyses indicate that there are strong ocean basin and regional differences that preclude the occurrence of identical cosmopolitan species in all habitats (Vanreusel *et al.* 2010; Williams *et al.* 2010). Regional patterns can differ from a summed global pattern. This is evident for deep-sea fishes in the North Atlantic, where key roles for the position of the thermocline, local water masses, resuspensed organic matter (OM) and seasonality create distinctive diversity patterns (Priede *et al.* 2010).

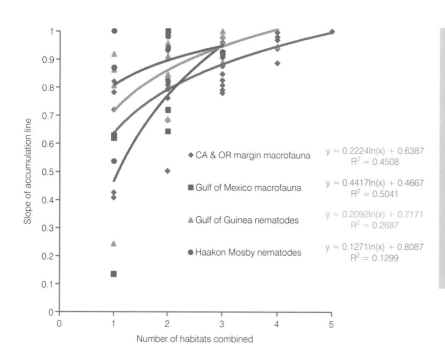

$y = 0.2224\ln(x) + 0.6387$
$R^2 = 0.4508$

$y = 0.4417\ln(x) + 0.4667$
$R^2 = 0.5041$

$y = 0.2092\ln(x) + 0.7171$
$R^2 = 0.2687$

$y = 0.1271\ln(x) + 0.8087$
$R^2 = 0.1299$

◆ CA & OR margin macrofauna

■ Gulf of Mexico macrofauna

▲ Gulf of Guinea nematodes

● Haakon Mosby nematodes

Fig. 5.5

Rate of species accumulation for macrofauna or genus accumulation for nematodes across habitats. The CA and OR margin macrofauna include species-level data from near-seep sediments, vesicomyid clam beds, oxygen minimum zones, bacterial mats, and background sediments. Gulf of Mexico macrofauna include species-level data from vestimentiferan tubeworm aggregations, mussel beds, and scleractinian coral habitats. Gulf of Guinea nematodes are genus-level data from seep, transition, canyon, and control sediments. Haakon Mosby mud volcano samples are also genus-level nematode meiofauna from bacterial mats, siboglinid-associated sediments from the outer rim of the volcano, and non-seep influenced sediments. Reproduced with permission from Cordes *et al.* 2010, copyright 2009 by Blackwell Publishing Ltd.

Fig. 5.4

Continental margin heterogeneity in images: **(A)** assemblages of mytilids, vesicomyids, and siboglinid tube-worms in a giant pockmark in the Gulf of Guinea (3,200 m depth) (copyright Ifremer, Biozaïre 2 cruise, 2002); **(B)** authigenic carbonates associated with a hydrocarbon seep are colonized by corals in the Gulf of Mexico (530 m depth) (courtesy of Derk Berquist and Charles Fisher, cruise sponsored by NOAA Ocean Exploration Program and US Mineral Management Service); **(C)** ophiuroids, antipatharians, and anemones are inhabiting Lophelia-reefs off Ireland (900 m depth) (copyright Ifremer, Caracole cruise, 2001); **(D)** A cloud of zooplankton around Lophelia reefs off Italy (600 m depth) (copyright Ifremer, Medeco cruise, 2007); **(E)** filter-feeding organisms such as Brinsing asteroids are dominant in the Nazare Canyon off Portugal (1,000 m) (copyright NOC Southampton and UK Natural Environment Research Council); **(F)** high sediment loading in the Var Canyon off France favors the sediment-dwelling or burrowing fauna such as squat lobsters (2,200 m) (copyright Ifremer, Medeco cruise, 2007); **(G)** the "featureless" muddy slope is actually punctuated with small-scale heterogeneities such as fecal pellets of large holothuroids *Benthodytes lingua* (35 cm in length), Alaminos Canyon, Northern Gulf of Mexico (2,222 m depth) (courtesy of Robert Carney, Louisiana State University).

In many instances sources of heterogeneity are superimposed on one another; this can create additional complexity or, if stress or disturbance is involved, it can impose local homogeneity. The influence of heterogeneity has proven to be context-dependent as well. Heterogeneity that adds structure or nutrients often has greater effect at deeper than shallower depths (Levin & Mendoza 2007) because deeper margins tend to be more structurally homogeneous and more food poor. Biotic interactions between substrate provider and epibionts (Dattagupta *et al.* 2007; Järnegren *et al.* 2007), between animals and sediment microbes (Bertics & Ziebis 2009), or predation and competition between taxa can generate additional sources of heterogeneity.

5.2.2 Models underlying the heterogeneity–diversity relation

Continued exploration of margins has revealed that any continental margin habitats (for example cold seeps, canyons, deep-water coral reefs) are distributed as patches in a sedimented slope matrix. The resident species are predicted to function as metapopulations and metacommunities. The species-sorting model, in which diversity and metacommunity structure is dictated by different niche requirements (Leibold *et al.* 2004), appears to explain community patterns for species that occupy methane seep habitats (Cordes *et al.* 2010) and hypoxic settings (Gooday *et al.* 2010). These niches are defined by substrate (abiotic, biotic), flow regimes, sulfide or methane requirements, and geochemical tolerances to sulfidic or hypersaline fluids (Brand *et al.* 2007; Levin & Mendoza 2007; Olu-Le Roy *et al.* 2007a; Levin *et al.* 2010; Sellanes *et al.* 2010; Van Gaever *et al.* 2010). In addition to chemoautotrophic symbioses, reduced compounds (methane and sulfide) also fuel a free-living microbial community that provides nutrition (and possibly settlement cues) for a vast array of smaller grazing, deposit feeding, and suspension feeding taxa, as well as for bacterivores that may specialize on microbes with specific metabolic pathways or morphologies (Levin & Mendoza 2007; Thurber *et al.* 2009; Van Gaever *et al.* 2010). Very localized, small-scale variations in geochemical settings may dictate diversity and evenness among meiofauna (Levin & Mendoza 2007; Thurber *et al.* 2009; Van Gaever *et al.* 2010; Vanreusel *et al.* 2010).

Other metacommunity models including mass effects (source-sink dynamics) or patch dynamics (succession based on tradeoffs between dispersal/colonization ability and competition) (Leibold *et al.* 2004) appear to apply to the canyon and deep-water coral reef settings where many species are not habitat endemics or obligate symbionts (Ramirez-Llodra *et al.* 2010; Vetter *et al.* 2010). The communities of deep-water coral reefs and vesicomyid tube worms exhibit clear successional stages on margins (Cordes *et al.* 2009).

5.2.3 The societal value of continental margin heterogeneity

The recent recognition of a high degree of heterogeneity on single margins and its influence on margin diversity offers new challenges to the assessment, management, and conservation of margin resources (Schlacher *et al.* 2010; Williams *et al.* 2010). It becomes essential that this heterogeneity is incorporated into planning for exploration, research, and monitoring (Levin & Dayton 2009). Habitat heterogeneity plays prominently in metapopulation and metacommunity theory, biodiversity–function relations, trophic dynamics, and in understanding roles of ecosystem engineers and invasive species.

Habitat heterogeneity unquestionably influences the key ecosystem services provided by the continental slope. Over 0.62 GtC y^{-1} settles to the seafloor on margins, of which 0.06 GtC y^{-1} may be buried in sediments (Muller-Karger *et al.* 2005). Sequestration occurs by margin biota and through carbonate precipitation (often microbially mediated). Hard bottoms, including those associated with methane seeps, seamounts, canyons, and coral and sponge reefs, are hot spots for fishes and invertebrates and provide major fisheries resource production on margins (Koslow *et al.* 2000). Oil and methane gas are linked to chemosynthetic environments on margins. The role of microbes and animals in transforming or consuming methane is of considerable interest, given that methane is a powerful greenhouse gas that contributes to global warming.

As we confront increasing pressures on margins from fishing, mineral resource extraction, and climate change, there is much to be gained by combining our newfound understanding of margin complexity with ecological theory into research and management solutions (Levin & Dayton 2009).

5.3 Spatial Trends in Biodiversity

Since the discovery of a diverse deep-sea fauna, deep-sea biologists have debated different hypotheses to explain depth-related patterns of the distribution and diversity of benthic and demersal organisms (Rex 1981), in particular that

- the diversity reaches a maximum at mid- to lower-slope depth,
- the fauna is zoned into bands according to water depth,
- faunal assemblages are dissimilar between depth zones but have a circum-margin distribution within a depth zone,
- and the width of the zones increases with depth.

Since these observations were initially made (1880s–1960s), numerous data have been collected on continental margins, but few syntheses have been attempted. The COMARGE project explored several ways to test those old but still unresolved hypotheses. A major issue has been the lack of taxonomic consistency across studies. Our first approach, thus, was to focus on two taxa that are widespread on continental margins and for which there is an active community of deep-sea taxonomists. For squat lobsters, we first compiled the literature and published a list of 800 known species (Baba *et al.* 2008), which we have now analyzed to address these questions. We also gathered, standardized, and analyzed individual datasets on deep-sea nematodes to decipher the processes that define global species distributions. The second approach was to undertake meta-analyses across taxa from data either mined from the literature and available in databases, such as OBIS and COMARGIS, or directly provided by members of the COMARGE network.

The role of multiple large-scale oceanographic features that change with latitude on diversity and zonation is more problematic than depth effects. Certainly when shallow-water data predominate analysis there are latitude changes in the ranges of individual species (Macpherson 2002) and in species diversity (Hillebrand 2004). Seeking such patterns below the permanent thermocline removes one of the major consequences of latitude. During the COMARGE project there has been an emphasis upon recognizing the high degree of local and regional heterogeneity on the margins. Until global-scale studies are undertaken, using a uniform design that examines both global and local factors, the actual role of latitude cannot be resolved.

5.3.1 Zonation and distribution on continental margins

Compared with the vast abyssal seafloor and the relatively wide continental shelf the continental margin lies in between as a narrow ribbon of ocean bottom characterized by dramatic transitions. The environment goes from upper slope regions where limited light may actually reach the seabed to a seafloor in total darkness. Except for polar and boreal regions, there is a sharp transition at the thermocline from warmer surface water to deep, cold water (typically less than 3 °C). Water pressure increases continuously with depth. Local bottom currents are usually weaker than and decoupled from upper ocean circulation. Importantly, photosynthetically derived food energy in the form of sinking detritus becomes progressively scarcer. Therefore, it is not remarkable that the margin also experiences major biotic transitions. The upper margin experiences a sharp decline in continental shelf fauna as few such species extend into the very different habitat of deeper water. The lower margin transitions to one dominated by abyssal species that extend out across the somewhat similar, larger, but much more food-poor seafloor habitat.

What is remarkable is that the narrow ribbon of margin also harbors a diverse suite of species that seem to be truly margin-endemic. These species occupy restricted bathymetric ranges along any given section of the margin, but often with basin-scale horizontal ranges. The overlap of within-margin species, shelf-to-slope, and slope-to-abyss transitions produces a vertical species change or turnover at specific depths that is known as bathymetric zonation (Carney *et al.* 1983). The process of describing this zonation is to develop a matrix of similarity values from some taxonomic component of the sampled fauna and then partition that similarity through multivariate analyses. The full process of numerical analysis has several very subjective steps that alter the results. Thus, the sampled depth is dividing into a series of zones that seem to have relatively homogenous biota.

At the beginning of the COMARGE project, a literature survey was undertaken to assess the level of knowledge about bathymetric zonation with three primary objectives (Carney 2005). These were to determine (1) if zonation was the most common distribution pattern found in studies since the 1960s, (2) if there were global similarities in the zonation found, and (3) whether global correlations of zonation help identify most likely causes for the phenomena. Six margin regions were identified as more extensively studied within the context of specific investigation of zonation (Fig. 5.6): Porcupine Sea Bight, Gulf of Mexico, Mediterranean, Cascadia Basin in the northeast Pacific, and Chatham Rise off New Zealand. Studies in these regions, as well as the results from a few single studies produced 33 regional descriptions of zonation. In a meta-analysis it was found that the number of zones reported increased with the depth range sampled (Fig. 5.7). Therefore, fauna underwent species turnover at specific depths in all studies, and zonation did not stop at any depth. The width of the deepest zone was greater than the shallowest zone in all except five cases (Fig. 5.8), suggesting some increased uniformity of faunal composition with depth in most regions. Except for the shelf-to-slope transition, the boundaries of zones did not coincide among the regional patterns. This might indicate the importance of local phenomena or simply be an artifact of inconsistent sampling design and analysis across multiple projects.

There was no indication that the temperature transition from shallow warm water to cold deeper water played a significant role in bathymetric zonation on a global scale. Deep slope species did not emerge extensively into cold shallow water in polar surveys. Similarly, shelf species did not descend into the unusually warm deep water of the Mediterranean. The surveys on Chatham Rise had been undertaken in part to examine the influence of different water masses and productivity regimes on zonation. Unfortunately, both the faunal and oceanographic data

Fig. 5.6

The results of thirty-four zonation studies around the world were examined for common patterns. Solid sections were considered homogenous by the authors. White sections were transition regions, and blank areas represent unresolved gaps.

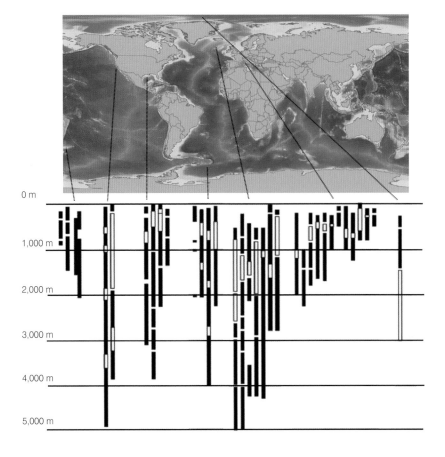

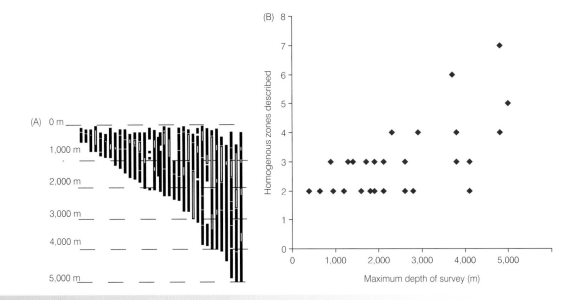

Fig. 5.7

(A) Although the execution of each zonation study differed greatly across locations and taxa, **(B)** the relation between maximum depth samples and homogenous zones recognized indicates faunal change occurs at all depths on the margin.

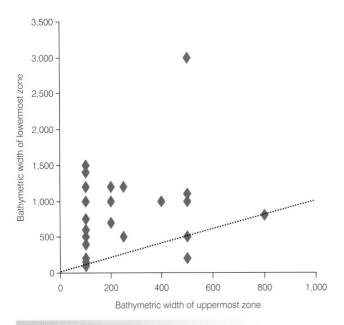

Fig. 5.8

A comparison of the depth width of the uppermost and lowermost zones in all surveys supports the older observation that faunal change slows somewhat with depth producing wider zones.

once in an extensive survey (Carney 1997). Even the most abundant species may represent a smaller proportion of the total fauna than is found in many other environments. When inferred from the distributions of species with a low frequency of occurrence, ranges appear to be narrow and the spatial change in fauna on the margin becomes exaggerated.

Developing a definitive zonation map for the global margins is of great practical as well as scientific value. The deep margins are already being exploited, but, because so many areas of the deep ocean are poorly sampled, the data available to regulators are limited. Regulatory agencies that are charged with developing science-based strategies must now rush to catch up with industry. One serious risk with this rush is oversimplification, whereby regulators may ignore the complex set of regulations that has evolved for the more data-rich shallow water environment and develop a single set of regulations for the entirety of the deep ocean. The one uniting theme of COMARGE is that the continental margins are complex and heterogeneous. Zonation studies clearly show that the biota of the upper slope is dissimilar from that of the middle and the lower slope. When zones have been mapped using appropriate sampling, expert taxonomy, and consistent analyses, then regulations can be developed that protect all of the zones present.

proved to be equivocal regarding the actual location of fronts and the environmental control of distribution.

The COMARGE literature review and meta-analysis of published conclusions confirmed the ubiquity of bathymetric zonation on all studied continental margins. Another key finding was severe limitations on the extent to which results from different studies can be compared. Some are obvious, such as inconsistent and possibly erroneous identification of specimens. Smaller meio- and macroinfauna comprise the most diverse and abundant metazoan components of these systems (Rex 1981; Rex *et al.* 2006). Many are new to science and too poorly characterized for consistent identification. This status makes it problematic to compile different datasets to produce accurate species ranges over basin- and global-scale areas. The hypothesis that individual margin species occupy narrow depth ranges over large (thousands of kilometers) horizontal distances requires considerable future study. A less obvious problem is the potential effect that sampling design and data analyses may have on data interpretation and conclusions. The placement and effort of sampling always impose artifacts into the patterns of distribution found. Boundaries between zones are often the result of uneven sampling effort at different depths, especially uneven depth intervals between sampling stations (Carney 2005). Furthermore, the margins and the abyss share a key characteristic of species diversity that demands consideration in future studies of zonation. Species inventories contain a high proportion of very rare organisms, where a given species may be collected only

5.3.2 Exploring depth–diversity trends along margins: from local patterns to global understanding

5.3.2.1 Expected depth–diversity trends and processes

The relation between diversity and depth is of long-standing interest to deep-sea ecologists, and unraveling the mechanisms underlying its origin and maintenance is of fundamental importance to understanding the determinants of deep-sea biodiversity. Rex (1981) was the first to show that the diversity of dominant macrofaunal and megafaunal groups was unexpectedly high but peaked at intermediate depths in the western North Atlantic, somewhere between 1,900 and 2,800 m depending on taxon. A similar parabolic trend was also observed in the eastern North Atlantic and tropical Atlantic for polychaetes (Paterson & Lambshead 1995; Cosson-Sarradin *et al.* 1998), thus supporting the hypothesis of a biodiversity peak at mid-slope depth along continental margins. Contradictory patterns have, however, also been found (Stuart & Rex 2009).

In the framework of the COMARGE project, our aim was to question the generality of this pattern and, if confirmed, to identify environmental variables that might explain the relation for well-studied taxa. We gathered data on diversity and sampling depth from 16 cross-margin datasets from the Arctic, Atlantic, Pacific, Indian, Southern

Oceans, and the Gulf of Mexico, each spanning a depth range of 1,000–4,000 m (Fig. 5.9 and Table 5.1). Most datasets represented single transects across a continental margin. The Deep Gulf of Mexico Benthos program (Rowe & Kennicutt 2008) sampled multiple transects to provide data at a regional scale, and Stuart & Rex (2009) compiled diversity values for Atlantic gastropods at the ocean basin scale. Each diversity value was computed from a single quantitative sample.

Once the ubiquity of the diversity–depth relation had been tested, we set out to determine its underlying cause(s). Three environmental variables were tested as significant correlates of diversity: organic carbon flux to the seafloor, temperature, and oxygen. The choice of these variables is based on theoretical considerations.

The relation between diversity and productivity is of central interest to ecologists in many ecosystems including the deep sea (Mittelbach *et al.* 2001; Stuart & Rex 2009). Past observations and models drawn primarily from the plant ecology literature suggest that a unimodal diversity–productivity relation is ubiquitous (Rosenzweig 1992; Huston & Deangelis 1994). Recent reviews, however, showed that other types of relations without definite maxima (linear increase or decrease) can be equally as common (Mittelbach *et al.* 2001). For our analysis, estimates for surface primary productivity were obtained

from the model of Behrenfeld & Falkowski (1997). The organic carbon that fuels deep-sea benthic communities is only a fraction of surface primary productivity, and it decreases exponentially with depth. Therefore, we used the empirical function given by Berger *et al.* (1987) to estimate organic carbon fluxes to seafloor from surface primary productivity values.

The metabolic theory of ecology provides a theoretical background for diversity-temperature trends and predicts an exponential increase in species richness with increasing temperature (Allen *et al.* 2002; Brown *et al.* 2004). The prediction has been both confirmed and contradicted by observations along latitudinal and longitudinal gradients (Allen *et al.*, 2002; Brown *et al.* 2004; Hawkins *et al.* 2007). Along a depth transect off Shetland Islands in the northeastern Atlantic, Narayanaswamy *et al.* (2005) found a peak in polychaete diversity at the boundary between two water masses. The relation between diversity and temperature was thus parabolic, though the maximum in diversity was explained by the high temperature range rather than temperature per se. For our broad-scale analysis, we did not have bottom temperature data for many of the sampling locations that we wished to include in the analysis. We, therefore, extracted data from the World Ocean Atlas 2005 for each sampling location. Because bottom temperatures in the Atlas are reported in 1° × 1° cells,

Fig. 5.9

Map of the location of datasets used in analyses of depth–diversity trends; see Table 5.1 for a description of the datasets.

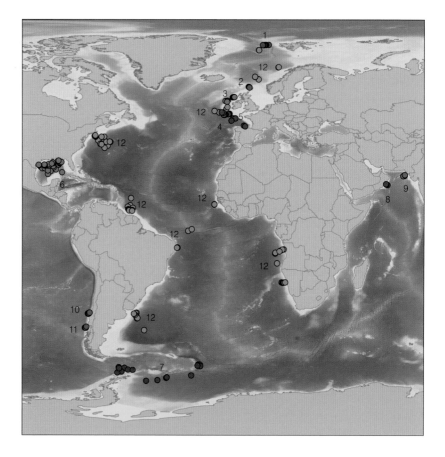

Table 5.1

Description of datasets used in the analyses of depth–diversity trends. Location numbering refers to numbers on the map (Fig. 5.9).

Location	Depth range (m)	Taxa	Sampling device	Diversity index	Reference
1. Arctic Ocean	1246–4200	Nematoda		ES(51)	E. Hoste, T. Soltwedel & A. Vanreusel, unpublished data
2. Faroe–Shetland channel	150–1000	Polychaeta	Megacorer, USNEL box-corer, Day grab	ES(26)	Narayanaswamy *et al.* 2005
3. Rockall Trough	401–2900	Polychaeta	USNEL box corer	ES(51)	Paterson & Lambshead 1995
4. Goban Spur	208–4470	Macrofaunal taxa	Circular box-corer	H′ (log *e*)	Flach & De Bruin 1999
5. Bay of Biscay	2035–4650	Echinodermata	Beam trawl	ES(57)	Sibuet 1987
6. Gulf of Mexico	213–3670	Polychaeta Bivalvia Isopoda	Box corer	ES(20) ES(20) ES(10)	Rowe & Kennicutt 2008
7. Southern Ocean	774–4976	Isopoda	Epibenthic sledge	H′ (log)	Brandt *et al.* 2004
8. Oman (OMZ)	400–3300	Macrofaunal taxa	Box corer	ES(100)	Levin *et al.* 2000
9. Pakistan (OMZ)	140–1850	Polychaeta Foraminifera	Megacorer Multicorer	ES(100) ES(100)	Hughes *et al.* 2009 Gooday *et al.* 2009
10. Off Conception (OMZ)	365–2060	Macrofaunal taxa	Multiple box-corer	ES(10)	Palma *et al.* 2005
11. Off Chiloe	160–1961	Macrofaunal taxa	Multiple box-corer	ES(10)	Palma *et al.* 2005
12. Atlantic Ocean	119–5216	Gastropoda	Epibenthic sledge	ES(50)	Stuart & Rex 2009

we selected the deepest nearby value as an estimate of temperature near the seafloor.

The effect of low oxygen on diversity is obvious in areas where hypoxia imposes a physiological stress on metazoans, such as in OMZs. Oxygen may have a major impact on deep-sea biodiversity below a threshold value of $0.45\,ml^{-1}$ (Levin & Gage 1998). We therefore predicted a positive, logarithmic relation between diversity and oxygen, and, to test this prediction, we extracted oxygen values from the World Ocean Atlas 2005 as described for temperature.

5.3.2.2 Observed depth–diversity trends and potential drivers

Plots of diversity as a function of depth for each dataset show that in most cases they were significantly correlated and, except in the Goban Spur region, a parabolic curve provided the best data fit (Table 5.2 and Fig. 5.10). Deviations from a common pattern of mid-slope depth maximum in diversity are particularly striking for oxygen minimum zones. Diversity and depth were not correlated off the coasts of Conception, Oman, and Pakistan. The parabolic

curves were actually inverted from the predicted pattern. Foraminiferans and polychaetes in these cases showed a diversity minimum at mid-slope depth, which coincided with the core of the OMZ (Gooday *et al.* 2009; Hughes *et al.* 2009). The unimodal relation between diversity and depth held true at a regional scale in the Gulf of Mexico for polychaetes, isopods, and bivalves but not at an ocean scale for gastropods. The discrepancy in depth–diversity trends at different spatial scales may simply reflect the fact that depth, or pressure, is not the main factor driving the pattern but rather provides a natural interaction term for many environmental variables that co-vary with depth among but not across regions (Levin *et al.* 2001; Stuart & Rex 2009). The fact that multiple environmental variables and depth are often strongly correlated raises the problem of multicollinearity in multiple regression models (Graham 2003). To limit potential artifacts, we correlated diversity with organic carbon fluxes, temperature, and oxygen at regional to ocean scales only (Table 5.3). For polychaetes, bivalves, and isopods in the Gulf of Mexico, as well as gastropods in the Atlantic Ocean, the diversity–productivity trend is unimodal and peaks at *ca.* 10–$15\,gC\,m^{-2}\,year^{-1}$, but it only explains a small portion of the

Table 5.2

Adjusted R^2 for linear, log-linear (ln) and quadratic regression models of diversity as a function of depth. NS, not significant.

Datasets	Diversity = f(depth)	Diversity = f(ln(depth))	Diversity = f(depth²)
1. Arctic Ocean	0.44***	0.35**	0.56***
2. Faroe–Shetland	0.42***	0.32***	0.42***
3. Rockall Trough	NS	NS	0.60***
4. Goban Spur	NS	0.45*	NS
5. Bay of Biscay	NS	NS	NS
6. Gulf of Mexico: Polychaeta	0.22***	0.12**	0.31***
6. Gulf of Mexico: Bivalvia	NS	NS	0.33***
6. Gulf of Mexico: Isopoda	NS	NS	0.36***
7. Southern Ocean	NS	NS	0.34*
8. Oman	NS	NS	NS
9. Pakistan: Foraminifera	0.34*	NS	0.41*
9. Pakistan:Polychaeta	NS	NS	0.93***
10. Off Conception	NS	NS	NS
11. Off Chiloe	NS	NS	NS
12. Atlantic Ocean	NS	NS	NS

*$p < 0.05$;
**$p < 0.01$;
***$p < 0.001$.

variance (Fig. 5.11 and Table 5.3). Diversity values are especially variable at the lower end of the productivity gradient. Interestingly, temperature was the best predictor of diversity at a global scale. The parabolic relation is heavily driven by low diversity values at temperatures below 0 °C in the Norwegian Sea (Fig. 5.12A); however, the depressed diversity in the deep Norwegian basin may be driven by isolation from the deep Atlantic that has slowed recolonization following the last glaciation and catastrophic slides 6,000–8,000 years ago (Bouchet & Warén 1979; Rex *et al.* 2005). When data from the Norwegian Sea are removed from the analysis, the parabolic relation between diversity and temperature is still statistically significant (Fig. 5.12B), but explains a much smaller portion of variance ($p = 0.02$, adjusted $R^2 = 0.08$). This type of collinearity and confounding among ecological, historical, and biogeographic drivers underlines interplay between local and regional processes that may prove the most difficult to control in quantitative models of large-scale diversity patterns.

5.3.2.3   Limitations and prospects for local and global analyses

As for other large-scale biodiversity patterns on Earth, deciphering depth–diversity trends may provide clues on which mechanisms structure biological communities, but patterns and processes are challenging to establish. The shape of the pattern is better described at local to regional scales because depth provides a natural interaction term for processes that act on benthic diversity. Although not the only pattern observed, a unimodal pattern is the most common outcome. Nonetheless, the collinearity of factors reduces the sensitivity of analyses designed to separate and quantify true predictors against confounding and potentially irrelevant variables. For this reason, meta-analyses at global scale might prove more useful. The single ocean-scale dataset included in this analysis suggests that the diversity of gastropods may be partly explained by geographic variations in food supply and temperature. It should be noted, however, that in addition to multicollinearity, there is also

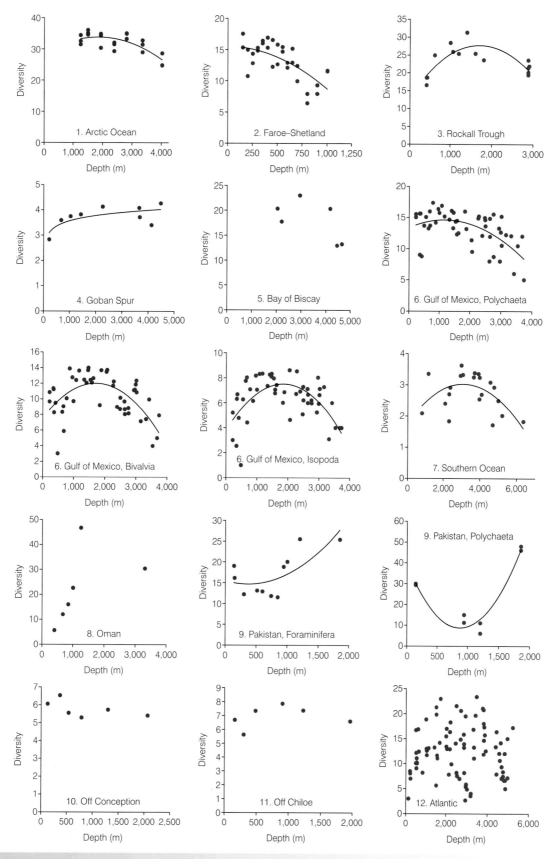

Fig. 5.10

Plots of diversity as a function of depth and best-fitted regression models when statistically significant; see Table 5.2 for results of the regression models.

Table 5.3

Adjusted R^2 for linear, loglinear (ln), exponential (exp), or quadratic regression models of diversity as a function of POC fluxes, oxygen, and temperature on the seafloor. NS, not significant.

Datasets	Diversity = f(POC flux)	Diversity = f(POC flux^2)	Diversity = f(O$_2$)	Diversity = f(ln(O$_2$))	Diversity = f(T)	Diversity = f(exp(T))	Diversity = f(T^2)	Diversity = f(POCflux^2 + ln(O$_2$)+T^2)
6. Gulf of Mexico – Polychaeta	$R^2 = 0.06*$	$R^2 = 0.14**$	NS	NS	NS	NS	NS	NS
6. Gulf of Mexico: Bivalvia		$R^2 = 0.20**$	NS	NS	NS	NS		0.18*
6. Gulf of Mexico: Isopoda	$R^2 = 0.20**$	$R^2 = 0.20**$	NS	NS	NS	NS	0.11*	0.43***
12. Atlantic Ocean	$R^2 = 0.05*$	$R^2 = 0.05*$	NS	NS	NS	0.06*	0.26***	0.28***

* $p < 0.05$;
** $p < 0.01$;
*** $p < 0.001$.

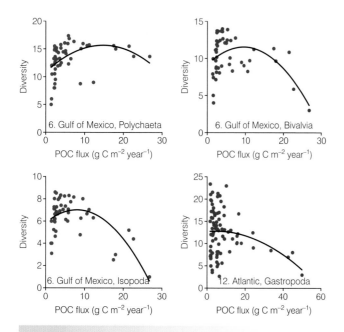

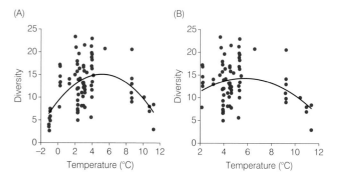

Fig. 5.12

Plots of gastropod diversity in the Atlantic Ocean as a function of temperature and quadratic regression model **(A)** with data from the Norwegian Sea, **(B)** without data from the Norwegian Sea. See text for details.

Fig. 5.11

Plots of diversity as a function of particulate organic carbon fluxes and best-fitted regression models when statistically significant; see Table 5.3 for results of the regression models.

concern that the spatial and temporal resolution of available environmental data may limit the utility of these meta-analyses. First, in the World Ocean Atlas, data are reported at grid cells of 1° × 1°. At this scale, a steep continental slope at a given locale is only spanned by one or at best two cells, thus potentially masking any down-slope gradient in oxygen or temperature. Second, whereas the spatial resolution of surface primary productivity is much better (*ca.*

20 km), the size and position of the benthic catchment areas that contribute food to particular benthic sites are unknown (Stuart & Rex 2009). Furthermore, lateral inputs of food by cross-slope currents are largely unknown and, therefore, ignored. Finally, the environmental data considered here represent mean annual values over one or two years. Seasonal, annual, or inter-annual variation is not considered. The advent of ocean biodiversity informatics (Costello & Vanden Berghe 2006), coupled with higher resolution global ocean models (see, for example, Chassignet *et al.* 2006) as well as refined estimates of particulate organic carbon fluxes to the seafloor (see, for example, Seiter *et al.* 2005) will eventually improve these first predictions on the extent to which these contemporary ecological processes drive diversity patterns.

5.4 Human Influence on Continental Margin Ecosystems

Fifty years ago, as the study of margin ecology began the transition from a descriptive low-resolution to a process-oriented, high-resolution endeavor, there were few recognized environmental threats to the deep sea. In contrast, today, many COMARGE participants are currently actively advising on the environmental management of the deep. Although deep-sea ecology has advanced significantly in the past 10 years, there is much that we still do not know about this vast realm. A great need remains to inform resource developers, policy makers, and even scientists about the diverse life and ecosystems in the deep once thought to be azoic. Deep-sea exploitation is being planned and pursued in the absence of adequate environmental knowledge to ensure protection.

5.4.1 Direct human intrusions in the deep-sea

The impacts of human land-based activities have generally been ignored in the deep sea, but continental slope environments are often closely coupled to events on land. For instance, the composition of sediments in deep water on continental margins is affected directly by drainage areas on land (see, for example, Soh 2003). Toxic wastes such as dichlorodiphenyltrichloroethane (DDT) and polychlorinated biphenyls (PCBs) may eventually concentrate into tissues of deep-sea organisms (Arima *et al.* 1979; Solé *et al.* 2001; Thiel 2003; Storelli *et al.* 2007) after accumulating in coastal deposits (Managaki & Takada 2005) that are transported to the deep sea through canyons and down slopes. Continental margins are, thus, depositional centers for pollutants produced by human activities. Additional contaminants in the deep sea include fluorescent whitening agents (FWAs) that have been detected in high concentrations in sediments at 1,450 m in Sagami Bay (Managaki & Takada 2005), industrial lead in canyon systems (Richter *et al.* 2009), and trace metals in deep-sea canyon sediment cores that have been correlated with metal inputs associated with industrial activities (Palanques *et al.* 2008).

Demersal fishing has significant impacts on deep-sea ecosystems. The depletion of shallow water fisheries has progressively pushed fishing activity into deeper water (Davies et al. 2007; Koslow et al. 2000). Compared with their shallow-water counterparts, the growth rates of several commercially valuable deep-sea fish species are much slower, they mature at a late age, their fecundity is much lower, and they exhibit extreme longevity. Collectively, these traits make them extremely sensitive to sustained fishery (Gordon 2005). Because many deep-water fishes,

both target and by-catch species, migrate over a wide depth range, demersal trawling affects not only the actual depths fished (*ca.* 1,600 m) but also extends into deeper waters up to at least 2,500 m (Bailey *et al.* 2009). Individual fishing vessels can impact a 100 km² area of the seabed in a single 10-day fishing expedition (Hall-Spencer *et al.*, 2002). Demersal fishing has a particularly significant effect on epifaunal coral and sponge habitats (Freiwald *et al.* 2004; Wheeler *et al.* 2005; Davies *et al.* 2007; Reed *et al.* 2007; UNEP 2007).

The mining of massive polymetallic sulfides and cobalt-rich ferromanganese crusts is becoming increasingly attractive from an economic perspective. Although many potential mineral reserves occur on distant mid-ocean ridges, some of the largest deposits, and those that might be developed first, occur in back-arc rifts and rocky terrain in continental slope settings (ISA 2002). These deposits can be rich in gold, copper, zinc, and silver. The demand for metals from developing nations is expected to rise, raising concerns about the environmental management of near-future mining operations (Halfar & Fujita 2008).

The extraction of hydrocarbons from the deep seafloor will have predictable impacts. Drill cutting spoil may smother organisms, cause organic enrichment, and release toxic chemicals (Currie & Isaac 2005; Jones *et al.* 2006, 2007; Santos *et al.* 2009). The effects of exposure to contaminants on deep-sea species is poorly known (Skadsheim *et al.* 2005). Drill cuttings may pose a greater local environmental hazard in the deep sea than in shallow water because recovery rates will be lower (Glover & Smith 2003).

Other commercial developments may have a profound effect on deep-sea ecosystems. These include the subseabed storage of CO_2, water column and sediment surface CO_2 disposal, the use of methane hydrates, submarine mine tailing disposal, bioprospecting and artificial iron fertilization of surface waters (Glover & Smith 2003; UNEP 2007). Experimental studies on the effects of CO_2 on benthic organisms have been equivocal (Bernhard *et al.* 2009; Ricketts *et al.* 2009; Sedlack *et al.* 2009). The disposal of mine waste from land into the deep sea, in some cases down canyon systems, has received surprisingly little study (Ellis 2001), although in shallow water the activity has significant smothering and trace metal effects on fauna.

5.4.2 Climate change

The increasing level of CO_2 in the atmosphere over the past century has created a cascading set of effects on ocean environments. Increased atmospheric CO_2 can act as a fertilizer or as a greenhouse gas that warms and increasingly stratifies and acidifies the ocean. Concomitant effects of warming include reduced turnover of oxygen and destabilization of gas (methane) hydrates. Elevated CO_2 in seawater lowers pH and leads to dissolution of carbonate. With its natural steep gradients in water temperature, salinity, density, and oxygen, the continental margins may exhibit the most pronounced effects of climate change.

5.4.2.1 Ocean warming

Global ocean temperature has risen by 0.1 °C from 1961 to 2003 and may rise by another 0.5 °C over the next century (IPCC 2007). Ocean warming leads to melting of polar ice caps and increased density stratification. Stratification of the world's oceans is increasing, with the greatest change in the North Pacific. Consequences include decreased surface productivity and organic matter flux to the seabed, as well as reduced mixing of oxygen from surface waters to the interior. Reduced mixing of oxygen is partly responsible for the expansion of OMZs in the world's tropical oceans (Stramma *et al.* 2008). The expansion of OMZs will lead to habitat compression in species of plankton and fishes and strong changes in species habitat availability on the upper margins. The stratification of the upper ocean and the increase in the extent of OMZs further alter the flux of materials to the seabed and, thus, impact benthic boundary layer communities indirectly.

Gas hydrates are present along many of the world's most productive continental margins. The stability of gas hydrate is determined by temperature and pressure (Hester & Brewer, 2009). Warming will decrease stability and may lead to massive dissolution and release of methane into the ocean, although much of this methane is likely to be oxidized by bacteria. Large-scale methane release events are considered to have modified past climate conditions (the clathrate gun hypothesis (Kennett *et al.* 2003)). Such releases could trigger slope instabilities but also might increase resource availability for slope chemosynthetic communities.

Additional effects of warming may include range expansions and contractions. Examples include the Humbolt (jumbo) squid (*Dosidicus gigas*), which follows low oxygen waters northward from Mexico in the eastern Pacific and are now routinely found off Oregon, Washington, and Alaska (Zeidberg & Robison 2007). Increasing temperatures may lead species to seek cooler refugia in canyons or down slope. The steep topography of some margins could facilitate range shifts in response to changes in temperature and oxygenation, given that only small distances are required to migrate to suitable conditions.

Less predictable, but also a likely effect of increasing surface temperatures, is a weakening or shifting of ocean circulation (Toggweiler & Russell 2008) leading to different connectivity between populations and, thus, faunal distributions. Certainly, long-term monitoring of deep-sea benthic communities has shown distinct changes in the population size of benthic Foraminifera in relation to changes in organic flux. For instance, in Sagami Bay during El Niño years between 1996 and 1998 average abundance decreased from 4,000 individuals per 10 cm^2 to 2,000 individuals per 10 cm^2 (Kitazato *et al.* 2003).

5.4.2.2 Ocean acidification on margins

Although the ocean is seen as a sink for atmospheric CO_2 that has absorbed one-third of the anthropogenic CO_2 added to the atmosphere, the ocean carbonate system is slowly being disrupted by this uptake. Enhanced pCO_2 has decreased pH in the water column by 0.1 unit since 1750 (IPCC 2007). This "acidification" effect leads to a lowered calcium carbonate saturation state and is particularly severe in colder waters. On northeast Pacific margins, some of this "corrosive" water is upwelled onto the shelf (Feely *et al.* 2008). Deep-water coral reefs are highly susceptible to changes in carbonate saturation because their aragonite skeletons are a particularly soluble form of $CaCO_3$ (Davies *et al.* 2007). Coral distributions already reflect the acidic conditions in the north Pacific (Guinotte *et al.* 2006), but the entire ecosystem could be threatened by acidification (Turley *et al.* 2007). Similarly the magnesium calcite in skeletons of echinoderms represents another easily dissolved form of carbonate. As acidification continues to worsen, mollusks, foraminiferans, and other calcifying taxa could also be affected. Notably, low oxygen waters are also low in pH since the respiration processes that deplete oxygen also elevate CO_2. Thus, expansion of hypoxic waters will also bring the stresses associated with acidification.

5.5 Achievements and Perspectives: The Unknown Aspects of Margin Biodiversity

Our new, higher-resolution view of the continental margins reveals ecosystems that are unique on Earth. Because continental margins are habitat ribbons that stretch almost continuously along all continents, a diverse array of environmental forcing factors occurs over very small distances. Not so surprisingly, they might harbor among the most diverse faunal communities on Earth. Continental margins cross wide latitudinal and longitudinal swaths, and, when considered in tandem with adjacent trenches, represent the highest variations in depth on Earth, thus crossing gradients in pressure, temperature, oxygen, currents, and food inputs. Moreover, these large-scale gradients are superimposed over a wide range of heterogeneous habitats such as cold seeps, cold corals, canyons, and OMZs, often with their own sets of environmental drivers. These hot spots provide exceptions to the rule of a detritus-based, muddy ecosystem and undoubtedly enhance biodiversity at regional scale. They also provide natural experiments for future study to decipher the processes underlying species distributions and diversity patterns across habitats. It is this integrated approach of continental margin ecosystems, both in terms of fauna, environmental drivers, and habitats that COMARGE has fostered. This is also the approach needed to describe, predict, and mitigate changes due to human activities.

Through COMARGE, its current participants and future marine ecologists are better prepared to increase knowledge of the system and meet the demands of science-based ocean management. We recognize four critical unknowns that could become knowns, and one major habitat, the trenches, that has been disregarded but is now within sight of new underwater vehicles (Jamieson *et al.* 2010). The first practical unknown is consistent and correct species identification. Some progress is being made toward synthesizing what is known. Species name compilation accelerated toward the end of the twentieth century through today, with most effort being directed toward WoRMS (World Register of Marine Species: www.marinespecies.org). In many cases, however, comparisons across margin datasets remain impossible, severely limiting integration across habitats and regions. This situation is caused by the taxonomic challenges of the high species richness of the margin combined with declining taxonomic expertise. In addition, the vast geographical coverage of continental margins and the difficulty in accessing them has limited sampling coverage. A large part of the margin taxa is new to science and remains undescribed. For example, surveys of the Australian margins have shown that 90% of 365 species of eastern slope isopods (Poore *et al.* 1994) and 30% of 524 western slope decapods (Poore *et al.* 2008) are undescribed. Research expeditions on the continental margin rarely target biodiversity studies or taxonomy specifically, especially as political considerations are increasingly important in funding decisions. The inventory of margin species remains far from complete.

A second key unknown is related to the first. The continental margin is among the least understood of the marine realms biogeographically. Horizontal or geographical distribution patterns are much less understood than depth-related patterns. Surprisingly little information is available about physical and environmental indicators and species composition on which to base biogeographic units on the continental slope. The Census data compilation project, OBIS, goes some way to overcoming this limitation. The Global Open Oceans and Deep Seabed (GOODS) biogeographic classification recognizes 14 lower bathyal provinces defined in terms of geographic or ocean current boundaries (within the depth range 800–3,000 m) on continental slopes and oceanic ridges (Fig. 7 in UNESCO 2009): Arctic, Northern North Atlantic, Northern North Pacific, North Atlantic, Southeast Pacific Ridges, New Zealand-Kermadec, Cocos Plate, Nazca Plate, Antarctic, Subantarctic, Indian, West Pacific, South Atlantic, and North Pacific. The GOODS biogeographic classification did not address upper bathyal depths of the slope (200–800 m), off-shelf areas within marginal seas, and

semi-enclosed ocean basins. The UNESCO report acknowledged the need for more species distribution data, improvement of the scientific basis for biogeographic classification, and greater integration of biodiversity data and independent datasets. The question of whether slope species are concentrated in biodiversity "hot spots" remains largely unanswered, but a correlation between species richness and habitat complexity is expected. Biogeographic studies of slope specialist taxa that incorporate an evolutionary component are rare. The evolutionary–historical legacy remains virtually unknown though there is increasing evidence that it may play an important role in present-day species distribution and diversity patterns. Evidence is emerging from taxonomic syntheses, some supported by COMARGE, of some taxa, such as bathyal squat lobsters (Machordom & Macpherson 2004), deep-water scleractinian corals (Cairns 2007), and nematodes (A. Vanreusel *et al.* unpublished data) that the centre of species richness for bathyal faunas is the same as that for shallow water, namely the Indo-West Pacific "coral triangle".

A third unknown is how bathyal communities change over ecological timescales. This need is particularly relevant to the assessment of human impacts on continental margins, noting that natural and anthropogenic signals cannot be dissociated. Therefore, there is a strong need for the development of observatory tools for continuous monitoring of bathyal environments and their biological communities.

A fourth unknown is the resilience of margin ecosystem functioning when impacted by natural and anthropogenic perturbations. Ecosystem functioning and dynamics in particular need to be addressed at population-community and ecosystem levels (Levin & Dayton 2009). A positive exponential relation between diversity and ecosystem functioning has recently been shown for deep-sea nematodes (Danovaro *et al.* 2008). So far, such a diversity–function relation is unique (Loreau 2008) and suggests that ecosystem functioning might quickly collapse in response to biodiversity loss in the deep sea. Further investigations and experiments are needed to corroborate these findings for various habitats and taxa in natural and stressed communities. They also stressed the need for an understanding of recovery processes after a system has collapsed, including the larval dispersal, supply, settlement, and recruitment processes that control the connectivity between populations.

To meet these and additional emerging needs, there must be more sampling undertaken in the context of optimal biogeographic design. There must be increased support of taxonomy including informatics and new technologies. Access to the ocean via HOV (human occupied vehicle), ROV, AUV, and cables must be increased on all margins. And, there must be more effective use of experimentation to understand the causes of patterns. International collaborations as those initiated by COMARGE and an increased partnership with the industry are desirable to achieve these goals.

A conclusion to COMARGE is yet premature. In just four years the project has underscored the complex interplay of large oceanographic features and habitat heterogeneities, acting at a hierarchy of spatial scales to shape distribution and diversity patterns. COMARGE has fostered a holistic understanding of continental margin ecosystems, enhancing international collaborations and increasing the standardization of methodologies that now translates in sampling schemes and cruises specifically addressing its overarching question. Beyond continental margins, the five Census deep-sea projects (Abyssal Plains (CeDAMar), Seamounts (CenSeam), Vents and Seeps (ChEss), Continental Margins (COMARGE), Mid-ocean Ridges (MAR-ECO); see Introduction for definitions) initiated a large data-mining endeavor in the framework of the synthesis project SYNDEEP (Towards a First Global Synthesis of Biodiversity, Biogeography, and Ecosystem Function in the Deep Sea) to tackle similar issues across all deep-sea habitats. In the years to come, these initiatives will further explore the cumulative and synergetic effects of species turnover among and across habitats to provide new insights on diversity maintenance and ecosystem functioning in the deep sea. Meanwhile, ongoing global analyses on the distribution of ecologically significant taxonomic groups will provide clues on the biogeography and phylogeography of the margin fauna thus placing contemporary diversity patterns in an historical context. A holistic view of continental margin ecosystems also has to consider anthropogenic impacts. By 2010, COMARGE will provide the first comprehensive map of the footprint of human activities on the northeast Atlantic margins. Altogether, a better understanding of habitat heterogeneity and assessment of human impacts will be brought to bear on management and conservation efforts on continental margins.

Acknowledgments

We gratefully acknowledge the support given by the Census of Marine Life, the Alfred P. Sloan Foundation, and the Total Foundation. The COMARGE project was managed from the Institut Oceanographic in Paris (coordination, secretariat, education, and outreach), the Louisiana State University (finances), and Ifremer (database and website). All three institutes are gratefully acknowledged for their contributions. Last but not least we thank all the scientists, students, institutes, and programs that have contributed to feed COMARGE syntheses through cruise funding, activities at sea, data, and expertise.

References

Allen, A.P., Brown, J.H. & Gillooly, J.F. (2002) Global biodiversity, biochemical kinetics, and the energetic-equivalence rule. *Science* 297, 1545–1548.

Arima, S., Marchand, M. & Martin, J.-L.M. (1979) Pollutants in deep-sea organisms and sediments. *Ambio Special Report* 6, 97–100.

Baba, K., Macpherson, E., Poore, G.C.B., *et al.* (2008) Catalogue of squat lobsters of the world (Crustacea: Decapoda: Anomura – families Chirostylidae, Galatheidae and Kiwaidae). *Zootaxa* 1905, 1–220.

Baguley, J.G., Montagna, P.A., Hyde, L.J. & Rowe, G.T. (2008) Metazoan meiofauna biomass, grazing, and weight-dependent respiration in the Northern Gulf of Mexico deep sea. *Deep-Sea Research II* 55, 2607–2616.

Bailey, D.M., Collins, M.A., Gordon, J.D.M., *et al.* (2009) Long-term changes in deep-water fish populations in the northeast Atlantic: a deeper reaching effect of fisheries? *Proceedings of the Royal Society of London B* 276, 1965–1969.

Behrenfeld, M.J. & Falkowski, P.G. (1997) Photosynthetic rates derived from satellite-based chlorophyll concentration. *Limnology and Oceanography* 42, 1–20.

Berger, W.H., Fischer, K., Lai, C. & Wu, G. (1987) Ocean productivity and organic carbon flux. I. Overview and maps of primary production and export production. San Diego, University of California.

Bernhard, J.M., Barry, J.P., Buck, K.R. & Starczak, V.R. (2009) Impact of intentionally injected carbon dioxide hydrate on deep-sea benthic foraminiferal survival. *Global Change Biology* 289, 197–22.

Bertics, V.J. & Ziebis, W. (2009) Biodiversity of benthic microbial communities in bioturbated coastal sediments is controlled by geochemical microniches. *ISME Journal* 3, 1269–1285.

Bouchet, P. & Warén, A. (1979) The abyssal molluscan fauna of the Norwegian Sea and its relation to others faunas. *Sarsia* 64, 211–243.

Brand, G.L., Horak, R.V., Bris, N.L., *et al.* (2007) Hypotaurine and thiotaurine as indicators of sulfide exposure in bivalves and vestimentiferans from hydrothermal vents and cold seeps. *Marine Ecology* 28, 208–218.

Brandt, A., Brökeland, W., Brix, S. & Malyutina, M. (2004) Diversity of Southern Ocean deep-sea Isopoda (Crustacea, Malacostraca) – a comparison with shelf data. *Deep-Sea Research II* 51, 1753–1768.

Brown, J.H., Gillooly, J.F., Allen, A.P., *et al.* (2004) Toward a Metabolic Theory of Ecology. *Ecology* 85, 1771–1789.

Cairns, S.D. (2007) Deep-water corals: an overview with special reference to diversity and distribution of deep-water scleractinian corals. *Bulletin of Marine Science* 81, 311–322.

Carney, R.S. (1997) Basing conservation policies for the deep-sea floor on current diversity concepts: a consideration of rarity. *Biodiversity and Conservation* 6, 1463–1485.

Carney, R.S. (2005) Zonation of deep biota on continental margins. *Oceanography and Marine Biology* 43, 211–278.

Carney, R.S., Haedrich, R.L. & Rowe, G.T. (1983) Zonation of fauna in the deep sea. In: *The Sea* (ed. G.T. Rowe), pp. 371–398. New York: John Wiley & Sons.

Chassignet, E.P., Hurlburt, H.E., Martin Smedstat, O., *et al.* (2006) Ocean prediction with the Hybrid Coordinate Ocean Model (HYCOM). In: *Ocean Weather Forecasting: An Integrated View of Oceanography* (eds. E.P. Chassignet & J. Verron), pp. 413–426. Springer.

Cordes, E.E., Bergquist, D.C. & Fisher, C.R. (2009) Macro-ecology of Gulf of Mexico cold seeps. *Annual Review of Marine Science* 1, 143–168.

Cordes, E.E., Ribeiro Da Cunha, M., Galéron, J., *et al.* (2010) The influence of geological, geochemical, and biogenic habitat heterogeneity on seep biodiversity. *Marine Ecology* 31, 51–65.

Corliss, B.H., Brown, C.W., Sun, X. & Showers, W.J. (2009) Deep-sea benthic diversity linked to seasonality of pelagic productivity. *Deep-Sea Research I* 56, 835–841.

Cosson-Sarradin, N., Sibuet, M., Paterson, G.L.J. & Vangriesheim, A. (1998) Polychaete diversity at tropical deep-sea sites: Environmental effects. *Marine Ecology Progress Series* **165**, 173–185.

Costello, M.J. & Vanden Berghe, E. (2006) 'Ocean biodiversity informatics': a new era in marine biology research and management. *Marine Ecology Progress Series* **316**, 203–214.

Currie, D.R. & Isaac, L.R. (2005) Impact of exploratory offshore drilling on benthic communities in the Minerva gas field, Port Campbell, Australia. *Marine Environmental Research* **59**, 217–233.

Danovaro, R., Gambi, C., Dell'anno, A., *et al.* (2008) Exponential decline of deep-sea ecosystem functioning linked to benthic biodiversity loss. *Current Biology* **18**, 1–8.

Dattagupta, S., Martin, J., Liao, S.-M., *et al.* (2007) Deep-sea hydrocarbon seep gastropod *Bathynerita naticoidea* responds to cues from the habitat-providing mussel *Bathymodiolus childressi*. *Marine Ecology* **28**, 193–198.

Davies, A.J., Roberts, J.M. & Hall-Spencer, J. (2007) Preserving deep-sea natural heritage: Emerging issues in offshore conservation and management. *Biological Conservation* **138**, 299–312.

De Mello E Sousa, S.H., Passos, R.F., Fukumoto, M., *et al.* (2006) Mid-lower bathyal benthic foraminifera of the Campos Basin, Southeastern Brazilian margin: biotopes and controlling ecological factors. *Marine Micropaleontology* **61**, 40–57.

Ellis, D.V. (2001) A review of some environmental issues affecting marine mining. *Marine Georesources and Geotechnology* **19**, 51–63.

Fabri, M.-C., Galeron, J., Larour, M. & Maudire, G. (2006) Combining the Biocean database for deep-sea benthic data with the online Ocean Biogeographic Information System. *Marine Ecology Progress Series* **316**, 215–224.

Feely, R.A., Sabine, C.L., Hernandez-Ayon, J.M., *et al.* (2008) Evidence for Upwelling of Corrosive "Acidified" Water onto the Continental Shelf. *Science* **320**, 1490–1492.

Flach, E. & De Bruin, W. (1999) Diversity patterns in macrobenthos across a continental slope in the NE Atlantic. *Journal of Sea Research* **42**, 303–323.

Forbes, E. (1844) Report on the Mollusca and Radiata of the Aegean Sea, and on their distribution, considered as bearing on geology. Report of the British Association for the Advancement of Science for 1843, 129–193.

Freiwald, A. (2002) Reef-forming cold-water corals. In: *Ocean Margin Systems* (eds. G. Wefer, D.S.M. Billett, D. Hebbeln, *et al.*), pp. 365–385. Berlin and Heildenberg: Springer.

Freiwald, A., Fossa, J.H., Grehan, A., *et al.* (2004) *Cold Water Coral Reefs – Out of Sight No Longer Out of Mind.* Cambridge, UK: United Nations Environment Program, World Conservation Monitoring Centre.

Fujiwara, Y., Kawato, M., Yamamoto, T., *et al.* (2007) Three-year investigations into sperm whale-fall ecosystems in Japan. *Marine Ecology* **28**, 219–232.

Glover, A.G. & Smith, C.R. (2003) The deep-sea floor: current status and prospects of anthropogenic change by the year 2025. *Environmental Conservation* **30**, 219–241.

Gooday, A.J., Bett, B., Escobar-Briones, E., *et al.* (2010) Habitat heterogeneity and its influence on benthic biodiversity in oxygen minimum zones. *Marine Ecology* **31**, 125–147.

Gooday, A.J., Levin, L.A., Aranda Da Silva, A., *et al.* (2009) Faunal responses to oxygen gradients on the Pakistan margin: A comparison of foraminiferans, macrofauna and megafauna. *Deep-Sea Research II* **56**, 488–502.

Gordon, J.D.M. (2005) Environmental and biological aspects of deepwater demersal fishes. *FAO Fisheries Proceedings* **3**, 70–88.

Graham, M.H. (2003) Confronting multicollinearity in ecological multiple regression. *Ecology* **84**, 2809–2815.

Guinotte, J., Orr, J., Cairns, S., *et al.* (2006) Will human-induced changes in seawater chemistry alter the distribution of deep-sea scleractinian corals? *Frontiers in Ecology and the Environment* **4**, 141–146.

Halfar, J. & Fujita, R.M. (2008) Danger of deep-sea mining. *Science* **316**, 987.

Hall-Spencer, J., Allain, V. & Fosså, J.H. (2002) Trawling damage to Northeast Atlantic ancient coral reefs. *Proceedings of the Royal Society of London B* **269**, 507–511.

Hawkins, B.A., Albuquerque, F.S., Araujo, M.B., *et al.* (2007) A global evaluation of metabolic theory as an explanation for terrestrial species richness gradients. *Ecology* **88**, 1877–1888.

Hester, K.C. & Brewer, P.G. (2009) Clathrate hydrates in nature. *Annual Review of Marine Science* **1**, 303–327.

Hillebrand, H. (2004) On the generality of the latitudinal diversity gradient. *American Naturalist* **163**, 192–211.

Hughes, D.J., Lamont, P.A., Levin, L.A., *et al.* (2009) Macrofaunal communities and sediment structure across the Pakistan margin oxygen minimum zone, north-east Arabian Sea. *Deep-Sea Research II* **56**, 434–448.

Huston, M.A. & Deangelis, D.L. (1994) Competition and coexistence: the effects of resource transport and supply rates. *American Naturalist* **144**, 954–977.

IPCC (2007) Climate Change 2007: The Physical Science Basis. Contribution of Working Group I to the Fourth Assessment Report of the Intergovernmental Panel on Climate Change. (eds. S. Solomon, D. Qin, M. Manning, *et al.*). Cambridge and New York: Cambridge University Press.

ISA (2002) Polymetallic massive sulphides and cobalt-rich ferromanganese crusts: status and prospects. *ISA Technical Study No. 2.*

Jamieson, A.J., Fujii, T., Mayor, D.J., *et al.* (2010) Hadal trenches: the ecology of the deepest places on Earth. *Trends in Ecology & Evolution* **25**, 190–197.

Järnegren, J., Rapp, H.T. & Young, C.M. (2007) Similar reproductive cycles and life-history traits in congeneric limid bivalves with different modes of nutrition. *Marine Ecology* **28**, 183–192.

Jones, D.O.B., Hudson, I.R. & Bett, B.J. (2006) Effects of physical disturbance on the cold-water megafaunal communities of the Faroe-Shetland Channel. *Marine Ecology Progress Series* **319**, 43–54.

Jones, D.O.B., Wigham, B.D., Hudson, I.R. & Bett, B.J. (2007) Anthropogenic disturbance of deep-sea megabenthic assemblages: a study with remotely operated vehicles in the Faroe-Shetland Channel, NE Atlantic. *Marine Biology Research* **151**, 1731–1741.

Kennett, J.P., Cannariato, K.G., Hendy, I.L. & Behl, R.J. (2003) *Methane Hydrates in Quaternary Climate Change: The Clathrate Gun Hypothesis.* Washington, DC: American Geophysical Union.

Kitazato, H., Nakatsuka, T., Shimanaga, M., *et al.* (2003) Long-term monitoring of the sedimentary processes in the central part of Sagami Bay, Japan: rationale, logistics and overview of results. *Progress in Oceanography* **57**, 3–16.

Koslow, J.A., Boehlert, G.W., Gordon, J.D.M., *et al.* (2000) Continantal slope and deep-sea fisheries: implications for a fragile ecosystem. *ICES Journal of Marine Science* **57**, 548–557.

Laubier, L. & Sibuet, M. (1979) Ecology of the benthic communities of the deep N.E. *Atlantic. Ambio* **6**, 37–42.

Le Danois, E. (1948) *Les Profondeurs de la Mer.* Paris: Payot.

Leibold, M.A., Holyoak, M., Mouquet, N., *et al.* (2004) The metacommunity concept: a framework for multi-scale community ecology. *Ecology Letters* **7**, 601–613.

Levin, L.A. (2003) Oxygen minimum zone benthos: Adaptation and community response to hypoxia. *Oceanography and Marine Biology* **41**, 1–45.

Levin, L.A. & Dayton, P.K. (2009) Ecological theory and continental margins: where shallow meets deep. *Trends in Ecology and Evolution* **24**, 606–617.

Levin, L.A., Etter, R.J., Rex, M.A., *et al.* (2001) Environmental influences on regional deep-sea species diversity. *Annual Review of Ecology and Systematics* 32, 51–93.

Levin, L.A. & Gage, J.D. (1998) Relationships between oxygen, organic matter and the diversity of bathyal macrofauna. *Deep-Sea Research II* 45, 129–163.

Levin, L.A., Gage, J.D., Martin, C. & Lamont, P.A. (2000) Macrobenthic community structure within and beneath the oxygen minimum zone, NW Arabian Sea. *Deep-Sea Research II* 47, 189–226.

Levin, L.A. & Mendoza, G.F. (2007) Community structure and nutrition of deep methane-seep macrobenthos from the North Pacific (Aleutian) Margin and the Gulf of Mexico (Florida Escarpment). *Marine Ecology* 28, 131–151.

Levin, L.A., Mendoza, G.F., Gonzalez, J., *et al.* (2010a) Diversity of bathyal macrofauna on the northeastern Pacific margin: the influence of methane seeps and oxygen minimum zones. *Marine Ecology* 31, 94–110.

Levin, L.A., Sibuet, M., Gooday, A.J., *et al.* (2010b) The roles of habitat heterogeneity in generating and maintaining biodiversity on continental margins: an introduction. *Marine Ecology* 31, 1–5.

Loreau, M. (2008) Biodiversity and ecosystem functioning: the mystery of the deep sea. *Current Biology* 18, R126–R128.

Machordom, A. & Macpherson, E. (2004) Rapid radiation and cryptic speciation in squat lobsters of the genus *Munida* (Crustacea, Decapoda) and related genera in the South West Pacific: molecular and morphological evidence. *Molecular Phylogenetics and Evolution* 33, 259–279.

Macpherson, E. (2002) Large-scale species-richness gradients in the Atlantic Ocean. *Proceedings of the Royal Society of London B* 269, 1715–1720.

Managaki, S. & Takada, H. (2005) Fluorescent whitening agents in Tokyo Bay sediments: molecular evidence of lateral transport of land-derived particulate matter. *Marine Chemistry* 95, 113–127.

Menot, L., Galéron, J., Olu, K., *et al.* (2010) Spatial heterogeneity of macrofaunal communities in and near a giant pockmark area in the deep Gulf of Guinea. *Marine Ecology* 31, 78–93.

Mills, E. (1983) Problems of deep-sea biology: an historical perspective. In: *The Sea*, Vol. 8 (ed. G.T. Rowe), pp. 1–79. New York: John Wiley & Sons.

Mittelbach, G.G., Steiner, C.F., Scheiner, S.M., *et al.* (2001) What is the observed relationship between species richness and productivity? *Ecology* 82, 2381–2396.

Muller-Karger, F.E., Varela, R., Thunell, R., *et al.* (2005) The importance of continental margins in the global carbon cycle. *Geophys. Res. Lett.* 32, L01602.

Narayanaswamy, B.E., Bett, B.J. & Gage, J.D. (2005) Ecology of bathyal polychaete fauna at an Arctic-Atlantic boundary (Faroe-Shetland Channel, North-east Atlantic). *Marine Biology Research* 1, 20–32.

Olu-Le Roy, K., Caprais, J.C., Fifis, A., *et al.* (2007a) Cold-seep assemblages on a giant pockmark off West Africa: spatial patterns and environmental control. *Marine Ecology* 28, 115–130.

Olu-Le Roy, K., Cosel, R.V., Hourdez, S., *et al.* (2007b) Amphi-Atlantic cold-seep *Bathymodiolus* species complexes across the equatorial belt. *Deep-Sea Research I* 54, 1890–1911.

Pailleret, M., Haga, T., Petit, P., *et al.* (2007) Sunken wood from the Vanuatu Islands: identification of wood substrates and preliminary description of associated fauna. *Marine Ecology* 28, 233–241.

Palanques, A., Masqué, P., Puig, P., *et al.* (2008) Anthropogenic trace metals in the sedimentary record of the Llobregat continental shelf and adjacent Foix Submarine Canyon (northwestern Mediterranean). *Marine Geology* 248, 213–227.

Palma, M., Quiroga, E., Gallardo, V., *et al.* (2005) Macrobenthic animal assemblages of the continental margin off Chile (22° to 42°S). *Journal of the Marine Biological Association of the United Kingdom* 85, 233–245.

Paterson, G.L.J. & Lambshead, P.J.D. (1995) Bathymetric patterns of polychaete diversity in the Rockall Trough, northeast Atlantic. *Deep-Sea Research II* 42, 1199–1214.

Poore, G.C.B., Just, J. & Cohen, B.F. (1994) Composition and diversity of Crustacea isopoda of the southeastern Australian continental slope. *Deep-Sea Research I* 41, 677–693.

Poore, G.C.B., Mccallum, A.W. & Taylor, J. (2008) Decapod Crustacea of the continental margin of south-western and central Western Australia: preliminary identifications of 524 species from FRV Southern Surveyor voyage SS10-2005. *Museum Victoria Science Reports* 11, 1–106.

Priede, M., Godbold, J.A., King, N.J., *et al.* (2010) Deep-sea demersal fish species richness in the Porcupine Seabight, NE Atlantic Ocean: global and regional patterns. *Marine Ecology* 31, 247–260.

Ramirez-Llodra, E., Company, J.B., Sarda, F. & Rotland, G. (2010) Megabenthic diversity patterns and community structure of the Blanes submarine canyon and adjacent slope in the Northwestern Mediterranean: a human overprint? *Marine Ecology* 31, 167–182.

Reed, J.K., Koenig, C.C. & Shepard, A.N. (2007) Impacts of bottom trawling on a deep-water Oculina coral ecosystem off Florida. *Bulletin of Marine Science* 81, 481–496.

Rex, M.A. (1981) Community structure in the deep-sea benthos. *Annual Review of Ecology and Systematics* 12, 331–353.

Rex, M.A., Crame, J.A., Stuart, C.T. & Clarke, A. (2005) Large-scale biogeographic patterns in marine mollusks: a confluence of history and productivity? *Ecology* 86, 2288–2297.

Rex, M.A., Etter, R.J., Morris, J.S., *et al.* (2006) Global bathymetric patterns of standing stock and body size in the deep-sea benthos. *Marine Ecology Progress Series* 317, 1–8.

Richter, T.O., De Stigter, H.C., Boer, W., *et al.* (2009) Dispersal of natural and anthropogenic lead through submarine canyons at the Portuguese margin. *Deep-Sea Research I* 56, 267–282.

Ricketts, E.R., Kennett, J.P., Hill, T.M. & Barry, J.P. (2009) Effects of carbon dioxide sequestration on California margin deep-sea foraminiferal assemblages. *Marine Micropaleontology* 72, 165–175.

Rogers, A.D., Billett, D.S.M., Berger, G.W., *et al.* (2002) Life at the edge: Achieving Prediction from Environmental Variability and Biological Variety. In: *Ocean Margin Systems* (eds. G. Wefer, D.S.M. Billett, D. Hebbeln, *et al.*), pp. 387–404. Berlin and Heidelberg: Springer.

Rosenzweig, M.L. (1992) Species diversity gradients: we know more and less than we thought. *Journal of Mammalogy* 73, 715–730.

Rowe, G.T. & Kennicutt, M.C. (2008) Introduction to the Deep Gulf of Mexico Benthos Program. *Deep-Sea Research II* 55, 2536–2540.

Sanders, H.L. (1968) Marine benthic diversity: a comparative study. *American Naturalist* 102, 243–282.

Sanders, H.L., Hessler, R.R. & Hampson, G.R. (1965) An introduction to the study of the deep-sea benthic faunal assemblages along the Gay Head–Bermuda transect. *Deep-Sea Research* 12, 845–867.

Santos, M.F.L., Lana, P.C., Silva, J., *et al.* (2009) Effects of non-aqueous fluids cuttings discharge from exploratory drilling activities on the deep-sea macrobenthic communities. *Deep-Sea Research II* 56, 32–40.

Schlacher, T.A., Williams, A., Althaus, F. & Schlacher-Hoenlinger, M.A. (2010) High-resolution seabed imagery as a tool for biodiversity conservation planning on continental margins. *Marine Ecology* 31, 200–221.

Sedlack, L., Thistle, D., Carman, K.R., *et al.* (2009) Effects of carbon dioxide on deep-sea harpacticoids revisited. *Deep-Sea Research I* 56, 1018–1025.

Seiter, K., Hensen, C. & Zabel, M. (2005) Benthic carbon mineralization on a global scale. *Global Biogeochemical Cycles* 19, 1–26.

Sellanes, J., Neira, C. & Quiroga, E. (2010) Diversity patterns along and across the Chilean margin: a continental slope encompassing oxygen gradients and methane seep benthic habitats. *Marine Ecology* **31**, 111–124.

Sibuet, M. (1987) *Structure des peuplements benthiques en relation avec les conditions trophiques en milieu abyssal dans l'ocean Atlantique.* Thèse de doctorat d'Etat, Sciences Naturelles, Université Pierre et Marie Curie, Paris 6, vol. 1, p. 280, Vol. 2, p. 413.

Sibuet, M. & Olu-Le Roy, K. (2003) Cold seep communities on continental margins: Structure and quantitative distribution relative to geological and fluid venting patterns. In: *Ocean Margin Systems* (eds. G. Wefer, D.S.M. Billett, D. Hebbeln, *et al.*), pp. 235–251. Berlin: Springer.

Sibuet, M. & Olu, K. (1998) Biogeography, biodiversity and fluid dependence of deep sea cold seeps communities at active and passive margins. *Deep-Sea Research II* **45**, 517–567.

Sibuet, M. & Vangriesheim, A. (2009) Deep-Sea Benthic Ecosystems of the Equatorial African Margin: the Multidisciplinary BIOZAIRE Program. *Deep-Sea Research II* **56**, special issue.

Skadsheim, A., Borseth, J.F., Bjornstad, A., *et al.* (2005) Hydrocarbons and chemicals: Potential effects and monitoring in the deep sea. In: *Offshore Oil and Gas Environmental Effects Monitoring: Approaches and Technologies* (eds S.L. Armsworthy, P.J. Cranford & K. Lee). Columbus, Ohio: Batelle Press.

Snelgrove, P.V.R. & Smith, C.R. (2002) A riot of species in an environmental calm; the paradox of the species-rich deep sea. *Oceanography and Marine Biology* **40**, 311–342.

Soh, W. (2003) Transport processes deduced from geochemistry and the void ratio of surface core samples, deep sea Sagami Bay, central Japan. *Progress in Oceanography* **57**, 109–124.

Solé, M., Porte, C. & Albaigés, J. (2001) Hydrocarbons, PCBs and DDT in the NW Mediterranean deep-sea fish *Mora moro*. *Deep-Sea Research I* **48**, 495–513.

Storelli, M.M., Perrone, V.G. & Marcotrigiano, G.O. (2007) Organochlorine contamination (PCBs and DDTs) in deep-sea fish from the Mediterranean Sea. *Marine Pollution Bulletin* **54**, 1968–1971.

Stramma, L., Johnson, G.C., Sprintall, J. & Mohrholz, V. (2008) Expanding oxygen-minimum zones in the tropical oceans. *Science* **320**, 655–658.

Stuart, C.T. & Rex, M.A. (2009) Bathymetric patterns of deep-sea gastropod species diversity in 10 basins of the Atlantic Ocean and Norwegian Sea. *Marine Ecology* **30**, 164–180.

Thiel, H. (2003) Anthropogenic impacts in the deep sea. In: *Ecosystems of the Deep Ocean*, Vol. **28**, Ecosystems of the World. (ed. P.A. Tyler). Amsterdam: Elsevier.

Thurber, A.R., Kröger, K., Neira, C., *et al.* (2009) Stable isotope signatures and methane use by New Zealand cold seep benthos. *Marine Geology* doi:10.1016/j.margeo.2009.06.001.

Toggweiler, J.R. & Russell, J. (2008) Ocean circulation in a warming climate. *Nature* **451**, 286–288.

Turley, C., Roberts, J. & Guinotte, J. (2007) Corals in deep-water: will the unseen hand of ocean acidification destroy cold-water ecosystems? *Coral Reefs* **26**, 445–448.

UNEP (2007) Deep-sea biodiversity and ecosystems: a scoping report for their socio-economy, management and governance.

UNESCO (2009) Global Open Oceans and Deep Seabed (GOODS) biogeographic classification. UNESCO-IOC, IOC Technical Series 84, 1–87.

Van Gaever, S., Raes, M., Pasotti, F. & Vanreusel, A. (2010) Spatial scale and habitat-dependent diversity patterns in nematode communities in three seepage related sites along the Norwegian margin. *Marine Ecology* **31**, 66–77.

Vanreusel, A., Fonseca, G., Danovaro, R., *et al.* (2010) The contribution of deep-sea macrohabitat heterogeneity to global nematode diversity. *Marine Ecology* **31**, 6–20.

Vetter, E.W., Smith, C.R. & De Leo, F. (2010) Hawaiian hotspots: enhanced megafaunal abundance and diversity in submarine canyons on the oceanic islands of Hawaii. *Marine Ecology* **31**, 183–199.

Weaver, P.P.E., Billett, D.S.M., Boetius, A., *et al.* (2004) Hotspot ecosystem research on Europe's deep-ocean margins. *Oceanography* **17**, 132–143.

Wefer, G., Billett, D.S.M., Hebbeln, D. *et al.* (eds.) (2003) *Ocean Margin Systems*. Berlin and Heildelberg: Springer.

Wheeler, A.J., Bett, B.J., Billett, D.S.M., *et al.* (2005) The impact of demersal trawling on Northeast Atlantic deepwater coral habitats: the case of the Darwin Mounds, United Kingdom. *American Fisheries Society Symposium* **41**, 807–817.

Williams, A., Althaus, F., Dunstan, P., *et al.* (2010) Scales of habitat heterogeneity and megabenthos biodiversity on an extensive Australian continental margin (100–1,000 m depths). *Marine Ecology* **31**, 222–236.

World Ocean Atlas (2005) NODC. Accessed May 18, 2010. http://www.nodc.noaa.gov/OC5/WOA05/pr_woa05.html

Zeidberg, L.D. & Robison, B.H. (2007) Invasive range expansion by the Humboldt squid, *Dosidicus gigas*, in the eastern north Pacific. *Proceedings of the National Academy of Science of the USA* **104**, 12948–12950.

Chapter 6

Biodiversity Patterns and Processes on the Mid-Atlantic Ridge

Michael Vecchione[1], Odd Aksel Bergstad[2], Ingvar Byrkjedal[3], Tone Falkenhaug[2], Andrey V. Gebruk[4], Olav Rune Godø[5], Astthor Gislason[6], Mikko Heino[7], Åge S. Høines[5], Gui M. M. Menezes[8], Uwe Piatkowski[9], Imants G. Priede[10], Henrik Skov[11], Henrik Søiland[5], Tracey Sutton[12], Thomas de Lange Wenneck[5]

[1]NMFS National Systematics Laboratory, National Museum of Natural History, Smithsonian Institution, Washington DC, USA
[2]Institute of Marine Research, Flødevigen, His, Norway
[3]University of Bergen, Bergen Museum, Department of Natural History, Bergen, Norway
[4]P.P. Shirshov Institute of Oceanology, Russian Academy of Sciences, Moscow, Russia
[5]Institute of Marine Research, Bergen, Norway
[6]Marine Institute of Iceland, Reykjavik, Iceland
[7]Department of Biology, University of Bergen, Bergen, Norway
[8]Departamento de Oceanografia e Pescas, Universidade dos Açores, Horta, Portugal
[9]Leibniz-Institut für Meereswissenschaften, IFM-GEOMAR, Forschungsbereich Marine Ökologie, Kiel, Germany
[10]Oceanlab, University of Aberdeen, Aberdeen, UK
[11]DHI, Hørsholm, Denmark
[12]Virginia Institute for Marine Science, College of William and Mary, Gloucester Point, Virginia, USA

6.1 Introduction

The network of mid-ocean ridges constitutes the largest continuous topographic feature on Earth, 75,000 km long (Garrison 1993). Some of the known chemosynthetic ecosystems (Chapter 9) in these deep seafloor habitats have been relatively well studied, but remarkably little is known about ridge-associated pelagic and benthic fauna that are sustained by photosynthetic production in association with mid-ocean ridges (Box 6.1). This knowledge gap inspired the initiation of the multinational field project "Patterns and Processes of the Ecosystems of the Northern Mid-Atlantic", MAR-ECO (Bergstad & Godø 2003; Bergstad et al. 2008c). Extensive investigations were conducted along the Mid-Atlantic Ridge between Iceland and the Azores (Fig. 6.1) with the aim to "describe and understand the patterns of distribution, abundance, and trophic relationships of the organisms inhabiting the mid-oceanic area of the North Atlantic, and to identify and model ecological processes that cause variability in these patterns". Compared with other mid-ocean ridge sections the Mid-Atlantic Ridge region under consideration is special in that it is shallow and emerges at both ends with islands, namely Iceland and the Azores. There have been fisheries since the

Box 6.1

Historical Context

In 1910, the R/V *Michael Sars* expedition across the North Atlantic (Murray & Hjort 1912) revealed markedly elevated abundance and species numbers in shallow mid-ocean areas, including approximately 45 fish species and well over 100 invertebrates new to science, many of which came from what later would be recognized as the Mid-Atlantic Ridge (MAR).

The general bathymetry of the North Atlantic mid-ocean ridge was mapped by the early 1960s and studies of oceanic circulation across the ridge and deep water flow through the Charlie Gibbs Fracture Zone (CGFZ) were well advanced by the start of MAR-ECO field work (see, for example, Krauss 1986; Rossby 1999; Bower *et al.* 2002). Gradually improved bathymetric data revealed the axial valley, numerous hills and valleys, and major fracture zones reaching abyssal depths. Circulation features are shown in Figure 6.1, including the Sub-Polar Front (SPF), which crosses the ridge in the vicinity of the CGFZ at around 52°N and may be significant to biogeography.

The SPF separates the Cold Temperate Waters Province (CTWP), and the Warm Temperate Waters Province (WTWP), defined by The Oslo–Paris Commission (OSPAR) based on extensive reviews of the regional biogeography data (Dinter 2001). Provinces defined by Longhurst (1998) were mainly based on surface features, one of them being an east–west asymmetry in the diversity patterns of zooplankton in the central North Atlantic (Beaugrand *et al.* 2000, 2002). Biogeography of the bathyal benthic fauna at the northern MAR was addressed in recent studies of the Reykjanes Ridge and seamounts south of the Azores (Mironov *et al.* 2006), but almost no data were available from the CGFZ-to-Azores section of the MAR. On the ocean-basin scale, Mironov (1994) proposed the concept of "meridional asymmetry": specifically, that some western Atlantic species are widely distributed in the Azorean-Madeiran waters whereas the eastern Atlantic benthic invertebrates are confined (with very rare exceptions) to the East Atlantic.

Pelagic and demersal nekton of the northern MAR were investigated by various historical expeditions that crossed the North Atlantic (see, for example, Murray & Hjort 1912; Schmidt 1931; Tåning 1944), and later by the Atlantic Zoogeography Program (Backus *et al.* 1977), and German expeditions to the mid-ocean and seamounts (see, for example, Post 1987; Fock *et al.* 2004). Information existed on the distribution of cephalopods at various specific locations in the Atlantic (Vecchione *et al.* 2010), revealing general latitudinal patterns and information from isolated seamounts, but none were focused on the MAR. Although the fish fauna and general distribution patterns of deepwater fishes of the

northern Atlantic Ocean had been described (see, for example, Whitehead *et al.* 1986; Haedrich & Merrett 1988; Merrett & Haedrich 1997), surprisingly few previous studies have focused specifically on the role of the mid-oceanic ridges in the distribution and ecology of either pelagic or demersal fishes. Studies from the Azores have shown very low endemism, and that most species have distributional affinities with the eastern Atlantic and the Mediterranean (Santos *et al.* 1997; Menezes *et al.* 2006). Considerable knowledge of fishes associated with ridge systems has been gained from fisheries-related research (Bergstad *et al.* 2008b, c), but most reports focused strongly on target species and usually on only the shallower parts of the ridge and specific seamounts. Only in exceptional cases have full species lists of the catches been published (see, for example, Hareide & Garnes 2001; Kukuev 2004). Areas of the northern MAR have been, and still are, exploited for fish species such as redfish (*Sebastes* spp.) (Clark *et al.* 2007). Pelagic fisheries of the open ocean have targeted tuna, swordfish, and sharks that tend to be found near fronts, eddies, and islands. Whales also occur in such areas (Sigurjónsson *et al.* 1991) and, like the epipelagic fishes, they migrate extensively, perhaps associated with the MAR.

Life-history strategies had not been studied for any species on the MAR, but information was available for some species on adjacent seamounts or continental slopes. These data constituted valuable comparative sources for new studies of the diversity of life-history strategies characterizing ridge-associated species.

Knowledge of large-scale distributions across and along the MAR was lacking for most pelagic and demersal macro- and megafaunal groups. Basin-wide population connections were also unknown. It was uncertain whether the MAR fauna was unique or composed of elements from the adjacent continental slopes.

MAR food webs were unknown, except for a few studies along the Reykjanes Ridge, and life-history information was only available for a very limited number of zooplankton taxa (copepods, mainly *Calanus* spp.), but lacking for most other species. The general trophic positions of some common zooplankton species, primarily copepods, amphipods, and euphausiids, inhabiting the epi- and upper mesopelagic layers above the Reykjanes Ridge have been described (Magnusson & Magnusson 1995; Petursdottir *et al.* 2008). Also, the spawning aggregations of redfish confined to the western slopes of the Reykjanes Ridge suggest that this is a productive area (Pedchenko & Dolgov 2005). However, no information existed on how the MAR affects productivity or abundance of mesopelagic organisms.

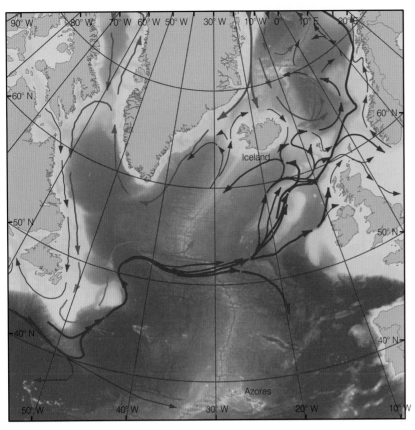

Fig. 6.1
Bathymetry and main circulation features of the North Atlantic.

Water mass

→ Arctic Water

→ Atlantic Water

→ Mixture of Atlantic and Arctic Water

→ Coastal Water

1970s, and the information available suggested high biodiversity and a strong potential for new discoveries.

6.1.1 The MAR-ECO project

MAR-ECO was conceived as the first comprehensive international exploration of a substantial section of the global mid-ocean ridge system. Working in mid-ocean waters at great depths and in rugged topography is technologically challenging and expensive. MAR-ECO's strategy was to mobilize a cadre of experts, using a variety of instruments and ships from several countries, to achieve the research capacity to meet the many and varied challenges. The "flagship" expedition for this project was conducted by R/V *G.O. Sars* during summer 2004, with concurrent longline fishing by F/V *Loran*, but several other cruises both before and after have contributed substantially as well (www.mar-eco.no). Using multiple technologies on the same platform provides more comprehensive results and enhances the potential for new discoveries. Our goal was to sample and/or observe organisms ranging in size from millimeters to meters (for example small zooplankton to whales), hence many types of sampler were used (Fig. 6.2). To sample all relevant depths, the technologies needed to function from surface waters to at least 3,500 m, preferably as deep as 4,500 m to reach the bottom of the deepest valleys. Along with the sampling of biota, hydrographic data were collected to characterize the physical and chemical environment. In addition to ships, other platforms such as manned and unmanned submersibles, moored instruments, and benthic landers were adopted. These instruments used optics and acoustics, and some were deployed for long periods to collect temporal data for certain taxa or selected features. Detailed accounts of technologies and methods and sampling strategies for the different taxa and functional groups were given by Wenneck *et al.* (2008), Gaard *et al.* (2008), and in many papers describing results of analyses (see, for example, several papers in Gebruk (2008a) and Gordon *et al.* (2008)). Those references also describe methods used in the post-cruise analyses of taxonomy and systematics, trophic ecology, and life-history strategies.

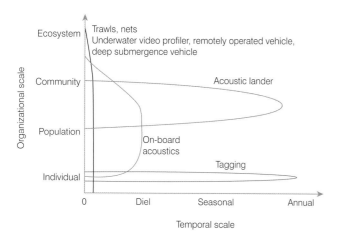

Fig. 6.2

Technologies and their spatiotemporal sampling scales.

6.2 Discoveries

6.2.1 Hydrography

Composite remote-sensing images were prepared to identify the location of the SPF in relation to location of the ridge (Søiland *et al.* 2008). From these and ship-board sampling, four different hydrographic regions were identified in the surface layers. North of 57°N on the Reykjanes Ridge, Modified North Atlantic Water dominated. Between 57°N and the SPF there was Sub Arctic Intermediate Water. South of the SPF, North Atlantic Central Water traverses the ridge in the general eastward flow of the North Atlantic Current but mixing with Sub Arctic Intermediate Water forms in a complex pattern of eddies to south of 50°N. The southern boundary of the SPF was thus very indistinct, containing many features with patches of high productivity and high abundances.

6.2.2 Identification and distribution of the fauna

6.2.2.1 Faunal composition and biodiversity

The number of species recorded in the samples from the two-month 2004 expedition by R/V *G.O. Sars* and F/V *Loran* illustrates the scale of diversity of the MAR-associated pelagic and epibenthic macro- and megafauna comprising animals of sizes from about 1 mm to several meters (Table 6.1). Examples include the 303 species from more than 60,000 fish specimens collected by net sampling during the *G.O. Sars* expedition. Of these fishes, two-thirds or more were pelagic (Sutton *et al.* 2008), the rest demersal

(that is, either benthic or benthopelagic (Bergstad *et al.* 2008b)). Many species were extremely rare and some were undescribed (see, for example, Orlov *et al.* 2006; Byrkjedal & Orlov 2007; Chernova & Møller 2008). The pelagic fish diversity was highest in the mesopelagic (200–1,000 m), whereas, surprisingly, biomass was highest in the bathypelagic (greater than 1,000 m). Numerically dominant families of pelagic fishes included Gonostomatidae, Melamphaidae, Microstomatidae, Myctophidae, and Sternoptychidae. The family Macrouridae was prominent among the demersal fishes, represented by 17 species (plus one that is probably new to science). *In situ* observations were also acquired: 22 fish taxa were photographed by a baited benthic lander, whereas bottom-dive segments with remotely operated vehicles (ROVs) found at least 36 taxa, including roundnose grenadier, orange roughy, oreos, halosaurs, codlings, and many additional macrourids. The longline catch comprised mainly large predatory fishes (mean weight 2.4 kg), dominated by the families Etmopteridae, Somniosidae, Ophidiidae, Macrouridae, Moridae, and Lotidae. This represented a different faunal composition from that of the demersal trawl catch.

A substantial cephalopod collection from the midwater and bottom trawls comprised 54 species in 29 families (Vecchione *et al.* 2010). The squid *Gonatus steenstrupi* was the most abundant cephalopod in the samples, followed by the squids *Mastigoteuthis agassizii* and *Teuthowenia megalops*. A multispecies aggregation of large cirrate octopods dominated the demersal cephalopods.

About 10% of species in the MAR-ECO epibenthic invertebrate species appeared to be new to science. The species richness of corals was high with a total of 40 taxa recorded. Octocorals dominated this coral fauna, with 27 taxa. *Lophelia pertusa* was one of the most frequently observed corals, present on five of the eight ROV-inspected sites. Massive live reef structures were not observed; only small colonies (less than 0.5 m across) were present. The number of megafaunal taxa was 1.6 times higher in areas where corals were present compared with areas without corals. Typical taxa that co-occurred with *Lophelia* were crinoids, sponges, the bivalve *Acesta excavata*, and squat lobsters.

Corresponding numbers for zooplankton taxa and top predators such as mammals and seabirds (from sightings along the ship's track) are given in Table 6.1. For all taxa, occurrence data were reported to the Ocean Biogeographic Information System (OBIS) as soon as identifications were validated.

6.2.2.2 Population structure

Many deepwater species have basin-wide distributions, and understanding potential sub-structuring is of substantial ecological and evolutionary interest with direct implications for management. Investigating underlying processes through a comparative assessment of species with differing

Table 6.1

Number of species recorded as of mid-2008.

Main taxa	Identified species	Described new species	Comments	New species references
Cetaceans	14			
Seabirds	22			
Fishes	303	2		Byrkjedal & Orlov 2007; Chernova & Møller 2008
Hemichordates	2	(2)	(Not described.) New species observed but not collected	Holland *et al.* 2005
Brachiopods	3			
Mollusks	75	2	Two new species of cephalopod	Vecchione & Young 2006; Young *et al.* 2006
Arthropods	306	2	One new genus	Brandt & Andres 2008; Crosnier & Vereshchaka 2008
Echinoderms	104	9	One new genus and one new family	Dilman 2008; Gebruk 2008; Martynov & Litvinova 2008; Mironov 2008
Annelids	3			
Chaetognaths	16			
Echiurans	2	1	New species of the genus *Jacobia* (Echiura)	Murina 2008
Sipunculids	2			Murina 2008
Ctenophores	3			
Cnidarians	112			
Sponges	35	13	One new genus	Menschenina *et al.* 2007; Tabachnick & Menshenina 2007; Tabachnik & Collins 2008
Fish parasites:				
Nematodes	11	2		Moravec *et al.* 2006; Moravec & Klimpel 2007
Monogeneans	18	1		Kritsky & Klimpel 2007
Cestodes	6			
Acanthocephalans	3			
Crustaceans	8			
Total	1048	34		

life-history characteristics (for example, duration of larval stages, fecundity, longevity, and habitat requirements) allows predictions about expected boundaries to gene flow, rates of gene flow, and demographic history. However, there have been some unexpected results. For example, the orange roughy (*Hoplostethus atlanticus*) has life-history characteristics that could promote population structure (for example, long life, comparatively low fecundity and larval duration), but genetic data suggest no structure in the North Atlantic study area (White *et al.* 2009; S. Stefanni, unpublished observations). On the other hand, the round-nose grenadier (*Coryphaenoides rupestris*), which has characteristics suggesting greater connectivity, showed considerable structure at the ocean-basin scale (H. Knutsen, P.E. Jorde & O.A. Bergstad, unpublished observations), some small-scale structure across a putative boundary (the sub-polar front), and evidence for selection associated with depth (White *et al.* 2010). In general the comparative studies highlight the importance of several key factors (Fig. 6.3): local habitat dependence (for example, tusk (*Brosme brosme*) in the MAR; Knutsen *et al.* 2009), isolation by geographic distance or along current pathways (see, for example, Knutsen *et al.* 2007), oceanic barriers to gene flow (see, for example, White *et al.* 2010), and the role of different life stages (see, for example, White *et al.* 2009).

Demersal fishes in general have low resilience to population disturbance, with a population doubling time on the order of 10 years. The existence of local, autonomous populations implies that local fishing areas may be sensitive to overexploitation. Our findings highlight the importance of considering population structure in deep-sea fishery management.

6.2.2.3 Taxonomy and phylogenetics

Extensive work on phylogenetic reconstruction for species discovery and to determine the origin of MAR radiations is ongoing. DNA barcoding of MAR species is proceeding. Most species of both pelagic and demersal nekton (fishes, cephalopods, and shrimps) will be barcoded. For example, over 190 fish species have been barcoded for the first time from MAR-ECO material. For zooplankton, species barcoding is being coordinated with the Census of Marine Zooplankton project (Chapter 13).

For two morphologically cryptic species (*Aphanopus carbo* and *A. intermedius*) with overlapping distributions (Stefanni & Knutsen 2007), a genetic marker suitable for routine discrimination has been developed (Stefanni *et al.* 2009). The huge collections of specimens and tissue samples including samples of very rare species not available elsewhere, motivated new revisions of difficult taxa. An example is the cusk-eel genus *Spectrunculus*, which was revised and split from one to two species based mainly on MAR-ECO material (Uiblein *et al.* 2008). The samples of rare deep-sea fishes are also very valuable in studies of the evolution of various groups. Modern standards for these phylogenetic reconstructions and studies of the interrelationships and origin of the fauna require tissue samples for DNA sequencing and corresponding voucher specimens for morphological characters and identification. Published studies based on MAR-ECO material include the slickhead and tubesholder fishes (Alepocephaliformes) (Lavoue *et al.* 2008, Poulsen *et al.* 2009), and several others are in progress (Ophidiiformes, Myctophidae).

6.2.3 Vertical distribution

6.2.3.1 Zooplankton

Copepod abundance was highest in upper layers (0–100 m) and decreased exponentially with depth (Gaard *et al.* 2008). Several species of copepods and decapods were observed to deepen their vertical distributions towards the south, following the isotherms (that is, equatorial submergence). Decapods peaked in the 200–700 m stratum north of the SPF, and at 700–2,500 m depth south of the SPF. The highest densities of euphausiids were found in the upper 200 m. The gelatinous fauna, dominated by cnidarians, siphonophores, and appendicularians, was most abundant at 400–900 m (Stemmann *et al.* 2008; Youngbluth *et al.* 2008). *In situ* observations of gelatinous zooplankton revealed that different taxa occurred in distinct, and often narrow (tens of meters), depth layers (Vinogradov 2005; Youngbluth *et al.* 2008). The most important contributors to the cnidarian biomass (wet mass) north of the SPF were the scyphomedusae *Periphylla periphylla* and *Atolla* spp. The vertical distributions of *P. periphylla* and *Atolla* spp. were deeper during the day than at night. The bulk of the *Atolla* spp. population usually resided deeper in the water column than *P. periphylla*. Appendicularians were generally abundant at 450–1,000 m and were observed to accumulate in the lowermost 50 m (Vinogradov 2005; Youngbluth *et al.* 2008), suggesting that these feeding specialists (extremely small particles) are a prominent component of the benthopelagic zooplankton.

6.2.3.2 Pelagic nekton (fishes and cephalopods)

Depth was by far the most important determinant of faunal composition for pelagic fish species, with along-ridge variation secondary. The most surprising finding was the water-column maximum fish biomass between 1,500 and 2,300 m (Sutton *et al.* 2008); this pattern stands in stark contrast to the typical exponential decline in fish biomass below 1,000 m seen in open oceanic ecosystems. Furthermore, evidence from acoustics and trawl catches suggests that in some locations, deep pelagic fish abundance and biomass peak within the benthic boundary layer, suggesting the possibility of predator–prey relationships between demersal fishes and migrating pelagic fishes as a mechanism underlying

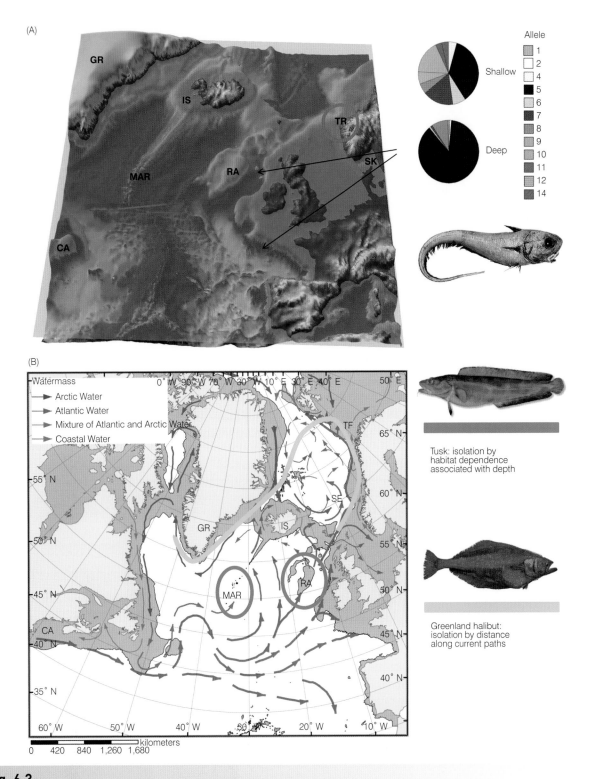

Fig. 6.3

(A) Barriers to gene flow (after H. Knutsen, P.E. Jorde & O.A. Bergstad, upublished observations; White *et al.* 2010) and evidence for local adaptation (after White *et al.* 2010) in *Coryphaenoides rupestris*. Barriers are shown in mid-Atlantic ridge (MAR); RA, Rockall; TR, Trondheim coastal site; SK, Skagerrak; IS, Iceland; GR, Greenland; CA, Canada. The allele frequency pie charts show how allele 5 (at a microsatellite DNA locus evidently linked to a relevant functional gene) is associated with depth of sample. MAR = 2,563–2,573. **(B)** Illustration of isolation by distance along current paths for Greenland halibut (yellow), and local differentiation of *Brosme brosme* populations (likely associated with depth) (green). SE, Storegga; TF, Tromsøflaket.

enhanced demersal fish biomass over the MAR (Bergstad *et al.* 2008b).

Acoustic data from both vessel and stationary systems revealed biophysical interaction of presumed importance to production and species interactions. In the epi- and meso-pelagic zones (0–1,000 m) of most areas along the MAR a clear diel vertically migrating community was observed by acoustics, with daytime depths of 500–1,000 m, and night-time occupation of surface layers (Opdal *et al.* 2008). The range and the patterns were affected by topography and light levels that may have been affected by phytoplankton density. Lanternfishes and pearlfishes (Sternoptychidae) were the dominant diel vertically migrating fishes. Seasonal information from an upward-looking acoustic lander showed abrupt changes in distribution and abundance of sound scatterers in early autumn and spring (Doksæter *et al.* 2009). Vertical migration was reduced to a minimum during mid-winter and peaked during summer.

Mesoscale eddies, validated by satellite sea level altimetry data, were recorded from the surface to 1,200 m. The acoustic lander observed extensive internal wave activity, mainly close to the seabed, but sometimes extending into the entire water column from 900 m to the surface. Occasionally, breaking internal waves apparently created turbulence in the near bottom zone resulting in disruption of scattering layers and chaotic distribution of individual acoustic scatterers.

Observations from submersibles have shown some cirrate octopods (*Grimpoteuthis* and *Opisthoteuthis*) to sit on the bottom and/or to float just above it (Vecchione & Roper 1991; Vecchione & Young 1997; Felley *et al.* 2008). All specimens of these genera, as well as *Cirroteuthis* and *Cirrothauma*, were collected in the bottom trawl. Conversely, many *Stauroteuthis syrtensis* were taken in midwater, including a specimen that had to have been at least 1,690 m above the bottom, although most specimens came from the bottom trawl. It therefore appears that this species aggregates near bottom but its distribution also extends far up into the deep water column.

6.2.3.3 Demersal fishes

Overall, demersal fish biomass and abundance declined with depth from the summit of the ridge to the middle rises on either side. Multivariate analyses of catch data from trawls and longlines (Bergstad *et al.* 2008b; Fossen *et al.* 2008) revealed that the species composition primarily changed with depth and that, as with pelagic fishes, variation by latitude was secondary. Species evenness was higher in deep slope and rise areas than on the slopes. Assemblages of species could be defined for different depth zones and sub-areas. *In situ* observations of scavenging fishes attracted to baited landers revealed three main assemblages: shallow (924–1,198 m), intermediate (1,569–2,355 m), and deep (2,869–3,420 m). These assemblages were dominated respectively by three species, *Synaphobranchus kaupii*,

Antimora rostrata, and *Coryphaenoides armatus*. Abyssal species were found in the axial valley region (*C. armatus*, *Histiobranchus bathybius*, and *Spectrunculus* sp.). Fishing by longlines in rugged terrain at all depths resulted in catches dominated by elasmobranchs (sharks and skates). Fishes were observed during the dives in 2003 of the manned submersibles MIR 1 and 2 in the CGFZ between 1,700 and 4,500 m (Felley *et al.* 2008). Perhaps the most remarkable observation was that of rich densities of small juvenile macrourids in the deep soft-bottom areas, presumably dominated by the abyssal grenadier *C. armatus*. MAR-ECO data formed a significant element of a global analyses of the depth distribution of elasmobranch and teleost fishes, demonstrating that elasmobranchs are uncommon or rare deeper than 3,000 m (Priede *et al.* 2006).

6.2.3.4 Benthos

For a range of taxa from many depths new species were described (Table 6.1). Corals were observed at all MAR-ECO sites inspected with ROVs at bottom depths between 800 and 2,400 m, but were most common shallower than 1,400 m. The deepest record of *Lophelia* was at 1,340 m, south of the CGFZ. Accumulations of coral skeleton debris were observed at several locations, indicating presence of former *Lophelia* reefs.

Manned submersible dives in the CGFZ from 1,700 to 4,500 m observed scattered rich sponge gardens. Dense aggregations of small elpidiid holothurians, *Kolga* sp., occurred at abyssal depths (4,500 m) in a sediment-filled depression. Abundance/biomass of giant protists (foraminiferan Syringamminidae, reaching the size of a golf ball) was noteworthy north of the SPF.

6.2.4 Variation along the ridge

The number of species recorded showed a clear latitudinal pattern for most taxa, with a discontinuity at about the location of the SPF (Fig. 6.4).

6.2.4.1 Zooplankton

The assemblages of copepods, cnidarians, chaetognaths, gelatinous zooplankton, and macrozooplankton were related to the distribution of three main water masses in the area (Gaard *et al.* 2008; Hosia *et al.* 2008; Pierrot-Bults 2008; Stemmann *et al.* 2008): a northern assemblage in Modified North Atlantic Water, a southern assemblage in North Atlantic Central Water, and a frontal assemblage influenced by North Atlantic Central Water and Sub-Arctic Intermediate Water. The species richness of most taxa was found to increase towards the south (Copepoda, Cnidaria, Decapoda, Euphausiacea, Amphipoda, Chaetognatha). Temperature appeared to be the most important factor in determining the structure of the copepod communities.

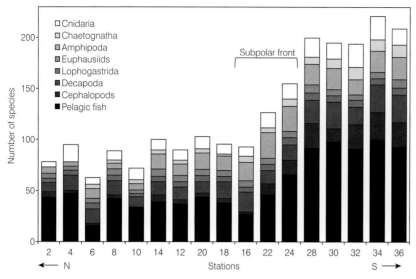

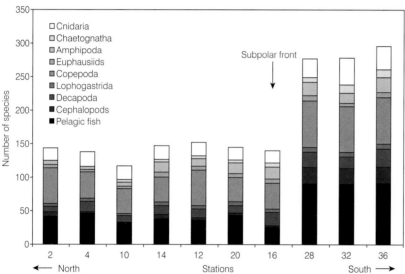

Fig. 6.4

Occurrence of species of various higher taxa at different stations from north (left) to south along the MAR. Top, midwater trawl stations; bottom, mesozooplankton stations.

6.2.4.2 Pelagic nekton

Among cephalopods, the squids *Mastigoteuthis agassizii* and *Teuthowenia megalops* were distributed throughout the whole area; *Gonatus steenstrupi* was most abundant in the northern and central regions (Reykjanes Ridge and CGFZ) (Vecchione *et al.* 2010). In contrast, the bobtail squid *Heteroteuthis dispar* was only common in the Azorean area, with a few specimens near the Faraday Seamounts and the CGFZ. Multivariate analysis revealed a clear separation of a southern cephalopod assemblage (Azorean area), an assemblage confined to the Reykjanes Ridge, and an assemblage concentrated at stations at the CGFZ. The Azorean assemblage was very similar to an assemblage recently described from samples along the Biscay–Azores Ridge and MAR north of the Azores (C. Warneke-Cremer, unpublished observations). Cephalopod species richness per station clearly increased from north to south (five species at a station on the Reykjanes Ridge, 28 species at a station near the Azores). Several benthic and one pelagic species, all taken in small numbers, were captured only in the CGFZ. Numbers of common bentho-pelagic species were highest in the CGFZ (Vecchione *et al.* 2010).

Within the top 750 m of the water column, there were two primary faunal groups of pelagic fishes (Sutton *et al.* 2008): a higher abundance, lower diversity assemblage from Iceland to the Faraday Seamount Zone (numerically dominated by the lanternfish *Benthosema glaciale*), and a lower abundance, higher diversity assemblage in the region of the Azores (29 lanternfish species contributed half of total abundance). Below 750 m there was a large assemblage of deep meso- to bathypelagic fishes that spanned from the Reykjanes Ridge all the way to the Azores (numerically dominated by the bristlemouth, *Cyclothone microdon*).

6.2.4.3   Demersal nekton

The latitudinal variation in occurrence of demersal fishes caught in demersal trawls was greater in shallow than in deep areas. The number of species was inversely related to latitude, but declined with depth below the slope depths. For example, the macrourids *Coryphaenoides rupestris*, *C. brevibarbis*, and *C. armatus* rank among the most abundant demersal fishes on the ridge or in the deep axial valleys or fracture zones, while other members of the family are uncommon or rare (Bergstad *et al.* 2008a). Whereas a few species in the family apparently have restricted northerly or southerly distributions, most are widespread, but showing definite depth-related patterns of distribution (Fig. 6.5). Similar patterns were observed in the demersal predators sampled by longlines.

6.2.4.4   The Sub-Polar Front: a biogeographic barrier

The SPF acted as a boundary for several zooplankton taxa (Falkenhaug *et al.* 2007). For copepods this delineation was asymmetrical: sub-tropical and warm-temperate species had limited dispersal northward, whereas cold-water species often extended south of the SPF (Gaard *et al.* 2008). The spatial distribution of the dominant copepods, *Calanus finmarchicus* and *C. helgolandicus*, was separated at the SPF, with the latter found only south of the SPF, at depths associated with Mediterranean water masses. The separation of Cnidaria at the SPF was found to be strongest in the upper 500 m but apparent down to 1,500 m (Hosia *et al.* 2008). Epi- and mesopelagic fish distributional trends mirrored those of zooplankton, with species diversity substantially higher near the Azores, but again with cold-water forms common in the south of the SPF (that is, a "fuzzy" southern limit to distribution of northern species). Overall, the strength of the SPF as a boundary to pelagic fauna varied vertically, with deeper-water samples showing less variation in species composition. For epibenthic fauna, changes were observed between the CGFZ and the Azores, particularly in the region of the SPF. The abundance of benthos is higher north of the SPF.

6.2.4.5   A site of enhanced biota

Chlorophyll *a* (Chl *a*) concentrations were elevated in the SPF/CGFZ area (approximately 50–100 mg Chl *a* m^{-2}, 0–30 m) compared with other regions along the ridge (approximately 10–50 mg Chl *a* m^{-2}, 0–30 m) (Gaard *et al.* 2008; Gislason *et al.* 2008; Opdal *et al.* 2008). Several zooplankton taxa were more abundant in the SPF region than elsewhere, for example *Calanus* (Gislason *et al.* 2008), *Pareuchaeta* (Falkenhaug *et al.* 2007), Decapoda (unpublished data), Chaetognatha (Pierrot-Bults 2008), and gelatinous megaplankton (Youngbluth *et al.* 2008). Interestingly, most these taxa are predatory. The area of elevated Chl *a* at the SPF corresponded with an area of elevated rates of egg production by *Calanus finmarchicus* (Gislason *et al.* 2008). Elevated bioluminescence at the SPF (Heger *et al.* 2008) also coincided with higher zooplankton abundances

Fig. 6.5

Depth distribution of macrourid fishes from the summit of the MAR to the lower slopes. Adapted from Bergstad *et al.* (2008a).

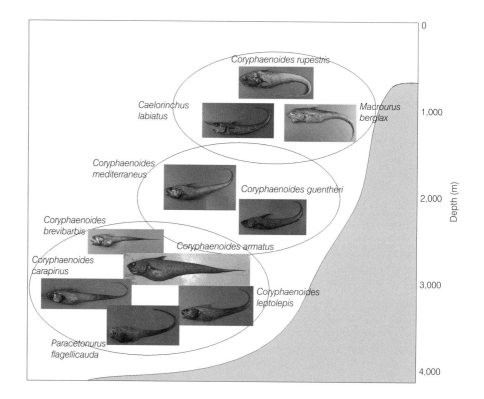

in this region. The most likely producers of the biolumi-nescence are crustaceans (decapods/euphausiids) and gelati-nous zooplankton. In contrast, lower abundances in the total copepods were recorded in the CGFZ (Gaard *et al.* 2008).

The density of sound scatterers, presumably dominated by fishes, was highest in the CGFZ area and in association with the SPF. In this area, higher densities tended to be associated with marked topographical features, that is, hills or troughs. There appeared to be a positive relation between horizontal distribution of standing stock of phy-toplankton as measured by Chl *a* and acoustic backscatter-ing. High abundances of demersal fishes were also found in the SPF (King *et al.* 2006; Bergstad *et al.* 2008b; Fossen *et al.* 2008).

Observations on seabirds and cetaceans unambiguously point to the surface manifestation of the SPF near the CGFZ as the most important large-scale habitat feature for several species of planktivores as well as nekton feeders along the MAR. Results indicate general co-occurrence between these top predators and concentrations of poten-tial prey like squid, pelagic fishes, and macrozooplankton in the CGFZ. Synoptic acoustic and visual transects across the ridge and individual seamounts indicated a segregation of sei whale (*Balaenoptera borealis*) and sperm whale (*Physeter macrocephalus*) in relation to topography. Multivari-ate analyses of the distribution of the two species with concurrent water current measurements (from acoustic Doppler current profilers) and hydrographic and topo-graphic data at various spatial scales indicated that small-scale frontal processes interacting with the topography in the surface and subsurface waters just north of the surface SPF are important for the transfer of energy to higher trophic levels in the MAR.

6.2.5 Cross-ridge (east–west) patterns of distribution

Cross-ridge patterns of zooplankton were less clear than along the ridge. Data on distributions of copepods and amphipods showed indications of east–west differences in species composition. More warm-water species (for example *C. helgolandicus*) were found east of the ridge (Gaard *et al.* 2008) than to the west. The northeastward trajectory of the North Atlantic Current may enable species of the warm-temperate association to be present east of the MAR and north to 50°N. As a result, the northern distribution limit of warm-water species is farther north on the eastern side than on the western side of the ridge.

Cross-ridge differences in pelagic fish distributions were not detectable by multivariate analysis of catch data (Sutton *et al.* 2008). Similarly, cross-ridge differences in demersal fish distributions were apparently minor, but pairwise com-parisons of trawl stations at the same depth and latitude on

either side of the central rift valley indicate that such differences may occur.

6.2.6 Food webs and carbon fluxes

Estimated phytoplankton ingestion rates by copepods were not higher at the SPF/CGFZ than in the other areas, indi-cating that the relatively high production of copepod eggs in the CGFZ is mainly fueled by sources of energy other than phytoplankton (for example microzooplankton) (Gislason *et al.* 2008). *Calanus* spp. are very important in the pelagic ecosystem over the northern MAR as food for organisms at higher trophic levels. Two main trophic path-ways were inferred over the Reykjanes Ridge. In one, *Calanus* spp. are important in the diet of *Maurolicus muelleri*, *B. glaciale*, and *Sergestes arcticus*, whereas in the other pathway *Meganyctiphanes norvegica* is the dominant food for the redfish *Sebastes mentella* and *Calanus* are of less importance (Petursdottir *et al.* 2008). For *S. arcticus*, *Calanus hyperboreus* is an important part of the food. Addi-tionally, *S. arcticus* probably has a benthic component in the diet (Petursdottir *et al.* 2008).

Two different types of food web appear to be important on the MAR (Fig. 6.6). One is a classic predatory trophic enrichment with higher levels feeding on those lower. Pre-liminary analysis of stable-isotopes in the pelagic food web show consumption/conversion of energy from shallow to deep, with a concomitant increase in trophic level: epipe-lagic fauna to vertically migrating mesopelagic fauna to non-migrating bathypelagic fauna to demersal fauna (trophic levels 2–6). The other, the benthic web, has organ-isms all essentially feeding on detritus in the sediment but elevated 'trophic' levels (as indicated by δN isotopic values) result from repackaging and remineralization of that mate-rial as it passes through the digestive tracts of other deposit feeders. Hence sub-surface feeders have greater δN values than surface feeders.

6.2.7 Life-history studies

The preponderance near the ridge of large, adult bathype-lagic fishes (Sutton *et al.* 2008; M. Heino *et al.*, unpub-lished observations), many in gravid condition (A. Stene, unpublished observations), suggests that the MAR, and perhaps other mid-ocean ridge systems, may be important spawning locations for otherwise widely (basin-wide) dis-tributed fish species. Abundance and distributional data from the MAR-ECO project further suggest that mid-ocean ridges may serve to concentrate deep-pelagic fishes, thereby enhancing reproductive and trophic interactions.

The shapes of length distributions of many common pelagic fish species are unusual, peaking near the maximum known size and with fewer small individuals than normally

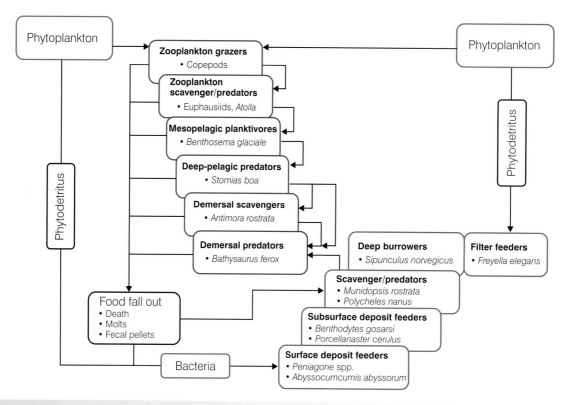

Fig. 6.6

A simplified MAR food web, showing consumption/conversion of energy from shallow to deep, with a concomitant increase in trophic level: epipelagic fauna to vertically migrating mesopelagic fauna to non-migrating bathypelagic fauna to demersal fauna (trophic levels 2–6). Example organisms are given for each trophic step. the benthic "chain" has organisms all feeding on essentially detritus in the sediment but the elevated "trophic" levels (as indicated by δN isotopic values) are a result of repackaging and remineralization of that material as it passes through the digestive tracts of other deposit feeders.

expected – opposite to 'typical' fish length distribution where most individuals are small or intermediate in size. If samples from the MAR resemble the overall population structure (that is, the MAR assemblages do not "overrepresent" adult fishes), then these unusual length distributions suggest that these species have fast growth rate relative to mortality rate. This does not imply that their growth is fast in absolute terms, only that growth outpaces mortality, leading to accumulation of individuals in size classes near the species' asymptotic size.

Small bobtail squids of the species *H. dispar* are the most pelagic members in the family Sepiolidae. During the 2004 *Sars* cruise, 46 specimens were collected in the southern region of the study area (Vecchione *et al.* 2010). All females, including immature specimens, were carrying sperm packages, indicating that the animals can take advantage of chance encounters between the sexes at any time (Hoving *et al.* 2008). Although all female cephalopods can carry or store sperm, the anatomy of the female reproductive system of *H. dispar* suggests that they also fertilize their eggs internally, a reproductive strategy so far unknown for squid or cuttlefish. This reproductive trait, and the small egg size, indicates an adaptation to an oceanic lifestyle, unique within the bobtail squids.

New life-history information was collected for selected demersal fishes. Orange roughy (*Hoplostethus atlanticus*) were not prominent in the MAR-ECO samples, but the few specimens collected comprised very young juveniles of ages 1–2 years and adults within the age range 78–139 years (Tyssebotn 2008). Previous studies have also shown the extended life cycle of this commercially important species, but small juveniles are rarely sampled. A more widespread slope species most abundant at 1,000–3,000 m was the blue hake, *Antimora rostrata*. This species was sampled from the Azores to Greenland (Fossen & Bergstad 2006). *A. rostrata* has intermediate longevity compared with other co-occurring deep-water fishes that have been studied sufficiently. It is neither short-lived, nor especially long-lived. Small juveniles are found in upper slope waters off Greenland, but were rare on the MAR. Hitherto unknown to science, postlarvae of this common species were found on the MAR, indicating that spawning does take place on the ridge (P.R. Møller, unpublished observations).

6.2.8 Management implications

MAR-ECO research has already had conservation implications. Based largely on the results of the *Sars* expedition, the North-East Atlantic Fisheries Commission (NEAFC) adopted measures that close more than 330,000 km² to bottom fisheries on the Mid-Atlantic Ridge, including the area of the CGFZ and SPF (Probert *et al.* 2007) "to protect Vulnerable Marine Ecosystems in the High Seas".

6.3 Knowledge Gaps

The primary remaining gap in current knowledge is the determination of temporal (diel, seasonal, and interannual) variation in faunal composition, abundance, and ecology. The species composition and distribution patterns now described are snapshots from summer seasons, namely after the presumed spring peak in epipelagic production. More extensive sampling into adjacent basins would be needed to determine definitively that observed patterns for pelagic biota are ridge-specific. Few samples with the same methods have ever been collected on adjacent continental slopes, islands, and seamounts; hence it is currently only feasible to assess basin-wide trends relative to impact of ridge for a few faunal components (for example demersal fishes, epibenthos, zooplankton, but not pelagic nekton). To what extent conclusions about structuring of communities by depth, water masses, and frontal zones from the North Atlantic MAR are valid for other sections of the mid-ocean ridge system has not yet been determined.

To enhance knowledge of food web and trophic ecology, the feeding, behavior, and physiology of more species need to be studied. Variables for calculating energy budgets and growth rates are unknown (rates of consumption, excretion, respiration, mortality). To estimate the predation impact of fishes on gelatinous prey, commensurate gelatinous and other zooplankton data on same spatial scales as nekton would be required. Information on the significance of microbial processes (microbial loop) is also needed.

The documented information on the significance of the mid-ocean ridge in the life cycles of nekton and zooplankton remains scattered. Although the occurrence of mature size classes of many nekton species suggests that the ridge is a significant reproduction area, studies of reproduction are lacking for most species. Comparative analyses across taxa would be needed to determine the diversity of life-history characteristics such as longevity, fecundity, and growth. For several demersal fish species, present connectivity between ridge and slope conspecifics appears to be limited, but similar studies for pelagic species are lacking.

Production estimates are few, and the relative significance of different physical and biological processes regulating production and transfer of energy horizontally and vertically remains unclear. Near-ridge zooplankton and pelagic nekton distributions in the entire water column appear to be affected by physical forcing. Without knowledge of the near-ridge flow field many of the potential biotic interactions cannot be substantiated.

6.4 Recommendations

6.4.1 Technologies

Many of the technological challenges for the future are common to studies in many habitats, but are especially pronounced in mid-ocean investigations of large volumes and depth ranges and at great distances from land. Although the analysis of abiotic factors can often be performed in near real-time, processing biological samples can take months and even years. This time lag especially applies to taxonomic studies, basic to studies of structure of assemblages and ecosystem processes. New methods to accelerate identification of organisms and processing of samples should be sought. The underwater video profiler used by MAR-ECO is an example of a major step in this direction. ROVs better adapted for pelagic studies have the potential to provide small-scale data at the individual level for many taxa. Despite advances made by ROV and submersible technology, sampling on rough rocky ground remains challenging. New gear also seems necessary to study the organisms of the benthic boundary layers.

To fill the major gap in temporal dynamics, traditional cruise-based sampling should be supplemented more widely with multiple long-term automatic recording devices using optics and acoustics. Enhancing the depth rating of transducers would be needed to extend the depth to which acoustic landers can be deployed. Current moored acoustic instruments have the potential to monitor sound-scattering zooplankton and nekton of all sizes, and their interactions, at several spatial scales (Fig. 6.7). Observation volume has to increase, however, and species identification has to improve, for example through visual observations or recognition of vocalization. Multiple-frequency sounders facilitate categorizing of the acoustic recordings to some degree, but concurrent targeted trawl sampling is needed to substantiate identification and to reveal details. Low-frequency acoustics may sample at subpopulation scales and high frequencies provide details of millimeter scales.

6.4.2 Strategies for continued exploration of the global ridge system

Mid-ocean ridges circle the globe, and to understand biodiversity patterns and ecological processes of this vast habitat a major multi-year and very costly field effort would be required. This effort would involve ships, observatories, and autonomous vehicles, not to mention the dedication and resources of scientists across the globe. Planning efforts should use new knowledge and experiences gained from these recent studies. Some major conclusions from the North Atlantic may provide guidance for strategic decisions on designs of new field efforts. It

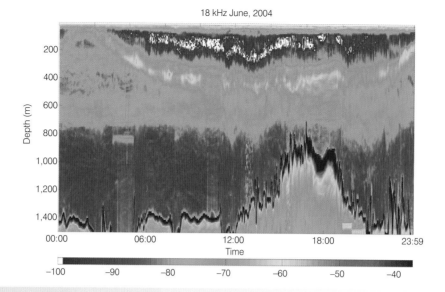

Fig. 6.7

Diel echogram from June 11, 2004, during crossing of MAR (time axis is in UTC). Between surface (top of graph) and bottom (wide red line at lower part of graph) are two major layers of biomass. In the mid-part of the echogram is a deep scattering layer of mesopelagic fishes from 600 to 800 m. Above that is a layer of fishes migrating from about 500 m depth during day to the surface at night. This layer seems to include various components with different migration and distribution patterns, probably caused by the variety of species and sizes included in this layer. At surface during day there are layers of planktonic organisms. The deeper part of the echogram (greater than 900 m) had fewer fish echoes. The echogram patterns also demonstrate that topography affect density and distribution of marine organisms, particularly obvious where bottom peaks to 800 m depth. The three thin lines going from surface to deep water and back again like elongated Vs are recordings of instruments lowered from surface to about 1,000 m. The thick and short line at about 850 m at the first super station is a false-bottom echo.

is likely that depth and topography play major structuring roles, and along-ridge patterns appear more associated with water-mass distributions and circulation features than with other features, for example, fracture zones. This suggests that new studies, inevitably resource limited, would have to be truly depth stratified both in the pelagic and near-bottom zones and would further benefit from targeting features such as hydrographic frontal zones. Within this framework, an exploratory effort would perhaps benefit from a random stratified design. The strength of using multiple gears and instruments has been demonstrated and any departure from multidisciplinarity should be discouraged. Accommodating more elements such as all benthos components including meiofauna and microbes, and so forth, would be beneficial.

Expansion of exploratory programs into the South Atlantic has considerable interest because of its comparability with the North Atlantic MAR, and the level of knowledge of its fauna is very limited indeed. However, there may also be benefits from focusing on frontal zones and ridges that are suspected to be substantially different from the MAR. For example, the faster-spreading East Pacific Rise might be of special interest for comparison and contrasting with the MAR. Selecting study areas in varying pelagic productivity regimes would probably enhance knowledge even further.

6.5   Conclusions

The mid-ocean habitats associated with the MAR have high diversity and considerable, if presently undetermined, abundance of macro-and megafauna. The first observations of elevated richness and catches made by the *Michael Sars* expedition in 1910 were thus confirmed and considerably expanded by the multi-vessel operations conducted by MAR-ECO.

Primary questions addressed to what extent faunal components from eastern and/or western slopes contributed to the MAR fauna, whether there was a latitudinal change in the fauna, and to what extent this was affected by two main hydrographic features: the Sub-Polar Front and the Charlie Gibbs Fracture Zone (see, for example, Krauss 1986; Rossby 1999; Reverdin *et al.* 2003). A general pattern identified was that the MAR supports high biodiversity which is not fundamentally different in structure and composition from the adjacent continental slopes. Furthermore, the results do not indicate a high degree of endemism. In this the MAR does not appear as a chain of isolated seamounts. The patterns emerging from the benthic invertebrate data showed that the MAR constitutes an interface between eastern and western continental slope fauna, with a slight overweight of the former (Mironov *et al.* 2006).

The SPF seems to be a more important feature than the CGFZ in affecting the composition of the benthic inverte-brate fauna (see, for example, Gebruk 2008a) and demersal fishes (King *et al.* 2006; Bergstad *et al.* 2008b). Similar conclusions were reached for zooplankton (Gaard *et al.* 2008; Hosia *et al.* 2008; Stemmann *et al.* 2008), although in zooplankton the SPF constitute more of a marked bound-ary for southern than northern species. Variation in the pelagic fauna is primarily related to depth, and the bathy-pelagic fish assemblages show a high degree of consistency from Iceland to the Azores (Sutton *et al.* 2008). Overall, the abundance of fishes declines with depth, but unexpect-edly the pelagic fishes showed a biomass maximum in the bathypelagic zone between 1,500 and 2,300 m and in the benthic boundary layer, 0–200 m above the bottom regard-less of depth.

For many taxa (see, for example, phytoplankton, crus-tacean mesozooplankton, pelagic decapods, bioluminescent organisms, mesopelagic acoustic scatterers, whales, and sea-birds), there appeared to be a maximum along-ridge abun-dance in association with the SPF, but this frontal zone spans a considerable latitudinal range (48–53°N) and varies temporally in character and configuration. Whales appeared associated with areas of steeply sloping bottom.

Several species were rare, especially benthic inverte-brates, and some were new to science (Fig. 6.8). Only

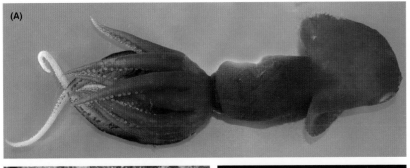

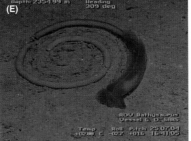

Fig. 6.8

Representative new species and remarkable observations. **(A)** Sloan squid, *Promachoteuthis sloani* Young *et al.* 2006; **(B)** giant protist family Syringamminidae; **(C)** mysterious track-like holes of unknown origin; **(D)** *Altelatipes falkenhaugae* Crosnier and Vereshchaka 2008 (© Publications Scientifiques du Muséum national d'Histoire naturelle, Paris); **(E)** undescribed enteropneust; **(F)** *Cottunculus tubulosus*, Byrkjedal & Orlov 2007 (courtesy of Ingvar Byrkjedal).

extended sampling can tell whether these represent species endemic to the MAR. The rich collection of MAR animals, now archived at the Bergen Museum, will certainly be useful for taxonomic, phylogenetic, and other types of study for many decades.

Sampling extended to the adjacent abyssal plains would be necessary to determine to what extent ridge-associated abundances are especially high. Such sampling would be of particular interest to gain more insight into the midwater biomass maximum found over the ridge for pelagic fishes.

The MAR-ECO sampling only provided a "snapshot" impression of the faunal composition and distribution. With the exception of some acoustic landers deployed over a longer time span, the sampling did not cover circadian or circumannual time series. Thus, the results tell nothing about faunal responses to periods of plankton blooms, and nothing new can be said about possible temporal life-history phenomena such as vertical and horizontal migration, reproduction periods, and special reproduction areas. Diel migration patterns could even change over the annual cycle, and possible seasonal changes in the occurrence of the benthic boundary layer are still unstudied.

The benthic boundary layer community is very difficult to sample adequately with the gear that was available to MAR-ECO. The species composition of this assemblage is therefore not well known. Such a study might be suspected to yield more insight into the abundance of nekton in the benthic boundary layer. Furthermore, studies of meiofaunal and microbial communities and processes were not included in the MAR-ECO project.

Food web structures of the MAR communities have been determined for the first time based on studies combining analyses of stable isotope ratios, fatty acid profiles, and traditional determination of diets from identification of gut contents. Some insight has also been gained from studies of fish parasites (Klimpel *et al.* 2008). Some inferences on the top predator feeding ecology were made by comparing occurrences of predators and prey (Doksæter *et al.* 2008). Major issues for the future are the estimation of rates, selectivity studies in relation to prey fields, and the predation on and by gelatinous organisms. MAR-ECO did not collect data to assess the significance of microbial loop pathways. Seasonal feeding and growth studies, and perhaps *in situ* experiments to determine rates would provide substantial advances.

At the time of year studied (June–August), highest primary production was observed in the region south of the CGFZ. Characterized by a well-defined northern boundary and an extended mixed area extending southwards, the frontal area and the North Atlantic Current have a major influence over a large sector of the ridge. We hypothesize that differences in organic carbon export, prevailing bottom water temperature, and topography combine to determine benthic species composition.

Future work should be directed to determine surface production and pelagic to benthic fluxes over a wider latitudinal range than is currently being investigated. The physical interaction of the pelagic biota with the ridge needs to be examined at spatial scales from meters to the whole ridge system. Progress will depend on technological developments, notably fixed and mobile autonomous instruments, new tagging and tracking systems, and so forth, that can overcome the limitations of shipborne measurements.

A major challenge is the spatial and temporal scale of phenomena in the pelagic realm. The bio-physical environment on the MAR region is very complex, and only remote sensing methods provide the regional and temporal scales required to detect seasonal and inter-annual changes. Frequent cruises at different times of the year would be necessary to reach a full understanding of events and processes.

Acknowledgments

MAR-ECO is grateful to numerous public institutions, funding agencies, private sponsors, and industry partners that provided essential resources to the project throughout its lifetime. The Institute of Marine Research and the University of Bergen (Bergen Museum), Norway, initiated and coordinated the project. As authors we are indebted to all other scientific colleagues who participated in the effort and contributed to the outcome.

References

Backus, R.H., Craddock, J.E., Haedrich, R.L. & Robison, B.H. (1977). Atlantic mesopelagic zoogeography. In: *Fishes of the Western North Atlantic*. (eds. R.H. Gibbs., F.N.H. Berry, J.E. Böhlke, *et al.*), Vol. 1, part 7, pp. 266–287. New Haven, Connecticut: Sears Foundation for Marine Research.

Beaugrand, G., Reid, P.C., Ibañez, F. & Planque, P. (2000) Biodiversity of North Atlantic and North Sea calanoid copepods. *Marine Ecology Progress Series* **204**, 299–303.

Beaugrand, G., Ibañez, F., Lindley, J.A. & Reid, P.C. (2002) Diversity of calanoid copepods in the North Atlantic and adjacent seas: species associations and biogeography. *Marine Ecology Progress Series* **232**, 179–195.

Bergstad, O.A. & Godø, O.R. (2003) The pilot project "Patterns and processes of the ecosystems of the northern Mid-Atlantic": aims, strategy and status. *Oceanologica Acta* **25**, 219–226.

Bergstad, O.A., Høines, Å.S., Orlov, A., *et al.* (2008a) Species composition and abundance patterns of grenadiers on the Mid-Atlantic Ridge between Iceland and the Azores. *American Fisheries Society Symposium* **63**, 65–80.

Bergstad, O.A., Menezes, G. & Høines, Å.S. (2008b) Demersal fish on a mid-ocean ridge: Distribution patterns and structuring factors. *Deep-Sea Research II* **55**, 185–202

Bergstad, O.A., Falkenhaug, T., Astthorsson, O.S., *et al.* (2008c) Towards improved understanding of the diversity and abundance patterns of the mid-ocean ridge macro- and megafauna. *Deep-Sea Research II* **55**, 1–5.

Bower, A.S., Le Cann, B., Rossby, T., *et al.* (2002) Directly measured mid-depth circulation in the northeastern North Atlantic Ocean. *Nature* **419**, 603–606.

Brandt, A. & Andres, H.G. (2008) Description of *Aega sarsae* sp.nov. and redescription of *Syscenus atlanticus* Kononeko, 1988 (Crustacea, Isopoda, Aegidae) from the Mid-Atlantic Ridge. *Marine Biology Research* **4**, 61–75.

Byrkjedal, I. & Orlov, A.M. (2007) A new species of *Cottunculus* (Teleostei: Psychrolutidae) from the Mid Atlantic Ridge. *Zootaxa* **1580**, 63–68.

Chernova, N.V. & Møller, P.R. (2008) A new snailfish, *Paraliparis nigellus* sp. nov. (Scorpaeniformes, Liparidae), from the northern Mid-Atlantic Ridge – with notes on occurrence of *Psednos* in the area. *Marine Biology Research* **4**, 369–375.

Clark, M.R., Vinnichenko, V.I., Gordon, J.D.M., *et al.* (2007) Large-scale Distant-water Trawl Fisheries on Seamounts. Chapter 17 In: *Seamounts: Ecology, Conservation and Management* (eds. T.J. Pitcher, T. Morato, P.J.B. Hart, *et al.*), Fish and Aquatic Resources Series, pp. 361–399. Oxford, UK: Blackwell.

Crosnier, A. & Vereshchaka, A. (2008) *Altelatipes falkenhaugae* n.gen., n.sp. (Crustacea, Decapoda, Benthesicymidae) de la ride médio-alantique nord. *Zoosystema* **30**, 399–411.

Dilman, A.B. (2008) Asteroid fauna of the northern Mid-Atlantic Ridge with description of a new species *Hymenasterides mironovi* sp.nov. *Marine Biology Research* **4**, 131–151.

Dinter, W.P. (2001) *Biogeography of the OSPAR maritime area.* Bonn, Germany: Federal Agency for Nature Conservation. 167 pp.

Doksæter, L., Olsen, E. Nøttestad, L. & Fernö, A. (2008) Distribution and feeding ecology of dolphins along the Mid-Atlantic Ridge between Iceland and the Azores. *Deep-Sea Research II* **55**, 243–253.

Doksæter, L., Godø, O.R., Olsen, E., Nøttestad, L. & Patel, R. (2009) Ecological studies of marine mammals using a seabed-mounted echosounder. *ICES Journal of Marine Science* **66**, 1029–1036.

Falkenhaug, T., Gislason, A. & Gaard, E. (2007) *Vertical distribution and population structure of copepods along the northern Mid-Atlantic Ridge.* ICES ASC Helsinki 17–21 September 2007. ICES CM 2007/F:07.

Felley, J.D., Vecchione, M. & Wilson, R.R. (2008) Small-scale distribution of deep-sea demersal nekton and other megafauna in the Charlie-Gibbs Fracture Zone of the Mid-Atlantic Ridge. *Deep-Sea Research II* **55**, 153–160.

Fock, H.O., Pusch, C. & Ehrich, S. (2004) Structure of deep-sea pelagic fish assemblages in relation to the Mid-Atlantic Ridge (45–501). *Deep-Sea Research I* **51**, 953–978.

Fossen, I. & Bergstad, O.A. (2006) Distribution and biology of blue hake, *Antimora rostrata* (Pisces: Moridae), along the mid-Atlantic Ridge and off Greenland. *Fisheries Research* **82**, 19–29.

Fossen, I., Cotton, C.F., Bergstad, O.A. & Dyb, J.E. (2008) Species composition and distribution patterns of fishes captured by longlines on the Mid-Atlantic Ridge. *Deep-Sea Research II* **55**, 203–217.

Gaard, E., Gislason, A., Falkenhaug, T., *et al.* (2008) Horizontal and vertical copepod distribution and abundance on the Mid-Atlantic Ridge in June 2004. *Deep-Sea Research II* **55**, 59–71.

Garrison, T. (1993) *Oceanography: An invitation to Marine Science.* Belmont, California: Wadsworth.

Gebruk, A.V. (ed.) (2008a) Benthic fauna of the northern Mid-Atlantic Ridge: results of the MAR-ECO expedition. *Marine Biology Research* **4**, 1–163.

Gebruk, A.V. (2008b) Holothurians (Holothuroidea, Echinodermata) of the northern Mid-Atlantic Ridge collected by the G.O. Sars MAR-ECO expedition with description of four new species. *Marine Biology Research* **4**, 48–60.

Gislason, A., Gaard, E., Debes, H. & Falkenhaug, T. (2008). Abundance, feeding and reproduction of *Calanus finmarchicus* in the Irminger Sea and on the northern Mid-Atlantic Ridge in June. *Deep-Sea Research II* **55**, 72–82.

Gordon, J.D.M., Bergstad, O.A. and Falkenhaug, T. (eds.) (2008) Mid-Atlantic Ridge Habitats and Biodiversity. *Deep-Sea Research II* **55**, 1–268.

Hareide, N.-R. & Garnes, G. (2001) The distribution and catch rates of deepwater fish along the Mid-Atlantic Ridge from 43 to 61° N. *Fisheries Research* **51**, 297–310.

Haedrich, R.L. & Merrett, N.R. (1988) Summary atlas of deep-living fishes in the North Atlantic. *Journal of Natural History* **22**, 1325–1362.

Heger, A., Ieno, E.N., King, N.J., *et al.* (2008) Deep-sea pelagic bioluminescence over the Mid-Atlantic Ridge. *Deep-Sea Research II* **55**, 126–136.

Holland, N.D., Clague, D.A., Gordon, D.P., Gebruk, A., Pawson, D.L., & Vecchione, M. (2005) 'Lophenteropneust' hypothesis refuted by collection and photos of new deep-sea hemichordates. *Nature* **434** (7031), 374–376.

Hosia, A., Stemmann, L. & Youngbluth, M. (2008) Distribution of net-collected planktonic cnidarians along the northern Mid-Atlantic Ridge and their associations with the main water masses. *Deep-Sea Research II* **55**, 106–118.

Hoving H.J.T., Laptikhovsky, V., Piatkowski, U. & Önsoy, B. (2008) Reproduction in *Heteroteuthis dispar* (Rüppell, 1844) (Mollusca: Cephalopoda): a sepiolid reproductive adaptation to an oceanic lifestyle. *Marine Biology* **154**, 219–230.

King, N., Bagley P.M. & Priede I.G. (2006) Depth zonation and latitudinal distribution of deep sea scavenging demersal fishes of the Mid-Atlantic Ridge, 42°–53° N. *Marine Ecology Progress Series.* **319**, 263–274.

Klimpel, S., Palm, H.W., Busch, M.W. & Kellermanns, E. (2008) Fish parasites in the bathyal zone: The halosaur *Halosauropsis macrochir* (Günther, 1878) from the Mid-Atlantic Ridge. *Deep-Sea Research II* **55**, 229–235.

Knutsen, H., Jorde, P.E., Albert, O.T., *et al.* (2007) Population genetic structure in the North Atlantic Greenland halibut: influenced by oceanic current systems? *Canadian Journal of Fisheries and Aquatic Sciences* **64**, 857–866.

Knutsen, H., Jorde, P.E., Sannæs, H., *et al.* (2009) Bathymetric barriers promoting genetic structure in the deepwater demersal fish tusk *Brosme brosme*. *Molecular Ecology* **18**, 3151–3162.

Krauss, W. (1986) The North Atlantic Current. *Journal of Geophysical Research C* **91**, 5061–5074.

Kritsky D.C. & Klimpel S. (2007) *Cyclocotyloides bergstadi* n. sp. (Monogenoidea: Diclidophoridae: Diclidophoropsinae) from the gills of grenadier, *Coryphaenoides brevibarbis* (Teleostei: Macrouridae), in the Northeast Atlantic Ocean. *Comparative Parasitology* **74**, 23–30.

Kukuev, E.I. (2004) 20 years of ichthyofauna research on seamounts of the North Atlantic Ridge and adjacent areas. A review. *Archive of Fishery and Marine Research* **51**, 215–232.

Lavoue, S., Miya, M., Poulsen, J.Y., *et al.* (2008) Monophyly, phylogenetic position and inter-familial relationships of the Alepocephaliformes (Teleostei) based on whole mitogenome sequences. *Molecular Phylogenetics and Evolution* **47**, 1111–1121.

Longhurst, A. (1998) *Ecological Geography of the Sea.* AcademicPress, London, 398 pp.

Magnusson, J. & Magnusson, J.V. (1995) Oceanic redfish (*Sebastes mentella*) in the Irminger Sea and adjacent waters. *Scientia Marina* **59**, 241–254.

Martynov, A.V. & Litvinova, N.M. (2008) Deep-water Ophiuroidea of the northern Atlantic with descriptions of three new species and taxonomic remarks on certain genera and species. *Marine Biology Research* **4**, 76–111.

Menezes, G.M., Sigler, M.F., Silva, H.M. & Pinho, M.R. (2006) Structure and zonation of demersal fish assemblages off the Azores archipelago (Mid Atlantic). *Marine Ecology Progress Series* **324**, 241–260.

Menschenina, L.L., Tabachnik, K.R., Lopes, D.A. & Hajdu, E. (2007) Revision of *Calycosoma* Schulze, 1899 and finding of *Lophocalyx* Schulze, 1887 (six new species) in the Atlantic Ocean (Hexactinellida, Rossellidae). In: *Porifera Research: Biodiversity, Innovation and Sustainability* (eds. M.R. Custódio, G. Lôbo-Hajdu, E. Hajdu & G. Muricy), pp. 449–465. Série Livros 28, Museo Nacional, Rio de Janeiro, Brazil.

Merrett, N.R. & Haedrich, R.L. (1997) *Deep-Sea Demersal Fish and Fisheries*. London: Chapman & Hall. 282 pp.

Mironov A.N. (1994) Bottom faunistic complexes of oceanic islands and seamounts. *Trudy Instituta Okeanologii AN USSR* **129**, 7–16. (In Russian with English summary.)

Mironov, A.N. (2008) Pourtalesiid sea urchins (Echinodermata: Echinoidea) of the northern Mid-Atlantic Ridge. *Marine Biology Research* **4**, 3–24.

Mironov, A.N., Gebruk, A.V. & Southward, A.J. (eds.) (2006) *Biogeography of the North Atlantic Seamounts*. Moscow: KMK Scientific Press. 196 pp.

Moravec F. & Klimpel S. (2007) A new species of *Comephoronema* (Nematoda : Cystidicolidae) from the stomach of the abyssal halosaur *Halosauropsis macrochir* (Teleostei) from the mid-Atlantic ridge. *Journal of Parasitology* **93**, 901–906.

Moravec, F., Klimpel, S. & Kara, E. (2006). *Neoascarophis macrouri* n. sp. (Nematoda: Cystidicolidae) from the stomach of *Macrourus berglax* (Macrouridae) in the eastern Greenland Sea. *Systematic Parasitology* **63**, 231–237.

Murina, V.V. (2008) New records of Echiura and Sipuncula in the North Atlantic Ocean, with description of a new species of *Jacobia*. *Marine Biology Research* **4**, 152–156.

Murray, J. & Hjort, J. (1912) *The Depths of the Ocean*. London: Macmillan.

Opdal, A.F., Godø, O.R., Bergstad, O.A. & Fiksen, Ø. (2008) Distribution, identity, and possible processes sustaining meso- and bathypelagic scattering layers on the northern Mid-Atlantic Ridge. *Deep-Sea Research II* **55**, 45–58.

Orlov, A., Cotton, C.F. & Byrkjedal, I. (2006) Deepwater skates collected during 2004 R/V G.O. Sars and M/V Loran cruises in the Mid-Atlantic Ridge area. *Cybium* **30**, 35–48.

Pedchenko A.P. & Dolgov, A.V. (2005) The results of the ecosystem investigations by PINRO in the Irminger Sea at beginning of XXI century. *XIII International conference on fisheries oceanology, Kaliningrad, Russia, September 2005*, pp. 216–217. Kaliningrad: AtlantNIRO Press. (In Russian.)

Petursdottir, H., Gislason, A., Falk-Petersen, S., *et al.* (2008) Trophic interactions of the pelagic ecosystem over the Reykjanes Ridge as evaluated by fatty acid and stable isotope analyses. *Deep-Sea Research II* **55**, 83–93.

Pierrot-Bults, A.C. (2008) A short note on the biogeographic patterns of the Chaetognatha fauna in the North Atlantic. *Deep-Sea Research II* **55**, 137–141.

Post, A. (1987) Stations lists and technical data of the pelagic transects of FRVs "Walther Herwig" and "Anton Dohrn" in the Atlantic Ocean 1966 to 1986. *Mitteilungen aus dem Institut für Seefischerei* **42**, 1–67.

Poulsen, J.Y., Møller, S., Lavoué, S., *et al.* (2009) Higher and lower-level relationships of the deep-sea fish order Alepocephaliformes (Teleostei: Otocephala) inferred from whole mitogenome sequences. *Biological Journal of the Linnean Society* **98**, 923–936.

Priede I.G., Froese R., Bailey D.M., *et al.* (2006) The absence of sharks from abyssal regions of the world's oceans. *Proceedings of the Royal Society of London B* **273**, 1435–1441.

Probert, P.K, Christiansen, S., Gjerde, K.M., *et al.* (2007) Management and conservation of seamounts. In *Seamounts: Ecology, Conservation and Management* (eds. T.J. Pitcher, T. Morato, P.J.B.

Hart, *et al.*), Fish and Aquatic Resources Series, pp. 442–475. Oxford: Blackwell.

Reverdin, G., Niiler, P.P. & Valdimarsson, H. (2003) North Atlantic Ocean surface currents. *Journal of Geophysical Research C* **108**, 3002.

Rossby, T. (1999) On gyre interaction. *Deep-Sea Research II* **46**, 139–164.

Santos, R.S., Porteiro, F.M. & Barreiros, J.P. (1997) Marine fishes of the Azores: annotated checklist and bibliography. *Arquipelago. Life and Marine Sciences* (Supplement 1), 1–244.

Schmidt, J. (1931) Oceanographic expedition of the Dana, 1928–1930. *Nature* **127**, 444–446, 487–490.

Sigurjónsson, J., Gunnlaugsson, T., Ensor, P., *et al.* (1991) North Atlantic sighting survey 1989 (NASS-89): shipboard surveys in Icelandic and adjacent waters July–August 1989. *Report on International Whale Communication* **41**, 559–572.

Søiland, H., Budgell, W.P. & Knutsen, Ø. (2008) The physical oceanographic conditions along the Mid-Atlantic north of the Azores in June–July 2004. *Deep-Sea Research II* **55**, 29–44.

Stefanni, S. & Knutsen, H. (2007) Phylogeography and demographic history of the deep-sea fish *Aphanopus carbo* (Lowe, 1839) in the NE Atlantic: vicariance followed by secondary contact or speciation? *Molecular Phylogeny and Evolution* **42**, 38–46.

Stefanni, S., Bettencourt, R., Knutsen, H. & Menezes G. (2009) Rapid polymerase chain reaction-restriction fragment length polymorphism method for discrimination of the two Atlantic cryptic deep-sea species of scabbardfish. *Moecular Ecology Resources* **9**, 528–530.

Stemmann, L., Hosia, A., Youngbluth, M.J., *et al.* (2008) Vertical distribution (0–1000 m) of macrozooplankton, estimated using the Underwater Video Profiler, in different hydrographic regimes along the northern portion of the Mid-Atlantic Ridge. *Deep-Sea Research II* **55**, 94–105.

Sutton, T.T., Porteiro, F.M., Heino, M., *et al.* (2008) Vertical structure, biomass and topographic association of deep-pelagic fishes in relation to a mid-ocean ridge system. *Deep-Sea Research II* **55** (1–2), 161–184.

Tabachnik, K.R. & Collins, A.G. (2008) Glass sponges (Porifera, Hexactinellida) of the northern Mid-Atlantic Ridge. *Marine Biology Research* **4**, 25–47.

Tabachnick, K.R. & Menshenina, L.L. (2007). Revision of the genus *Asconema* (Porifera: Hexactinellida: Rossellidae). *Journal of the Marine Biological Association of the United Kingdom* **87**, 1403–1429.

Tåning, A.V. (1944). List of supplementary pelagic stations in the Pacific Ocean and the Atlantic. *Dana-Report* **26**, 1–15.

Tysseboth, I.M. (2008) *Contamination in deep sea fish: toxic elements, dioxins, furans and dioxin-like PCBs in the orange roughy. Master of Science Thesis*, University of Bergen, Norway.

Uiblein, F., Nielsen, J.G. & Møller, P.R. (2008) Systematics of the ophidiid genus *Spectrunculus* (Teleostei: Ophidiiformes) with resurrection of *S. crassus*. *Copeia* **2008**, 542–551.

Vecchione, M. & Roper, C.F.E. (1991) Cephalopods observed from submersibles in the western North Atlantic. *Bulletin of Marine Science* **49**, 433–445.

Vecchione, M. & Young, R.E. (1997) Aspects of the functional morphology of cirrate octopods: locomotion and feeding. *Vie et Milieu* **47**, 101–110.

Vecchione, M. & Young, R.E. (2006) The squid family Magnapinnidae (Mollusca: Cephalopoda) in the Atlantic Ocean, with a description of a new species. *Proceedings of the Biological Society of Washington* **119**, 365–372.

Vecchione, M., Young, R.E. & Piatkowski, U. (2010) Cephalopods of the northern Mid-Atlantic Ridge. *Marine Biology Research* **6**, 25–52.

Vinogradov, G.M. (2005) Vertical distribution of macroplankton at the Charlie-Gibbs Fracture Zone (North Atlantic), as observed from the manned submersible "Mir-1". *Marine Biology* **146**, 325–331.

Wenneck, T. de L., Falkenhaug, T. & Bergstad, O.A. (2008) Strategies, methods, and technologies adopted on the R.V. *G.O. Sars* MAR-ECO expedition to the Mid-Atlantic Ridge in 2004. *Deep-Sea Research II* **55**, 6–28.

White, T.A., Stefanni, S., Stamford, J. & Hoelzel, A.R. (2009) Unexpected panmixia in a long-lived, deep-sea fish with well-defined spawning habitat and relatively low fecundity. *Molecular Ecology* **18**, 2563–2573.

White, T.A., Stamford, J. & Hoelzel, A.R. (2010) Local selection and population structure in a deep-sea fish, the roundnose grenadier (*Coryphaenoides rupestris*). *Molecular Ecology* **19**, 216–226.

Whitehead, P.J.P., Bauchot, M.-L., Hureau, J.-C., Nielsen, J. and Tortonese E. (eds.) (1986) *Fishes of the North-eastern Atlantic and the Mediterranean*, Vols. **I–III**, 1473 pp. Paris: UNESCO.

Young, R.E., Vecchione, M. & Piatkowski, U. (2006) *Promachoteuthis sloani*, a new species of the squid family Promachoteuthidae (Mollusca: Cephalopoda). *Proceedings of the Biological Society of Washington* **119**, 287–292.

Youngbluth, M., Sørnes, T., Hosia, A. & Stemmann, L. (2008) Vertical distribution and relative abundance of gelatinous zooplankton, *in situ* observations near the Mid-Atlantic Ridge. *Deep-Sea Research II* **55**, 119–125.

Chapter 7

Life on Seamounts

Mireille Consalvey[1], Malcolm R. Clark[1], Ashley A. Rowden[1], Karen I. Stocks[2]

[1]*National Institute of Water and Atmospheric Research, Wellington, New Zealand*
[2]*San Diego Supercomputer Center, University of California San Diego, La Jolla, California, USA*

7.1 Introduction: A History of Seamount Research

The rugged terrain and vast mountain ranges that rise from our continents inspire a strong passion, perhaps epitomized by the first climbing of Mount Everest by Sir Edmund Hillary and Sherpa Tenzing Norgay in 1953. In the same decade marine scientists Bruce Heezen and Marie Tharp were looking down, deep into our oceans, mapping the Atlantic seafloor. Painstakingly assembling echo-sounded data, they revealed for the first time the extent of the mid-Atlantic ridge. At 20,000 km, it easily surpasses the length of the Himalayas, Andes, and Rockies combined, and is the longest mountain range on Earth. On this ridge and elsewhere in the oceans stand undersea mountains, or seamounts (Box 7.1), the largest of which rise many kilometers from the sea floor.

The bathymetry of our oceans is now resolved at a scale and detail unimaginable by early pioneers. Yet despite advances in ocean mapping we are still unable to answer seemingly simple questions such as how many seamounts there are. To even begin to estimate the global number of seamounts requires advanced computational technologies. In 2007 Hillier & Watts took 40 million kilometers' worth of echosounder depth measurements and predicted the occurrence of around 40,000 seamounts over 1,000 m tall, most not yet discovered. Widening their scope to include seamounts >100 m high, the authors predicted about 200,000, and speculated there could be as many as 3 million seamounts (Fig. 7.1).

Biological research on seamounts has been limited, and we have a poor understanding of global seamount biodiversity. So far, fewer than 300 seamounts have been studied in sufficient biological detail to describe adequately the assemblage composition of seabed organisms. Furthermore, sampling has been biased toward larger fauna such as fishes, crustaceans, and corals (SeamountsOnline; Stocks 2009).

Carl L. Hubbs (1959) was one of the first biologists to work on seamounts, and the questions he posed in 1959 remain relevant half a century later. What species inhabit seamounts and with what regularity and abundance? How did these species disperse to, and establish on, seamounts? What bearing may the determined constitution of these isolated faunas have on our ideas concerning past and present oceanic circulation and temperatures? Do banks and seamounts provide stepping stones for trans-oceanic dispersal? To what degree has isolation led to speciation? What factors are responsible for the abundance of life over seamounts?

However, one of Hubbs' questions was to be answered quickly: are demersal or pelagic fishes sufficiently abundant on seamounts to provide profitable fisheries? Seamounts host significant commercial fisheries in many parts of the world. Traditional handline fisheries were likely the first fisheries associated with seamounts (Marques da Silva & Pinho 2007) as far back as the fourteenth century (Brewin *et al.* 2007), and continue to the present day. In the 1970s deep-sea trawling began in earnest, targeting large seamount associated fish aggregations (Clark *et al.* 2007a) with nations sending hundreds of vessels around the world's oceans. So far, at least 77 commercially valuable fish species have been fished on seamounts (Rogers 1994). Since the

Life in the World's Oceans, edited by Alasdair D. McIntyre
© 2010 by Blackwell Publishing Ltd.

Box 7.1

Seamounts

Seamounts are prominent features of the world's underwater topography, found in every ocean basin (Fig. 7.1). They are generally volcanic in origin, and often conical in shape. Over geological time seamounts sink (through isostatic adjustment) and erode to become less regular. The topography of seamounts can be complex and within any seamount one may find terraces, canyons, pinnacles, crevices, and craters.

Seamounts are traditionally defined by geologists as having an elevation greater than 1,000m above the seabed (Menard 1964). Biologists now widely include peaks less than 1,000m in their definitions, for there is no known ecological reason for this cutoff height. Pitcher *et al.* (2007) defined a seamount as any topographically distinct seafloor feature that is greater than 100m but which does not break the sea surface to become an island. This definition excludes large banks and shoals (as they differ in size) and topographic features on continental shelves (because of their proximity to other shallow topography).

1960s the total international catch of demersal fishes on seamounts by distant-water fishing fleets is estimated to be over 2.25 million tonnes (Clark *et al.* 2007a), although the true extent of trawling on seamounts may never be known through a combination of catches not being reported, or catches coming from wider areas than just seamounts (Watson *et al.* 2007).

Historically seamount ecosystems have not been well protected (Probert *et al.* 2007) and have been affected by fishing activities that can cause declines in fish stocks and

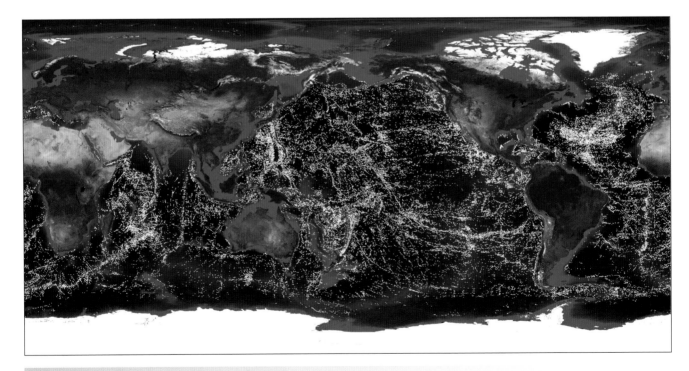

Fig. 7.1
Location of 63,000 seamounts collated from verified regional datasets, or estimated from satellite altimetry or vessel track sounding data (CenSeam 2009).

visible damage to benthic habitat (Davies *et al.* 2007). A great deal of fishing effort has, and continues to, occur on the high seas and many fisheries proceeded largely unregulated, falling outside of any nation's jurisdiction. Although the United Nations and regional fisheries management organizations are becoming more effective, enforcement of regulations on the high seas remains a challenge.

Emergent threats such as deep-sea mineral extraction and indirect threats to all deep-sea habitats are also increasingly being considered, such as rising CO_2 (Guinotte *et al.* 2006). High-profile governmental and non-governmental initiatives have elevated the position of seamounts in the public eye.

Recognizing it is not feasible to sample all of the world's seamounts, research efforts needed to be coordinated to assess the current state of knowledge, fill critical knowledge gaps, and target understudied regions and seamount types. The Global Census of Marine Life on Seamounts (CenSeam) has provided a focal point for coordinating global research and for communicating research results to stakeholders seeking scientific advice and guidance. To mark the end of the first Census of Marine Life, this chapter addresses some of the core research questions that have faced seamount researchers, including those of the CenSeam project, over the past five years. It also indicates where seamount research is likely to be directed in the future.

7.2 A Global Census of Marine Life on Seamounts (CenSeam)

The field of seamount biology has grown in recent decades, as shown by the increasing number of scientific publications each year (Brewin *et al.* 2007). The Census field project CenSeam started in 2005, and has served to bring together more than 500 seamount researchers, policy makers, environmental managers, and conservationists from every continent. At the outset of CenSeam, our understanding of seamount ecosystems was hampered by significant gaps in global sampling, a variety of approaches and sampling methods, and a lack of large-scale synthesis; scientific attention was not yet consistent with their potential biological and ecological value (Stocks *et al.* 2004). CenSeam has aimed to do the following: (1) synthesize and analyze existing data (Box 7.2); (2) coordinate and expand existing and planned research (Box 7.3); (3) communicate the findings through public education and outreach; and (4) identify priority areas for research and foster scientific expeditions to these regions. CenSeam researchers have augmented sampling efforts and analyses in the well studied Southwest Pacific and Northern Atlantic. CenSeam has also identified three key undersampled regions: the Indian Ocean, the South Atlantic, the Western

Box 7.2

SeamountsOnline: Providing Researchers and Managers with Tools for Finding and Accessing Information on the Biological Communities that live on Seamounts

Since 2005, SeamountsOnline (Stocks 2009) has been collecting data on species that have been recorded from seamounts globally, and making them available through a free online data portal (Fig. B7.2). By bringing together global seamount data into a standardized, searchable, electronic format, SeamountsOnline facilitates research and management objectives looking at patterns across different seamounts and regions.

Through a map interface, users can select seamounts of interest, or see the distribution of taxa globally. Users can search for information by management boundaries, such as Exclusive Economic Zones, and biogeographic regionalizations, such as Longhurst Provinces. Seamounts can also be searched by summit depth. Taxonomic searching has options for searching by phylum, class, order, or family, in addition to genus or species. All species observations in SeamountsOnline are also contributed to the Ocean Biogeographic Information System (www.iobis.org), which integrates data from across all the Census of Marine Life projects.

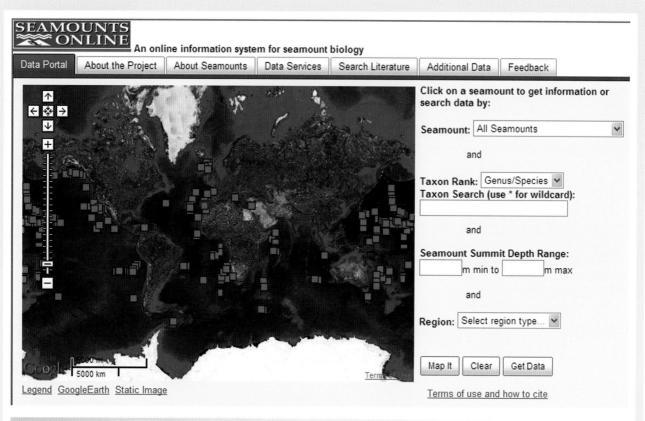

Fig. B7.2

SeamountsOnline is a free online data portal that makes available global data on species that have been recorded from seamounts (seamounts.sdsc.edu, see also its sister database SeamountCatalog at earthref.org/cgi-bin/er.cgi?s=sc-s0-main.cgi).

Box 7.3

Sampling Seamounts

On seamount voyages researchers will typically conduct a bathymetric survey (usually using multibeam sonar) of the target seamount. The resulting baythmetric map provides the basis for more detailed planning of the sampling program: plans that will take into account factors such as seamount size, shape, and depth. Echosounder information can also be used to identify substrate type and can guide sampling to target soft and hard bottoms.

The sampling gear and methodology used will depend on the nature of the research and the *in situ* conditions, for example weather and substrate. For biodiversity surveys,

camera transects (undertaken using towed camera platforms, remotely operated vehicles, or submersibles) should be performed where possible. Remote methodologies have the advantage of being non-destructive and enabling researchers to view intact community composition, and to potentially gain valuable information on animal behavior. However, to quantify biodiversity fully, "ground-truthing" is required, and physical collection is vital for the completion of a full taxonomic inventory. A combination of sampling gears (for example grabs, corers, dredges, sleds, trawls) may be deployed on seamount surveys, but the hard and

Fig. B7.3

Researchers will typically conduct **(A)** a bathymetric survey followed by **(B)** camera transects e.g., Deep Towed Imaging System (pictured) and then collect physical specimens using a **(C)** beam trawl and/or **(D)** epibenthic sled (National Institute of Water and Atmospheric Research).

rough ground that frequently prevails on seamounts may limit researchers to the use of towed dredges or sleds (Fig. B7.3).

The sample from each gear type deployed is sorted on board and separated out as close to species or putative species (that is, apparently morphologically distinct organisms, sometimes called operational taxonomic units) as possible. The samples are then chemically fixed or frozen (following taxa-specific recommendations, as well as taking into account genetic sampling requirements). At the end of the voyage the samples will be delivered

to taxonomists who will complete the final faunal inventory. This assessment of biodiversity can take many years, based on the high numbers of samples and low numbers of taxonomists.

Two working groups have helped drive the CenSeam research effort; the Data Analysis (DAWG) and Standardization Working Group (SWG). Members of each group have convened several workshops to tackle specific research questions and challenges, for example standardizing survey design, sampling, and analysis techniques (where possible) to facilitate geographic comparisons.

and Southern Central Pacific, and researchers have worked toward securing funding to sample these regions. So far, CenSeam-linked scientists have participated in over 20 voyages.

To help focus global research efforts, the CenSeam community identified two overarching priority themes: (1) What factors drive community composition and diversity on seamounts, including any differences between

Fig. 7.2

The CenSeam field program aims to investigate what factors drive community composition and diversity on seamounts, and to understand better the impacts of human activities such as fishing on seamounts (Erika Mackay, National Institute of Water and Atmospheric Research).

Seamount ecosystem

seamounts and other habitat types? (2) What are the impacts of human activities on seamount community structure and function? (Fig. 7.2). Within these themes key questions were developed to address where more science was needed to improve our understanding of the structure and functioning of seamount ecosystems and to inform management and conservation objectives. These questions will be used below to present what is known about seamounts so far, and how CenSeam has contributed to this knowledge. The CenSeam research effort has focused on seamount mega- and macroinvertebrates.

7.3 What Factors Drive Community Composition and Diversity on Seamounts?

Effective management of any seamount ecosystem must be based on a solid understanding of the seamount community, and associated physical and biological processes. Furthermore, it is important to determine the interactions of seamount communities with those in the wider deep-sea realm.

7.3.1 Seamount community composition and diversity

The dominant large fauna of hard substrate on many deep-sea seamounts are attached, sessile organisms that feed on particles of food suspended in the water (Fig. 7.3). These suspension feeders are predominantly from the phylum Cnidaria, which includes stony corals, gorgonian corals,

black corals, sea anemones, sea pens, and hydroids. Deep-sea (or cold water) corals are one of the most studied groups, and CenSeam has promoted research on their global distributions (Clark *et al.* 2006; Rogers *et al.* 2007; Tittensor *et al.* 2009). Corals can grow as individual colonies, or can coalesce as reefs; potentially providing complex three-dimensional habitat for a wide range of other animals, providing more refuge, an enhanced supply of food, surface area for settlement, and microhabitat variability to support a greater faunal diversity than less complex habitat. However, the role of biogenic habitat in the deep sea has only recently emerged as an area of both academic and conservation interest, and only a few quantitative studies have been made of the relationship between biogenic habitats and the composition of seamount fauna (see, for example, O'Hara *et al.* 2008).

In the literature there are many studies that describe the fish that live on seamounts. The most recent review by Morato *et al.* (2004) identified a total of 798 species of seamount fish, though exactly how to define a "seamount fish" is not straightforward. Commonly cited examples include the orange roughy (*Hoplostethus atlanticus*), alfonsino (*Beryx splendens*), Patagonian toothfish (*Dissostichus eleginoides*), oreos (*Pseudocyttus maculatus*, *Allocyttus niger*), and pelagic armourhead (*Pseudopentaceros wheeleri*) (Table 7.1). Sharks and tuna are also reported as occurring on seamounts, and in the waters above some shallow seamounts serranids (including sea basses and the groupers) and jacks are observed to spawn (Morato & Clark 2007).

Seamounts are popularly referred to as hot spots of high species richness in the deep sea. However, many researchers are failing to find support for this premise. Stocks & Hart (2007) report variability but no overall trend of elevated species richness across approximately 18 studies

Fig. 7.3

Corals on seamounts can grow as reefs or as individual colonies. **(A)** *Solenosmilia variabilis* on Ghoul Seamount (approximately 1,000 m; New Zealand; National Institute of Water and Atmospheric Research). **(B)** *Paragorgia arborea* and a dense population of basket stars *Gorgonocephalus* sp. on San Juan Seamount (USA; courtesy of the Monterey Bay Aquarium Research Institute). **(C)** *Viminella* on the summit of Condor Seamount (200 m; Azores, North Atlantic; © Greenpeace/Gavin Newman). **(D)** Paragorgiid, acanthogorgiid, and chrysogorgiid corals on Pioneer Seamount (approximately 1,700–1,800 m; Northwestern Hawaiian Islands; 2003 NWHI exploration team: Amy Baco-Taylor, Chris Kelley, John Smith, and pilot Terry Kerby, NOAA Office of Ocean Exploration and Hawaii Undersea Research Laboratory).

comparing seamounts to either surrounding deep sea or nearby continental margins. However, sampling-related issues complicate such comparisons, an issue that has been addressed within the CenSeam Data Analysis Working Group (DAWG). Taking sampling factors into account, DAWG member O'Hara (2007) compared levels of ophiuroid species richness between seamount and non-seamount areas (for the latter by randomly generating populations from areas and depth ranges that reflected the typical sampling profile of seamounts) and concluded that seamounts do not show elevated levels of species richness.

At macro-ecological scales, the fauna of individual seamounts have been found to broadly reflect the species pools present on neighboring seamounts and continental margins (see, for example, Samadi *et al.* 2006; Stocks & Hart 2007; McClain *et al.* 2009; Clark *et al.* 2010; Brewin *et al.* 2009). Although the main body of evidence suggests

that broad assemblage composition may be similar to surrounding deep-sea environments, community structure may differ between habitats. For example McClain *et al.* (2009) present preliminary evidence that the faunal communities of Davidson Seamount (off the west coast of the USA), although similar in composition to adjacent canyon habitat, are structurally different, particularly in the frequency of occurrence of particular species.

Seamounts are hypothesized to serve as biogeographical "islands" that could also function as shallow stepping stones across the abyssal plains. The isolated nature of many seamounts has fueled the hypothesis that seamounts can support high levels of endemicity, and numerous studies have supported this hypothesis (Richer de Forges *et al.* 2000; see review by Stocks & Hart 2007). However, so far, it is unlikely that we have identified enough of the regional or global deep-sea fauna to use the term endemic with any confidence, and

Table 7.1

Distribution of main commercial fish species on seamounts (North Atlantic (NA); South Atlantic (SA); North Pacific (NP); South Pacific (SP); Indian Ocean (IO); Southern Ocean (SO)) and the depth range commonly fished.

Common name	Scientific name	Distribution	Main depth range (m)
Alfonsino	*Beryx splendens*	NA, SA, NP, SP, IO	300–600
Black cardinalfish	*Epigonus telescopus*	NA, SA, SP, IO	500–800
Rubyfish	*Plagiogenion rubiginosum*	SA, SP, IO	250–450
Black scabbardfish	*Aphanopus carbo*	NA	600–800
Redbait	*Emmelichthys nitidus*	SA, SP, IO	200–400
Sablefish	*Anoplopoma fimbria*	NP	500–1,000
Pink Maomao	*Caprodon* spp.	NP, SP	300–450
Southern boarfish	*Pseudopentaceros richardsoni*	SA, SP, IO	600–900
Pelagic armorhead	*Pseudopentaceros wheeleri*	NP	250–600
Orange roughy	*Hoplostethus atlanticus*	NA, SA, SP, IO	600–1,200
Oreos	*Pseudocyttus maculatus, Allocyttus niger*	SA, SP, IO, SO	600–1,200
Bluenose	*Hyperoglyphe antarctica*	SA, SP, IO	300–700
Redfish	*Sebastes* spp. (*S. marinus, S. mentella, S. proriger*)	NA, NP	400–800
Roundnose grenadier	*Coryphaenoides rupestris*	NA	800–1,000
Toothfish	*Dissostichus* spp.	SA, SP, IO, SO	500–1,500
Notothenid cods	*Notothenia* spp.	SA, SP, IO, SO	200–600

apparently high levels of endemism may be an artifact of sampling species-rich communities or uneven sampling effort. Furthermore, many seamount fauna are recorded to have global or near-global distributions including reef-building scleractinian corals (*Lophelia pertusa, Solenosmilia variabilis,* and *Madrepora oculata*) (Roberts *et al.* 2006) and fish species such as orange roughy (Francis & Clark 2005).

In summary, seamounts can host abundant and diverse benthic communities. However, in many instances community composition is similar to those of adjacent habitats including continental slope. Today the concept of seamounts being islands in the sea has little support, but more sampling is required to be able to address this idea fully.

7.3.2 Connectivity of seamount populations

Differences in the connectivity of faunal populations among seamounts is almost certainly an important determinant of community composition on seamounts, a potential driver of

endemicity, and a major consideration for the management of seamount ecosystems. The dispersal capabilities of deep-sea fauna depend primarily on whether species can disperse as adults, or only as eggs, larvae, and/or post-larvae; however, one cannot fully predict the distribution of a species based on larval life history alone (see Johannesson's (1988) paradox of Rockall). Relatively little is known about the life-history traits of deep-sea organisms, including those found on seamounts. Studies so far have been restricted to a single seamount (Parker & Tunnicliffe 1994) or limited taxa (see, for example, Calder (2000) on hydroids). Distance from shore, or degree of isolation from other seamounts, is widely proposed as an important factor determining community composition and richness. Leal & Bouchet (1991) report a significant decline in species richness of prosobranch gastropods moving offshore along the Vitória-Trindade seamount chain, but could not attribute this to any differences in larval life histories and posed that species may be passively dispersed along the chain through "island hopping". In fact a suite of physical and biological factors also influence dispersal, and hence connectivity, over space and time. In a major CenSeam review paper, Clark *et al.* (2010) break these factors down: (1) physical ocean structure (for example hydrographic retention, large-scale and local currents), (2) factors influencing larval development time (for example temperature, food availability, predation), (3) habitat availability for larval settlement, and (4) post-settlement survival; with interactions thereof driving variations in the dispersal capabilities of fauna among seamounts.

Genetic studies are essential to our understanding of connectivity, but so far have been limited to few fauna and by the sensitivity of the current techniques. Historically, seamount genetic connectivity studies focused on commercially fished fauna but in recent years efforts have expanded to non-commercial fauna. No consistent pattern has emerged, with mixed results indicating both genetic differentiation and genetic homogeneity between some commercially fished fauna on seamounts, and those on oceanic islands and the continental margins at both oceanic and regional scales (Aboim *et al.* 2005; Stockley *et al.* 2005).

Baco & Shank (2005) discovered relatively high levels of genetic diversity, as well as low yet significant levels of population differentiation, for the precious coral *Coralium lauuense* among several Hawaiian seamounts and islands. They suggested that *C. lauuense* are primarily self-recruiting with occasional long-distance dispersal events maintaining genetic connectivity between sites. In contrast, Smith *et al.* (2004) provided evidence of widespread distribution of bamboo coral species in the Pacific which were not endemic to seamounts. However, the authors could not rule out that the mitochondrial markers they used in their analysis were insensitive to recent speciation events. Samadi *et al.* (2006) determined that populations of a gastropod with dispersive larvae were more similar than populations of non-planktotrophic gastropod species. Samadi *et al.* (2006) also determined that dispersive squat lobster species were genetically similar among populations on seamounts and the adjacent island slope.

The potential ability of certain seamount fauna to disperse widely is perhaps not surprising when compared with the finding of dispersive fauna at other isolated deep-sea habitats, such as hydrothermal vents or cold seeps (Samadi *et al.* 2007; see Chapter 9).

As well as understanding the dispersal characteristics of seamount fauna, it is vital to set these in the context of both large-scale oceanic circulation and localized hydrological phenomena. For example, Taylor cones have been cited as a possible trapping mechanism that may drive endemism. Mullineaux & Mills (1997) recorded larval concentrations above and around Fieberling Guyot to be consistent with modeled tidally rectified recirculation over the seamount. Parker & Tunnicliffe (1994) proposed the presence of a modified Taylor cap on Cobb Seamount was important for trapping short-lived larvae, but because water mass is replaced approximately every 17 days, concluded that

medium and long-lived larvae would be dispersed. Recent research by DAWG members has concluded, for some faunal groups, that seamount-scaled oceanographic retention is weak compared with other ecological drivers of community diversity on seamounts (Brewin *et al.* 2009).

To conclude, current understanding of the dispersal capabilities of seamount fauna, and the role of large and smaller-scale oceanographic processes, is limited and as such we cannot assess the role that dispersal may play in producing spatial differences between communities. The premise that seamounts may be dominated by short-lived or non-planktonic larval phased fauna has not been widely tested.

7.3.3 Environmental factors driving differences in diversity and species composition of seamount fauna

Seamount communities, as with slope and abyssal faunas, may exhibit latitudinal turnover in species composition. For example, O'Hara (2007) reports a clear biogeographical gradient for both seamount and non-seamount ophiuroid fauna from the tropics to the sub-Antarctic. Though incomplete, research so far has demonstrated that environmental factors can vary at large spatial scales, hence, can have the potential to influence community composition of deep-sea fauna.

Seamounts differ in their location, depth, elevation, and geological history (Rowden *et al.* 2005), all factors that may alter environmental conditions on large and small spatial scales and, in turn, influence seamount biodiversity and species composition. Clark *et al.* (2010) list as among the main factors that may determine the character of seamount benthic assemblages: seamount geomorphology, geological origin and age; local hydrodynamic regime (all preceding influence substrate type); light levels; water chemistry (for example oxygen); productivity of the overlying water (which relates to food availability); as well as the presence of volcanic/hydrothermal activity (see Chapter 9), temperature, and pressure. All these factors may operate in tandem and can create a unique set of conditions for a region, for any given seamount, and within a seamount.

Most marine animals are restricted to a limited bathymetric range (see, for example, Rex *et al.* 1999), and recent work by CenSeam-linked researchers has demonstrated that seamount assemblages can be depth stratified (O'Hara 2007; Lundsten *et al.* 2009). Work on the very deep slopes of seamounts has been limited but new research indicates that these can support distinct assemblages (see, for example, Baco 2007). However, the depth-related patterns, and the drivers thereof, remain largely unexplored for seamounts, but environmental gradients that correlate with depth such as temperature, oxygen concentration, food availability, and pressure are likely to be as important as they are for other deep-sea habitats (Clark *et al.* 2010).

Large seamount chains can divert major currents, and individual seamounts can affect localized hydrographic events including the formation of a rotating body of water retained over the summit of a seamount (Taylor cone, which may flatten to a cap) and the generation or interaction with internal waves (White *et al.* 2007). These may influence the faunal composition through larval transport (section 7.3.2). Additionally currents can be amplified around seamounts creating favorable conditions for suspension feeders as the waters bring an increased particle supply, as well as removing sediments (Genin 2004). Suspension feeders (for example corals, sponges, hydroids, crinoids, anemones, sea pens, feather stars, and brittlestars) have been found to dominate some seamounts (particularly their peaks) (Genin *et al.* 1986; Wilson & Kaufmann 1987), and large sessile fauna can, in turn, form structural habitat for a diverse range of smaller, mobile fauna (section 7.3.1).

Exposed rock surfaces are limited in the deep sea and seamounts represent a significant source of this substrate (Gage & Tyler 1991). Soft sediments can also dominate some seamounts, particularly flat-topped seamounts, called guyots, and in these circumstances community composition can switch from suspension to deposit feeders, similar to neighboring continental slopes (see, for example, Ávila & Malaquias 2003; Lundsten *et al.* 2009). The composition of infaunal communities on seamounts have not been well studied but include a wide diversity of polychaetes, crustaceans, mollusks, ribbon worms, peanut worms, and oligochaetes, as well as meiofaunal organisms such as nematode worms, loriciferans, and kinorhynchs (Samadi *et al.* 2007).

Although depth-related factors and substrate type (including biogenic habitat) are important drivers of community composition on seamounts, more research is required to describe and explore large-scale biogeographic patterns on seamounts.

7.3.4 Productivity on seamounts

Enhanced productivity at seamounts is a widely cited phenomenon, and seamounts appear to support relatively large planktonic and higher consumer (fish) biomass when compared with surrounding ocean waters, particularly so in oligotrophic oceans (Genin & Dower 2007). Seamounts can each have their own local oceanographic regimes (section 7.3.3), which could influence seamount productivity.

Elevated phytoplankton concentrations have been observed on some seamounts (see, for example, Genin & Boehlert 1985; Dower *et al.* 1992; Mouriño *et al.* 2001), and it has been theorized that nutrient-rich upwelled waters and eddies around a seamount enhance surface primary productivity which leads to an energy transfer to higher trophic levels. However, the persistence of upwelling does not generally seem sufficient for such a transfer, making this an unlikely explanation for the elevated zooplankton and fish biomasses found over seamounts (Genin & Dower 2007).

It is now proposed that the high biomass on some seamounts may be fueled not by enhancement of primary production but instead by a trophic subsidy to carnivores (Genin & Dower 2007). These authors proposed food inputs by the following routes: (1) bottom trapping of vertically migrating zooplankton, (2) greatly enhanced horizontal fluxes of suspended food through current acceleration, and (3) amplification of internal waves increasing horizontal fluxes of planktonic prey. Porteiro & Sutton (2007) proposed that fish behavior may have evolved to capitalize on the regular planktonic food supply passing over a seamount, enabling them to convert mid-trophic level biomass efficiently to higher trophic levels.

The importance of biogenic structure in supporting higher fish densities has been widely cited but work so far has yielded mixed results (see, for example, Husebø *et al.* 2002; Krieger & Wing 2002; Ross & Quattrini 2007). Hydrographic factors around deep-sea coral reefs may increase zooplankton density (Dower & Mackas 1996; Husebø *et al.* 2002) in turn benefiting planktivorous fish, but there is no consistent explanation for enhanced fish productivity over seamounts. Fish aggregations may occur independently of biogenic fauna; for example, orange roughy are observed to spawn over seamounts but do not feed during spawning (Morato & Clark 2007). Clearly, more information on the life histories of many seamount fish (for example larval stages, juvenile fish grounds) is needed.

In summary, enhanced secondary productivity over seamounts is most likely attributable to a food supply exported from elsewhere, and not locally enhanced primary productivity. The influence of bio-physical coupling, and the potentially complicated and varied interactions of forcing mechanisms and seamount topographies, is uncertain. Vital to future research on seamount-related productivity will be the establishment of long-term monitoring programs, in concert with the development of physical and trophic models.

7.3.5 Trophic architecture of seamount communities

Large suspension feeders such as corals, sponges, and crinoids can dominate the biomass of the seamount megabenthos on hard substrates. A dominance of suspension feeders may suggest that most animals are consuming similar resources at a low trophic level; hence, it might be thought seamounts have short food chains and low guild complexity.

CenSeam-linked researchers Samadi *et al.* (2007) used isotopic analysis to report that the benthic food webs on seamounts of the Norfolk Ridge have a diverse trophic architecture, with food-chain lengths (four to five trophic levels) toward the upper end of reported values from other aquatic systems, and broadly similar to other deep-sea food webs. Samadi *et al.* (2007) excluded the larger predatory fish which would further elongate seamount food chain length.

Pelagic food webs are likely to be a key part of the seamount ecosystem. Midwater fishes represent an important link from zooplankton to higher trophic level predators such as seabirds, squids, piscivorous fishes (for example tunas), sharks, and marine mammals (see Morato & Clark 2007; Porteiro & Sutton 2007). Many benthopelagic and demersal fish also feed on zooplankton, hence, there exists the possibility of numerous benthopelagic couplings in the water column around a seamount.

Food supply may vary as a result of the complex topographic and oceanographic patterns around seamounts, and feeding flexibility is also probably instrumental in enhancing trophic complexity on seamounts; for example, sponges are highly efficient at capturing ultraplankton (Pile & Young 2006) but also include carnivorous forms that prey on copepods (Watling 2007).

To conclude, despite a dominance of filter feeders, seamount communities are no less complex than other marine communities. However, considerably more research is required to gain a broad-scale understanding of the trophic architecture of seamounts.

7.4 What are the Impacts of Human Activities on Seamount Community Structure and Function?

Anthropogenic impacts in the deep sea are indisputable and this environment is more sensitive to human and natural impacts than previously thought (Davies *et al.* 2007). Human-induced changes are likely to be more intense, and occur over a shorter time period, than natural events, especially in the deep sea.

7.4.1 Vulnerability of seamount communities to fishing

In contrast to nearshore communities there have been limited studies investigating the impacts of deep-sea fishing on seamounts (Koslow *et al.* 2001; Clark & O'Driscoll 2003; Althaus *et al.* 2009; Clark & Rowden 2009). As distinct geological features, seamounts can provide a focal point for both fish to aggregate, and for intensive fishing effort (Clark 1999). The concentration of trawling on a seamount can be much higher than on the continental slope where activities might be more diffuse (O'Driscoll & Clark 2005).

The aggregating nature of seamount fish can facilitate large catch volumes (Fig. 7.4A). Seamount fisheries have often been boom and bust (Clark *et al.* 2007a) and under current management practice most deep-sea fisheries are not sustainable in the long term (Glover & Smith 2003). Deep-sea fish are typically less productive than shallow

Fig. 7.4
(A) Orange roughy catch on deck. **(B)** Orange roughy swimming over a trawled area of a seamount (Malcolm Clark/New Zealand's National Institute of Water and Atmospheric Research).

water shelf species and are highly vulnerable to overfishing. Deep-sea fish are often slow growing, long lived (for example orange roughy have been aged at 100 years or more (Andrews *et al.* 2009)), and slow to mature with low fecundity and sporadic reproduction (see, for example, reviews by Morato & Clark 2007; Clark 2009). Furthermore, deep-sea fish have low natural rates of mortality. The aforementioned are all factors that can limit recovery and resilience. So far many deep-sea fish populations have shown no signs of recovery and indeed it is uncertain if they can while commercial fisheries remain (Clark 2001; Dunn 2007).

The effects of fishing on seamounts must be considered beyond the impact on the target catch (Koslow *et al.* 2000). Longlines, gillnets, traps, and pots can all have some effect on non-target fauna, but bottom trawling is widely recognized as the primary threat to the seabed communities of seamounts (Fig. 7.4B).

The initial composition of the community will largely affect the scale of impact, with attributes of the component fauna such as fragility, size, and mobility as key traits determining the potential resistance (ability to withstand change and/or avoid trawl damage) of the community to trawling (Probert *et al.* 1997). Habitat-forming fauna such as stony corals are particularly vulnerable to trawling damage. Comparing eight seamounts on the Chatham Rise, New Zealand, Clark & Rowden (2009) found unfished seamounts to possess a relatively large amount of stony coral habitat (*Solenosmilia variabilis* and *Madrepora oculata* predominantly on the seamount peaks) compared with fished seamounts with relatively little coral habitat, and they reported significant differences in assemblage composition between fished and unfished seamounts. Overall, fishing can impact benthic species composition, abundance, age composition, size structure, and overall structural complexity of the benthic habitat (Clark & Koslow 2007).

Researchers agree that communities on seamounts are vulnerable to human activities, with fishing the major impact so far. The effects of fishing are now widely acknowledged to extend beyond the target species. A better understanding of the entire seamount ecosystem is vital if we are to mitigate human impacts.

7.4.2 Recovery of seamount communities from human-induced disturbance

The life-history characters of many seamount species (such as limited mobility, long generation time, low larval output, and limited dispersal capability) predispose many seamount ecosystems to recover over a scale of decades to centuries. Recovery will also be influenced by substrate type (and any changes associated with trawling), seamount location (for

example proximity to seed populations), and prevailing oceanographic conditions.

Althaus *et al.* (2009) reported no evidence of the structure forming taxa (primarily stony corals) recovering a decade after cessation of trawling on seamounts off Tasmania, Australia. Furthermore, the long-term loss of biogenic faunas such as corals and sponges may ultimately mean that impacted communities never return to their pre-disturbance state.

However, some fauna have shown either resistance to damage or capacity for rapid recovery and have been reported as more abundant on trawled seamounts (e.g., hydrocorals, gold corals, bryozoans, and some anemones) (Althaus *et al.* 2009; Clark & Rowden 2009). Within a given seamount there can exist refuges, such as areas that are too rugged for nets, which may help conserve biodiversity on an individual seamount and serve as source populations for recovery.

So far, there has been limited research on the recovery of seamount communities after disturbance. However, seamount biological communities have been considered among the least resilient in the marine environment (Clark *et al.* 2010) and timescales of recovery for some taxa exceed human life expectancies. The closures of previously fished seamounts to fishing operations will present future opportunity to monitor recovery, and provide valuable information to help guide future management initiatives.

7.5 Knowledge Transfer to Stakeholders

The scientific basis necessary for the successful management, protection, and restoration of deep-sea habitats such as seamounts is limited at national and international levels (Davies *et al.* 2007), and the variability in seamounts and their associated communities dictates that no single management model can be applicable to all seamounts. Crucial to increasing our understanding of seamount ecosystems is an open, honest dialogue and a free exchange of information between all seamount stakeholders.

Future research must focus on addressing urgent management needs. Seamount fisheries on the high seas remain poorly regulated with no unified single managing authority or mechanism in place. Despite gaps in global coverage and inconsistent measures to prevent damage or destruction to vulnerable habitats such as seamounts, regional fisheries management organizations provide the best option for the management/protection of seamount ecosystems. CenSeam researchers have contributed extensively to recent guidelines from the Food and Agriculture Organization of the United Nations (see Rogers *et al.* 2008) to assist in making future management of deep-sea seamount habitats more effective. These cover the need for careful and controlled development of any fishery, and Rogers *et al.* (2008) outline options for controlling initial exploitation levels (encompassing effort and catch limits). The instruments to protect seamounts are available (for example marine protected areas, closed areas, site-based effort control, licensing, gear restrictions) but examples of their implementation are rare (Alder & Wood 2004; Probert *et al.* 2007). The most effective practice for seamounts is thought to be closure of areas to trawling (Clark & Koslow 2007), and many regions are moving forward to designate marine protected areas in the deep sea (Davies *et al.* 2007). This places increasing pressure on scientists to deliver information to aid the selection process. Limitations to knowledge should not restrict efforts toward seamount conservation and, recognizing that seamounts are under-sampled, it is vital to develop means of predicting what communities might occur where (for example CenSeam publications: Clark *et al.* 2006; Tittensor *et al.* 2009) as well as robust means of classifying seamounts (Rowden *et al.* 2005) that may ultimately lead to the design of networks of marine protected areas balancing conservation and exploitation (Leathwick *et al.* 2008).

CenSeam has facilitated the first global seamount classification based on "biologically meaningful" physical variables (M.R. Clark *et al.* personal communication). This global classification of seamounts first uses a general biogeographic classification for the bathyal depth zone (near-surface to 3,500 m; UNESCO 2009) and then four key environmental variables (overlying export production, summit depth, oxygen levels, and seamount proximity) to group seamounts with similar characteristics. The classification determined 194 seamount classes throughout the world's oceans. The development of such classifications is vital for enabling the transparent selection of seamounts as candidates for protection, as well as to help guide researchers in strategically targeting seamounts for research.

7.6 Moving Beyond 2010: Emerging Issues

So far, the major threat to seamount ecosystems has been deep-sea bottom fishing (Probert *et al.* 2007). The maximum depths which can be fished are 2,000 m for trawling and 3,000 m for longline, but future generations may see technological advances in the depths and bottom types that can be fished. Hence the current "footprint" of fishing may expand into totally new areas.

The search for minerals in the deep sea has emerged as a potential threat to deep-sea biodiversity, including that of seamounts. Seamounts can have thick cobalt-rich ferromanganese crusts, manganese nodules, and polymetallic sulfide accumulations, which could be exploited for base metals, such as copper, zinc, and lead, or for precious metals such as gold and silver (see, for example, Hein 2002; Glover & Smith 2003; Davies *et al.* 2007). Regions of interest for seamount mining have been identified in the Pacific Ocean; cobalt-rich ferromanganese crusts off Hawaii, Micronesia, and the Marshall Islands, and polymetallic sulfide deposits around Papua New Guinea, New Zealand, and Vanuatu.

Direct physical disturbance of the seabed by mining equipment, and associated indirect and direct effects of sediment suspension and deposition, have been shown to influence the composition of macrofaunal assemblages that live on or in seabed substrates (Baker *et al.* 2001; Glover & Smith 2003). Glover & Smith (2003) cite our limited knowledge of the taxonomy, species structure, biogeography, and basic natural history of deep-sea animals as preventing any accurate assessment of the risk of species extinctions from large-scale seabed mining. Research is critical in this current era of prospecting to inform the formulation of environmental guidelines effectively before commercial mining begins. Researchers from CenSeam and ChEss (Biogeography of Deep-Water Chemosynthetic Ecosystems) have provided input to the environmental guidelines produced by the International Seabed Authority and those initiated by the mining industry itself (Clark *et al.* 2007b).

Rising levels of carbon dioxide in our atmosphere may lower the pH and calcium carbonate saturation of the oceans. A reduction in carbonate saturation inhibits the ability of marine organisms to build calcium carbonate skeletons, shells, and tests. Guinotte *et al.* (2006) predicted that within this century we might see substantial changes in the distribution of deep-sea corals, a dominant component fauna of communities on seamounts.

7.7 Moving Forward: The Next Decade of Seamount Research

Half a century after Hubbs (1959) posed his initial seamount questions, most remain unanswered. However, researchers have moved from a census of seamount fauna to examining the structure and function of seamount communities, recognizing the need for cross-habitat and multidisciplinary research.

So far, limited sampling prevents broad statements being made about seamount ecosystems, and it is most likely that high variability will ultimately prevent broad generalizations. However, expanding the global sampling effort will increase our chances of being better able to understand seamount ecosystems, and to predict global patterns in faunal diversity and distribution. At the outset of CenSeam, seamount researchers identified the Indian Ocean, the South Atlantic, and the Western and Southern Central Pacific as priority undersampled regions, and in the CenSeam lifetime new expeditions have been launched that will sample seamounts in most of these regions.

Just as we have recognized the geographical confines of our research, we must also recognize bathymetric limitations. The summits of seamounts remain much more intensively sampled than their slopes and bases. Technical limitations have constrained our abilities to sample deep seamounts; sampling deeper than 2,000 m requires specialized equipment such as remotely operated vehicles and submersibles. In the coming decades new sampling capabilities will start to extend the depth boundaries of what we currently think possible.

Virtually every seamount voyage that sails will return with species new to science, and a common challenge across the entire deep-sea realm is to overcome the so-called "taxonomic impediment" (Giangrande 2003). An ever-declining number of taxonomic experts face the challenge of updating faunal inventories, and the considerable time and effort taken to formally name and describe a species limits the rate at which we can census the marine realm. However, new methods, such as barcoding and metagenomics, may make inventories of a broader range of taxa, as well as taxonomically difficult organisms, more tractable. Advances in genetic technologies can also help us better determine the extent to which species, populations, and communities on seamounts are isolated.

So far, the specific relations between environmental drivers and faunal distributions are generally unresolved. CenSeam researchers have been at the forefront of developing new modeling methodologies to extrapolate from the known to the unknown, and expand our knowledge of the distributions of seamount communities. Rigorous testing of the prevailing hypotheses about what environmental factors may affect species survival is still required and, one day, improvements in aquarium design may enable long-term maintenance of deep-sea populations to test future climate change scenarios.

There is a growing need to understand temporal as well as spatial changes in species distributions, recognizing both natural and anthropogenic drivers of such changes, and to feed this information into management approaches (Probert *et al.* 2007). The effectiveness of marine protected areas in the deep sea is largely untested, and researchers need to be afforded the opportunities to study the recovery of deep-sea communities and feed the results back into future management strategies. Long-term observation programs are no longer a pipe dream with recent methodological advances. The Global Ocean Observing System (GOOS) and its national counterparts are instrumenting the ocean with a suite of physical, chemical, geological, and biological sensors that will provide valuable new information to explain drivers and patterns of marine biodiversity better.

Economics is a major driver in seamount research, and it is likely that, just as fishing operations provided the foundations of much of our current understanding, activities of the future, such as seabed mining, may drive future research direction. Although the negative impacts of fishing activities on some seamount communities are undisputed, fisheries have also increased our knowledge of seamount communities. Industries such as oil and mining are recognizing that they can provide a valuable platform for science. Initiatives such as SERPENT (Scientific and Environmental ROV Partnership using Existing Industrial Technology; www.serpentproject.com) have already proven their worth in providing a research capacity not previously available

One of the principal legacies of CenSeam will be to have challenged and changed the seamount paradigms that reigned at its inception. The project has played a role in moving deep-sea ecology and conservation issues into the mainstream, and it is likely that the call for better management of our deep-sea resources will grow ever louder. At the global scale, researchers must work together to align and standardize sampling and analysis approaches, ultimately strengthening our capacity to conduct cross-regional analyses. Our increased understanding of seamount ecosystems has highlighted the crucial need for seamount research to be set in a broad ecological context and include other habitats to increase our understanding of the deep sea in general.

Acknowledgments

The authors are the secretariat of CenSeam. We thank and acknowledge the many seamount researchers who have contributed to CenSeam as well as the entire CenSeam Steering Committee and Data Analysis Working Group participants for their role throughout the life of the project; for having given so freely of their time and ideas toward developing the core themes and questions. Without their enthusiasm the success achieved so far would not have been possible. The National Institute of Water and Atmospheric Research (NIWA) hosts the CenSeam secretariat. SeamountsOnline has been supported by NSF grants DBI 0074498 and OCE 0340839.

References

Aboim, M.A., Menezes, G.M., Pinho, M.R., *et al.* (2005) Genetic structure and history of populations of the deep-sea fish *Helicolenus dactylopterus* (Delaroche 1809) inferred from mtDNA sequence analysis. *Molecular Ecology* 14, 1343–1354.

Alder, J. & Wood, L. (2004) Managing and protecting seamounts ecosystems In: *Seamounts: Biodiversity and Fisheries* (eds. T. Morato & D. Pauly). *Fisheries Centre Research Reports* **12**(5), 67–73.

Althaus, F., Williams, A., Schlacher, T.A., *et al.* (2009) Impacts of bottom trawling on deep-coral ecosystems of seamounts are long-lasting. *Marine Ecology Progress Series* **397**, 279–294.

Andrews, A.H., Tracey, D.M. & Dunn, M.R. (2009) Lead-radium dating of orange roughy (*Hoplostethus atlanticus*): validation of a centenarian life span. *Canadian Journal of Fisheries and Aquatic Sciences* **66**, 1130–1140.

Ávila, S.P. & Malaquias, M.A. (2003) Biogeographical relationships of the molluscan fauna of the Ormonde Seamount (Gorringe Bank, Northeast Atlantic Ocean). *Journal of Molluscan Studies* **69**, 145–150.

Baco, A.R. (2007) Exploration for deep-sea corals on North Pacific seamounts and islands. *Oceanography* **20**, 109–117.

Baco, A.R. & Shank, T.M. (eds.) (2005) Population genetic structure of the hawaiian precious coral *Corallium lauuense* (Octocorallia: Coralliidae) using microsatellites. In: *Cold-water Corals and Ecosystems* (eds. A. Freiwald & J.M. Roberts), pp. 663–678. Berlin and Heidelberg: Springer.

Baker, M.C., Bett, B.J., Billett, D.S.M., *et al.* (2001) An environmental perspective. In: WWF/IUCN (eds.) *Status of Natural Resources on the High Seas*. Gland, Switzerland: WWF/IUCN.

Brewin, P.E., Stocks, K.I., Haidvogel, D.B., *et al.* (2009) Effects of oceanographic retention on decapod and gastropod community diversity on seamounts. *Marine Ecology Progress Series* **383**, 225–237.

Brewin, P.E., Stocks, K.I. & Menezes, G. (2007) A history of seamount research. In: *Seamounts: Ecology, Fisheries, and Conservation* (eds. T.J. Pitcher, T. Morato, P.J.B. Hart, M.R. Clark, N. Haggan & R.S. Santos), pp. 41–61. Oxford: Blackwell.

Calder, D.R. (2000) Assemblages of hydroids (Cnidaria) from three seamounts near Bermuda in the western North Atlantic. *Deep-Sea Research I* **47**, 1125–1139.

Clark, M. (1999) Fisheries for orange roughy (*Hoplostethus atlanticus*) on seamounts in New Zealand. *Oceanologica Acta* **22**, 593–602.

Clark, M.R. (2001) Are deepwater fisheries sustainable? The example of orange roughy. *Fisheries Research* **51**, 123–135.

Clark, M.R. (2009) Deep-sea seamount fisheries: a review of global status and future prospects. *Latin American Journal of Aquatic Research* **37**, 501–512.

Clark, M.R. & Koslow, J.A. (2007) Impacts of fisheries on seamounts. In: *Seamounts: Ecology, Fisheries, and Conservation* (eds. T.J. Pitcher, T. Morato, P.J.B. Hart, *et al.*), pp. 413–441. Oxford: Blackwell.

Clark, M.R. & O'Driscoll, R.L. (2003) Deepwater fisheries and aspects of their impact on seamount habitat in New Zealand. *Journal of Northwest Atlantic Fisheries Science* **31**, 441–458.

Clark, M.R. & Rowden, A.A. (2009) Effect of deepwater trawling on the macro-invertebrate assemblages of seamounts on the Chatham Rise, New Zealand. *Deep-Sea Research I* **56**, 1540–1554.

Clark, M.R., Rowden, A.A., Schlacher, T., *et al.* (2010) The ecology of seamounts: structure, function, and human impacts. *Annual Review of Marine Science* **2**, 253–278.

Clark, M.R., Rowden, A.A. & Stocks, K.I. (2007b) The proposed Census of Marine Life Seamounts Project: towards a global baseline and synthesis of seamount community data – its applicability in minimizing impacts from crusts mining. *Polymetallic sulphides and cobalt-rich ferromanganese crusts deposits: establishment of environmental baselines and an associated monitoring programme during exploration*. Kingston, Jamaica, International Seabed Authority.

Clark, M.R., Tittensor, D.P., Rogers, A.D., *et al.* (2006) *Seamounts, deep-sea corals and fisheries: vulnerability of deep-sea corals to fishing on seamounts beyond areas of national jurisdiction*. UNEP-WCMC.

Clark, M.R., Vinnichenko, V.I., Gordon, J.D.M., *et al.* (2007a) Large-scale distant-water trawl fisheries on seamounts. In: *Seamounts: Ecology, Fisheries, and Conservation* (eds. T.J. Pitcher, T. Morato, P.J.B. Hart, *et al.*), pp. 361–399. Oxford: Blackwell.

Davies, A.J., Roberts, J.M. & Hall-Spencer, J.M. (2007) Preserving deep-sea natural heritage: emerging issues in offshore conservation and management. *Biological Conservation* **138**, 299–312.

Dower, J.F., Freeland, H. & Juniper, K. (1992) A strong biological response to oceanic flow past Cobb seamount. *Deep-Sea Research I* **39**, 1139–1145.

Dower, J.F. & Mackas, D.L. (1996) "Seamount effects" in the zooplankton community near Cobb Seamount. *Deep-Sea Research* **43**, 837–858.

Dunn, M.R. (2007) Orange roughy. What might the future hold? *New Zealand Science Review* **63**, 70–75.

Francis, R.I.C.C. & Clark, M.R. (2005) Sustainability issues for orange roughy fisheries *Bulletin of Marine Science* **76**, 337–351.

Gage, J.D. & Tyler, P.A. (1991) *Deep-Sea Biology: A Natural History of Organisms at The Deep-Sea Floor*. Cambridge, UK: Cambridge University Press.

Genin, A. (2004) Bio-physical coupling in the formation of zooplankton and fish aggregations over abrupt topographies. *Journal of Marine Systems* **50**, 3–20.

Genin, A. & Boehlert, G.W. (1985) Dynamics of temperature and chlorophyll structures above a seamount. An oceanic experiment. *Journal of Marine Research* **43**, 907–924.

Genin, A., Dayton, P.K., Lonsdale, P.F., *et al.* (1986) Corals on seamount peaks provide evidence of current acceleration over deep-sea topography. *Nature* **322**, 59–61.

Genin, A. & Dower, J.F. (2007) Seamount plankton dynamics In: *Seamounts: Ecology, Fisheries, and Conservation* (eds. T.J. Pitcher, T. Morato, P.J.B. Hart, *et al.*), pp. 85–100. Oxford: Blackwell.

Giangrande, A. (2003) Biodiversity, conservation, and 'taxonomic impediment'. *Aquatic Conservation: Marine and Freshwater Ecosystems* **13**, 451–459.

Glover, A.G. & Smith, C.R. (2003) The deep-sea floor ecosystem: current status and prospects of anthropogenic change by the year 2025. *Environmental Conservation* **30**, 219–241.

Guinotte, J.M., Orr, J., Cairns, S., *et al.* (2006) Will human-induced changes in seawater chemistry alter the distribution of deep-sea scleractinian corals? *Frontiers in Ecology and the Environment* **4**, 141–146.

Hein, J. (2002) Cobalt-rich ferromanganese crusts: global distribution, composition, origin and research activities. *Workshop on Polymetallic massive sulphides and cobalt-rich ferromanganese crusts: status and prospects*. Jamaica: International Seabed Authority, ISA Technical Study.

Hillier, J.K. & Watts, A.B. (2007) Global distribution of seamounts from ship-track bathymetry data. *Geophysical Research Letters* **34**, doi:10.1029/2007GL029874.

Hubbs, C. (1959) Initial discoveries of fish faunas on seamounts and offshore banks in the eastern Pacific. *Pacific Science* **12**, 311–316.

Husebø, Å., Nøttestad, L., Fosså, J.H., *et al.* (2002) Distribution and abundance of fish in deep-sea coral habitats. *Hydrobiologia* **471**, 91–99.

Johannesson, K. (1988) The paradox of Rockall: why is a brooding gastropod (*Littorina saxatilis*) more widespread than one having a planktonic larval dispersal stage (*L. littorea*)? *Marine Biology* **99**, 507–513.

Koslow, J.A., Boehlert, G.W., Gordon, J.D.M., *et al.* (2000) Continental slope and deep-sea fisheries: implications for a fragile ecosystem. *ICES Journal of Marine Science* **57**, 548–557.

Koslow, J.A., Gowlett-Holmes, K., Lowry, J.K., *et al.* (2001) Seamount benthic macrofauna off southern Tasmania: community structure and impacts of trawling. *Marine Ecology Progess Series* **213**, 111–125.

Krieger, K.J. & Wing, B.L. (2002) Megafauna associations with deep-water corals (*Primnoa* spp.) in the Gulf of Alaska. *Hydrobiologia* **471**, 83–90.

Leal, J.H. & Bouchet, P. (1991) Distribution patterns and dispersal of prosobranch gastropods along a seamount chain in the Atlantic Ocean. *Journal of the Marine Biological Association of the United Kingdom* **71**, 11–25.

Leathwick, J., Moilanen, A., Francis, M., *et al.* (2008) Novel methods for the design and implementation of marine protected areas in offshore waters. *Conservation Letters* **1**, 91–102.

Lundsten, L., Barry, J.P., Caillet, G.M., *et al.* (2009) Benthic invertebrate communities on three seamounts off southern and central California, USA. *Marine Ecology Progress Series* **374**, 23–32.

Marques Da Silva, H. & Pinho, M.R. (2007) Small-scale fishing on seamounts. In: *Seamounts: Ecology, Fisheries, and Conservation* (eds. T.J. Pitcher, T. Morato, P.J.B. Hart, *et al.*), pp. 335–360. Oxford: Blackwell.

McClain, C.R., Lundsten, L., Ream, M., *et al.* (2009) Endemicity, biogeography, composition and community structure on a northeast Pacific seamount. *PLoS ONE* **4** e4141.

Menard, H.W. (1964) *Marine Geology of the Pacific* New York: McGraw-Hill.

Morato, T., Cheung, W.W.L. & Pitcher, T.J. (2004) Additions to Froese and Sampang's checklist of seamount fishes. In: *Seamounts: Biodiversity and Fisheries* (eds. T. Morato & D. Pauly). *Fisheries Centre Research Reports* **12**(5), 51–60.

Morato, T. & Clark, M.R. (2007) Seamount fishes: ecology and life histories. In: *Seamounts: Ecology, Fisheries, and Conservation* (eds. T.J. Pitcher, T. Morato, P.J.B. Hart, *et al.*), pp. 170–188. Oxford: Blackwell.

Mouriño, B., Fernandez, E., Serret, P., *et al.* (2001) Variability and seasonality of physical and biological fields at the Great Meteor Tablemount (sub-tropical NE Atlantic). *Oceanologica Acta* **24**, 1–20.

Mullineaux, L.S. & Mills, S.W. (1997) A test of the larval retention hypothesis in seamount-generated flows. *Deep-Sea Research I* **4**, 745–770.

O'Hara, T.D. (2007) Seamounts: centres of endemism or species richness for ophiuroids? *Global Ecology and Biogeography* **16**, 720–732.

O'Hara, T.D., Rowden, A.A. & Williams, A. (2008) Cold-water coral habitats on seamounts: do they have a specialist fauna? *Diversity and Distributions* **14**, 925–934.

O'Driscoll, R.L. & Clark, M.R. (2005) Quantifying the relative intensity of fishing on New Zealand seamounts. *New Zealand Journal of Marine and Freshwater Research* **39**, 839–850.

Parker, T. & Tunnicliffe, V. (1994) Dispersal strategies of the biota on an oceanic seamount: implications for ecology and biogeography. *Biological Bulletin* **187**, 336–345.

Pile, A.J. & Young, C.M. (2006) The natural diet of a hexactinellid sponge: benthic-pelagic coupling in a deep-sea microbial food web. *Deep-Sea Research I* **53**, 1148–1156.

Pitcher, T.J., Morato, T., Hart, P.J.B., *et al.* (2007) Preface. In: *Seamounts: Ecology, Fisheries, and Conservation* (eds. T.J. Pitcher, T. Morato, P.J.B. Hart, *et al.*). Oxford: Blackwell.

Porteiro, F.M. & Sutton, T. (2007) Midwater fish assemblages and seamounts. In: *Seamounts: Ecology, Fisheries, and Conservation* (eds. T.J. Pitcher, T. Morato, P.J.B. Hart, *et al.*), pp. 101–116. Oxford: Blackwell.

Probert, P.K., Christiansen, B., Gjerde, K.M., *et al.* (2007) Management and conservation of seamounts. In: *Seamounts: Ecology, Fisheries, and Conservation* (eds. T.J. Pitcher, T. Morato, P.J.B. Hart, *et al.*), pp. 442–475. Oxford: Blackwell.

Probert, P.K., Mcknight, D.G. & Grove, S.L. (1997) Benthic invertebrate bycatch from a deep-water trawl fishery, Chatham Rise, New Zealand. *Aquatic Conservation: Marine and Freshwater Ecosystems* **7**, 27–40.

Rex, M.A., Etter, R.J., Clain, A.J., *et al.* (1999) Bathymetric patterns of body size in deep-sea gastropods. *Evolution* **53**, 1298–1301.

Richer De Forges, B., Koslow, J.A. & Poore, G.C.B. (2000) Diversity and endemism of the benthic seamount fauna in the south-west Pacific. *Nature* **405**, 944–947.

Roberts, J.M., Wheeler, A.J. & Freiwald, A. (2006) Reefs of the deep: the biology and geology of cold-water coral ecosystems. *Science* **312**, 543–547.

Rogers, A.D. (1994) The biology of seamounts. *Advances in Marine Biology* **30**, 305–350.

Rogers, A.D., Baco, A., Griffiths, H., *et al.* (2007) Corals on seamounts. In: *Seamounts: Ecology Fisheries and Conservation* (eds. T.J. Pitcher, T. Morato, P.J.B. Hart, *et al.*), pp. 141–169. Oxford: Blackwell.

Rogers, A.D., Clark, M.R., Hall-Spencer, J.M., *et al.* (2008) *The science behind the guidelines: a scientific guide to the FAO Draft International Guidelines (December 2007) for the management of Deep-Sea Fisheries in the high Seas and examples of how the guidelines may be practically implemented.* Switzerland: IUCN. 39 pp.

Ross, S.W. & Quattrini, A.M. (2007) The fish fauna associated with deep coral banks off the southeastern United States. *Deep-Sea Research I* **54**, 975–1007.

Rowden, A.A., Clark, M.R. & Wright, I.C. (2005) Physical characterisation and a biologically focused classification of "seamounts" in the New Zealand region. *New Zealand Journal of Marine and Freshwater Research* **39**, 1039–1059.

Samadi, S., Bottan, L., Macpherson, E., *et al.* (2006) Seamount endemism questioned by the geographical distribution and population genetic structure of marine invertebrates. *Marine Biology* **149**, 1463–1475.

Samadi, S., Schlacher, T. & Richer De Forges, B. (2007) Seamount benthos. In: *Seamounts: Ecology, Fisheries, and Conservation* (eds. T.J. Pitcher, T. Morato, P.J.B. Hart, *et al.*), pp. 119–140. Oxford: Blackwell.

Smith, P.J., Mcveagh, S.M., Mingoia, J.T., *et al.* (2004) Mitochondrial DNA sequence variation in deep-sea bamboo coral (Keratoisidinae) species in the southwest and northwest Pacific Ocean. *Marine Biology* **144**, 253–261.

Stockley, B.M., Menezes, G., Pinho, M.R., *et al.* (2005) Genetic population structure of the black-spot sea bream (*Pagellus bogaraveo*) from the NE Atlantic. *Marine Biology* **146**, 793–804.

Stocks, K.I. (2009) *SeamountsOnline: an online information system for seamount biology.* Version 2009-1. Available at http://seamounts.sdsc.edu.

Stocks, K.I., Boehlert, G.W. & Dower, J.F. (2004) Towards an international field programme on seamounts within the Census of Marine Life. *Archive of Fishery and Marine Research* **51**, 320–327.

Stocks, K.I. & Hart, P.J.B. (2007) Biogeography and biodiversity of seamounts. In: *Seamounts: Ecology, Fisheries, and Conservation* (eds. T.J. Pitcher, T. Morato, P.J.B. Hart, *et al.*), pp. 255–281. Oxford: Blackwell.

Tittensor, D.P., Baco, A.R., Brewin, P.E., *et al.* (2009) Predicting global habitat suitability for stony corals on seamounts. *Journal of Biogeography* **36**, 1111–1128.

UNESCO (2009) Global Open Oceans and Deep Seabed (GOODS) bioregional classification. In: *UNESCO-IOC* (eds. M. Vierros, I. Cesswell, E. Escobar, *et al.*). *IOC Technical Series.*

Watling, L. (2007) Predation on copepods by an Alaskan cladorhizid sponge. *Journal of the Marine Biological Association of the United Kingdom* **87**, 1721–1126.

Watson, R., Kitchingman, A. & Cheung, W. (2007) Catches from world seamount fisheries. In: *Seamounts: Ecology, Fisheries, and Conservation* (eds. T.J. Pitcher, T. Morato, P.J.B. Hart, *et al.*), pp. 400–412. Oxford: Blackwell.

White, M., Bashmachnikov, I., Arístegui, J., *et al.* (2007) Physical processes and seamount productivity. In: *Seamounts: Ecology, Fisheries, and Conservation* (eds. T.J. Pitcher, T. Morato, P.J.B. Hart, *et al.*), pp. 65–84. Oxford: Blackwell.

Wilson, R.R. & Kaufmann, R.S. (1987) Seamount biota and biogeography. *Geophysics Monographs* **43**, 355–377.

Chapter 8

Diversity of Abyssal Marine Life

Brigitte Ebbe[1], David S. M. Billett[2], Angelika Brandt[3], Kari Ellingsen[4], Adrian Glover[5],
Stefanie Keller[1], Marina Malyutina[6], Pedro Martínez Arbizu[1], Tina Molodtsova[7], Michael Rex[8],
Craig Smith[9], Anastasios Tselepides[10]

[1]*Senckenberg Institute, Deutsches Zentrum für Marine Biodiversitätsforschung, Wilhelmshaven, Germany*
[2]*National Oceanography Centre, Southampton, UK*
[3]*Zoologisches Museum und Biozentrum Grindel, Hamburg, Germany*
[4]*Norwegian Institute for Nature Research, Polar Environmental Centre, Tromsø, Norway*
[5]*Natural History Museum, London, UK*
[6]*A.V. Zhirmunsky Institute of Marine Biology, Vladivostok, Russia*
[7]*P.P. Shirshov Institute of Oceanology, Russian Academy of Sciences, Moscow, Russia*
[8]*Department of Biology, University of Massachusetts, Boston, Massachusetts, USA*
[9]*Marine Sciences Building, University of Hawaii at Manoa, Honolulu, Hawaii, USA*
[10]*Thalassocosmos, Heraklion, Crete, Greece*

8.1 Introduction

The Census of the Diversity of Abyssal Marine Life (CeDAMar) was devoted to the study of the largest and remotest ecosystem on Earth, the major deep basins stretching between continental margins and the mid-ocean ridge system. Abyssal plains and basins account for about half of Earth's surface (Tyler 2003) and harbor a great variety of life forms. As part of the overall Census of Marine Life, the field project CeDAMar was designed to study the diversity, distribution, and abundance of organisms living in, on, or directly above the seafloor. Prominent features such as ridges, seamounts, trenches, and chemosynthetic environments were covered by other Census projects.

Life in the World's Oceans, edited by Alasdair D. McIntyre
© 2010 by Blackwell Publishing Ltd.

8.2 Abyssal Plains

Until the late nineteenth century, abyssal sediments were believed to be azoic deserts owing to a lack of sunlight and primary production. This view changed dramatically with the British *Challenger* expedition (1872–1876), which found deep-sea life throughout the world ocean. Modern marine diversity research began in the 1960s when Sanders, Hessler, and co-workers were able to show that the abundance of macrobenthic organisms decreased with depth whereas the number of species increased (Sanders *et al.* 1965; Hessler & Sanders 1967; Sanders & Hessler 1969). Pivotal in the development of the scientific interest in marine diversity patterns was a study by Grassle & Maciolek (1992) of a series of box corer samples collected along a 176 km transect on the northwest Atlantic continental slope. Species turnover rates along the transect suggested that the number of species at the deep-ocean floor may rival that of tropical rainforests. This study led to broad debate about the number of marine species and the distribution of diversity along bathymetric and latitudinal gradients (Poore & Wilson 1993; Rex *et al.* 1993, 1997; Thomas & Gooday 1996; Culver & Buzas 2000).

Before the year 2000, biological research in the abyss had been conducted only sporadically as part of the classic worldwide expeditions aboard American, German, Danish, and Swedish vessels around the turn of the century into the mid-1990s. More recently, between 1948 and 2000, the P.P. Shirshov Institute sampled more than 1,700 stations below 3,000 m including abyssal plains, basins, and trenches down to 9,000 m. Studies of abyssal diversity and biogeography were complicated by the logistic challenges of deep-sea exploration. When the first CeDAMar expeditions were planned, the total sampled area of deep-sea floor was equal to no more than a few football fields, and by the year 2005 the total sampled area below 4,000 m amounted to about $1.4 \times 10^{-9}\%$ (Stuart *et al.* 2008).

8.3 The CeDAMar Rationale

When CeDAMar was initiated, published results suggested that deep-sea sediments supported low biotic abundance and biomass, but potentially high species richness depending on taxon. All expeditions to abyssal plains and basins showed that regardless of the location, roughly 90% of the infaunal species collected in a typical abyssal sample were new to science.

8.3.1 Open questions in deep-sea research

One fundamental gap in our knowledge of the abyss was the existence of vast geographic areas that had not been sampled, for example, the central Pacific Ocean and oceans of the southern hemisphere, because they were so remote from oceanographic institutions. CeDAMar expeditions were specifically designed to explore both sides of the southern Atlantic, southern Indian Ocean, and the Southern Ocean; the Northeast Atlantic; the central Pacific; and, as an example for a warm, ultra-oligotrophic deep sea, the eastern Mediterranean Sea (Fig. 8.1).

The occurrence of high biodiversity in the extreme habitat conditions that characterize the abyss, such as low temperature, very high hydrostatic pressure, little habitat complexity, and extremely low food availability, was perceived to be one of the major biogeographic puzzles of our time. Despite the potential importance of this vast ecosystem as a reservoir for genetic diversity and evolutionary novelty, very little was known about the factors regulating deep-sea species richness (Gage & Tyler 1991; Gray 2002). CeDAMar therefore aimed to collect new reliable data on species assemblages of ocean basins and determine

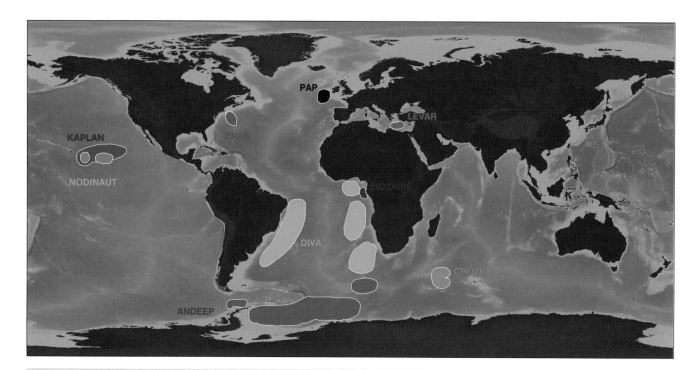

Fig. 8.1

Study areas of CeDAMar. For explanations of project names see sections 8.4.1 through 8.4.8.

the large-scale distribution of species among these basins. Documentation of the actual species diversity of abyssal plains provided a baseline for global-change research and for a better understanding of historical causes and ecological factors regulating biodiversity.

Even less is known about the biology of abyssal organisms. One of the unanswered questions in this context was the relation between food supply and the number of species present in a given deep-sea area. The deep-sea benthos depends ultimately on surface production that sinks through the water column. Although it seems evident that the biomass of deep-sea organisms should be positively correlated with food availability (Rowe 1971; C.R. Smith *et al.* 1997; Brown 2001), the productivity–biodiversity relationship is less clear.

8.3.2 Specific CeDAMar questions

Considering our lack of knowledge, CeDAMar focused research efforts in a way that would produce tangible results within a set timeframe of less than ten years. Deep-sea biologists identified the most urgent questions to be addressed by CeDAMar expeditions, keeping in mind the overarching Census themes of diversity, abundance, and distribution.

8.3.2.1 Questions concerning diversity

- How does diversity vary at different geographic scales, between different size classes of organisms, and with differences in food supply?
- Are there centers of high diversity (hot spots of diversity) in the deep sea?
- What is the role of evolutionary-historic processes in determining diversity levels?
- How do manganese nodules or drop stones influence benthic diversity?

8.3.2.2 Questions concerning abundance

- How do organisms of different size classes respond to environmental factors?
- What is the relation between food availability and benthic standing stock?

8.3.2.3 Questions concerning distribution

- Do biogeographic barriers affect the distribution of abyssal fauna? How endemic is the abyssal fauna?
- How common are cosmopolitan species in the abyss? Is there gene flow between distant abyssal communities of the same species?

- Are there latitudinal gradients in species richness? Is the diversity of a given basin similar to the diversity of basins in other oceans at similar latitudes?

8.4 Finding Answers: Methods and Programs of CeDAMar

The most prominent reason why the abyss has been explored to such a small degree is the difficulty of reaching it. Apart from the scarcity of research vessels, there are many logistic challenges, the time required for sampling great ocean depths not being the least. To lower sampling gear to the seafloor some 4,500 m below the surface and retrieve it back to the ship, several hours are necessary for each single sampling. The control of the actual sampling process on the bottom is limited by the great depth and the amount of wire between ship and gear. The methodology that CeDAMar used was more traditional than hi-tech, consisting of coring devices (box corer and multi-corer), epibenthic sledges, Agassiz trawls, and, when possible, a sediment profiling camera with or without a video camera. This set of gear was used in a standardized way to ensure (1) collection of organisms of all size classes from bacteria to large epifauna such as corals, sea anemones, sponges, holothurians, and stalked crinoids, and (2) comparability of results among CeDAMar projects and with the existing literature. The Time Series study of the seafloor in the Porcupine Abyssal Plain used a time-lapse camera and sediment traps to monitor processes on the seafloor.

8.4.1 Project DIVA

DIVA (diversity gradients in the Atlantic) is the seed project of CeDAMar, with the main focus on the question of latitudinal gradients in biodiversity in the southern Atlantic. Sampling locations were the abyssal basins off west Africa from the Cape to the equator and the Argentine and Brazil basins off the east coast of South America.

8.4.2 Project ANDEEP

ANDEEP (Antarctic benthic deep-sea biodiversity – colonization and recent community patterns) was dedicated to the abyssal waters in the Atlantic sector of the Southern Ocean. This region is one of the least investigated and it closed the gap between the two study areas of DIVA. It is also the location closest to the pole and farthest away from the equator, which made it very suitable to prove or disprove that a decline in marine biodiversity is present from the equator to the poles.

8.4.3 Projects KAPLAN and NODINAUT

The study area of KAPLAN and NODINAUT was the manganese nodule field in the Clarion-Clipperton Fracture Zone (CCZ), with the main focus centered on the question of the impact of nodules on biodiversity at different scales. Results were used for recommendations concerning marine protected areas (MPAs) to protect the fauna in case of nodule mining. In light of increasing demand for minerals, deep-sea mining has become a realistic possibility.

8.4.4 Project Biozaire

Biozaire was conducted off West Africa, just inshore of the DIVA area, encompassing the deep slope, abyssal plain, and a chemosynthetic site (a so-called pockmark). The objective was to characterize the "benthic community structure in relation with physical and chemical processes in a region of oil and gas interest" (Sibuet & Vangriesheim 2009).

8.4.5 Project LEVAR

LEVAR (Levantine Basin Biodiversity Variability) was one of the younger projects of CeDAMar, the study area being the eastern Mediterranean Sea with its comparatively shallow abyss (around 3,000 m), warm water at depth, and extremely poor food supply. Stations near Crete were sampled during one cruise. The aim was to determine whether proximity to shore or the depth was more important in influencing community composition and the distributions of abyssal biota.

8.4.6 Project CROZEX

The relation between surface primary production and benthic community composition was also explored during three cruises of the CROZEX (Crozet circulation iron fertilization and export production experiment) expedition off the sub-Antarctic Crozet Islands (Indian Ocean). The background of this study was a proposal put forward by biogeochemists suggesting that natural iron fertilization might enhance algal growth, which would sink to the abyssal seafloor, thus sequestering CO_2 and taking it out of the atmosphere. By observing processes driven by natural fertilization through iron eroded from the islands, CROZEX was designed to assess whether artificial iron fertilization might be a feasible option to fight global warming.

8.4.7 Project Time Series

A time-lapse camera system and moorings including sediment traps have been used to observe the deep ocean floor in the Porcupine Abyssal Plain since 1989, changing our perception of the quiescent, stable abyss to that of a very dynamic environment with sometimes radical changes in communities. One incident, the so-called *Amperima* Event named after the sea cucumber *Amperima rosea*, has become famous because of substantial changes in abundance related to changes in food supply.

8.4.8 Project ENAB

Evolution in the deep sea was the focus of ENAB (Evolution in the North Atlantic Basin), with a sampling cruise conducted along the famous Gay Head–Bermuda transect that in the early 1960s had started biodiversity research in the deep sea. The program was dedicated to assessing spatial population genetic structure in deep-sea mollusks to determine patterns of population differentiation, speciation, and phyletic evolution.

8.4.9 CeDAMar database

One of the legacies that may prove to be highly valuable to deep-sea researchers today and in the future is a freely accessible database that will be maintained and updated beyond the life of CeDAMar. So far, some 12,000 records, representing more than 3,000 species from nearly 4,800 locations distributed in all oceans can be queried. These records are made available to Ocean Biogeographic Information System (see Chapter 17), the database of the Census, from where they can also be accessed by anyone. With a special tool, maps can be created with different resolutions. Figure 8.2 shows the number of abyssal records per area, in this case a grid of 10 degree × 10 degree squares (roughly 100 km × 100 km). There are four areas with relatively extensive sampling on which much of our knowledge of the abyssal fauna is based: (1) the northwest Atlantic off the US east coast sampled in the 1980s, including stations on the continental slope that led to the estimates of deep-sea species richness by Grassle & Maciolek (1992); (2) the manganese nodule area off Peru, where the German DISCOL disturbance experiment was performed in the 1980s and 1990s to assess recovery of abyssal benthic fauna after massive disturbance mimicking possible effects of nodule mining; (3) the Porcupine Abyssal Plain and Gulf of Gascogne where British and French deep-sea investigations were concentrated; and (4) the Kurile–Kamchatka Trench, which was a main study area of Russian deep-sea research. The remaining area of the abyssal plains is still unsampled or poorly sampled, showing that even the substantial effort put into abyssal expeditions during CeDAMar has relatively little effect on sample coverage from a worldwide perspective.

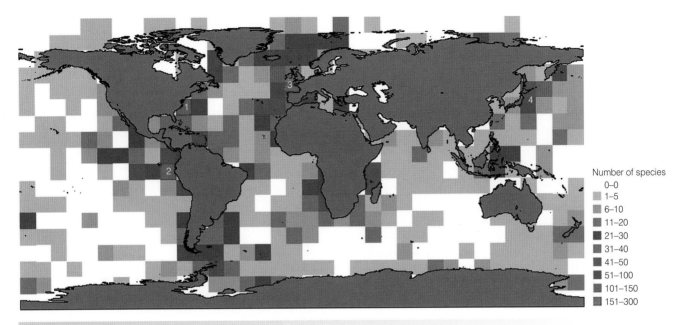

Fig. 8.2

Records of abyssal benthic species from the CeDAMar database, shown as species per 10-degree square (see color code on the right). The numbers indicate major research areas with the most extensive sampling activities. (1) US Atlantic slope and rise, (2) Peru Basin, (3) Porcupine Abyssal Plain, (4) Kurile–Kamchatka Trench.

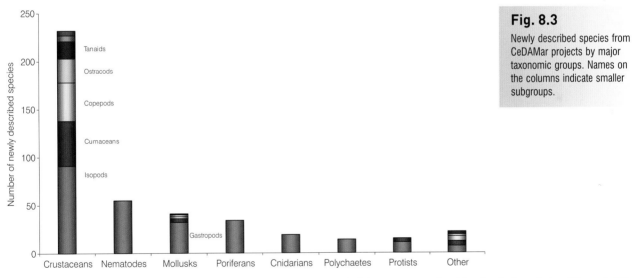

Fig. 8.3

Newly described species from CeDAMar projects by major taxonomic groups. Names on the columns indicate smaller subgroups.

8.4.10 Overcoming the taxonomic impediment

As all knowledge about ecosystems is based on knowing the identity of species in a particular system, much effort has been put into overcoming the so-called taxonomic impediment. The term means the general lack of specialists for identification of marine animals. Workshops and short-term stays at participating institutions (taxonomic exchanges) have helped to foster communication and intercalibration of the numerous personal databases from a broad range of projects. For polychaetes, a platform (www.polychaetes. info) was created with the help of the Natural History Museum (London) to exchange information by the Internet on yet unpublished but already well-defined "working species", allowing specialists to share information on an additional 50–90% of their respective taxa. A more visible outcome for the entire scientific community was CeDAMar's goal to deliver formal descriptions of 500 new abyssal species by the end of the first Census in October 2010. The goal will have been reached by the time this book is published (Fig. 8.3). Nearly half of all newly described or redescribed species are crustaceans (243 species, 91 of which are isopods), followed by nematodes (55 species) and mollusks (41 species, including 32 gastropods).

8.5 Major Results

Through the results generated by the CeDAMar project our perception of the abyss has changed fundamentally. This change in perception may be condensed into two statements which, although they may seem trivial at first glance, are significant changes in how scientists view the abyss: (1) extreme is normal; (2) rare is common.

8.5.1 Extreme is normal

Quite surprisingly, scientists even in the twentieth century viewed remote habitats on Earth from an anthropocentric perspective. The richness of life on abyssal seafloors showed quite convincingly that this habitat, which is extreme or even "inhospitable" to us, is highly habitable for a remarkable range of organisms. Even though we still know very little about the biology of abyssal organisms, it has become very apparent that many are well adapted to "extreme" conditions; reproduction takes place as well as speciation, and observations of a single site over time, such as the Porcupine Abyssal Plain (PAP) Time Series project, revealed that the abyssal seafloor can be unexpectedly dynamic. The massive bloom of the holothurian *Amperima rosea* in the PAP observed in the late 1990s was followed by a significant shift in the communities of several other deep-sea invertebrates that was documented over a period of 20 years (Billett *et al.* 2009). Not all other organisms seem to be affected by the alterations of the environment. Some of the polychaete populations, for example, did not react in any visible way, whereas others showed a significant increase in the number of individuals which could be related to increased nutrient input.

8.5.2 Rare is common

In terms of the general structure of benthic communities, there are large differences between the abyss and shallower environments. Nearly all species found in the abyss are rare, at least to our current knowledge. In practical terms it means that most species have been recorded as one or two individuals from one or two sampling sites, even in large programs during which thousands of animals were collected (Fig. 8.4). With very few exceptions, none of the communities sampled during CeDAMar expeditions were characterized by one or a few numerically dominating species as is typically the case in shelf communities.

8.5.3 Diversity of abyssal benthos

One of the ways to measure diversity is to look at the number of species at one particular site (alpha diversity), in addition species turnover along a certain distance (beta

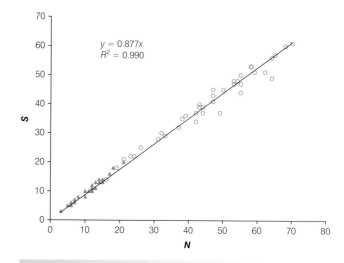

Fig. 8.4

Copepod species numbers (*S*) and corresponding numbers of specimens (*N*) collected by multicorer at two stations in the Angola Basin (triangles, station 325; circles, station 346) with line of linear regression. The number of individuals nearly equals the number of species. From Rose *et al.* 2005.

diversity) may also be assessed. Both measures of diversity were found to be much higher than expected. For example, copepods in the southeast Atlantic occurred everywhere in high abundances, but most species were undescribed (DIVA cruises): 98% of these species had never been seen before. Even smaller animals, the unicellular foraminiferans, showed high species turnover rates in the manganese nodule fields in the Pacific. At sampling sites no more than roughly 600 miles apart, different communities of foraminiferans were found. However, not all foraminiferan distributions appear to be restricted. In another study, including the ANDEEP material, other foraminiferans were discovered that are distributed from pole to pole, obviously coping with many very different habitat conditions.

Habitat heterogeneity is considered to be one of the major drivers of biodiversity because it provides a greater range of niches for the formation of new species. The abyssal seafloor was found to be as heterogeneous as shallower areas, perhaps most obviously in manganese nodule fields of the equatorial Pacific and in the Southern Ocean where stones drop out of melting icebergs and provide greater heterogeneity in substrata. The community structure of abyssal megafauna and macrofauna in manganese nodule fields was found to differ not only due to the availability and quality of food but also because of the heterogeneity in physical and chemical properties of the habitat (nodules and superficial sediment). Studies undertaken at the local scale (1–5 km in distance) with the manned submersible *Nautile* showed for the first time that nodule fields constitute a distinct habitat for infaunal communities, and that macrofauna and meiofauna components differ in abundance depending on the presence of nodules (Miljutina *et al.* 2009).

The geologic history of a basin can play an important role for biodiversity as well. A good example is the Southern Ocean. Its history includes not only periods of anoxia in the late Jurassic and cooling in the late Eocene/early Oligocene, but also cycles of glaciation and deglaciation which led to migration of shallow-water species into bathyal and abyssal depths (submergence) as well as recolonization of shallow sea bottoms from the deep (emergence). Applying molecular methods, Raupach *et al.* (2004, 2009) showed that shallow-water isopods colonized the deep sea at least on four separate occasions. Several isopod families

underwent spectacular radiation events in the abyss, resulting in an exceptionally high number of species and species complexes (Fig. 8.5). The Scotia and Weddell Seas, the geographic focus of the ANDEEP investigations, are characterized by a complex tectonic history related to the Middle Jurassic break-up of the Gondwana supercontinent which began around 180 million years (Ma) ago (Storey 1995). The Scotia Sea is much younger and formed during the past approximately 40 Ma (Thomson 2004). However, it is unknown whether the great biodiversity documented for many taxa in the deep Weddell Sea can be explained by the age of the ocean floor.

Another example is the generally low diversity of the benthos in the deep Mediterranean Sea, which is related to,

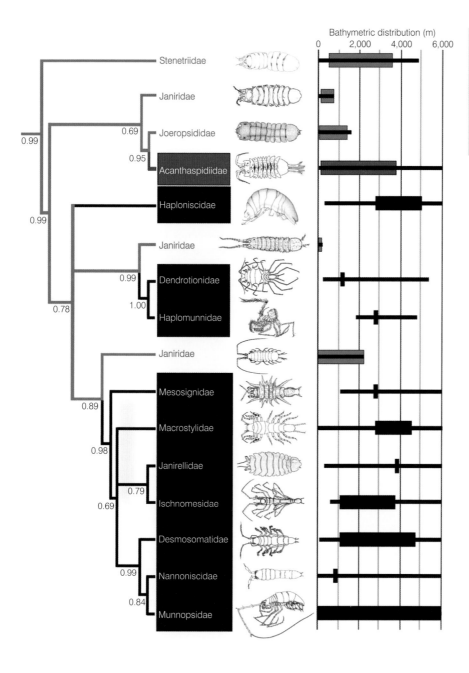

Fig. 8.5

Phylogenetic tree and distributional patterns of deep-sea isopod families based on molecular investigations. Families marked in blue are found in bathyal and abyssal depths but possess eyes, indicating that they invaded the deep sea from the shelf. From Raupach *et al.* 2004.

among other reasons, the complex paleoecological history characterized by the Messinian salinity crisis and the almost complete desiccation of the basin.

8.5.3.1 Spatial and temporal variability in primary productivity in the world's oceans and its effects on abyssal communities

Changes in primary productivity in the surface waters of the world's oceans are mirrored in abyssal communities in both space and time (C.R. Smith *et al.* 2008a). Organic matter created by photosynthetic production provides the food for most deep-sea life. Changes in food production at the sea surface, therefore, and the subsequent transport of organic matter into the ocean's interior through the biological carbon pump, have a profound effect on life on the abyssal seafloor.

It is well known that in regions where seasons are evident in surface waters, seasonal changes occur on the deep-sea floor within a matter of weeks (Billett *et al.* 1983; C.R. Smith *et al.* 1997; Beaulieu 2002). Large-scale biogeographical provinces in surface waters are reflected in broad changes in the structure of abyssal communities (Smith C.R. *et al.* 2008a). Decadal-scale shifts in primary production, caused by climate-related oscillations, produce long-term radical changes in deep-sea communities (Billett *et al.* 2001, 2009; Ruhl & Smith 2004; Ruhl 2007; C.R. Smith *et al.* 2008a; Smith K.L. *et al.* 2009). The fall of the carcasses of whales and fish (C.R. Smith & Baco 2003) and the mass deposition of jellyfish (Billett *et al.* 2006) provide additional, if localized, organic inputs. The abyss is linked intimately to processes at the sea surface.

CeDAMar projects have contributed significantly to recent advances made in our understanding of how surface water productivity affects abyssal ecosystems. Spatial variations in the distribution of species have been related to changes in surface water productivity in the Kaplan, DIVA, and CROZEX projects. In addition, radical changes in abyssal communities with time have been documented at the PAP in the Northeast Atlantic Ocean. Similar large-scale changes with time have been noted in the northeast Pacific Ocean (K.L. Smith *et al.* 2009).

At the PAP, CeDAMar has documented how over a 20-year time series (1989 to 2009) the abyssal megafauna changed in total abundance by two orders of magnitude in 1996 (Billett *et al.* 2009). This was mainly due to the increase in the holothurian species *Amperima rosea* and became known as the "*Amperima* Event" (Billett *et al.* 2001). Significant changes in the abundances of several megafaunal taxa occurred, including ophiuroids, actiniarians, pycnogonids, tunicates, and holothurians other than *A. rosea*. The changes were evident over a vast area of the abyssal plain (Billett *et al.* 2001) and had a significant effect on the recycling of organic matter at the sediment surface (Bett *et al.* 2001). During the

CeDAMar project it has been determined that protozoan and metazoan meiofauna (Gooday *et al.* 2010; Kalogeropoulou *et al.* 2010) and polychaete macrofauna (Soto *et al.* 2009) also increased significantly in abundance during the "*Amperima* Event". All elements of the benthic community showed a simultaneous change indicative of a large environmental event.

Protozoan phytodetritus indicator species showed a sharp decrease in abundance, whereas trochamminaceans, which previously had been comparatively rare, became dominant, potentially because of the increased disturbance caused by the megafauna (Gooday *et al.* 2009). In the metazoan meiofauna increases in abundance were seen in the nematode and the meiofaunal polychaetes, but not in the copepods. Ostracods decreased in abundance. The three dominant macrofaunal polychaete families, Cirratulidae, Spionidae, and Opheliidae, all increased in abundance but no major changes occurred in the community structure and dominant species (Soto *et al.* 2009), unlike the megafauna.

These results show that abyssal benthic communities change significantly with time. Similar results in the northeast Pacific Ocean indicate that such phenomena are widespread in productive regions of the world's oceans (K.L. Smith *et al.* 2009). The flux of organic matter may change by about an order of magnitude from one year to the next (Lampitt *et al.* 2010) and abundances in fauna have been shown to be correlated to climate indices that influence the biological carbon pump on regional scales (K.L. Smith *et al.* 2006, 2009).

Although many elements of the benthic community change at the same time in the Time Series studies, the scale of the response is not the same in all taxa or size classes. Larger changes in abundance are apparent in the megafauna and there are greater changes in the dominant species. This has important implications for interpreting geographic variations in the distributions of species in the different size classes of the benthic community.

Annual particulate organic carbon (POC) flux and benthic parameters have been measured together at only a few sites in the abyssal ocean. However, where POC flux has been measured directly, there are strong linear relations between POC flux and the abundance and/or biomass of specific biotic size classes, including megafauna, macrofauna, and microbes (C.R. Smith *et al.* 1997; C.R. Smith *et al.* 2008a; K.L. Smith *et al.* 2009). Average biomass of megafauna (Lampitt *et al.* 1986) and macrofauna (Rowe 1971) decline significantly with increasing water depth (and hence decreasing POC flux), resulting in the smaller size classes (bacteria and meiofauna) dominating community biomass at abyssal water depths (greater than 3,000 m) (Rex *et al.* 2006). Despite this, experimental results (Witte *et al.* 2003) and time-lapse photography (Bett *et al.* 2001) indicate that larger organisms play important functional roles in energy flow through food-limited abyssal ecosystems by outcompeting the smaller size classes for freshly deposited detritus. Changes in the spatial distribution of abyssal fauna therefore not only reflect the total input of organic matter, but also the periodicity and predictability in its supply. In addition, changes may be related to the quality of the organic matter (Ginger *et al.* 2001; Wigham *et al.* 2003; FitzGeorge-Balfour *et al.* 2010).

In another CeDAMar study around the Crozet Islands in the southern Indian Ocean, the distributions of protozoan and metazoan meiofauna, and of megafauna, were studied in relation to an area of natural iron fertilization in the oceans (Pollard *et al.* 2009). Iron carried off the volcanic islands of Crozet leads to seasonal phytoplankton blooms to the north of the Crozet plateau, as opposed to the south of the islands where iron is limiting. The eutrophic site had a greater diversity of live foraminiferans, and the phytodetritus indicator species *Epistominella exigua* was more abundant at this locality (Hughes *et al.* 2007). In contrast, the megafaunal communities in the two areas were radically different (Wolff *et al.* personal communication). The most abundant species *Peniagone crozeti* (Cross *et al.* 2009), found only at the seasonally productive site, was new to science. This indicates that megafaunal communities may be the most sensitive to changes in surface water productivity, whereas the smaller size fractions may show broader distributions, depending on the taxon. However, broad generalizations are difficult to make because certain macrofaunal species, including isopods and polychaetes, are restricted to productive areas of the ocean, such as the Southern Ocean (Brandt *et al.* 2007a, b, c).

8.5.3.2 Latitudinal/depth gradients of biodiversity in the Atlantic Ocean

Latitudinal gradients are the most conspicuous and ubiquitous biogeographic patterns in terrestrial and coastal ecosystems, but their explanation remains elusive. They were long assumed not to occur in the deep sea because the deep overlying water column buffered communities from the climatic phenomena thought to ultimately shape large-scale patterns of diversity. However, there is evidence that latitudinal gradients of diversity do exist in several macrofaunal taxa and foraminiferans in bathyal communities (Rex *et al.* 1993; Sun *et al.* 2006). They have not been examined previously at abyssal depths, largely because there are so few abyssal samples. The comprehensive DIVA datasets are being used to test whether latitudinal gradients do exist at abyssal depths. The results will be especially interesting because it is unclear whether latitudinal gradients in macrofaunal taxa exist in the southern hemisphere (Rex *et al.* 2000).

Results from the ANDEEP expeditions have shown that the impact of depth on species richness is not consistent among taxonomic groups. Ellingsen *et al.* (2007) examined general macrofaunal response to water depth in the Atlantic sector of the deep Southern Ocean using data on poly-

chaetes, isopods, and bivalves collected during the EASIZ II (Ecology of the Antarctic Sea-Ice Zone, 1998) and ANDEEP I and II cruises (2002), ranging from 774 to 6,348 m depth. They found that the isopods displayed higher species richness in the middle depth range (216 species in 3,000 m depth) and lower in the shallower and deeper parts of the area (Brandt *et al.* 2005), as reported for other deep-sea areas (see, for example, Gage & Tyler 1991). However, the number of bivalve species showed no clear relation to depth, and polychaetes showed a negative relation to depth (Ellingsen *et al.* 2007) (Fig. 8.6). Although the data were collected over a wide geographical area (58°14′–74°36′ S,

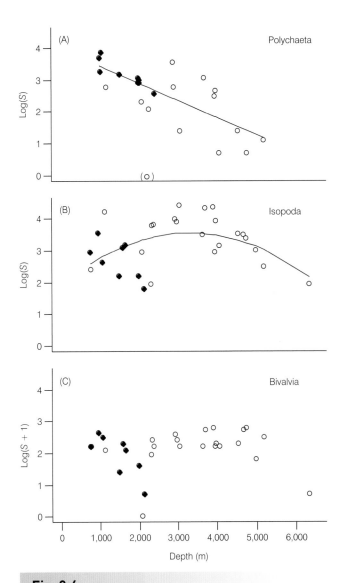

Fig. 8.6

Depth distributions of major taxa in the bathyal and abyssal Southern Ocean. Species richness of polychaetes declines with depth **(A)**, that of isopods peaks at about 3,000 m **(B)**, whereas no relation with depth can be seen for bivalves **(C)**.

22°08′–60°44′ W), the numbers of isopod, polychaete, and bivalve species did not show any consistent relation to latitude or longitude. Gastropods and bivalves show a variety of diversity–depth patterns among deep-sea basins (Allen 2008; Stuart & Rex 2009). Brandt *et al.* (2009) investigated the bathymetric distribution patterns of bivalves, gastropods, isopods and polychaetes in the Southern Ocean from 0 to 5,000 m, and found that the patterns differed between the different taxonomic groups.

8.5.3.3 Diversity and biogeography of Antarctic deep-sea fauna

Within the Southern Ocean, the abyssal benthic realm is the largest ecosystem and covers 27.9 million km² (Clarke & Johnston 2003). The Southern Ocean is characterized by some unique environmental features, which include a very deep continental shelf and a weakly stratified water column. It is also the source for the deep-water production influencing the deep circulation throughout the world. These physical characteristics led to the assumption that the Southern Ocean deep-sea fauna may be related both to adjacent shelf communities and to those living in other deep oceans. In the past century, Antarctic benthic shelf communities have been investigated extensively and are known to be characterized by high levels of endemism, gigantism, slow growth, longevity, and late maturity. Some amphipod, isopod, and fish families have adaptive radiations which have led to considerable novel biodiversity in these groups. Contrary to the Southern Ocean shelf, little was known about life in the vast Southern Ocean deep-sea region before the ANDEEP project. ANDEEP was a multidisciplinary international project which involved two expeditions to the Weddell and Scotia Seas in 2002 (Brandt & Hilbig 2004) and a third expedition (ANDEEP III) in 2005 to the Cape and Agulhas Basins, Weddell Sea, Bellingshausen Sea, and Drake Passage. In total, 40 stations were sampled between 748 and 6,348 m water depth with a focus on the abyss (Brandt & Hilbig 2004; Brandt & Ebbe 2007; Brandt *et al.* 2007a, b, c). The analyses revealed an astonishingly high biodiversity of several different taxa. From the material analyzed, more than 1,400 species were identified, and of these, more than 700 were new to science. In some groups of organisms, such as nematodes and isopods, greater than 90% of the species collected were new to science. Among the most important isopod families, over 95% of the species collected were unknown (Brandt *et al.* 2007a; Malyutina & Brandt 2007). Although we know that some species complexes have radiated in the deep Southern Ocean (Brökeland & Raupach 2008; Raupach & Wägele 2006; Raupach *et al.* 2007), it is unclear whether they have evolved here and subsequently spread into other ocean basins. Many species (>50%) were rare or patchy and occurred at only one station. Many species were singletons.

Biogeographic and bathymetric trends varied between groups and were probably related to differences in the reproductive mode (Brandt *et al.* 2007b, 2009; Pearse *et al.* 2009). In the isopods and polychaetes, slope assemblages included species that have invaded from either the shelf or the abyss through emergence or submergence, respectively, whereas in other taxa such as bivalves and gastropods, the shelf and slope assemblages were more distinct. Abyssal faunas tended to have stronger biogeographic links to other oceans, particularly the Atlantic, but mainly for organisms with good dispersal capabilities such as the foraminiferans (Brandt *et al.* 2007b; Pawlowski *et al.* 2007) and polychaetes (Schüller & Ebbe 2007; Schüller *et al.* 2009). The isopods, ostracods, and nematodes, which are poor dispersers, include many species currently known only from the Southern Ocean. In some groups, such as the Munnopsidae (Isopoda), the highest number of species (219) was reported in a worldwide biogeographical analysis (Malyutina & Brandt 2007). The ANDEEP results challenge the hypothesis that deep-sea diversity is depressed in the Southern Ocean and provide a sound basis for future explorations of the evolutionary significance of the varied biogeographic patterns observed in this remote environment.

8.5.3.4 The Mediterranean Sea: Diversity patterns in a warm deep sea

The Mediterranean region is characterized by the presence of both low and very high biodiversity, high levels of endemism are apparent, and in some areas strong energetic gradients in primary production and food supply to the deep occur decreasing from the western to the eastern basins and from shallower to deeper sites. The deep Mediterranean has generally been considered to have lower diversity than other deep-sea regions. Faunal exchange with the Atlantic Ocean is impaired by differences in deep-sea temperatures (approximately 10 °C higher in the Mediterranean than in the Atlantic Ocean at the same depth), which makes the establishment of incoming deep Atlantic fauna difficult. In particular, the abyssal basins of the Eastern Mediterranean are extremely unusual deep-sea systems with water temperatures at 4,000 m in excess of 14 °C. Barriers to colonization from the Atlantic also include salinity gradients and differences in food supply, as well as the existence of shallow sills. The deep Mediterranean is thus generally considered a "biological desert", although certain areas display such high benthic activity as to be characterized as "benthic hot spots". These areas are in most cases located at or near the mouth of submarine canyons that transport, through flash flooding, sediment failures, and dense shelfwater cascading, large amounts of sediment and organic material to the deep-sea floor (Canals *et al.* 2006). Abyssal trenches act as traps of organic matter of either terrestrial or pelagic origin (Tselepides & Lampadariou

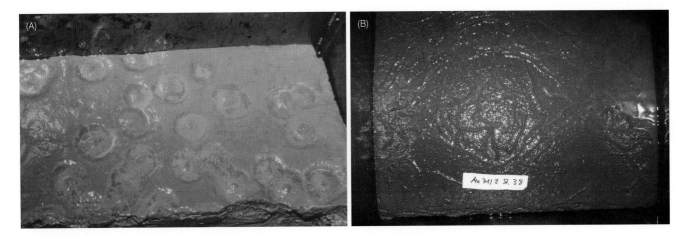

Fig. 8.7

(A) Box corer sample taken in 1998 in the Ierapetra Basin at 4,300 m depth. Circular shaped surface structures are "lebenspuren", made by the highly dominant polychaete *Myriochele fragilis*. **(B)** Sample from the same site taken in 2006 during LEVAR expedition. *Myriochele fragilis* was no longer found.

2004; Boetius *et al.* 1996). Large-scale hydrographic changes (Eastern Mediterranean Transient) have also been implicated in enhancing the productivity of the euphotic zone and indirectly structuring the underlying deep benthic communities (Danovaro *et al.* 2004).

The Mediterranean differs from other deep-sea ecosystems in terms of its megafaunal species composition (Jones *et al.* 2003). Typical deep-water groups, such as echinoderms, glass sponges, and macroscopic Foraminifera (Xenophyophora) are absent in the deep Mediterranean, whereas other faunistic groups (fishes, decapod crustaceans, mysids, and gastropods) are represented poorly compared with the Northeast Atlantic.

Although the low-diversity pattern is based on the analysis of macro- and megabenthos, recent evidence (Danovaro *et al.* 2008) suggests that the Mediterranean deep-sea nematode fauna is rather diverse and cannot be considered "biodiversity depleted". In fact, it was suggested that meiofaunal benthic biodiversity in the deep Atlantic and Mediterranean basins is similar.

A detailed analysis of food availability in the deep Mediterranean revealed that organic matter composition differed between the east and the west Mediterranean. Organic matter in the east was dominated by a high fraction of proteins and lipids. Therefore, although there were reduced amounts of organic matter in the east, this was to a certain extent compensated for by higher food quality and bioavailability. It seems that biodiversity patterns are not controlled by the amounts of food resources alone but also by the availability of the organic matter.

The project LEVAR explored not only the composition of benthic communities, but also environmental factors such as distance from shore, that is, supply of nutrients from shallower areas nearby, versus primary production in surface waters right above the sampling site and their respective influence on diversity. Preliminary results show that the benthic fauna at abyssal sites of the eastern Mediterranean is extremely poor in terms of abundance during normal oligotrophic periods, but can quickly develop high biomass when pulses of organic material settle down to the seafloor after unpredictable phytoplankton bloom events in surface waters (Figs. 8.7A and B).

8.5.3.5 Abyssal diversity hot spots

The diversity of life in the Southern Ocean (Brandt & Hilbig 2004) and the central Pacific Ocean (Glover *et al.* 2002) is high enough to characterize these areas as abyssal biodiversity hot spots. Glover *et al.* (2002) stated, "Local polychaete species diversity beneath the equatorial Pacific upwelling (measured by rarefaction) appears to be unusually high for the deep sea, exceeding by at least 10 to 20% that measured in abyssal sites in the Atlantic and Pacific, and on the continental slopes of the North Atlantic, North Pacific, and Indian Oceans." The use of molecular genetic methods will likely reveal an even higher diversity as many organisms looking alike under the microscope turn out to belong to different species, discernible only by differences in their genes.

8.5.4 Abundance of abyssal benthos

Studies of the CeDAMar project Biozaire on the continental slope of the Gulf of Guinea, adjacent to the DIVA 1 study area in the abyss, revealed that benthic communities living closer to shore are influenced by a very complex system of environmental parameters. Nevertheless, as in the PAP, the megafauna seemed to respond most directly to the influence of the organic material supplied by the Congo channel, whereas densities of smaller organisms – macrofauna and meiofauna – were subject to changes in environmental parameters, particularly in trophic inputs, at regional scale beyond the effects of the Congo channel (Sibuet & Vangriesheim 2009). Two of three study sites were located in approximately 4,000 m depth, 15 and 150 km south of the Congo channel. They were sampled during three cruises that were roughly one and two years apart.

Abundance of macro- and meiofauna increased substantially between 2001 and 2003, but interestingly not near the Congo river fan where increased input of organic matter was observed but rather at the station away from the fan. Obviously it was the quality of the food rather than the quantity that had the most profound effect on abundance, the organic matter near the Congo channel being mostly terrigeneous and, thus, of lower value for the deep benthos. These results agree well with findings from the deep eastern Mediterranean Sea.

8.5.5 Distributional patterns in the abyss: endemism versus cosmopolitanism

The traditional view of an abyssal cosmopolitan fauna has been strongly favored given the enormous, contiguous nature of abyssal environments, and the isolated records of apparently conspecific animals in separate ocean basins. Recent CeDAMar field projects such as the ANDEEP cruises in the Southern Ocean, the KAPLAN cruises in the central Pacific, and the DIVA cruises in the south Atlantic have created new opportunities to re-assess degrees of cosmopolitanism, which are reviewed here.

CeDAMar scientists have focused on a range of dominant abyssal taxa, which exhibit a range of reproductive strategies. These include peracarid crustaceans, copepods, polychaete worms, mollusks, holothurians, and foraminiferans. Peracarids generally brood young in their marsupium and there is no distinct larval stage (Brandt *et al.* personal communication). Copepods are direct-developing, with juveniles and adults both probably distributed by ocean currents. Polychaetes include species that either brood or display a bi-phasic life cycle with free-swimming planktotrophic or lecithotrophic larvae: both modes are thought to occur in abyssal species (Beesley *et al.* 2000). Deep-sea gastropods and bivalves generally reproduce by planktotrophic or lecithotrophic larval dispersal (Rex *et al.* 2005). Deep-sea holothurians have a broad range of egg sizes, from 180 to 4,000 μm (Billett 1991). The largest egg sizes are thought to lead to direct development of free-swimming juvenile holothurians within the abyssopelagic zone allowing for wide dispersal (Billett *et al.* 1985). Abyssal foraminiferans are thought to reproduce asexually (Murray 1991).

A study of cosmopolitanism in 45 deep-sea peracarid species has revealed only 11 species which occur in all oceans studied (the North Atlantic, South Atlantic, Southern Ocean, North and South Pacific, and Indian Oceans) (Brandt *et al.* personal communication). However, 33 species have distributions across more than one ocean basin, and 16 species are shared between the North Atlantic and North Pacific. Molecular-based studies of asellote isopods have revealed cryptic species, but these studies have so far been limited to a small range of taxa (Raupach *et al.* 2009). For benthic harpacticoid copepods, a study in the South Atlantic and Southern Ocean recorded 19 species of which 11 were restricted to particular regions, and eight widespread between ocean basins (Gheerardyn & Veit-Köhler 2009).

In polychaetes, sampling and analysis projects associated with CeDAMar have revealed both cosmopolitanism and cryptic speciation. Several species of small infaunal deposit-feeding spionids from abyssal depths are apparently distributed globally, based on examination of gross and ultra-structural morphology using scanning electron microscopy (Mincks *et al.* 2009; A. Glover unpublished data). Conversely, specimens of *Aurospio dibranchiata* Maciolek, 1981 from two central Pacific abyssal plain sites appear to be cryptic species based on 18S rRNA sequences, a normally highly-conserved gene (Mincks *et al.* 2009). A study of the distribution of multiple species of Southern Ocean abyssal polychaetes has revealed similar trends in terms of broad distributions of several species, based on morphology. Out of 70 Southern Ocean species studied in detail, 17 were shown to be cosmopolitan and only 13 apparently locally restricted to particular Southern Ocean sites (Schüller & Ebbe 2007). The remainder were at the very least broadly distributed, some between ocean basins (for example the Southern Ocean and North Atlantic).

A review of the distribution of protobranch bivalves in the east and west North Atlantic has revealed broadly distributed species at multiple bathymetric levels (McClain *et al.* 2009). Forty-three percent of the species studied were shared between the two ocean basins, of which 88% had overlapping depth ranges. The degree of apparent cosmopolitanism increased with depth, from 40% in bathyal regions to 60% in abyssal.

Systematic studies of deep-sea holothurians from the *Galathea* expedition revealed several cosmopolitan species in the abyss (Hansen 1975). Few taxa have been studied yet in detail using molecular methods, but the cosmopolitan species *Oneirophanta mutabilis* Théel, 1879 (Fig. 8.8A) and *Psychropotes longicauda* Théel, 1882 have been recovered from multiple ocean basins. These species are characterized by large egg sizes up to 1 mm, which suggests lecithotrophic larvae or direct development (Ramirez-Llodra *et al.* 2005).

One of the more enigmatic abyssal groups is the Komokiacea, a group of soft-bodied foramaminfera that produce large branching tests. A recent systematic review

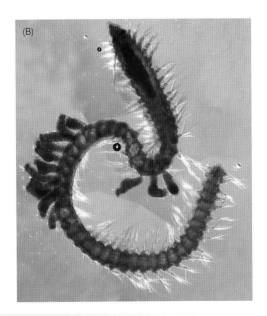

Fig. 8.8

(A) *Oneirophanta mutabilis*, a cosmopolitan abyssal elasipod holothurian recovered from 5,000 m on the central Pacific abyssal plain. **(B)** *Sphaerosyllis* sp. B, a polychaete recovered from a central Indian site, apparently conspecific with specimens from the North Pacific and North Atlantic, with direct-developing juveniles visible budding off mid-body segments. Photographs: A. Glover.

of komokiaceans from the Southern Ocean has revealed nine species, of which five are also present in the North Atlantic (Gooday *et al.* 2007).

Some foraminiferans apparently are truly cosmoplitan as they cannot be discriminated even with molecular genetic methods, indicating that gene flow is taking place from pole to pole (Lecroq *et al.* 2009). This global gene flow is difficult to imagine at first glance, and it may be confined to organisms with certain traits in their biology. Body size, which is inversely related to population size (that is, the smaller the organism is, the more individuals there are), plays an important role, and so do planktonic dispersal capabilities and the ability to survive long periods of famine. For example, the cosmopolitan species *Epistominella exigua* can live in substrata with organic carbon concentrations spanning orders of magnitude and episodic flux to small ephemeral patches on a seemingly homogeneous seabed (Lecroq *et al.* 2009). This flexibility is thought to facilitate gene flow even under marginal conditions.

In summary, available data are sparse yet support the view that both cosmopolitanism and basin endemism occur across a wide range of taxa in the abyss. These include species that exhibit direct development and bi-phasic life cycles where larvae can be carried by ocean currents. Evidence from molecular genetics is now starting to challenge some of these apparent cosmopolitan distributions, but even if many abyssal species are cryptic, it is clear that gross morphologies, and in some cases fine ultrastructure, are highly conserved in the abyss. This pattern may be a result of relatively rapid colonization of the abyss from bathyal depths and subsequent slow rates of adaptive radiation, in response to relatively similar environmental conditions.

Studies of reproductive biology are extremely rare, and are required to find independent lines of evidence for species ranges. Polychaetes with clear direct-developing offspring have recently been recovered from an isolated, oligotrophic central Indian Ocean abyssal site that are apparently conspecific with specimens from both the north Atlantic and north Pacific (Fig. 8.8B). The simplification of a pattern where only species with larval stages are likely to be broadly dispersed is clearly being challenged, future studies involving physiological data (see, for example, Hall & Thatje 2009) and modeling of available habitats may yet provide the additional lines of evidence required to resolve the paradox of cosmopolitan abyssal species.

However, as so many animals in the abyss are rare, any distributional patterns have to be interpreted with great caution. "Endemic" species may just not have been found again in other locations, and all newly described species are by default "endemics". Conversely, many species considered to be cosmopolitan may have been misidentified, for example, through the use of identification keys not pertaining to the area. There is some indication that generally, distributional patterns as we interpret them from samples taken so far may represent extreme patchiness. The scale of this patchiness may be rather small (Kaiser *et al.* 2007), and we may have to change sampling strategies from large-scale coverage of entire ocean basins to concentrated sampling at a single site.

8.5.6 Evolution and speciation in the abyss

During the past several decades, much has been learned about patterns of species diversity in the deep sea and their potential ecological causes. However, we are only now beginning to explore the evolutionary processes that generated this rich and distinctive fauna. How and where did all these species originate? Currently, our entire understanding of evolution is based on patterns in other ecosystems.

Deep-sea mollusks were chosen for a study of deep-sea evolution because their basic taxonomy and biogeography is particularly well known. The ENAB project is testing models of evolution based originally on analyses of shell form within species arrayed along depth gradients (Etter & Rex 1990). This research suggested that most population differentiation occurred at intermediate depths in the narrow bathyal zone along continental margins, and that the abyss played only a minor role in promoting deep-sea biodiversity. However, it was not possible to determine whether bathymetric ranges in shell form represented evolved genetic differences or simply environmentally caused morphological differences.

New laboratory methods were developed to extract, amplify, and sequence mitochondrial DNA from specimens that had been fixed in formalin and then preserved in alcohol, sometimes for decades. The resulting patterns of genetic differentiation tended to confirm that the bathyal zone was an evolutionary hot spot (Etter *et al.* 2005). This research has now been expanded to examine very large-scale geographic variation in mollusks among deep-sea basins in the North and South Atlantic (Zardus *et al.* 2006). A variety of patterns has emerged including differentiation at great depths.

In the summer of 2008, the first deep-sea sampling expedition devoted exclusively to studying evolutionary patterns in the deep sea was performed. The objective was to collect fresh material in order to sequence both nuclear and mitochondrial genes. A broad range of genes is essential to verifying geographic patterns of differentiation. Fresh material also enables us to develop better primers to more effectively sequence genes in the vast amount of archived preserved material. Being able to use multiple genes adds a new dimension to evolutionary studies in the deep sea. Except for foraminiferans, there is virtually no fossil record of deep-sea assemblages to assist us in unraveling long-term adaptive radiation and the global spread of higher taxa. Instead, phylogenetic evolution must be inferred from molecular genetic data. For the first time, we now have broadly distributed material that is amenable

to phylogeographic analysis. This will allow us to answer very fundamental questions, adding an evolutionary-historical perspective to our understanding of life in the deep sea. One of the most puzzling discoveries of this research so far is an apparent genetic break within eurybathic species at about 3,300 m, indicating that there is limited gene flow around this depth. This phenomenon not only occurs in mollusks, but was also reported for a widely distributed amphipod (France & Kocher 1996).

8.5.7 Nodule Mining and MPAs in the Pacific abyss

Manganese nodules, or polymetallic concretions of iron and manganese hydroxides, can be abundant at the abyssal seafloor beneath regions of low to moderate ocean primary productivity (Ghosh and Mukhopadhyay 2000). In some regions, nodules may cover more than 50% of the seafloor (Fig. 8.9) and are potential mineral sources of copper, nickel, and cobalt. Manganese nodule mining is expected to occur in the abyss by the year 2025 and could ultimately be the largest scale human activity to directly impact the deep-sea floor (C.R. Smith *et al.* 2008b). Thirteen pioneer investor countries and consortia have conducted hundreds of prospecting cruises to investigate areas of high manganese nodule coverage in the Pacific and Indian Oceans, especially in the area between the Clarion and the Clipperton fracture zones, which covers roughly 6 million km² and may contain 340 million tonnes of nickel and 265 million tonnes of copper (Ghosh and Mukhopadhyay 2000; Morgan 2000). Eight contractors are now licensed by the International Seabed Authority (ISA) to explore nodule resources and to

Fig. 8.9

The yellow elasipod holothuroid *Psychropotes longicauda*, here shown on a dense bed of manganese nodules, is a widely distributed deposit feeder and uses its upright "sail" to use current energy for transport along the seafloor. It was collected at 4,900 m in the Clarion-Clipperton Fracture Zone. Photograph: IFREMER.

test mining techniques within individual claim areas, each covering 75,000 km² (Fig. 8.10) (C.R. Smith *et al.* 2008b; www.isa.org.jm/en/home). In addition to harboring mineral resources, abyssal Pacific sediments in the CCZ may also be major reservoirs of biodiversity (Glover *et al.* 2002). However, it has been extremely difficult to predict the threat of nodule mining to biodiversity (in particular, the likelihood of species extinctions) because of very limited knowledge of (1) the number of species residing within areas likely to be perturbed by single mining operations, and (2) the typical geographic ranges of species within the nodule provinces (Glover & Smith 2003). During the CeDAMar field projects KAPLAN and NODINAUT, we used state-of-the-art molecular and morphological methods to begin to evaluate biodiversity and species ranges of three key faunal groups in the abyssal Pacific nodule province: polychaete worms, nematode worms, and foraminiferans. Together, these groups can constitute more than 50% of faunal abundance and species richness in abyssal sediments (Smith & Demopoulos 2003), and represent a broad range of ecological and life-history types.

CeDAMar results indicate high, unanticipated levels of species diversity for all three sediment-dwelling faunal components studied at our individual sites E, C, and W (Fig. 8.10). Based on morphological analyses, the Foraminifera contain at least 252 species at site E and at least 180 species at site C (Nozawa *et al.* 2006). Many of these species are new to science and appear not to have been collected elsewhere (Nozawa *et al.* 2006; C.R. Smith *et al.* 2008c). Based on DNA sequencing studies, the nematode worms also exhibit very high within-site diversity, with 73 molecular operational taxonomic units (or putative species) from only 97 sequenced individuals (C.R. Smith *et al.* 2008c). Because of a high ratio of one new species for every 1.3 individuals sequenced, the total nematode species richness is still grossly undersampled; we can be certain that far more species remain to be collected at each of our abyssal Pacific sites.

The polychaetes also exhibit very high within-site diversity for the families studied in detail; for example, Site E contains at least 48 polychaete species within 16 polychaete families (C.R. Smith *et al.* 2008c). A high abundance

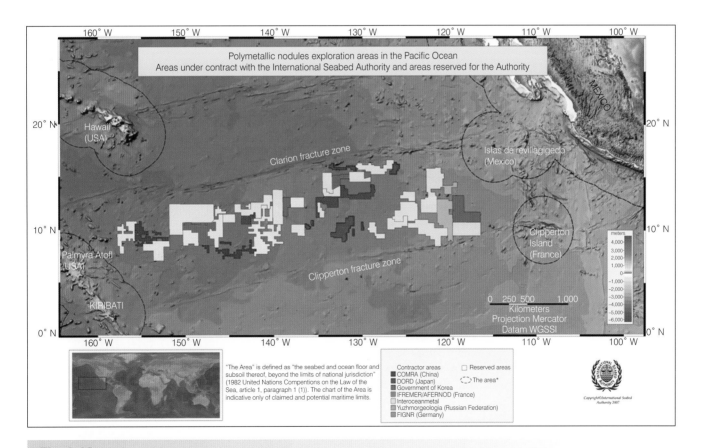

Fig. 8.10

The region of maximum commercial interest in the Pacific nodule province (box in inset) and claim areas licensed to exploration contractors in 2008. The Kaplan sites where samples were collected for CeDAMar were located at 15°N, 119°W (Site E), 14°N, 130°W (Site C), and 9°30′N, 150°W (Site W).

of apparently cryptic species found with our molecular studies indicates that earlier estimates of polychaete species richness within abyssal Pacific sites based on morphological studies, for example the 170 species from $3\,m^2$ by Glover *et al.* (2002), are likely to be low by at least a factor of two. We speculate that, even based on the relatively limited number of samples we have been able to analyze thus far, the total species richness of sediment-dwelling foraminiferans, nematodes, and polychaetes (a subset of the total fauna) at a single site in the CCZ could easily exceed 1,000 species (C.R. Smith *et al.* 2008c).

Our combined results for the foraminferans, nematodes, and polychaetes suggest that there is a characteristic fauna of the Pacific abyss, indicating that the abyss is not merely a sink of non-reproducing individuals transported from the continental margins (Rex *et al.* 2005; C.R. Smith *et al.* 2008a). Many of the hundreds of species of Foraminifera identified from our samples appear to be restricted to, or at least characteristic of, the abyss (Nozawa *et al.* 2006; C.R. Smith *et al.* 2008c). Seventy of the 73 molecular operational taxonomic units (MOTUs) of nematodes appear to be new genera distinct from shallow-water genera, and thus may well have evolved in the abyss (C.R. Smith *et al.* 2008c). The molecular data for the polychaetes also indicate numerous cryptic new species in our KAPLAN abyssal samples, again suggesting that the abyssal polychaete fauna contains higher species diversity than previously appreciated, and may include numerous species evolved in the abyss. All of these results suggest that the central Pacific abyss harbors a specially adapted, diverse fauna distinct from the fauna of the continental margins. It seems very unlikely that all, or even many, species found in the CCZ abyss are protected from extinction by populations residing many thousands of kilometers away at much shallower depths on the continental margins (C.R. Smith *et al.* 2008a).

Although the data are still limited, there is significant evidence that community structure of the Foraminifera and polychaetes differ substantially on scales of 1,000–3,000 km across the CCZ. These apparent patterns of faunal turnover seem likely to be driven in part by the east to west decline in primary productivity thus the flux of food to the seafloor across the CCZ , but may also be driven in part by varying habitat heterogeneity (C.R. Smith *et al.* 2008c).

Using results from the KAPLAN and NODINAUT projects, CeDAMar helped to convene a workshop of experts to draft recommendations to ISA for the design of MPAs in the CCZ to conserve marine biodiversity and ecosystem structure and function in the region in the face of nodule mining. Based on sound scientific principles, it was recommended that a network of nine $400\,km \times 400\,km$ protected areas (or "areas of particular environmental interest") be set up within the CCZ where mining would be prohibited (Fig. 8.11) (International Seabed Authority 2008, 2009). This network of protected areas would be stratified by regional variations in primary productivity and protect a total area of $1,440,000\,km^2$, placing roughly 25% of the total CCZ management area under protection (International Seabed Authority 2008). The ISA is currently considering these recommendations. If implemented, these CeDAMar recommendations would initiate scientifically based conservation management in international waters, would establish the ISA as a leader in the application of modern conservation management, and would set a precedent for protecting seabed biodiversity, a common heritage of mankind, before the initiation of exploitive activities (International Seabed Authority 2008).

8.6 Remaining Challenges and New Questions

8.6.1 Natural history and environmental factors

Although we learned much about the faunal elements of abyssal benthos communities, we still know almost nothing about the natural history of abyssal animals or environmental factors structuring abyssal communities. To the human eye an abyssal plain looks uniform over hundreds of kilometers. Nonetheless, benthic communities are not nearly as homogeneous as originally thought. To abyssal animals, the habitat bears enough heterogeneity to cause species turnover even within a single ocean basin. However, we are just beginning to understand the scale of species turnover in abyssal plains.

In the deep Southern Ocean, the ANDEEP project has revealed patterns of biodiversity within different faunal groups, but we still do not know anything about the processes behind these biodiversity patterns. The ANDEEP follow-up International Polar Year project SYSTCO (system coupling) therefore focuses on coupling processes between atmosphere, water column, and deep-sea floor near the Polar Front and in the abyssal Weddell Sea and includes ecological questions and investigations of the role of deep-sea fauna in trophodynamic coupling and nutrient cycling in oceanic ecosystems.

8.6.2 Speciation in the abyss

On an evolutionary scale, the same gap in our knowledge becomes apparent. We know very little about speciation in the abyss, and we are just now beginning to gain insights into the origin of the abyssal fauna and the very high diversity of abyssal benthic communities. Especially for soft-bodied organisms that leave no fossil record, molecular clocks have to be developed to reconstruct their evolutionary history. ENAB has developed novel techniques which are promising for future research.

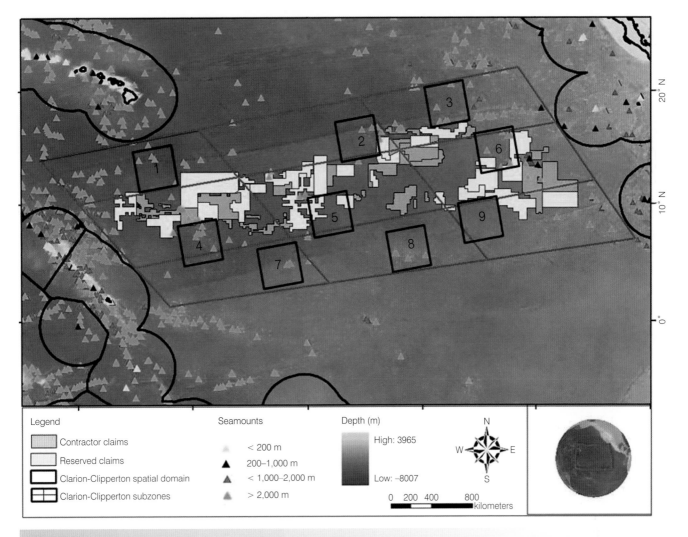

Fig. 8.11

Clarion-Clipperton Zone divided into nine management subregions, with one 400 × 400 km protected "area of particular environmental interest" centered in each subregion. This figure shows one of many options for location of preservation reference areas within the management subregions (International Seabed Authority 2008).

8.6.3 Abyssal species numbers and taxonomy

We will probably never know the true number of species in the abyss. The research area is far too large to be sampled adequately considering how heterogeneous this habitat turns out to be and how high the percentage is of rare species which have been recorded from just one site, often also by just one individual among thousands. Nevertheless, with knowledge gained continuously, scientists continue to try to reach better and better estimates.

The remarkable gain of knowledge about the abyssal benthos, notwithstanding the taxonomic impediment which brought about the birth of CeDAMar, is still apparent. We are still facing an overwhelming amount of species awaiting formal description and a scarcity of specialists to do the task. Taxonomic intercalibration, which has come a long way during CeDAMar, will have to continue as we have just scratched the surface. Molecular genetic and morphological methods will have to be integrated in a continuing effort to understand each other and communicate.

8.7 Moving On

Although public awareness about the deep sea has risen a great deal during CeDAMar, the abyss is still perceived by most people as a somewhat remote part of the planet, not affecting humankind in any way worth mentioning, and the research is still felt to be somewhat academic.

However, the abyss is on its way to become a resource for human exploitation very quickly. Industrial harvesting of manganese nodules may become a reality before most of us notice. Necessary technology is far advanced, largely unnoticed by anybody other than those directly involved. Even before man-made gear enters this still pristine environment, it is quite possible that the abyssal seafloor, which accounts for the largest area on the planet, may warrant our close attention because biogeochemical cycles of the seafloor have a strong influence on the global climate and climate change.

Climate warming is expected to increase regional sea surface temperatures and thermal stratification in low to mid-latitudes, yielding reductions in nutrient upwelling (C.R. Smith *et al.* 2008b; K.L. Smith *et al.* 2009). These changes will in turn alter the quantity and quality of food flux from the euphotic zone to the abyssal seafloor (Fig. 8.12). CeDAMar studies suggest that resulting long-term declines in POC flux to the abyss will cause reductions in the abundance and biomass of benthic fauna, and yield reductions in species diversity and body size over large regions, such as in the equatorial Pacific. Substantial shifts in the taxonomic composition of abyssal assemblages, especially the megafauna, are also expected, as well as changes

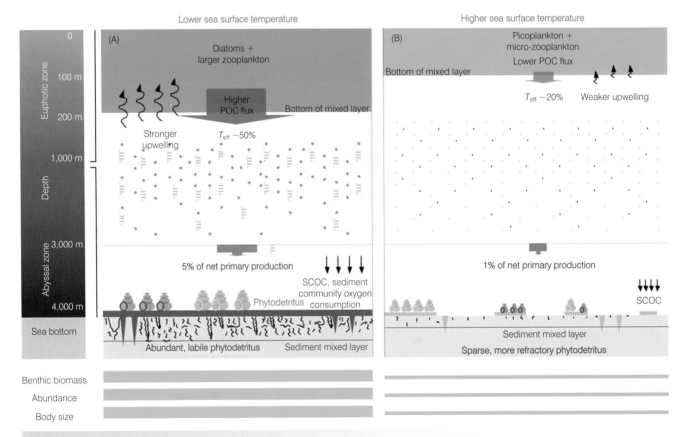

Fig. 8.12

Predictions of the effects of rising atmospheric $p\mathrm{CO_2}$ and climate change on abyssal benthic ecosystems. By increasing mean sea-surface temperature (SST) and ocean stratification, and by reducing upwelling, global warming has the potential to shift pelagic ecosystems from **(A)** diatom- and large zooplankton-dominated assemblages with higher export efficiencies to **(B)** picoplankton- and microzooplankton-dominated assemblages with lower export efficiencies. Such pelagic community shifts will reduce overall primary production and the efficiency of organic-carbon export from the euphotic zone into the deep ocean, and thus will substantially reduce POC flux to large areas of the abyssal seafloor. Reductions in POC flux will in turn reduce sediment community oxygen consumption (SCOC), bioturbation intensities, sediment mixed-layer depths, faunal biomass, and body sizes of invertebrate taxa (for example gastropods), and alter a variety of other abyssal ecosystem parameters. Shifts in the quality of sinking POC, for example in fatty acid composition, caused by changes from diatoms to picoplankton, will alter the nutritional quality of this food material, favoring reproductive success of some abyssal species and reducing reproductive success of others.

in basic ecosystem functions at the seafloor, such as organic carbon burial and calcium carbonate mineralization. Climate induced reductions in abyssal food flux over large areas, such as the equatorial Pacific biodiversity hot spot, have the potential to cause regional species extinctions as populations are reduced below reproductively viable levels (Rex *et al.* 2005; C.R. Smith *et al.* 2008b). Because abyssal ecosystems are so sensitive to the quantity and quality of sinking food material from the upper ocean (C.R. Smith *et al.* 2008b; K.L. Smith *et al.* 2009), impacts on the abyss must be considered in predicting the effects of climate warming and eco-engineering (for example ocean fertilization to mitigate climate change) on the biodiversity and ecological functioning of ocean ecosystems.

8.7.1 What needs to be done?

When the first Census has ended, keeping the momentum of global collaboration has to become our first action item. One idea might be to establish an international consortium supported by national funding agencies to identify important questions that most urgently need answers. Funding for taxonomists and molecular biologists needs to be secured in the long term to truly overcome the taxonomic impediment. Sampling strategies need the same global perspective as the Census to avoid falling back into competition among nations or institutions for the most attractive results.

Innovative methods will have to be adopted for the exploration of life in the abyss, for example, *in situ* experiments that might tell us something about the biology of abyssal organisms, and autonomous vehicles that can travel along abyssal plains to collect data over large distances and areas. The technically challenging development of suitable instruments and research with such methods will require substantial additional funding which will be granted only if the general public gets involved and educated. Societal acceptance of deep-sea research is still measured by that of astronomy. Allocating public funds to investigate other planets, stars, and even galaxies, immeasurably farther away from human reach, is questioned by few, in contrast to investigating the portion of surface of our own planet which happens to be covered with water.

Exhibitions and trade fairs related to boating and diving lately included small individual submarines for pleasure, designed to dive to about 100 m, driven by the owners themselves. Although these submarines are targeted for a very wealthy clientele, they may perhaps raise awareness for the benthic environment in a different and more direct way than anything we can offer through the media.

8.7.2 Outlook and conclusions

The return to a more holistic perspective is perhaps a logical process following nearly a century of specialization and focus on smaller and smaller details of an ecosystem which, as we gathered more and more facts, seemed to become more and more difficult to comprehend. We may have reached a time that is right for taking a step back and looking at whole systems from different viewpoints, realizing how they all overlap and complement each other. If one could understand which factors regulate the presence of species in a given area and which factors regulate the absolute and relative abundance of these species, then one would understand much of the functioning of the ecosystem as a whole. The evaluation of biodiversity – defined as the variety and variability of genomes, populations, species, communities, and ecosystems in space and time (Heywood 1995) – continues to be a central theme in biology and conservation.

When the scientific scope of CeDAMar was planned, exclusion of continental margins, seamounts, and chemosynthetic environments was deliberate. Only through focusing on a few of the major abyssal basins of the global ocean was it possible to achieve any tangible results in the limited timeframe of the Census. Exploring the relations of the ecosystem "abyssal benthos" with neighboring systems is a logical second step to be undertaken in the future. Several habitats possibly interacting with the abyssal benthos come to mind, most obviously the continental margins (see Chapter 5); on an even larger scale, an integration of water column and benthos research is a desirable goal. To be able to gain more complete insights both spatially and temporally, the abyss must be integrated into ocean observing systems.

Although the rate of discovery of new species is intimidating, it is not equally large for all organisms. Specialists do not expect much beyond 10% new species, for example, of mollusks (whereas for others such as nematodes the rate may be about 90%). Several organisms have been found to be widespread, for example, on either side of the Atlantic Ocean or in both Polar seas. Although genetic investigations have to confirm these patterns based on morphology, we may eventually come to a realistic estimate of the number of species in the abyss. "Singletons", those species known from only one specimen, may eventually be recaptured at the original site or even elsewhere, and the recapture rate may be a good proxy for species richness.

Within the past 150 years, we have learned to look at the abyss through different lenses. The unfathomed depths turned from a mythical place inhabited, at best, by fearsome creatures waiting to attack the unwary seaman, to an integral part of our planet filled with a dazzling variety of life, well adapted to its environment and of unsuspected beauty and grace. There are still many more questions than answers. CeDAMar research has lifted some of the mysteries, and the facts are even more fascinating than the myths, to scientists as well as the general public. There is much hope among deep-sea scientists that CeDAMar, together with other deep-sea projects within the Census, acted as a spark for ongoing research in the decades to follow.

Acknowledgments

The field project CeDAMar, like the entire Census of Marine Life, would not have happened without the vision, inspiration, and continuous support of Jesse Ausubel, Sloan Foundation, and J. Frederick Grassle, Institute of Marine and Coastal Sciences, Rutgers University. National science foundations in many nations are thanked for their financial support of the research done during CeDAMar projects, and there are countless scientists and technicians, well-seasoned and young, who have to be thanked for their tireless work that produced the data. Captains and crews of many research vessels helped to collect the material on which the data are based, and heartfelt thanks are due to them as well.

References

Allen, J.A. (2008) The Bivalvia of the deep Atlantic. *Malacologia* **50**, 57–173.

Beaulieu, S.E. (2002) Accumulation and fate of phytodetritus on the sea floor. *Oceanography and Marine Biology* **40**, 171–232.

Beesley, P.L., Ross, G.J.B. & Glasby, C.J. (2000) *Polychaetes and Allies: The Southern Synthesis.* CSIRO Publishing.

Bett, B.J., Malzone, M.G., Narayanaswamy, B.E. & Wigham, B.D. (2001) Temporal variability in phytodetritus and megabenthic activity at the seabed in the deep Northeast Atlantic. *Progress in Oceanography* **50**, 349–368.

Billett, D.S.M. (1991) Deep-sea holothurians. *Oceanography and Marine Biology* **29**, 259–317.

Billett, D.S.M., Bett, B.J., Jacobs, C.L., et al. (2006) Mass deposition of jellyfish in the deep Arabian Sea. *Limnology and Oceanography* **51**, 2077–2083.

Billett, D.S.M., Bett, B.J., Rice, A.L., et al. (2001) Long-term change in the megabenthos of the Porcupine Abyssal Plain (NE Atlantic). *Progress in Oceanography* **50**, 325–348.

Billett, D.S.M., Bett, B.J., Reid, W.D.K., et al. (2009) Long-term change in the abyssal NE Atlantic: The "Amperima Event" revisited. *Deep-Sea Research II* doi:10.1016/j.dsr2.2009.02.001.

Billett, D.S.M., Hansen, B. & Huggett, Q.J. (1985) Pelagic Holothurioidea (Echinodermata) of the northeast Atlantic. In: *Echinodermata: Proceedings of the 5th International Echinoderms Conference, Galway* (eds. B.F. Keegan & B.D.S. O'Connor) pp. 399–411.

Billett, D.S.M., Lampitt, R.S., Rice, A.L. & Mantoura, F. (1983) Seasonal sedimentation of phytoplankton to the deep-sea benthos. *Nature* **302**, 520–522.

Boetius, A., Scheibe, S., Tselepides, A. & Thiel, H. (1996) Microbial biomass and activities in deep-sea sediments of the Eastern Mediterranean: trenches and benthic hotspots. *Deep-Sea Research* **43**, 1439–1460.

Brandt, A., Brökeland, W., Choudhury, M., et al. (2007c) Deep-sea isopod biodiversity, abundance and endemism in the Atlantic sector of the Southern Ocean – results from the ANDEEP I – III expeditions. *Deep-Sea Research II* **54**, 1760–1775.

Brandt, A., De Broyer, C., De Mesel, I., et al. (2007a) The deep benthos. *Philosophical Transactions of the Royal Society of London B* **362**, 39–66.

Brandt, A. & Ebbe, B. (2007) ANDEEP III ANtarctic benthic DEEP-sea biodiversity: colonisation history and recent community patterns. *Deep-Sea Research II* **54**, 1645–1904.

Brandt, A., Ellingsen, K.E.E., Brix, S., et al. (2005) Southern Ocean deep-sea isopod species richness (Crustacea, Malacostraca): influences of depth, latitude and longitude. *Polar Biology* **28**, 284–289.

Brandt, A., Gooday, A.J., Brix S.B., et al. (2007b) The Southern Ocean deep sea: first insights into biodiversity and biogeography. *Nature* **447**, 307–311.

Brandt, A. & Hilbig, B. (2004) ANDEEP (Antarctic benthic DEEP-sea biodiversity: colonization history and recent community patterns) – a tribute to Howard L. Sanders. *Deep-Sea Research II* **51**, 1457–1919.

Brandt, A., Linse, K. & Schüller, M. (2009) Bathymetric distribution patterns of Southern Ocean macrofaunal taxa: Bivalvia, Gastropoda, Isopoda and Polychaeta. *Deep-Sea Research I* **56**, 2013–2025.

Brökeland, W. & Raupach, M. (2008) A species complex within the isopod genus Haploniscus (Crustacea: Malacostraca) from the Antarctic deep sea. *Zoological Journal of the Linnean Society* **152**, 655–706.

Brown, B. (2001) Biomass of deep-sea benthic communities: polychaetes and other invertebrates. *Bulletin of Marine Science* **48**, 401–411.

Canals, M., Puig, P., Durrieu de Madron, X., et al. (2006) Flushing submarine canyons. *Nature* **444**, 354–357.

Clarke, A. & Johnston, N.M. (2003) Antarctic marine benthic diversity. *Oceanography and Marine Biology* **41**, 47–114.

Cross, I.A., Gebruk, A., Billett, D.S.M. & Rogacheva, A. (2009) *Peniagone crozeti*, a new species of elasipodid holothurian from abyssal depths off the Crozet Isles in the Southern Indian Ocean. *Zootaxa* **2096**, 484–488.

Culver, S.J. & Buzas, M.A. (2000) Global latitudinal species diversity gradient in deep-sea foraminifera. *Deep-Sea Research I* **47**, 259–275.

Danovaro, R., Dell'Anno, A. & Pusceddu, A. (2004) Biodiversity response to climate change in a warm deep sea. *Ecology Letters* **7**, 821–828.

Danovaro, R, Gambi, C., Lampadariou, N. & Tselepides, A. (2008) Deep-sea nematode biodiversity in the Mediterranean basin: testing for longitudinal, bathymetric and energetic gradients. *Ecography* **31**, 231–244.

Ellingsen, K., Brandt, A., Hilbig, B. & Linse, K. (2007) The diversity and spatial distribution of polychaetes, isopods and bivalves in the Atlantic sector of the deep Southern Ocean. *Polar Biology* **30**, 1265–1273.

Etter, R.J. & Rex, M.A. (1990) Population differentiation decreases with depth in deep-sea gastropods. *Deep-Sea Research* **37**:1251–1261.

Etter, R.J., Rex, M.A., Chase, M.R. & Quattro, J.M. (2005) Population differentiation decreases with depth in deep-sea bivalves. *Evolution* **59**, 1479–1491.

FitzGeorge-Balfour, T., Billett, D.S.M., Wolff, G.A., et al. (2010) Phytopigments as biomarkers of selectivity in abyssal holothurians; inter-species differences in response to a changing food supply. *Deep-Sea Research II* doi:10.1016/j.dsr2.2010.01.013.

France, S.C. & Kocher, T.D. (1996) Geographic and bathymetric patterns of mitochondrial 16S rRNA sequence divergence among deep-sea amphipods, *Eurythenes gryllus. Marine Biology* **126**, 633–643.

Gage, J.D. & Tyler, P.A. (1991) *Deep-Sea Biology: a Natural History of Organisms at the Deep-Sea Floor.* Cambridge, UK: Cambridge University Press.

Gheerardyn, H. & Veit-Köhler, G. (2009) Diversity and large-scale biogeography of Paramesochridae (Copepoda, Harpacticoida) in South Atlantic Abyssal Plains and the deep Southern Ocean. *Deep-Sea Research I* **56**, 1804–1815.

Ghosh, A.K. & Mukhopadhyay, R. (2000) *Mineral Wealth of the Ocean*. Rotterdam, the Netherlands: A.A. Balkema.

Ginger, M.L. *et al.* (2001) Organic matter assimilation and selective feeding by holothurians in the deep sea: Some observations and comments. *Progress in Oceanography* 50, 407–421.

Glover, A.G. & Smith, C.R. (2003) The deep-sea floor ecosystem: current status and prospects of anthropogenic change by the year 2025. *Environmental Conservation* 30, 219–241.

Glover, A.G., Smith, C.R., Paterson, J., *et al.* (2002) Polychaete species diversity in the central Pacific abyss: local and regional patterns, and relationships with productivity. *Marine Ecology Progress Series* 240, 157–170.

Gooday A.J., Kamenskaya, O.E. & Cedhagen, T. (2007) New and little-known Komokiacea (Foraminifera) from the bathyal and abyssal Weddell Sea and adjacent areas. *Zoological Journal of the Linnean Society* 151, 219–251.

Gooday, A.J., Malzone, M.G., Bett, B.J. & Lamont, P.A. (2010) Decadal-scale changes in shallow-infaunal foraminiferal assemblages at the Porcupine Abyssal Plain, NE Atlantic. *Deep-Sea Research II* doi:10.1016/j.dsr2.2010.01.012.

Grassle, J.F. & Maciolek, N.J. (1992) Deep-sea species richness: regional and local diversity estimates from quantitative bottom samples. *American Naturalist* 139, 313–341.

Gray, J.S. (2002) Species richness of marine soft sediment. *Marine Ecology Progress Series* 244, 285–297.

Hall, S. & Thatje, S. (2009) Global bottlenecks in the distribution of marine Crustacea: temperature constraints in the family Lithodidae. *Journal of Biogeography*, 36, 2125–2135.

Hansen, B. (1975) Systematics and biology of the deep-sea holothurians I. Elasipoda. *Galathea Report* 13, 1–262.

Hessler, R.R. & Sanders, H.L. (1967) Faunal diversity in the deep-sea. *Deep-Sea Research* 14, 65–78.

Heywood, V.H. (ed.) (1995) Global Biodiversity Assessment. *United Nations Environment Programme*. Cambridge, UK: Cambridge University Press.

Hughes, J.A., Smith, T., Chaillan, F., *et al.* (2007) Two abyssal sites in the Southern Ocean influenced by different organic matter inputs: Environmental characterization and preliminary observations on the benthic foraminifera. *Deep-Sea Research II* 54, 2275–2290.

International Seabed Authority (2008) Rationale and recommendations for the establishment of preservation reference areas for nodule mining in the Clarion-Clipperton Zone. ISBA/14/LTC/2*, Kingston, Jamaica, 12 pp.

International Seabed Authority (2009) Proposal for the designation of certain geographical areas in the Clarion-Clipperton Fracture Zone. ISBA/15/LTC/4, Kingston, Jamaica, 8 pp.

Jones, E.G., Tselepides, A., Bagley, P.M. & Priede, I.G. (2003) Bathymetric distribution of some benthic and benthopelagic species attracted to baited cameras and traps in the deep Eastern Mediterranean. *Marine Ecology Progress Series* 251, 75–86.

Kalogeropoulou, V., Bett, B.J., Gooday, A.J., *et al.* (2010) Temporal changes (1989–1999) in deep-sea metazoan meiofaunal assemblages on the Porcupine Abyssal Plain, NE Atlantic. *Deep-Sea Research II* doi:10.1016/j.dsr2.2009.02.002.

Kaiser, S., Barnes, D.K.A. & Brandt, A. (2007) Slope and deep-sea abundance across scales: Southern Ocean isopods show how complex the deep sea can be. *Deep-Sea Research II* 54, 1776–1789.

Lampitt, R.S., Billett, D.S.M. & Rice, A.L. (1986) The biomass of the invertebrate megabenthos from 500 to 4100 m in the North East Atlantic. *Marine Biology* 93, 69–81.

Lampitt, R.S., Salter, I., de Cuevas, B.A., *et al.* (2010) Long-term variability of downward particle flux in the deep Northeast Atlantic: causes and trends. *Deep-Sea Research II* doi:10.1016/j.dsr2.2010.01.011.

Lecroq, B., Gooday, A.J. & Pawlowski, J. (2009) Global genetic homogeneity in the deep-sea foraminiferan *Epistominella exigua* (Rotaliidae: Pseudoparrellidae). *Zootaxa* 2096, 23–32.

Malyutina, M. & Brandt, A. (2007) Diversity and zoogeography of Antarctic deep-sea Munnopsidae (Crustacea, Isopoda, Asellota). *Deep-Sea Research II* 54, 1790–1805.

McClain, C.R., Rex, M.A. & Etter, R.J. (2009) Patterns in deep-sea macroecology. In: *Marine Macroecology* (eds. J. Witman & K. Roy). Chicago: University of Chicago Press.

Miljutina, M.A., Miljutina, D.M., Mahatma, R. & Galéron, J. (2009) Deep-sea nematode assemblages of the Clarion-Clapperton Nodule Province (Tropical North-Eastern Pacific). *Marine Biodiversity* 40, 1–15.

Mincks, S.L., Dyal, P.L., Paterson, G.L.J., *et al.* (2009) A new species of *Aurospio* (Polychaeta, Spionidae) from the Antarctic shelf, with analysis of its ecology, reproductive biology and evolutionary history. *Marine Ecology* 30, 181–197.

Morgan, C.L. (2000) Resource estimates of the Clarion-Clipperton manganese nodule deposits. In: *Handbook of Marine Mineral Deposits* (ed. D.S. Cronan), pp. 145–170. Boca Raton, Florida: CRC Press

Murray, J.W. (1991) *Ecology and Palaeoecology of Benthic Foraminifera*. New York: Longman Scientific and Technical. 397 pp.

Nozawa, F., Kitazato, H., Tsuchiya, M. & Gooday, A.J. (2006) 'Live' benthic foraminifera at an abyssal site in the equatorial Pacific nodule province: abundance, diversity and taxonomic composition. *Deep-Sea Research I* 53, 1406–1422.

Pawlowski, J., Fahrni, J.F., Lecroq, B., *et al.* (2007) Bipolar gene flow in deep-sea benthic foraminifera. *Molecular Ecology* 16, 4089–4096.

Pearse, J.S., Mooi, R., Lockhart, S.J. & Brandt, A. (2009) Brooding and species diversity in the southern ocean: selection for brooders or speciation within brooding clades? In: *Smithsonian at the Poles: Contributions to International Polar Year Science* (eds. I. Krupnik, I.M.A. Lang & S.E. Miller) pp. 181–196.Proceedings of Smithsonian at the Poles Symposium, Smithsonian Institution, Washington, DC, 3–4 May 2007. Washington, DC: Smithsonian Institution Scholarly Press.

Pollard, R.T., Salter, I., Sanders, R., *et al.* (2009) Southern Ocean deep-water carbon export enhanced by natural iron fertilization. *Nature* 457, 577–580.

Poore, G.C.B. & Wilson, G.D.F. (1993) Marine species richness. *Nature* 361, 597–598.

Ramirez-Llodra, E., Reid, W.D.K. & Billett, D.S.M. (2005) Long-term changes in reproductive patterns of the holothurian *Oneirophanta mutabilis* from the Porcupine Abyssal Plain. *Marine Biology* 146, 683–693.

Raupach, M.J., Held, C. & Wägele, J.-W. (2004) Multiple colonization of the deep sea by the Asellota (Crustacea: Peracarida: Isopoda). *Deep-Sea Research II* 51, 1787–1795.

Raupach, M.J., Malyutina, M., Brandt, A. & Wägele, J.W. (2007) Molecular data reveal a highly diverse species flock within the deep-sea isopod *Betamorpha fusiformis* (Crustacea: Isopoda: Asellota) in the Southern Ocean. *Deep-Sea Research II* 54, 1820–1830.

Raupach, M.J., Mayer, C., Malyutina, M. & Wägele, J.-W. (2009) Multiple origins of deep-sea Asellota (Crustacea: Isopoda) from shallow waters revealed by molecular data. *Proceedings of the Royal Society B* 276, 799–808.

Raupach, M. & Wägele, J.-W. (2006) Distinguishing cryptic species in Antarctic Asellota (Crustacea: Isopoda) – a preliminary study of mitochondrial DNA in *Acanthaspidia drygalskii*. *Antarctic Science* 18, 191–198.

Rex, M.A., Etter, R.J. & Stuart, C.T. (1997) Large-scale patterns of species diversity in the deep-sea benthos. In: *Marine biodiversity* (eds. R.F.G. Ormond, J.D. Gage & M.V. Angel), pp. 94–121. Cambridge, UK: Cambridge University Press.

Rex, M.A., McClain, C.R., Johnson, N.A., *et al.* (2005) A source-sink hypothesis for abyssal diversity. *American Naturalist* **165**, 163–178.

Rex, M.A., Stuart, C.T. & Coyne, G. (2000) Latitudinal gradients of species richness in the deep-sea benthos of the North Atlantic. *Proceedings of the National Academy of Sciences of the USA* **97**, 4082–4085.

Rex, M.A., Stuart, C.T., Hessler, R.R., *et al.* (1993) Global-scale latitudinal patterns of species diversity in the deep-sea benthos. *Nature* **365**, 636–639.

Rex, M.A. *et al.* (2006) Global bathymetric patterns of standing stock and body size in the deep-sea benthos. *Marine Ecology Progress Series* **317**, 1–8.

Rose, A., Seifried, S., Willen, E., *et al.* (2005) A method for comparing within-core alpha diversity values from repeated multicorer samplings, shown for abyssal Harpacticoida (Crustacea: Copepoda) from the Angola Basin. *Organisms Diversity and Evolution* **5** (Suppl. 1), 3–17.

Rowe, G.T. (1971) Observations on bottom currents and epibenthic populations in Hatteras Canyon. *Deep-Sea Research* **18**, 569–581.

Ruhl, H.A. (2007) Abundance and size distribution dynamics of abyssal epibenthic megafauna in the northeast Pacific. *Ecology* **88**, 1250–1262.

Ruhl, H.A. & Smith, K.L. (2004) Shifts in deep-sea community structure linked to climate and food supply. *Science* **305**, 513–515.

Sanders, H.L., Hessler, R.R. & Hampson, G.R. (1965) An introduction to the study of deep-sea benthic faunal assemblages along the Gay Head–Bermuda transect. *Deep-Sea Research* **12**, 845–867.

Sanders, H.L. & Hessler, R.R. (1969) Diversity and composition of abyssal benthos. *Science* **166**, 1033–1034.

Schüller, M. & Ebbe, B. (2007) Global distributional patterns of selected deep-sea Polychaeta (Annelida) from the Southern Ocean. *Deep-Sea Research II* **54**, 1737–1751.

Schüller, M., Ebbe, B. & Wägele, J.-W. (2009) Community structure and diversity of polychaetes (Annelida) in the deep Weddell Sea (Southern Ocean) and adjacent basins. *Marine Biodiversity* **39**, 95–108.

Sibuet, M. & Vangriesheim, A. (2009) Deep-sea environment and biodiversity of the West African Equatorial margin. *Deep-Sea Research II* **56**, 2156–2168.

Smith C.R. & Baco, A.R. (2003) Ecology of whale falls at the deep-sea floor. *Oceanography and Marine Biology* **41**, 311–354.

Smith, C.R., Berelson, W., Demaster, D.J., *et al.* (1997) Latitudinal variations in benthic processes in the abyssal equatorial Pacific: control by biogenic particle flux. *Deep-Sea Research II* **44**, 2295–2317.

Smith, C.R., De Leo, F.C., Bernardino, A.F., *et al.* (2008a) Abyssal food limitation, ecosystem structure and climate change. *Trends in Ecology and Evolution* **23**, 518–528.

Smith, C.R. & Demopoulos, A.W.J. (2003) Ecology of the deep Pacific Ocean floor. In: *Ecosystems of the World*, Volume **28**, *Ecosystems of the Deep Ocean* (ed. P.A. Tyler), pp. 179–218. Elsevier, Amsterdam.

Smith, C.R., Levin, L.A., Koslow, A., *et al.* (2008b) The near future of deep seafloor ecosystems. In: *Aquatic Ecosystems: Trends and global prospects* (ed. N. Polunin), pp. 334–351. Cambridge University Press.

Smith, C.R., Paterson, G., Lambshead, J., *et al.* (2008c). Biodiversity, species ranges, and gene flow in the abyssal Pacific nodule province: predicting and managing the impacts of deep seabed mining. ISA Technical Study: No.3, International Seabed Authority, Kingston, Jamaica, 38 pp.

Smith K.L., *et al.* (2006) Climate effect on food supply to depths greater than 4,000 meters in the northeast Pacific. *Limnology and Oceanography* **51**, 166–176.

Smith, K.L., Ruhl, H.A., Bett, B.J., *et al.* (2009) Climate, carbon cycling, and deep-ocean ecosystems. *Proceedings of the National Academy of Sciences of the USA* **106**, 19211–19218.

Soto, E., Paterson, G.L.J., Billett, D.S.M., *et al.* (2009) Temporal variability in polychaete assemblages of the abyssal NE Atlantic Ocean. *Deep-Sea Research II* doi:10.1016/j.dsr2.2009.02.003.

Storey, B.C. (1995) The role of mantle plumes in continental breakup: case histories from Gondwanaland. *Nature* **337**, 301–308.

Stuart, C.T., Martinez Arbizu, P., Smith, C.R., *et al.* (2008) CeDAMar global database of abyssal biological sampling. *Aquatic Biology* **4**, 143–145.

Stuart, C.T. & Rex, M.A. (2009) Bathymetric patterns of deep-sea gastropod species diversity in 10 basins of the Atlantic Ocean and Norwegian Sea. *Marine Ecology* **30**, 164–180.

Sun, X., Corliss, B.H., Brown, C.W. & Showers, W.J. (2006) The effect of primary productivity and seasonality on the distribution of deep-sea benthic Foraminifera in the North Atlantic. *Deep-Sea Research I* **53**: 28–47.

Thomas, E. & Gooday, A.J. (1996) Cenozoic deep-sea benthic foraminifers: tracers for changes in oceanic productivity? *Geology* **24**, 355–358.

Thomson, M.R.A. (2004) Geological and palaeoenvironmental history of the Scotia Sea region as a basis for biological interpretation. *Deep-Sea Research II* **51**, 1467–1487.

Tselepides, A. & Lampadariou, N. (2004) Deep-sea meiofaunal community structure in the Eastern Mediterranean: are trenches benthic hot spots? *Deep-Sea Research I* **51**, 833–847.

Tyler, P.A. (2003) (ed.) *Ecosystems of the World*, Vol. 28 *Ecosystems of the Deep Oceans*, pp 1–569. Amsterdam: Elsevier.

Wigham, B.D., Hudson, I.R., Billett, D.S.M. & Wolff, G.A. (2003) Is long-term change in the abyssal Northeast Atlantic driven by qualitative changes in export flux? Evidence from selective feeding in deep-sea holothurians. *Progress in Oceanography* **59**, 409–441.

Witte, U., Wenzhöfer, F., Sommer, S., *et al.* (2003) *In situ* experimental evidence of the fate of a phytodetritus pulse at the abyssal sea floor. *Nature* **424**, 763–766.

Zardus, J.D., Etter, R.J., Chase, M.R., *et al.* (2006) Bathymetric and geographic population structure in the pan-Atlantic deep-sea bivalve *Deminucula atacellana* (Schenck, 1939). *Molecular Ecology* **15**, 639–651.

Chapter 9

Biogeography, Ecology, and Vulnerability of Chemosynthetic Ecosystems in the Deep Sea

Maria C. Baker[1], Eva Z. Ramirez-Llodra[2], Paul A. Tyler[1], Christopher R. German[3], Antje Boetius[4], Erik E. Cordes[5], Nicole Dubilier[4], Charles R. Fisher[6], Lisa A. Levin[7], Anna Metaxas[8], Ashley A. Rowden[9], Ricardo S. Santos[10], Tim M. Shank[3], Cindy L. Van Dover[11], Craig M. Young[12], Anders Warén[13]

[1]School of Ocean and Earth Science, University of Southampton, National Oceanography Centre, Southampton, UK
[2]Institut de Ciències del Mar, Barcelona, Spain
[3]Woods Hole Oceanographic Institution, Woods Hole, Massachusetts, USA
[4]Max Planck Institute for Microbiology, Bremen, Germany
[5]Department of Biology, Temple University, Philadelphia, Pennsylvania, USA
[6]Muller Laboratory, Pennsylvania State University, Pennsylvania, USA
[7]Integrative Oceanography Division, Scripps Institution of Oceanography, La Jolla, California, USA
[8]Department of Oceanography, Dalhousie University, Halifax, Nova Scotia, Canada
[9]National Institute of Water and Atmospheric Research, Wellington, New Zealand
[10]Department of Oceanography and Fisheries, University of the Azores, Horta, Portugal
[11]Nicholas School of the Environment and Earth Sciences, Duke University Marine Laboratory, Beaufort, North Carolina, USA
[12]Oregon Institute of Marine Biology, Charleston, Oregon, USA
[13]Department of Invertebrate Zoology, Swedish Museum of Natural History, Stockholm, Sweden

9.1 Life Based on Energy of the Deep

9.1.1 A spectacular discovery

This chapter is based upon research and findings relating to the Census of Marine Life ChEss project, which addresses the biogeography of deep-water chemosynthetically driven ecosystems (www.noc.soton.ac.uk/chess). This project has been motivated largely by scientific questions concerning phylogeographic relationships among different chemosynthetic habitats, evidence of conduits and barriers to gene flow among those habitats, and environmental factors that control diversity and distribution of chemosynthetically driven fauna. Investigations of chemosynthetic environments in the deep sea span just three decades, owing to their relatively recent discovery. Despite the excitement of many discoveries in the deep ocean since the early nineteenth century, nothing could have prepared the scientific community for the discovery made in the late 1970s, which would challenge some fundamental principles of our understanding of life on Earth. Deep hot water venting was observed for the first time in 1977 on the Galápagos Rift, in the eastern Pacific. To the astonishment of the deep-sea explorers of the time, a prolific community of bizarre

Life in the World's Oceans, edited by Alasdair D. McIntyre
© 2010 by Blackwell Publishing Ltd.

animals were seen to be living in close proximity to these vents (Corliss *et al.* 1979). Giant tubeworms and huge white clams were among the inhabitants, forming oases of life in the otherwise apparently uninhabited deep seafloor (Figs. 9.1A, B, and C). Most of the creatures first observed on vents were totally new to science, and it was a complete mystery as to what these animals were using for an energy source in the absence of sunlight and in the presence of toxic levels of hydrogen sulfide and heavy metals.

9.1.2 Chemosynthetic ecosystems: where energy from the deep seabed is the source of life

Until the discovery of hydrothermal vents, benthic deep-sea ecosystems were assumed to be entirely heterotrophic, completely dependent on the input of sedimented organic matter produced in the euphotic surface layers from photosynthesis (Gage 2003) and, in the absence of sunlight, completely devoid of any *in situ* primary productivity. The deep sea is, in general, a food-poor environment with low secondary productivity and biomass. In 1890, Sergei Nikolaevich Vinogradskii proposed a novel life process called chemosynthesis, which showed that some microbes have the ability to live solely on inorganic chemicals. Almost 90 years later the discovery of hydrothermal vents provided stunning new insight into the extent to which microbial primary productivity by chemosynthesis can maintain biomass-rich metazoan communities with complex trophic structure in an otherwise food-poor deep sea (Jannasch & Mottl 1985). Hydrothermal vents are found on mid-ocean ridges and in back arc basins where deep-water volcanic chains form new ocean floor (reviewed by Van Dover 2000; Tunnicliffe *et al.* 2003). The super-heated fluid (up to 407°C) emanating from vents is charged with metals and sulfur. Microbes in these habitats obtain energy from the oxidation of hydrogen, hydrogen sulfide, or methane from the vent fluid. The microbes can be found either suspended in the water column or forming mats on different substrata, populating seafloor sediments and ocean crust, or living in symbiosis with several major animal taxa (Dubilier *et al.* 2008; Petersen & Dubilier 2009). By microbial mediation, the rich source of chemical energy supplied from the deep ocean interior through vents allows the development of densely populated ecosystems, where abundances and biomass of fauna are much greater than on the surrounding deep-sea floor.

Eight years after the discovery of hydrothermal vent communities, the first cold seep communities were described in the Gulf of Mexico (Paull *et al.* 1984). Cold seeps occur in both passive and active (subduction) margins. Seep habitats are characterized by upward flux of cold fluids enriched in methane and often also other hydrocarbons, as well as a high concentration of sulfide in the sediments (Sibuet & Olu 1998; Levin 2005). The first observations of seep communities showed a fauna and trophic ecology similar to that of hydrothermal vents at higher taxonomical levels (Figs. 9.1D and E), but with dissimilarities in terms of species and community structure.

The energetic input to chemosynthetic ecosystems in the deep sea can also derive from photosynthesis as in the case of large organic falls to the seafloor, including kelp, wood, large fish, or whales. After a serendipitous discovery of a whale fall in 1989, the first links between vents, seeps, and the reducing ecosystems at large organic falls were made (Smith & Baco 2003). Bones of whales consist of up to 60% lipids that, when degraded by microbes, produce reduced chemical compounds similar to those emanating from vents and seeps (Fig. 9.1F) (Treude *et al.* 2009). Another deep-water reducing environment is created where oxygen minimum zones (OMZs, with oxygen concentrations below 0.5 ml l^{-1} or 22 µM) intercept continental margins, occurring mainly beneath regions of intensive upwelling (Helly & Levin 2004) (Figs. 9.1G and H). Only in the second half of the twentieth century was it understood that OMZs support extensive autotrophic bacterial mats (Gallardo 1963, 1977; Sanders 1969; Fossing *et al.* 1995; Gallardo & Espinoza 2007) and, in some instances, fauna with a trophic ecology similar to that of vents and seeps (reviewed in Levin 2003).

9.1.3 Adaptations to an "extreme" environment

Steep gradients of temperature and chemistry combined with a high disturbance regime, caused by waxing and waning of fluid flow and other processes during the life cycle of a hydrothermal vent, result in low diversity communities with only a few mega- and macrofauna species dominating any given habitat (Van Dover & Trask 2001; Turnipseed

---

Fig. 9.1

Hydrothermal vent (**A, B,** and **C**), cold seep (**D** and **E**), whale fall (**F**), and OMZ communities (**G** and **H**). (**A**) Zoarcid fish over a *Riftia pachyptila* tubeworm community in EPR vents; (**B**) *Bathymodiolus* mussel community in EPR vents (© Stephen Low Productions, Woods Hole Oceanographic Institution, E. Kristof, the National Geographic Society, and R. A. Lutz, Rutgers University). (**C**) Dense aggregations of the MAR vent shrimp *Rimicaris exoculata* (© Missao Sehama, 2002 (funded by FCT, PDCTM 1999/MAR/15281), photographs made by VICTOR6000/IFREMER). (**D**) *Lamellibrachia* tubeworms from the Gulf of Mexico cold seeps (© Charles Fisher, Penn State University). (**E**) *Bathymodiolus* mussel bed by a brine pool in the Gulf of Mexico cold seeps (© Stéphane Hourdez, Penn State University/Station Biologique de Roscoff). (**F**) Skeleton of a whale fall covered by bacteria (© Craig Smith, University of Hawaii). (**G**) Ophiuroids on an OMZ in the Indian margin (© Hiroshi Kitazato, JAMSTEC, and NIOO). (**H**) Galatheid crabs on an OMZ on the upper slopes of Volcano 7, off Acapulco, Mexico (© Lisa Levin, Scripps Institution of Oceanography).

Chapter 9 Biogeography, Ecology, and Vulnerability of Chemosynthetic Ecosystems in the Deep Sea

et al. 2003; Dreyer *et al.* 2005). The proportion of extremely rare species (fewer than five individuals in pooled samples containing tens of thousands of individuals from the same vent habitat) is typically high, in the order of 50% of the entire species list for a given quantitative sampling effort (C.L. Van Dover, unpublished observation).

Deep-water chemosynthetic habitats have also been shown to have a high degree of species endemicity in each habitat: 70% in vents (Tunnicliffe *et al.* 1998; Desbruyères *et al.* 2006a), about 40% in seeps both for mega epifauna (Bergquist *et al.* 2005; Cordes *et al.* 2006) and macro infauna (Levin *et al.* 2009a). In OMZs, the percentage of endemism is relatively low (Levin *et al.* 2009a), but has yet to be quantified. Some of the most conspicuous of the endemic species of reducing environments have developed unusual physiological adaptations for the extreme environments in which they live. These include symbiotic relationships with bacteria, organ and body modifications, and reproductive and novel adaptations for tolerating thermal and chemical fluctuations of great magnitude. Because chemosynthetic habitats are naturally fragmented and ephemeral habitats, successful species must also be specially adapted for dispersal to and colonization of isolated "chemosynthetic islands" in the deep sea (Bergquist *et al.* 2003; Neubert *et al.* 2006; Vrijenhoek 2009a).

9.1.4 Chemosynthetic islands: a biogeographic puzzle with missing pieces

Since their discovery just over 30 years ago, more than 700 species from vents (Desbruyères *et al.* 2006a) and 600 species from seeps have now been described and are listed on *ChEssBase* (Ramirez-Llodra *et al.* 2004; www.noc.soton.ac.uk/chess/database/db_home.php). This rate of discovery is equivalent to one new species described every two weeks, sustained over approximately one-quarter of the past century (Lutz 2000; Van Dover *et al.* 2002). Furthermore, geomicrobiologists have explored microbial diversity of chemosynthetic ecosystems, revealing a plethora of interesting and novel metabolisms, but also signature compositions for the different types of reduced habitat, and symbiotic organism (Jørgensen & Boetius 2007; Dubilier *et al.* 2008).

Although several hundred hydrothermal vent and cold seep sites have now been located worldwide (see ChEss webpages), only approximately 100 have been studied so far with respect to their faunal and microbial composition, and even for their ecosystem function. Nevertheless, through such investigations, scientists soon noticed the differences and in some cases similarities among the animal communities from different vent and seep sites. For example, why is the giant tubeworm *Riftia pachyptila* only found at Pacific vents whereas shrimp species in the genus *Rimicaris* are only found at Atlantic and Indian Ocean vents? Why is the mussel

genus *Bathymodiolus* generally widespread at vents and seeps but largely absent from seeps and vents in the northeastern Pacific Ocean? In 2002, at the onset of the ChEss project, biological investigations of known vent sites provided enough data to describe six biogeographic provinces for vent species (Van Dover *et al.* 2002) and identified several gaps that needed to be closed to complete the "biogeographical puzzle of seafloor life" (Shank 2004) (Fig. 9.2). In contrast, cold seep and whale fall communities appear to share many of the key taxa across all oceans. The ChEss project developed a major exploratory program to address and explain global patterns of biogeography in deep-water chemosynthetic ecosystems and the factors shaping them.

9.2 Finding New Pieces of the Puzzle (2002–2010)

9.2.1 Technological developments for exploration

One of the most significant advances in deep-sea investigations of chemosynthetic ecosystems, developed and implemented as a new international state of the art technique within the lifetime of the ChEss project, has been the use of deep-sea autonomous underwater vehicles (AUVs) to trace seafloor hydrothermal systems to their source or to map cold seep systems in the necessary resolution to quantify the distribution of chemosynthetic habitats. This approach (Baker *et al.* 1995; Baker & German 2004; Yoerger *et al.* 2007) was sufficient for geological investigations of global-scale heat-flux and chemical discharge to the oceans. However, the ChEss hypotheses concerning global-scale biogeography required more precise location of hydrothermal venting and hydrocarbon seepage on the seafloor; ideally with preliminary characterization of not only the vent and seep site itself but also a first-order characterization of the dominant species present.

So far, the method has been applied on seven separate hydrothermal vent cruises, from 2002 to 2009, throughout the Southern hemisphere, the least explored part of the global deep ocean. These expeditions have located 16 different new sites on the Galápagos Rift (Shank *et al.* 2003), in the Lau Basin (southwest Pacific; German *et al.* 2008a), the Mid-Atlantic Ridge (MAR) (South Atlantic; German *et al.* 2008b; Melchert *et al.* 2008; Haase *et al.* 2009), the southwest Indian Ridge (Southern Indian Ocean; C. Tao, personal communication), the East Pacific Rise (southeast Pacific; C. Tao, personal communication), and the Chile margin (C. German, unpublished observation). For cold seep mapping, a major success was the combined AUV and remotely operated vehicle (ROV) deployment in the Nile Deep Sea Fan, leading to the description of several new types of hydrocarbon seep in depths between 1,000 and 3,500 m

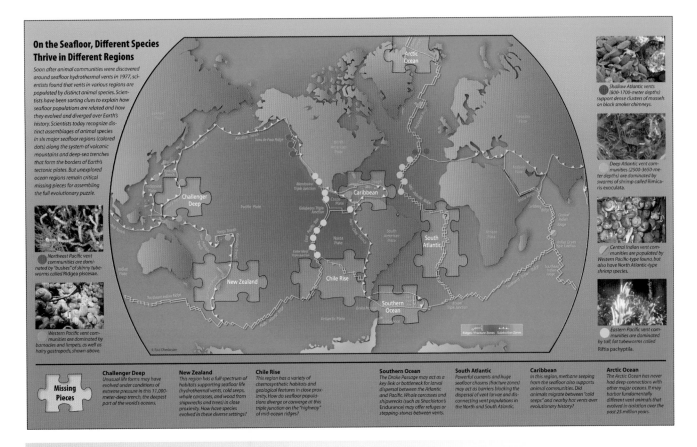

On the Seafloor, Different Species Thrive in Different Regions

Soon after animal communities were discovered around seafloor hydrothermal vents in 1977, scientists found that vents in various regions are populated by distinct animal species. Scientists have been sorting clues to explain how seafloor populations are related and how they evolved and diverged over Earth's history. Scientists today recognize distinct assemblages of animal species in six major seafloor regions (colored dots) along the system of volcanic mountains and deep-sea trenches that form the borders of Earth's tectonic plates. But unexplored ocean regions remain critical missing pieces for assembling the full evolutionary puzzle.

Northeast Pacific vent communities are dominated by "bushes" of skinny tubeworms called Ridgea piscesae.

Western Pacific vent communities are dominated by barnacles and limpets, as well as hairy gastropods, shown above.

Shallow Atlantic vents (800–1700-meter depths) support dense clusters of mussels on black smoker chimneys.

Deep Atlantic vent communities (2500–3650-meter depths) are dominated by swarms of shrimp called Rimicaris exoculata.

Central Indian vent communities are populated by Western Pacific-type fauna, but also have North Atlantic-type shrimp species.

Eastern Pacific vent communities are dominated by tall, fat tubeworms called Riftia pachyptila.

Missing Pieces

Challenger Deep Unusual life forms may have evolved under conditions of extreme pressure in this 11,000-meter-deep trench, the deepest part of the world's oceans.

New Zealand This region has a full spectrum of habitats supporting seafloor life (hydrothermal vents, cold seeps, whale carcasses and wood from shipwrecks and trees) in close proximity. How have species evolved in these diverse settings?

Chile Rise This region has a variety of chemosynthetic habitats, geological features in close proximity. How do seafloor populations diverge or converge at this triple junction on the "highway" of mid-ocean ridges?

Southern Ocean The Drake Passage may act as a key link or bottleneck for larval dispersal between the Atlantic and Pacific. Whale carcasses and shipwrecks (such as Shackleton's Endurance) may offer refuges or stepping-stones between vents.

South Atlantic Powerful currents and huge seafloor chasms (fracture zones) may act as barriers blocking the dispersal of vent larvae and disconnecting vent populations in the North and South Atlantic.

Caribbean In this region, methane seeping from the seafloor also supports animal communities. Did animals migrate between "cold seeps" and nearby hot vents over evolutionary history?

Arctic Ocean The Arctic Ocean has never had deep connections with other major oceans. It may harbor fundamentally different vent animals that evolved in isolation over the past 25 million years.

Fig. 9.2

Global map showing the mid-ocean ridge system, the recognized hydrothermal vent biogeographic provinces (colored dots) and the unexplored regions that are critical missing pieces of the full evolutionary puzzle. Reproduced from Shank 2004 with permission of the Woods Hole Oceanographic Institution.

(Foucher *et al.* 2009; technical details described in Dupré *et al.* (2009)).

The way the AUV technique works for the exploration of vents is described in detail by German *et al.* (2008a). Perhaps most surprising to us, and of widest long-term significance, is that, when flying close to the seafloor, the techniques have not only been sufficiently sensitive to locate high-temperature "black-smoker" venting, but also sites of much more subtle lower-temperature diffuse flow (Shank *et al.* 2003). Building on these successes, future investigations will be reliant upon the new generation of exploratory vehicles such as a new hybrid AUV–ROV vehicle (Bowen *et al.* 2009), which has already been applied in ChEss studies (see below) as a technological precursor to future under-ice investigations (Jakuba *et al.* 2008; German *et al.* 2009).

9.2.2 Finding new species

In the past decade, we have seen a significant increase in molecular tools for studies to understand species evolution, metapopulations, and gene flow in chemosynthetic regions (Shank & Halanych 2007; Johnson *et al.* 2008; Plouviez

et al. 2009; Vrijenhoek 2009b). New high-resolution and high-throughput methods will result in the first insight into the structure and biogeography of microbial communities of chemosynthetic ecosystems in the Census International Census of Marine Microbes (ICoMM) project (see Chapter 12). However, a major concern today for marine biodiversity analysis is the paucity of taxonomists using morphological methods, and in particular taxonomists specializing in deep-sea species. Both morphological and molecular taxonomy are essential to develop fundamental knowledge and sustainable management of our marine resources. In an effort to raise the profile of taxonomy once more, ChEss set up an annual program of Training Awards for New Investigators (TAWNI). These awards have been made to a total of 10 scientists from around the globe to develop further their taxonomic skills relating to chemosynthetic organisms (www.noc.soton.ac.uk/chess/science/sci_tawni. php). As a result, they have collectively achieved impressive outputs where many meio-, macro-, and megafauna species have been described and new records identified from different sites (Table 9.1). These descriptions have been added to the approximately 200 species that have been described and published from vents, seeps, and whale falls

Table 9.1

Species new to science described or identified by TAWNI awardees during the ChEss project.

Group	Family	Species	Location	References	TAWNI
Anomura	Kiwaidae	*Kiwa* sp. nov.	Costa Rica seeps	Thurber *et al.* in preparation	Andrew Thurber
Polychaete	Spionidae	Gen. & sp. nov.	New Zealand seeps	Thurber *et al.* in preparation	Andrew Thurber
Polychaete	Ampharetidae	Gen. & sp. nov.	New Zealand seeps	Thurber *et al.* in preparation	Andrew Thurber
Polychaete	Ampharetidae	Gen. & sp. nov.	New Zealand seeps	Thurber *et al.* in preparation	Andrew Thurber
Harpacticoid copepod	Tegastidae	*Smacigastes barti*	9°50′N EPR vents	Gollner *et al.* 2008	Sabine Gollner
Nematoda	Monhysteridae	*Thalassomonhystera fisheri* n. sp.	9°50′N EPR vents	Zekely *et al.* 2006	Julia Zekely
Nematoda	Monhysteridae	*Halomonhystera hickeyi* n. sp.	9°50′N EPR vents	Zekely *et al.* 2006	Julia Zekely
Nematoda	Monhysteridae	*Thalassomonhystera vandoverae* n. sp.	Mid-Atlantic Ridge vents	Zekely *et al.* 2006	Julia Zekely
Nematoda		*Anticoma* sp. 1	9°50′N EPR vents		Julia Zekely
Nematoda		*Chromadorita* sp. 1	9°50′N EPR vents		Julia Zekely
Nematoda		*Daptonema* sp. 1	9°50′N EPR vents		Julia Zekely
Nematoda		*Daptonema* sp. 2	9°50′N EPR vents		Julia Zekely
Nematoda		*Euchromadora* sp.	9°50′N EPR vents		Julia Zekely
Nematoda		*Eurystomina* sp. 1	9°50′N EPR vents		Julia Zekely
Nematoda		*Halomonhystera hickeyi*	9°50′N EPR vents		Julia Zekely
Nematoda		*Halomonhystera* sp. 1	9°50′N EPR vents		Julia Zekely
Nematoda		*Leptolaimus* sp. 1	9°50′N EPR vents		Julia Zekely
Nematoda		*Metoncholaimus* sp. 1	9°50′N EPR vents		Julia Zekely
Nematoda		*Microlaimus* sp. 1	9°50′N EPR vents		Julia Zekely
Nematoda		*Molgolaimus* sp. 1	9°50′N EPR vents		Julia Zekely
Nematoda		*Paracantonchus* sp. 1	9°50′N EPR vents		Julia Zekely
Nematoda		*Paralinhomeus* sp. 1	9°50′N EPR vents		Julia Zekely
Nematoda		*Rhabdocoma* sp. 1	9°50′N EPR vents		Julia Zekely
Nematoda		*Prooncholaimus* sp. 1	9°50′N EPR vents		Julia Zekely
Actiniaria		*Amphianthus* sp. nov.	Lau Basin vents		Kevin Zelnio
Actiniaria		*Anthosactis* sp. nov.	Lau Basin vents		Kevin Zelnio
Actiniaria		*Bathydactylus* sp. nov.	Lau Basin vents		Kevin Zelnio
Actiniaria		*Chondrophellia* sp. nov.	Lau Basin vents		Kevin Zelnio

Group	Family	Species	Location	References	TAWNI
Actiniaria		*Sagartiogeton* sp. nov.	Lau Basin vents		Kevin Zelnio
Actiniaria		Gen. et sp. nov.?	Lau Basin vents		Kevin Zelnio
Zoanthid		Sp. nov?	Lau Basin vents		Kevin Zelnio
Frenulate polychaete	Siboglinidae	*Bobmarleya gadensis* gen. et sp. nov.	Gulf of Cadiz mud volcanoes	Hilário & Cunha 2008	Ana Hilário
Frenulate polychaete	Siboglinidae	*Spirobrachia tripeira* sp. nov.	Gulf of Cadiz mud volcanoes	Hilário & Cunha 2008	Ana Hilário
Frenulate polychaete	Siboglinidae	*Lamellisabella denticulata* (new record in Gulf of Cadiz)	Gulf of Cadiz mud volcanoes	Hilário & Cunha 2008	Ana Hilário
Frenulate polychaete	Siboglinidae	*Lamellisabella* sp. nov.	Gulf of Cadiz mud volcanoes	Hilário *et al.* in prep	Ana Hilário
Frenulate polychaete	Siboglinidae	*Polybrachia* sp. nov.	Gulf of Cadiz mud volcanoes	Hilário *et al.* in prep	Ana Hilário
Frenulate polychaete	Siboglinidae	*Polybrachia* sp. nov.	Gulf of Cadiz mud volcanoes	Hilário *et al.* in prep	Ana Hilário
Frenulate polychaete	Siboglinidae	*Siboglinum poseidoni* (new record in Gulf of Cadiz)	Gulf of Cadiz mud volcanoes	Hilário *et al.* submitted	Ana Hilário

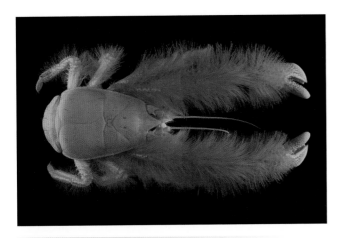

Fig. 9.3

The yeti crab, *Kiwa hirsuta*, from the Easter Island microplate hydrothermal vents. © Ifremer/A. Fifis.

since the onset of the ChEss project in 2002. One of the most extraordinary animals that has consequently received much media attention was discovered on southeast Pacific vents in 2005: the yeti crab *Kiwa hirsuta* (Fig. 9.3). This is not only a species new to science, but also represents a new genus and new family (Macpherson *et al.* 2005). Recently, a close relative of the vent yeti crab was discovered from Costa Rican cold seeps and is being described

with the aid of a TAWNI grant (A. Thurber, personal communication).

9.2.3 Global biogeography patterns in deep-water chemosynthetic ecosystems

Addressing global biogeographic patterns for species from all deep-water chemosynthetic ecosystems and the phylogenetic links among habitats needed a coordinated international effort, with shared human and infrastructure resources, that no single nation could attempt alone. In 2002, ChEss outlined a field program for the strategic exploration and investigation of chemosynthetic ecosystems in key areas that would provide essential information to close some of the main gaps in our knowledge (Tyler *et al.* 2003). The ChEss field program was motivated by three scientific questions. (1) What are the taxonomic relationships among different chemosynthetic habitats? (2) What are the conduits and barriers to gene flow among those habitats? (3) What are the environmental factors that control diversity and distribution of chemosynthetically driven fauna? To address these questions at the global scale, four key geographic areas were selected for exploration and investigation: the Atlantic Equatorial Belt (AEB), the New Zealand Region (RENEWZ), the Polar Regions (Arctic and

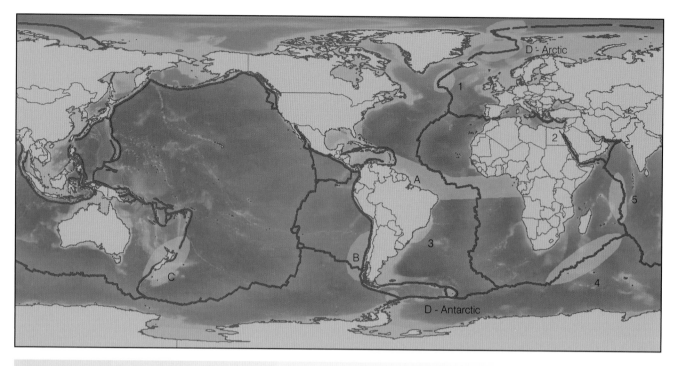

Fig. 9.4

ChEss field program study sites. Light blue and pink areas are key priority regions: **(A)** Equatorial Atlantic Belt region; **(B)** Southeast Pacific region; **(C)** New Zealand region; **(D)** Polar regions. Yellow areas are important for ChEss-related activities: 1, northern MAR between the Iceland and Azores hot-spots; 2, Eastern Mediterranean; 3, Brazilian continental margin; 4, southwest Indian Ridge; 5, Central Indian Ridge.

Antarctic), and the southeast Pacific off Chile region (INSPIRE) (Fig. 9.4). Below, we describe the issues addressed and main findings in each area.

9.2.3.1 The Atlantic Equatorial Belt: barriers and conduits for gene flow

The AEB is a large region expanding from Costa Rica to the West Coast of Africa that encloses numerous seep (for example Costa Rica, Gulf of Mexico, Blake Ridge, Gulf of Guinea) and vent (e.g., northern MAR (NMAR), southern MAR (SMAR), Cayman Rise) habitats. This region is particularly significant for investigating connectivity among populations and species' maintenance across large geographic areas. Potential gene flow across the Atlantic (west to east) is subject to the effects of deep-water currents (Northeast Atlantic deep water), equatorial jets, and topographic barriers such as the MAR. When considering a north–south direction, gene flow along the MAR may be affected by mid-ocean ridge offsets such as the Romanche and Chain fracture zones. These fracture zones are significant topographic features 60 million years old, 4 km high and 935 km ridge offset, which cross the equatorial MAR prominently, affecting both the linearity of the ridge system and large-scale ocean circulation in this region. North Atlantic Deep Water flows south along the East coasts of

North and South America as far as the Equator before being deflected east, crossing the MAR through conduits created by these major fracture zones (Speer *et al.* 2003). Circulation within these fracture zones is turbulent and may provide an important dispersal pathway for species from west to east across the Atlantic (Van Dover *et al.* 2002), for example between the Gulf of Mexico and the Gulf of Guinea. The cold seeps in the Pacific Costa Rican margin were included in this study to address questions of isolation between the Pacific and the Atlantic faunas after the closure of the Isthmus of Panama 5 million years ago. The fauna from methane seeps on the Costa Rica margin, just now being explored, are yielding surprising affinities, which suggests that this site operates as a crossroads. Some animals appear related to the seep faunas in the Gulf of Mexico and off West Africa, whereas others show phylogenetic affinities with nearby vents at 9°N on the East Pacific Rise and with more distant vents at Juan de Fuca Ridge and the Galápagos (L. Levin, unpublished observations). Furthermore, recent investigations have shown (C. German, C.L. Van Dover & J. Copley, unpublished observation) there is active venting in the ultra-slow Cayman spreading ridge in the Caribbean at depths of 5,000 m (CAYTROUGH 1979), and investigations are underway to determine how the animals colonizing these vents are related to vent and seep faunas on either side of the Isthmus of Panama. The first

Fig. 9.5

Cold seep communities from different Atlantic Equatorial Belt areas. **(A)** Gulf of Mexico; **(B)** Costa Rica; **(C)** Barbados Prism; **(D)** Congo margin. © Erik Cordes, Temple University (photographs A and B); © Ifremer (photographs C and D courtesy of Karine Olu).

plumes were located in November 2009 at depths below 4,500 m, suggestive of active venting, and these plumes were further explored by ChEss scientists in 2010 who located the source of active venting at 5,000 m – the deepest known vent ever found. Exploration on the MAR has also led to the discovery of the hottest vents (407 °C) (Kochinsky 2006; Kochinsky *et al.* 2008), as well as another deep vent (4,100 m), named Ashadze (Ondreas *et al.* 2007; Fouquet *et al.* 2008).

In the AEB, the connections across the Atlantic have been relatively clearly defined. The seeps of the African margin contain a fauna with very close affinities to the seep communities in the Gulf of Mexico (Cordes *et al.* 2007; Olu-Le Roy *et al.* 2007; Warén & Bouchet 2009) and the seep communities on the Blake Ridge and Barbados accretionary wedge (Fig. 9.5). The communities are dominated by vestimentiferan tubeworms and bathymodioline mussels and the common seep-associated families of galatheid crabs

and alvinocarid shrimp. The bathymodioline species complexes on both sides of the Atlantic sort out among the same species groupings within the genus *Bathymodiolus*, with *B. heckerae* from the Gulf of Mexico and Blake Ridge and *Bathymodiolus* sp. 1 from the African Margin in one grouping, and *B. childressi* from the Gulf of Mexico and *Bathymodiolus* sp. 2 from Africa in another group (Cordes *et al.* 2007). However, our understanding of the biogeographic puzzle beyond this is less clear and requires further investigation (E. Cordes, unpublished observation).

The discovery of vent sites on the southern MAR (Haase *et al.* 2007, 2009; German *et al.* 2008b) and the morphological similarity of their fauna to that of NMAR vents, suggest that the Chain and Romanche fracture zones are less of an impediment to larval dispersal than previously hypothesized (Shank 2006; Haase *et al.* 2007). Further support for unhindered dispersal of vent fauna along the MAR comes from molecular analyses of the two dominant invertebrates of MAR vents, *Rimicaris* shrimp and *Bathymodiolus* mussels. These studies showed recent gene flow across the equatorial zone for these key host species and their symbionts (Petersen & Dubilier 2009; Petersen *et al.* 2010). In addition to finding known species, new species, including the shrimp *Opaepele susannae* (Komai *et al.* 2007), have been described, and now more than 17 (morpho-) species have been identified from the SMAR. Many of these have been genetically compared with taxonomically similar fauna on the NMAR and reveal significant genetic divergence among species considered "the same" in both regions (T. Shank, unpublished observation).

9.2.3.2 New Zealand region: phylogenetic links among habitats

The New Zealand region hosts a wide variety of chemosynthetic ecosystems, all in close geographic proximity. During the ChEss/COMARGE New Zealand field program (RENEWZ), more than 10 new seep sites were discovered off the New Zealand North Island (Baco-Taylor *et al.* 2009). One of these sites (Builder's Pencil) covers 135,000 m², making it one of the largest known seep sites in the world. These initial and ongoing research activities aim at locating the sites, describing their environmental characteristics, investigating their fauna, and determining potential phylogeographic relationships among species from vents, seeps, and whale falls found in close proximity to one another.

In the New Zealand region, sampling and description of chemosynthetic communities is in its infancy. Baco-Taylor *et al.* (2009) have now provided an initial characterization of cold seep faunal communities of the New Zealand region. Preliminary biological results indicate that, although at higher taxonomic levels (family and above) faunal composition of vent and seep assemblages in the New Zealand region is similar to that of other regions, at the species level, several taxa are apparently endemic to the region. Bathymodiolin mussels and an eolepadid barnacle dominate (in number and biomass) at vent sites on the seamounts of the Kermadec volcanic arc. Genetic analysis of mussels from chemosynthetic habitats by Jones *et al.* (2006) revealed the New Zealand vent mussel *Gigantidas gladius* to be closely related to species from New Zealand and Atlantic cold seeps.

New Zealand vents are often characterized by the barnacle *Vulcanolepus osheai* (Buckeridge 2000), found at very high densities which is different from those found farther north in the Pacific, and is most similar to an undescribed species found on the Pacific–Antarctic Ridge (Southward & Jones 2003). The most abundant motile species at Kermadec vent sites are caridean shrimp, including two species of endemic alvinocarids (*Alvinocaris niwa*, *A. alexander*), and one hippolytid (*Lebbeus wera*) (Webber 2004; Ahyong 2009), as well as two species of alvinocarid found elsewhere in the western Pacific (*A. longirostris*, *Nautilocaris saintlaurentae*; Ahyong 2009).

Only two species of low abundance and sparsely distributed vestimentiferan worms have been sampled so far from Kermadec vent sites (Miura & Kojima 2006). Of these species, *Lamellibrachia juni* has been found elsewhere in the western Pacific, whereas the other species, *Oasisia fujikurai*, is closely related to *O. alvinae* from the eastern Pacific (Kojima *et al.* 2006). Other species of macro- and megafauna found associated with Kermadec vent sites (Glover *et al.* 2004; Anderson 2006; Schnabel & Bruce 2006; McLay 2007; Munroe & Hashimoto 2008; Buckeridge 2009) suggest that levels of species endemism in the New Zealand region are relatively high, although some species are either closely related to species, or are found, elsewhere in the wider Pacific region. Community-level analysis (an update of the analysis of Desbruyères *et al.* (2006b)) suggests that although the New Zealand region does apparently contain a vent community with a distinct composition, there is a degree of similarity with communities from elsewhere in the western Pacific (A. Rowden, unpublished observation).

In total, the analysis of samples either compiled or collected as part of the ChEss/COMARGE project provide some support for the hypothesis that the region may represent a new biogeographic province for both seep and vent fauna. However, there is clearly a need for further sampling.

9.2.3.3 Exploring remote polar regions

The Polar Regions have received an increasing interest in the first decade of the twenty-first century, facilitated by new AUV technologies being developed to work in these remote areas of difficult access caused by ice coverage (Shank 2004; Jakuba *et al.* 2008). The exploration of the Arctic Ocean revealed, in 2003, evidence for abundant hydrothermal activity on the Gakkel Ridge (Edmonds *et al.* 2003). The Gakkel Ridge is an ultra-slow spreading ridge, which lies beneath permanent ice cover within the bathymetrically isolated Arctic Basin. The deep Arctic water is isolated from deep-water in the Atlantic by sills between Greenland and Iceland and between Iceland and Norway. This has important implications for the evolution and ecology of the deep-water Arctic vent fauna. In July/August 2007, the AGAVE (Arctic GAkkel Vent Exploration) project investigated the Gakkel Ridge using AUVs and a video-guided benthic sampling system (Camper). Investigations suggested "recent" and explosive volcanic activity (Sohn *et al.* 2008). Extensive fields and pockets of yellow microbial mats dominated the landscape. Microbial samples revealed highly diverse chemolithotrophic microbial communities fueled by iron, hydrogen, or methane (E. Helmke, personal communication). Macrofauna associated with these mats included shrimp, gastropods, and amphipods with hexactinellid sponges peripherally attached to "older lavas". These communities may be sustained by weak fluid discharge from cracks in the young volcanic surfaces (T. Shank, unpublished observation).

So far, the northernmost vent sites that have been investigated by ROV are at 71°N on the Mohns Ridge (Schander et al. 2009). The shallow (500–750 m) sites located there support extensive mats of sulfur-oxidizing bacteria. However, of the 180 species described from two fields explored, the only taxon that is potentially symbiont-bearing is a small gastropod, *Rissoa* cf. *griegi*, also known from seeps and wood falls in the North Atlantic. Arctic cold seeps have also been investigated at the Haakon Mosby Mud Volcano (HMMV) on the Barents Sea slope (72°N) at 1,280 m depth (Niemann et al. 2006; Vanreusel et al. 2009). This site has large extensions of bacterial mats and is dominated by siboglinid tubeworms (Lösekann et al. 2008), with many small bivalves of the family Thyasiridae living among them. In terms of macrofauna, the HMMV is dominated by polychaetes, with higher abundances and diversity at the siboglinid fields compared to the bacterial mats. The meiofauna is dominated by benthic copepods in the active centre, whereas the nematode *Halomonhystera disjuncta* dominates in the bacterial mats (Van Gaever et al. 2006). The other Nordic margin cold seeps of the Storegga and Nyegga systems are also characterized by a high abundance of potentially endemic siboglinid tubeworms in association with methane seepage, as well as the occurrence of diverse mats of giant sulfide-oxidizing bacteria, attracting large numbers of meio- and macrofauna (Vanreusel et al. 2009).

In the Southern Ocean, vent exploration in the East Scotia Arc and seep investigations on the Weddell Sea have addressed the role of the Circumpolar Current in dispersal of deep-water fauna as a conduit between the Pacific and the Atlantic, or as a barrier between these two oceans and the Southern Ocean. The ChEsSo (ChEss in the Southern Ocean) project explored the East Scotia Ridge in 2009 and 2010, providing further detail to the vent plume data described by German et al. (2000). A follow-up cruise is planned for 2011 to investigate further and locate the vent source and any potential vent fauna.

On the continental margin of the Antarctic Peninsula in the Weddell Sea, cold seep communities were discovered in 800 m water beneath what was the Larsen B ice shelf. The site was once covered by extensive areas of bacterial mat and beds of live vesicomyid clams (Domack et al. 2005), but hydrocarbon seepage appeared extinct only a few years later (Niemann et al. 2009). The discovery of vesicomyid clams is evidence that the hydrographic boundary between the southern Atlantic and Pacific Oceans with Southern Ocean is not a biogeographic barrier, at least for this taxon, though it remains to be determined if the Weddell Sea vesicomyid has been sufficiently isolated to be genetically distinct from any other vesicomyid species. An international team has recently returned to the Larsen B seep sites to determine the phylogeographic alliance of the Weddell Sea clams with other vesicomyids from the Atlantic and Pacific Basins.

9.2.3.4 Southeast Pacific off Chile: a unique place on Earth

The southeast Pacific region off Chile is of high interest for deep-water chemosynthetic studies, especially for the study of inter-habitat connectivity through migration and colonization. Only here can we expect to find every known form of deep-sea chemosynthetic ecosystem in very close proximity to one another. A key reason for this unique juxtaposition of chemosynthetic habitats is the underpinning plate-tectonic setting. The Chile Rise is one of only two modern sites where an active ridge crest is being swallowed by a subduction zone and the only site where such subduction is taking place beneath a continental margin (Cande et al. 1987; Bangs & Cande 1997). Consequently, one would expect to find hydrothermal vent sites along the East Chile Rise, cold seeps associated with subduction along the Peru–Chile trench at the intersection with the Chile Rise, and an oxygen minimum zone that abuts and extends south along the Peru and Chile margins (Helly & Levin 2004). Along with these geologic/oceanographic occurrences, significant whale feeding grounds and migration routes occur on the southwest American margin (Hucke-Gaete et al. 2004), and there is strong potential for wood-fall from the forests of southern Chile as the Andes slope steeply into the ocean south off approximately 45°S (V. Gallardo, personal communication). To what extent will the same chemosynthetic organisms be able to take advantage of the chemical energy available at all of these diverse sites? Alternatively, will each type of chemosynthetic system host divergent fauna based on additional factors (for example depth, longevity of chemically reducing conditions, extremes of temperature, and/or fluid compositions)? Some seep sites are already known further north along the margin (Sellanes et al. 2004) and first evidence for hydrothermal activity on the medium-fast spreading Chile Rise was suggested by metalliferous input to sediments in this region (Marienfeld & Marching 1992). Systematic exploration at the very intersection of the ridge-crest and adjacent margin has recently been conducted during a joint ChEss–COMARGE cruise (February–March 2010) and sources of venting were recorded, along with evidence of at least one cold seep site relatively close by, thereby confirming expectations of the scientists on board (A. Thurber, personal communication).

The Peru–Chile margin and subduction zone contain hydrate deposits and seep sites venting methane-rich fluids (Brown et al. 1996; Grevemeyer et al. 2003; Sellanes et al. 2004). Until recently these habitats had only been sampled remotely by trawl. A recent expedition provided the first Chile seep images and quantitative samples using a video-guided multicorer (A. Thurber, personal communication). Based on trawl collections, seeps of the Chilean Margin appear to have evolved in relative isolation from other chemosynthetic communities. There are at least eight species of symbiotic bivalves, including vesicomyids, thyasirids, solemyids, and lucinids, and at least one species of the tubeworm *Lamellibrachia* (Sellanes *et al.* 2008). The bivalve species do not appear to have close affinities to the chemosynthetic fauna of other seeps, in particular the seeps off the coast of Peru (Olu *et al.* 1996) or New Zealand (Baco-Taylor *et al.* 2009). The general composition of the community, including a high diversity of vesicomyids (four species) is similar to that of other seep sites of the eastern Pacific. Further taxonomic resolution of this key fauna along with a complete analysis of the Costa Rica fauna will help to refine further the location of this biogeographic puzzle piece.

9.2.4 Larval ecology: shaping faunal distribution under ecological timescales

The biological communities that inhabit chemosynthetic environments face several challenges that arise from the peculiarities of the habitat. Firstly, relatively few species are specifically adapted to the physical and chemical characteristics of these habitats. Secondly, these habitats are generally ephemeral at decadal scales (with the notable exception of seeps and OMZs) either because they are geologically unstable (vents) or because they are short lived (large organic falls). Thirdly, these habitats are patchy and can be separated by hundreds to thousands of kilometers of habitat unsuitable for the organisms that are adapted to chemosynthetic conditions. An additional challenge is that most of the organisms that inhabit chemosynthetic environments are either sessile (being attached to a substratum) or show limited mobility in their adult life; they rely solely on planktonic propagules, mainly larvae (Mills *et al.* 2009) (Fig. 9.6) to maintain existing populations and to colonize newly opened areas (e.g., after an eruption at a vent or when a whale lands on the ocean floor). Given the patchy distribution and ephemeral nature of their habitat, adaptations during larval life can have pronounced implications for the success of these species. Yet, our knowledge of the larval ecology of these species, and of deep-sea species in general, remains extremely limited (reviewed in Young 2003; Mills *et al.* 2009).

Most of our current understanding of larval ecology is based on species that inhabit hydrothermal vents, and information from other chemosynthetic habitats is sorely lacking. Larval populations are being increasingly sampled to assess their abundance and distribution relative to the hydrothermal vent where they most likely originated. In general, larvae are found in greater abundance near the ocean floor than near the plume at hundreds of meters above the bottom, suggesting that they may be dispersing along the ocean floor, taking advantage of the along-axis currents there. Although these types of study were initiated in the 1990s (e.g., Kim *et al.* 1994; Kim & Mullineaux 1998),

they accelerated in the 2000s. However, they have only focused at a handful of sites on the Juan de Fuca Ridge (Metaxas 2004), the East Pacific Rise (Mullineaux *et al.* 2005; Adams & Mullineaux 2008), and the Mid-Atlantic Ridge (Khripounoff *et al.* 2001; 2008). Similar studies were initiated in the 2000s and are ongoing at vents in Lau Basin and volcanically active seamounts on arcs in the southern (Kermadec) and western (Mariana) Pacific. These studies suggested that hydrodynamics can provide a mechanism of both larval retention to re-seed existing populations, as well as along-axis transport and dispersal to colonize newly opened areas within hundreds of kilometers. Larval abundance in cold seeps has been measured for two species in the Gulf of Mexico (Van Gaest 2006; Arellano 2008), indicating that, unlike most species at vents (except some crustaceans), larval migration from the seep of origin to surface waters likely occurs.

Larval colonization is better understood and has received more attention than larval dispersal. Since 2002, studies have focused on vents, seeps, whale and wood falls (see, for example, Mullineaux *et al.* 2003; Govenar *et al.* 2004; Pradillon *et al.* 2005; Levin *et al.* 2006; Braby *et al.* 2007; Kelly *et al.* 2007; Fujiwara *et al.* 2007; Tyler *et al.* 2007; Arellano 2008). In all habitats, larvae of different species settle and colonize areas in a particular sequence that appears to be related to chemical and biological cues of the environment. We now know that colonization and succession at vents can be quite rapid and communities can recover from catastrophic disturbances within 2–5 years (Shank *et al.* 1998). It appears that the spatial and temporal patterns of colonists are primarily related to the physicochemical environment and secondarily to biological interactions. However, the evidence on the latter is still scant, and both experimental manipulations and numerical modeling are being used increasingly to address this gap (Neubert *et al.* 2006; Shea *et al.* 2008; N. Kelly, personal communication).

The largest gap in our understanding of larval life is the factors that affect larval growth and development in all chemosynthetic environments. The main challenge is larval rearing for species that inhabit deep-water environments with very particular chemical and physical characteristics. Only a few studies have succeeded in rearing larvae of only a handful of species from vents (Marsh *et al.* 2001; Pradillon *et al.* 2001), seeps (Young *et al.* 1996; Van Gaest 2006; Arellano & Young 2009), and whale falls (Rouse *et al.* 2009), but none were successful in following larvae through to the end of that life stage. Larval rearing *in situ* has been attempted and has been partly successful (Marsh *et al.* 2001; Pradillon *et al.* 2001; Brooke & Young 2009).

A key unknown aspect of the question of larval development is the duration of the larval stage, the period that larvae spend in the water column, and thus their potential dispersal distance. Some specific exceptions include

(A)

(B)

(C)

(D)

Fig. 9.6

Photographs of vent larvae and juveniles collected from the 9°50′N EPR vent field. **(A)** larva of the gastropod genus *Lepetodrilus*; **(B)** juvenile of a polynoid polychaete; **(C)** megalopa larval stage of decapod crustacean *Bythograea thermydron*; **(D)** larva of the bivalve *Bathymodiolus thermophilus*. © Stace Beaulieu (photographs A, B, and D) and Susan Mills (photograph C), Woods Hole Oceanographic Institution.

experiments to estimate dispersal time in the vent tube-worm *Riftia pachyptila* (Marsh *et al.* 2001), the whale-fall polychaete *Osedax* (Rouse *et al.* 2009), the vent gastropod *Bathynerita* (Van Gaest 2006), and the mussel *Bathymodiolus* (Arellano & Young 2009). Other major gaps in knowledge include larval growth rates in relation to temperature, pressure, and food availability; the cues that induce them to stop swimming and settle onto a suitable habitat (or avoid an unsuitable one); and the role of larval behavior in vertical positioning while in the water column or near bottom. Increased knowledge of larval connectivity and colonization processes will add conceptual understanding of reducing ecosystems as metapopulations (Leibold *et al.* 2004; Neubert *et al.* 2006) and their response or resilience in the face of natural and anthropogenic disturbance (Levin *et al.* 2009b).

9.3 Limits to Knowledge

9.3.1 Understanding remote and dynamic ecosystems

Discovery of deep-water chemosynthetically driven communities is relatively recent. Thus the investigation of these habitats has a very strong exploration component. Biologically, the unknowns exceed the knowns. How many species are there? What is their distribution and why? How do species reproduce, disperse, and colonize new sites? What are the evolutionary history and phylogenetic relationships of species from different chemosynthetic habitats? The remoteness and abrupt topography of deep-water chemosynthetic ecosystems make observation, sampling, and experimentation difficult in these habitats. There is a strong need for international collaboration and sharing of resources, which has been accomplished through projects such as the Census, contributing greatly to our knowledge of chemosynthetically driven ecosystems. The ongoing learning process allows identifying gaps and provides a driver for science to transform the unknowns into knowns (Gomory 1995; Marchetti 1998).

Chemosynthetic research depends on our capacity to sample, observe, and experiment at great depths in "extreme" physicochemical conditions. Therefore research in chemosynthetic ecosystems closely follows technological developments. For example, the use of submersibles and ROVs made possible direct observation and *in situ* precise sampling of the ecosystem. AUVs are useful for investigation of regions that are not accessible from the surface (i.e. oceans under ice) and in revolutionizing the efficiency of deep-sea hydrothermal exploration (German *et al.* 2008a; Sohn *et al.* 2008). Deep-towed sidescan sonar instruments can produce detailed acoustic images of the deep seafloor. Submersible-mounted multibeam bathymetry can achieve centimeter-scale resolution maps of the sea floor. High-resolution, high-definition cameras are being used to produce photo-mosaics of remote habitats, providing a comprehensive overview of the habitat and allowing for first interpretations of the relationships between habitat and fauna. New types of biogeochemical sensor module and incubation instrument allow quantification of the transport of energy and the benthic community activities *in situ* (Boetius & Wenzhöfer 2009). Technology is also advancing rapidly in laboratory techniques, for example molecular biology, which has aided our understanding of processes that, until recently, were hidden from our senses. However, are there limits to knowledge? This question can be considered in terms of different timescales. At the ecological timescale, one of the major limits to knowledge is imposed by society

itself. The first barrier is one of economics and human resources.

In the case of the discrete and dynamic deep-water chemosynthetic ecosystems, some aspects might be unknown and unknowable at any given time. German & Lin (2004) have estimated that, for fast- and intermediate-spreading ridges, there should be a volcanic eruption approximately every 50 years for any given 100 km of ridge section. Slow-spreading ridges may exhibit much greater irregularity (German & Lin 2004). It would take the equivalent of one 1- to 2-year expedition to explore all of the southern MAR from north to south. Taking into account the episodic volcanic activity of mid-ocean ridges, we would expect three or four major new eruptions to have occurred over the time it would take to explore the whole 7,000 km of the MAR. By that reckoning (approximately one new vent area on the MAR each year), our current rate of discovery (nine sites on the northern MAR since 1986) is not even keeping pace with the rate of new production. Here, the limit to knowledge is a consequence of the dynamic characteristics of mid-ocean ridges. Cold seeps at passive margins are more stable ecosystems than vents (Sibuet & Olu 1998), but those on active margins (e.g., Chile, Costa Rica) are subject periodically to some of the most violently destructive forces on Earth. We can study these systems in their current "dormant" state and investigate their geophysics and geochemistry interactions and associated fauna. However, the episodicity of the great earthquakes (an approximate 100-year cycle predicted for the magnitude 9.5 great earthquake) may render the responses to such events impossible to observe and the extent to which the associated organisms are impacted by – or even anticipate – major tectonic events unknowable. Similar issues apply to stochastic, discrete, and ephemeral "habitats" such as whale falls and large organic falls. The life cycle of vents on fast-spreading ridges (i.e. East Pacific Rise) or cold seeps on active margins is equivalent, in time, to the construction of a cathedral in the Middle Ages (approximately 100 years), whereas on slower-spreading ridges they can extend to more than 10,000 years (Cave *et al.* 2002), longer than the oldest known European prehistoric constructs such as Stonehenge in the southern United Kingdom. These are unknowables for ChEss.

Furthermore, the fauna from deep-water vents, seeps, and whale falls is often new to science. Although the diversity of megafauna is low and the new species are described at a pace that keeps with discovery, this is not the case for the more diverse meiofauna, where up to 90% of the species collected can be new to science. The inability to keep up the rate of taxonomic identification is increasingly affected by the continuous decrease in taxonomic expertise among the new generations of marine biologists.

9.3.2 Technology: pushing backwards the limits to knowledge

Despite the best possible combinations of sensors and technologies, a limit to progress will remain related to accessing the appropriate sections of the global ocean using suitable research ships. As we develop AUV-based approaches we can improve the rates of exploration and discovery. Because AUVs are decoupled from ships' positions and motions (unlike ROVs) and do not put human lives in harm's way at the bottom of the ocean (unlike manned submersibles), we will be able to expand our systematic exploration surveys to higher and higher latitudes. Furthermore, a new generation of AUVs is now under construction that will have ranges in excess of 1,000 km and could ultimately span entire ocean basins. At this level, and with artificially intelligent control systems, we can imagine a day when long-range AUVs will be able to conduct first pass investigations along entire sections of the global ridge crest, interrogating, processing, and interpreting their data underway, to allow second pass mapping and, potentially, even photographic surveys of the seafloor. Although such approaches may seem like so much science fiction, preliminary algorithms to interpret hydrothermal field data on-the-fly have already been trialed (for example on the MAR) and a first long-range mid-ocean ridge ship-free AUV cruise has already reached the planning stage.

9.4 Human Footprints in Deep-Water Chemosynthetic Ecosystems

Deep-sea ecosystems support one of the highest biodiversities of the Planet as well as important natural and mineral resources. In the last decades, the depletion of resources in the upper layers of the oceans together with technological development has fueled the increasing interest of industry to explore and exploit deep-sea resources. Industries such as mining, hydrocarbon extraction, and fishing are working at great depths (below 1,000 m), and some of these activities affect chemosynthetic ecosystems. We briefly discuss below only anthropogenic impacts that have been addressed by ChEss scientists (i.e. mining, hydrocarbon extraction, and trawling), but other impacts such as litter accumulation or climate change are also important.

9.4.1 Major known impacts

Probably the most important industry potentially affecting chemosynthetically driven habitats and their fauna is mining for precious metals on seafloor massive sulfide deposits (SMS) from vents. SMS contain significant quantities of commercially valuable metals, such as gold, silver, copper, and zinc (Baker & German 2009). Although at a very early stage of exploration, the SMS industry is already extremely active, with two major companies (Nautilus Minerals and Neptune Minerals) developing plans for exploitation in national waters of Papua New Guinea and New Zealand. ChEss scientists have participated in the environmental impact assessment conducted by Nautilus Minerals, who have produced an environmental impact statement for mining activities in Papua New Guinea, and considerable baseline research has been done (Levin et al. 2009b; Erickson et al. 2009; see EIS at www.cares.nautilusminerals.com/Downloads.aspx). Although it seems highly likely that economically viable extraction of sulfides from the deep-sea floor may begin within the next decade, the true nature of mining impact on the ecosystem is still mostly unknown. There are several potential environmental effects of mining that are of concern to some stakeholders. These include direct physical damage to the seabed at the operation site and the surrounding area; potential extinction of isolated populations; production of sediment plumes and deposition of sediment, which will affect marine life by smothering or inhibiting filter feeders; alteration of fluid-flow paths at a vent, on which the benthic, often sessile, vent fauna rely; noise pollution; wastewater disposal; and equipment failure which may result in leakage (ISA 2004; Van Dover 2007). ChEss has been instrumental in instigating key collaborations including all stakeholders, conducting baseline research, participating in workshops and discussion fora, and contributing to the "Code for Environmental Management of Marine Mining" (IMMS 2009) and the International Seabed Authority's guidelines for environmental baselines and monitoring programs (International Seabed Authority 2004; Van Dover 2007), which both serve to monitor and mitigate the potential effects of deep-sea mining.

Another important extractive industry is oil and gas exploitation. Seep communities often coincide with large subsurface hydrocarbon reservoirs and the outcropping of gas hydrates. Thus they may also be susceptible to damage from oil and methane exploration and extraction activities in the future. One of the most intensive areas of deep-water oil industry activity is in the Gulf of Mexico, leased and monitored by the US Minerals Management Service. Collaborations between scientists and the Minerals Management Service in the Gulf of Mexico are well established (Fisher et al. 2007; Roberts et al. 2007), and exploration for new sources of hydrocarbons often coincides with the discovery of new biological communities. As new high-density biological communities are discovered, the Minerals Management Service establishes "mitigation areas" to prevent oil industry activity from impacting these sensitive areas. ChEss has also established strong collaborations with Fondation Total to promote science in chemosynthetic ecosystems.

Finally, ChEss scientists have been involved in issues related to deep-water trawling that affect chemosynthetic sites. Recent exploration of the eastern margin of New Zealand provided not only the first descriptions of seep communities in the region but evidence that seep sites have been subjected to bottom fishing (Baco-Taylor *et al.* 2009), with visible accumulations of coral or vesicomyid shell debris, lost trawl gear, or trawl marks in seep adjacent sediments. Similarly, signs of trawling have been found on the California margin at 500 m depth (Levin *et al.* 2006) and on the Chilean margin, where the commercially fished Patagonian toothfish is closely associated with seeps (Sellanes *et al.* 2008).

9.4.2 Marine protected areas

As vent sites have become the focus of intensive, long-term investigation, both governmental and non-governmental organizations have been discussing the need to introduce appropriate measures that combine preservation of habitat with scientific interests, tourism, and, potentially, mining (Mullineaux *et al.* 1998; Dando & Juniper 2000; Santos *et al.* 2003). The ChEss project has been an active participant in the planning of biogeographically representative networks of sites of interest for conservation and continued scientific research. ChEss scientists have contributed also to the Convention of Biological Diversity effort, to produce scientific criteria and guidance for the identification of ecologically or biologically significant marine areas and the designation of representative networks of marine protected areas (MPAs) (Convention on Biological Diversity 2009). So far, ecological reserves and/or MPAs have been proposed at the Mid-Atlantic Ridge and the East Pacific Rise, and have already been established on the Juan de Fuca Ridge (http://www.dfo-mpo.gc.ca/oceans/marineareas-zonesmarines/mpa-zpm/pacific-pacifique/factsheets-feuillets/endeavour-eng.htm). At the MAR, three sites – Lucky Strike, Menez Gwen, and Rainbow – were proposed to be included on the network of MPAs of The Convention for the Protection of the Marine Environment of the North-East Atlantic (OSPAR) maritime region V (the wider Atlantic). In the New Zealand region, some hydrothermal vent sites are currently protected from bottom trawling by a Benthic Protected Area. Although licenses that have been issued for exploratory mineral mining overlap with this area, it is expected that collaborations among stakeholders will instigate new prospects for conservation. The scientific community has recognized also the potential impact of continuous research activities at certain sites. This led to the "Statement of Responsible Research Practices at Hydrothermal Vents", developed by InterRidge with the collaboration of ChEss scientists (Devey *et al.* 2007). Following on those steps, the OSPAR convention

also proceeded with the design of a Code of Conduct for Research on Hydrothermal Vents, which included the usefulness of MPAs (OSPAR draft background report 2009).

9.5 Conclusions

Under the ChEss project, the scientific community has expanded our knowledge beyond the vent biogeographic regions recognized in the early twenty-first century (Van Dover *et al.* 2002). New technologies have been developed, making available the necessary tools to explore and investigate remote habitats of difficult access in a very efficient way, resulting in the discovery of new sites in all oceans on Earth. The number of species described from vents, seeps, whale falls, and OMZs is constantly growing, providing essential data to understand global biodiversity and phylogenetic links among habitats to refine existing biogeographic provinces and define new ones.

Indeed, Bachraty *et al.* (2009) have recently addressed the biogeographic relationships among deep-sea vent faunas at a global scale using statistical analysis of comprehensive vent data. They delineate six major hydrothermal provinces and identify possible dispersal pathways. Furthermore, detailed ecological investigations (for example trophic structure, reproduction, larval ecology) of known sites have resulted in a better understanding of ecosystem functioning and the role played by the environment in shaping deep-water chemosynthetic communities. Table 9.2 summarizes some of the key contributions the ChEss project participants have made toward a better global understanding of the distribution of chemosynthetic environments, the species that inhabit them, and some underlying ecological processes. Nevertheless, there are important geographic gaps in the global chemosynthetic biogeographic puzzle that remain to be explored and gaps to knowledge that remained unanswered in 2010. The momentum created by the Census and ChEss initiatives has promoted major international collaborations and strengthened existing ones. This synergy between laboratories around the globe, sharing expertise and resources in joint research projects, will continue beyond 2010 as one of the main legacies of ChEss. The scientific results obtained are crucial for the development of conservation and management options in an ecosystem that is already affected by human activities.

Acknowledgments

We thank the officers, crews, and technical support from all the cruises associated to the ChEss project. ChEss has been funded by the Alfred P. Sloan Foundation (2002–2010) and has received support from Fondation Total

Table 9.2

Summary of key discoveries, new technological developments and conclusions from the ChEss project.

Key Discovery/Insight/Tools	References
Hottest known hydrothermal vent discovered on MAR (407 °C).	Kochinsky 2006; Kochinsky *et al.* 2008
Deep vent (Ashadze) discovered at 4,100 m on MAR. Fauna here is strikingly different from that of other Atlantic sites.	Ondreas *et al.* 2007; Fouquet *et al.* 2008
Deepest known vent discovered at 5,000 m on Cayman Rise in 2010.	German, Copley, Connelly, Tyler, unpublished observation
First vent sites discovered south of the equator on the southern Mid-Atlantic Ridge (SMAR).	Haase *et al.* 2007; German *et al.* 2008b
New species discovered from SMAR vents.	Komai *et al.* 2007
Morphological similarity of fauna from SMAR and NMAR suggestive of unhindered larval dispersal.	Shank 2006; Haase *et al.* 2007; Petersen, personal communication; van der Heijden, personal communication
Faunal affinities discovered between seep species across Atlantic Equatorial Belt. Communities dominated by vestimentiferan tubeworms, bathymodioline mussels, and the common seep-associated families of galatheid crabs and alvinocarid shrimps.	Cordes *et al.* 2007; Olu-Le Roy *et al.* 2007; Warén & Bouchet 2009
More than 10 new seep sites discovered off of the New Zealand North Island including one of largest cold seeps known covering about 135,000 m² (approximately 33 acres) (Builder's Pencil).	Baco *et al.* 2009
Initial characterization of cold seep faunal communities of the New Zealand region.	Baco *et al.* 2009
Species of macro- and megafauna found associated with Kermadec vent sites suggest that levels of species endemism in the New Zealand region are relatively high.	Glover *et al.* 2004; Anderson 2006; Schnabel & Bruce 2006; McLay 2007; Munroe & Hashimoto 2008; Buckeridge 2009
Overall, the analysis of samples either compiled or collected as part of the ChEss project, provide some support for the hypothesis that the New Zealand region may represent a new biogeographic province for both seep and vent fauna.	Rowden, unpublished observation
Abundant hydrothermal activity revealed on Gakkel Ridge, Arctic Ocean, bathymetrically isolated from Atlantic Ocean. Yellow microbial mats supporting shrimps and amphipods discovered.	Edmonds *et al.* 2003; Sohn *et al.* 2008; Helmke, personal communication; Shank, unpublished observation
The northernmost vent sites investigated by ROV discovered at 71° N on the Mohns Ridge.	Schander *et al.* 2009
Arctic cold seeps investigated at the Haakon Mosby Mud Volcano on the Barents Sea slope (72° N) at 1280 m depth revealed large extensions of bacterial mats, siboglinid tubeworms, small bivalves, polychaetes, benthic copepods, and nematodes.	Niemann *et al.* 2006; Van Gaever *et al.* 2006; Lösekann *et al.* 2008; Vanreusel *et al.* 2009
ChEss has begun to explore the Southern Ocean for chemosynthetic communities during the ChEsSo programme in 2009 and 2010. Investigations to continue in 2011.	Tyler, unpublished observation; Connelly, personal communication
Initial studies of vesicomyid clams from Weddell Sea cold seeps suggest that the hydrographic boundary between the southern Atlantic and Pacific Oceans with Southern Ocean is not a biogeographic barrier at least for this taxon. Further work is planned for 2010.	Domack *et al.* 2005; Niemann *et al.* 2009; Van Dover, unpublished observation
Based on trawl collections, seeps of the Chilean Margin appear to have evolved in relative isolation from other chemosynthetic communities. There are at least eight species of symbiotic bivalves including vesicomyids, thyasirids, solemyids, and lucinids, four seep gastropods, and at least one species of the tubeworm *Lamellibrachia.*	Sellanes *et al.* 2008; Waren, personal observation
Studies of vent larval abundance suggest that hydrodynamics can provide a mechanism of both larval retention to re-seed existing populations, as well as along-axis transport and dispersal to colonize newly opened areas within hundreds of kilometers. In cold seeps, larval migration likely occurs from the seep of origin to surface waters. Some estimates of larval dispersal times have been made.	Metaxas 2004; Mullineaux *et al.* 2005; Van Gaest 2006; Adams & Mullineaux 2008; Arellano 2008; Khripounoff *et al.* 2008; Arellano & Young 2009; Rouse *et al.* 2009

Key Discovery/Insight/Tools	References
Since 2002, larval colonization studies have focused on vents, seeps, whale and wood falls. In all habitats, larvae of different species settle and colonize areas in a particular sequence that appears to be related to chemical and biological cues of the environment.	Mullineaux *et al.* 2003; Govenar *et al.* 2004; Pradillon *et al.* 2005; Levin *et al.* 2006; Neubert *et al.*, 2006; Braby *et al.* 2007; ; Fujiwara *et al.* 2007; Kelly *et al.* 2007; Tyler *et al.* 2007; Arellano 2008; Shea *et al.* 2008; Kelly, personal communication
Development and implementation of AUV and AUV–ROV hybrid technology for vent and seep exploration and investigation (including under-ice investigations).	Baker *et al.* 1995; Baker & German 2004; Shank 2004; Yoerger *et al.* 2007; German *et al.* 2008a; German *et al.* 2008b; Jakuba *et al.*2008; Melchert *et al.* 2008; Dupre *et al.* 2009; Foucher *et al.* 2009; German *et al.* 2009; Haase *et al.* 2009
Significant increase in molecular tools for studies to understand species evolution, metapopulations, and gene flow in chemosynthetic regions.	Shank & Halanych 2007; Johnson *et al.* 2008; Plouviez *et al.* 2009; Vrijenhoek 2009b
New types of biogeochemical sensor module and incubation instrument to quantify the transport of energy and the benthic community activities *in situ*.	Boetius & Wenzhöfer 2009
Approximately 200 species have been described and published from vents, seeps, and whale falls since the onset of the ChEss programme in 2002.	ChEssBase: www.noc.soton.ac.uk/database
TAWNI outputs: many meio-, macro-, and megafauna species have been described and new records identified from different vent and seep sites.	Zekely *et al.* 2006; Gollner *et al.* 2008; Hilário & Cunha 2008; Hilário, personal communication; Thurber, personal communication
The much publicized yeti crab *Kiwa hirsuta* discovered on the Pacific vents represents a new species, genus, and family. A close relative of this crab has also been discovered from Costa Rican cold seeps and is being described with the aid of a TAWNI grant.	Macpherson *et al.* 2005; Thurber, personal communication
Cold seep species from Costa Rica Margin are related to those from cold seeps in the Gulf of Mexico and West Africa, and to seep species in the Pacific.	Levin & Waren, unpublished observation
Novel reproductive adaptations found in vent animals, including seasonality.	Young 2003; Hilário *et al.* 2005; Tyler *et al.* 2007; Rouse *et al.* 2009
Adaptations to hypoxia and thermal tolerance in vent and seep animals.	Hourdez & Lallier 2007
ChEss scientists have provided leadership on global hydrothermal vent conservation.	ISA 2004; Van Dover 2007; Fisher *et al.* 2007; Roberts *et al.* 2007; CBD 2009; IMMS 2009

(2007–2010), which are kindly acknowledged. Maria Baker is funded by the A.P. Sloan Foundation; Eva Ramirez-Llodra is funded by the A.P. Sloan Foundation and Fondation Total.

References

Adams, D.K. & Mullineaux, L.S. (2008) Supply of gastropod larvae to hydrothermal vents reflects transport from local larval sources. *Limnology and Oceanography* **53**, 1945–1955.

Ahyong, S.T. (2009) New species and new records of hydrothermal vent shrimps from New Zealand (Caridea: Alvinocarididae, Hippolytidae). *Crustaceana* **82**, 775–794.

Anderson, M.E. (2006) Studies on the Zoarcidae (Teleostei: Perciformes) of the Southern Hemisphere. XI. A new species of *Pyrolycus* from the Kermadec Ridge. *Journal of the Royal Society of New Zealand* **36**, 63–68.

Arellano, S.M. (2008) Embryology, larval ecology, and recruitment of *Bathymodiolus childressi*, a cold-seep mussel from the Gulf of Mexico. PhD thesis, University of Oregon, USA.

Arellano, S.M. & Young, C.M. (2009) Spawning, development, and the duration of larval life in a deep-sea cold-seep mussel. *Biological Bulletin* **216**, 149–162.

Bachraty, C., Legendre, P. & Desbruyères, D. (2009) Biogeographic relationships among deep-sea hydrothermal vent faunas at global scale. *Deep-Sea Research I* **56**, 1371–1378.

Baco-Taylor, A., Rowden, A., Levin, L., *et al.* (2009) Initial characterization of cold seep faunal communities on the New Zealand Hikurangi Margin. *Marine Geology* (in press).

Baker, E.T., German, C.R. & Elderfield, H. (1995) Hydrothermal plumes: global distributions and geological inferences. In: *Seafloor Hydrothermal Systems: Physical, Chemical, Biological and Geological Interactions*, Geophysical Monograph **91**, 47–71.

Baker E.T. & German, C.R. (2004) On the global distribution of mid-ocean ridge hydrothermal vent-fields. In: *The Thermal Structure of the Oceanic Crust and the Dynamics of Seafloor Hydrothermal Circulation*, Geophysical Monograph **148**, 245–266.

Baker, M.C. & German, C.R. (2009) Going for gold! Who will win in the race to exploit ores from the deep sea? *Ocean Challenge* **16**, 10–17.

Bangs N.L. & Cande, S.C. (1997) Episodic development of a convergent margin inferred from structures and processes along the southern Chile margin. *Tectonics* **16**, 489–503.

Bergquist, D.C., Ward, T., Cordes, E.E., *et al.* (2003) Community structure of vestimentiferan-generated habitat islands from Gulf of Mexico cold seeps. *Journal of Experimental Marine Biology and Ecology* **289**, 197–222.

Bergquist, D.C., Fleckenstein, C., Knisel, J., *et al.* (2005) Variations in seep mussel bed communities along physical and chemical environmental gradients. *Marine Ecology Progress Series* **293**, 99–108.

Boetius, A. & Wenzhöfer, F. (2009) *In situ* technologies for studying deep-sea hotspot ecosystems. *Oceanography* **22**, 177–177.

Bowen, A.D., Yoerger, D.R., Taylor, C., *et al.* (2009) Field trials of the *Nereus* hybrid underwater robotic vehicle in the challenger deep of the Mariana Trench. In: *Proceedings of the IEEE/MTS OCEANS Conference, Biloxi.*

Braby, C.E., Rouse, G.W., Johnson, S.B., *et al.* (2007) Bathymetric and temporal variation among *Osedax* boneworms and associated megafauna on whale-falls in Monterey Bay, California. *Deep-Sea Research I* **54**, 1773–1791.

Brooke, S.D. & Young, C.M. (2009) Where do the embryos of *Riftia pachyptila* develop? Pressure tolerances, temperature tolerances, and buoyancy during prolonged embryonic dispersal. *Deep-Sea Research II* **56**, 1599–1606.

Brown, K.M., Bangs, N.L., Froelich, P.N., *et al.* (1996) The nature, distribution and origin of gas hydrate in the Chile Triple Junction region. *Earth and Planetary Science Letters* **139**, 471–483.

Buckeridge, J.S. (2000) *Neolepas osheai* sp. nov., a new deep-sea vent barnacle (Cirripedia: Pedunculata) from the Brothers Caldera, south-west Pacific Ocean. *New Zealand Journal of Marine and Freshwater Research* **34**, 409–418.

Buckeridge, J.S. (2009) *Ashinkailepas kermadecensis*, a new species of deep-sea scalpelliform barnacle (Thoracica: Eolepadidae) from the Kermadec Islands, southwest Pacific. *Zootaxa* **2021**, 57–65.

Cande, S.C., Leslie, R.B., Parra, J.C., *et al.* (1987) Interaction between the Chile Ridge and Chile Trench: geophysical and geothermal evidence. *Journal of Geophysical Research* **92**, 495–520.

Cave, R.R., German, C.R., Thomson, J., *et al.* (2002) Fluxes to sediments from the Rainbow hydrothermal plume, 36°14′N on the MAR. *Geochimica et Cosmochimica Acta* **66**, 1905–1923.

CAYTROUGH (1979) Geological and geophysical investigation of the Mid-Cayman Rise spreading centre: initial results and observations. In *Maurice Ewing Series 2* (eds. M. Talwani, C.G. Harrison & D.E. Hayes). Washington DC, American Geophysical Union, pp. 66–95.

Convention on Biological Diversity (2009) Azores scientific criteria and guidance for identifying ecologically or biologically significant marine areas and designing representative networks of marine protected areas in open ocean waters and deep sea habitats. Montréal, Canada: Secretariat of the Convention on Biological Diversity. 12 pp.

Cordes, E.E., Bergquist, D.C., Predmore, B.L., *et al.* (2006) Alternate unstable states: convergent paths of succession in hydrocarbon-seep tubeworm-associated communities. *Journal of Experimental Marine Biology and Ecology* **339**, 159–176.

Cordes, E.E., Carney, S.L., Hourdez, S., *et al.* (2007) Cold seeps of the deep Gulf of Mexico: Community structure and biogeographic comparisons to Atlantic equatorial belt seep communities. *Deep-Sea Research I* **54**, 637–653.

Corliss, J.B., Dymond, J. Gordon, L.I., *et al.* (1979) Submarine Thermal Springs on the Galápagos Rift. *Science* **203**, 1073–1083.

Dando, P. & Juniper, K. (2000) Management and conservation of hydrothermal vent ecosystems. Report from an InterRidge Workshop. *InterRidge News* **18**, 35 pp.

Desbruyères, D., Segonzac, M. & Bright, M. (2006a) *Handbook of Deep-Sea Hydrothermal Vent Fauna*, second edition. Linz: Denisia. 544 pp.

Desbruyères, D., Hashimoto, J & Frabri, M.-C. (2006b) Composition and biogeography of hydrothermal vent communities in western Pacific Back-Arc Basins. *Geophysical Monograph Series* **66**, 215–234.

Devey, C.W., Fisher, C.R. & Scott, S. (2007) Responsible science at hydrothermal vents. *Oceanography* **20**, 162–171.

Domack, E., Duran, D., Leventer, A., *et al.* (2005) Stability of the Larsen B ice shelf on the Antarctic Peninsula during the Holocene epoch. *Nature* **436**, 681–685.

Dreyer, J.C., Knick, K.E., Flickinger, W.B., *et al.* (2005) Development of macrofaunal community structure in mussel beds on the northern East Pacific Rise. *Marine Ecology Progress Series* **302**, 121–134.

Dubilier, N., Bergin C. & Lott, C. (2008) Symbiotic diversity in marine animals: the art of harnessing chemosynthesis. *Nature Reviews Microbiology* **6**, 725–740.

Dupré, S., Buffet, G., Mascle, J., *et al.* (2009) High-resolution mapping of large gas emitting mud volcanoes on the Egyptian continental margin (Nile Deep Sea Fan) by AUV surveys. *Marine Geophysical Researches* **29**, 275–290.

Edmonds, H.N., Michael, P.J., Baker, E.T., *et al.* (2003) Discovery of abundant hydrothermal venting on the ultraslow-spreading Gakkel ridge in the Arctic Ocean. *Nature* **421**, 252–256.

Erickson, K.L., Macko, S.A. & Van Dover, C.L. (2009) Evidence for a chemoautotrophically based food web at inactive hydrothermal vents (Manus Basin). *Deep-Sea Research II* **56**, 1577–1585.

Fisher, C.R., Roberts, H.H., Cordes, E.E. *et al.* (2007) Cold seeps and associated communities of the Gulf of Mexico. *Oceanography* **20**, 118–129.

Fossing. H., Gallardo, V.A., Jørgensen, B.B., *et al.* (1995) Concentration and transport of nitrateby the mat-forming sulphur bacterium *Thioploca. Nature* **374**, 714–715.

Foucher, J.P., Westbrook, G.K., Boetius, A., *et al.* (2009) Structure and drivers of cold seep ecosystems. *Oceanography* **22**, 92–109.

Fouquet, Y., Cherkashov, G. Charlou, J.L. *et al.* (2008) Serpentine cruise-ultramafic hosted hydrothermal deposits on the Mid-Atlantic Ridge: First submersible studies on Ashadze 1 and 2, Logatchev 2, and Krasnov vent fields. *InterRidge News* **17**, 15–19.

Fujiwara, Y., Kawato, M., Yamamoto, T., *et al.* (2007) Three-year investigations into spermwhale-fall ecosystems in Japan. *Marine Ecology* **28**, 219–232.

Gallardo, V.A. (1963) Notas sobre la densidad de la fauna bentónica en el sublittoral del norte de Chile. *Gayana* **10**, 3–15.

Gallardo, V.A. (1977) Large benthic microbial communities in sulfide biota under Peru–Chile subsurface countercurrent. *Nature* **268**, 331–332.

Gallardo, V.A. & Espinoza, C. (2007) New communities of large filamentous sulfur bacteria in the eastern South Pacific. *International Microbiology* **10**, 97–102.

Gage, J. (2003) Food inputs, utilisation, carbon flow and energetics. In: *Ecosystems of the World: Ecosystems of the Deep Ocean.* (ed. P.A. Tyler), pp. 313–426. Amsterdam: Elsevier.

German, C.R. & Lin, J. (2004) The thermal structure of the oceanic crust, ridge spreading, and hydrothermal circulation: How well do we understand their inter-connections? In: *Mid-ocean Ridges: Hydrothermal Interactions between the Lithosphere and Oceans* (eds. C.R. German, J. Lin & L.M. Parson), *American Geophysical Union Geophysical Monograph* 148, 1–18.

German, C.R., Livermore, R.A., Baker, E.T., *et al.* (2000) Hydrothermal plumes above East Scotia Ridge: an isolated high-latitude back-arc spreading centre. *Earth and Planetary Science Letters* 184, 241–250.

German, C.R., Bowen, A., Cooleman, M.L., *et al.* (2009) Hydrothermal exploration of the Mid-Cayman Spreading Centre: isolated evolution on Earth's deepest Mid-Ocean Ridge? EOS Transactions, American Geophysical Union (abstract #OS21B-08).

German, C.R., Yoerger, D.R., Jakuba, M., *et al.* (2008a) Hydrothermal exploration using the *Autonomous Benthic Explorer*. *Deep-Sea Research I* 55, 203–219.

German, C.R., Bennett, S.A., Connelly, D.P. *et al.* (2008b) Hydrothermal activity on the southern Mid-Atlantic Ridge: tectonically- and volcanically-controlled venting at 4–5°S. *Earth and Planetary Science Letters* 273, 332–344.

Glover, E.A., Taylor, J.D. & Rowden, A.A. (2004) *Bathyaustriella thionipta*, a new lucinid bivalve from a hydrothermal vent on the Kermadec Ridge, New Zealand and its relationship to shallow-water taxa (Bivalvia: Lucinidae). *Journal of Molluscan Studies* 70, 283–295.

Gollner, S., Ivanenko, V.N. & Martinez Arbizu, P. (2008) A new species of deep-sea Tegastidae (Crustacea: Copepoda: Harpacticoida) from 9°50′N on the East Pacific Rise, with remarks on its ecology. *Zootaxa* 1866, 323–336.

Gomory, R.E. (1995) An essay on the known, the unknown and the unknowable. *Scientific American* 272, 120.

Govenar, B., Freeman, M., Bergquist, D.C., *et al.* (2004) Composition of a one-year-old *Riftia pachyptila* community following a clearance experiment: insight to succession patterns at deep-sea hydrothermal vents. *Biological Bulletin* 207, 177–182.

Grevemeyer, I., Diaz-Naveas, J.L., Ranero, C.R., *et al.* (2003) Heat flow over the descending Nazca plate in central Chile, 32°S to 41°S: observations from ODP Leg 202 and the occurrence of natural gas hydrates. *Earth and Planetary Science Letters* 213, 285–298.

Haase, K.M., Petersen, S., Koschinsky, A., *et al.* (2007) Young volcanism and related hydrothermal activity at 5S on the slow-spreading Mid-Atlantic Ridge. *Geochemistry, Geophysics, Geosystems* 8, 17.

Haase, K.M., Koschinsky, A. Petersen, S., *et al.* (2009) Diking, young volcanism and diffuse hydrothermal activity on the southern Mid-Atlantic Ridge: the Lilliput fields at 9°33′S. *Marine Geology* 266, 52–64.

Helly, J. & Levin, L.A. (2004) Global distribution of naturally occurring marine hypoxia on continental margins. *Deep-Sea Research I* 51, 1159–1168.

Hilario, A. & Cunha, M.R. (2008) On some frenulate species (Annelida: Polychaeta: Siboglinidae) from mud volcanoes in the Gulf of Cadiz (NE Atlantic). *Scientia Marina* 72(2), 361–371.

Hilario, A., Young, C.M. & Tyler, P.A. (2005). Sperm storage, internal fertilization and embryonic dispersal in vent and seep tubeworms (Polychaeta: Siboglinidae: Vestimentifera). *Biological Bulletin* 208, 20–28.

Hourdez, S. & Lallier, F.H. (2007) Adaptations to hypoxia in hydrothermal-vent and cold-seep invertebrates. *Reviews in Environmental Science and Biotechnology* 6, 143–159.

Hucke-Gaete, R., Osman, L.P., Moreno, C.A., *et al.* (2004) Discovery of a blue whale feeding and nursing ground in southern Chile. *Proceedings of the Royal Society of London B* 271 (Supplement), S170–S173.

International Marine Minerals Society (2009) Code for Environmental Management of Marine Mining. Revised draft version from the International Marine Minerals Society adopted code (2001). http://

www.immsoc.org/IMMS_downloads/PAV_Code_082109_KM_082509.pdf.

International Seabed Authority (2004) Polymetallic sulphides and cobalt rich ferromanganese crust deposits: establishment of environmental baselines and an associated monitoring program during exploration. *Proceedings of the International Seabed Authority's workshop held in Kingston, Jamaica, 6–10 September 2004.*

Jakuba, M.V., Whitcomb, L.L., Yoerger, D.R., *et al.* (2008) *Toward Under-Ice Operations with Hybrid Underwater Robotic Vehicles.* Proceedings of the IEEE/OES conference on Polar AUVs (AUV2008), Woods Hole, MA, USA.

Jannasch, H.W. & Mottl, M.J. (1985) Geomicrobiology of deep-sea hydrothermal vents. *Science* 229, 717–725.

Johnson, S.B., Warén, A. & Vrijenhoek, R.C. (2008) DNA barcoding of *Lepetodrilus* limpets reveals cryptic species. *Journal of Shellfish Research* 27, 43–51.

Jones, W.J., Won, Y-J., Maas, P.A.Y., *et al.* (2006) Evolution of habitat use by deep-sea mussels. *Marine Biology* 148, 841–851.

Jørgensen, B.B. & Boetius, A. (2007) Feast and famine – microbial life in the deep-sea bed. *Nature Microbiology Reviews* 5, 770–781.

Kelly, N., Metaxas, A. & Butterfield, D. (2007) Spatial and temporal patterns in colonization of deep-sea hydrothermal vent invertebrates on the Juan de Fuca Ridge, NE Pacific. *Aquatic Biology* 1, 1–16.

Khripounoff, A., Vangriesheim, A., Crassous, P., *et al.* (2001) Particle flux in the Rainbow hydrothermal vent field (Mid-Atlantic Ridge): dynamics, mineral and biological composition. *Journal of Marine Research* 59, 633–656.

Khripounoff, A., Vangriesheim, A., Crassous, P., *et al.* (2008) Temporal variation of currents, particulate flux and organism supply at two deep-sea hydrothermal fields of the Azores Triple Junction. *Deep-Sea Research I* 55, 532–551.

Kim, S.L. & Mullineaux, L.S. (1998) Distribution and near bottom transport of larvae and other plankton at hydrothermal vents. *Deep-Sea Research II* 45, 423–440.

Kim, S.L., Mullineaux, L.S. & Helfrich, K. (1994) Larval dispersal via entrainment into hydrothermal vent plumes. *Journal of Geophysical Research C* 99, 12655–12665.

Kojima, S., Watanabe, H., Tsuchida, S., *et al.* (2006) Phylogenetic relationships of a tube worm (*Lamellibrachia juni*) from three hydrothermal vent fields in the South Pacific. *Journal of the Marine Biological Association of the United Kingdom* 86, 1357–1361.

Komai, T., Giere, O. & Segonzac, M. (2007) New record of Alvinocaridid shrimps (Crustacea: Decapoda: Caridea) from hydrothermal vent fields on the Southern Mid-Atlantic Ridge, including a new species of the genus *Opaepele*. *Species Diversity* 12, 237–253.

Kochinsky, A. (2006) Discovery of new hydrothermal vents on the southern Mid-Atlantic Ridge (4–10′S) during cruise M68/1. *Inter-Ridge News* 15, 9–15.

Kochinsky, A., Garbe-Schönberg, D., Sander, S., *et al.* (2008) Hydrothermal venting at pressure–temperature conditions above the critical point of seawater, 5°S on the Mid-Atlantic Ridge. *Geology* 36, 615–618.

Leibold, M.A., Holyoak, M., Mouquetn N., *et al.* (2004) The metacommunity concept: a framework for multi-scale community ecology. *Ecology Letters* 7, 601–613.

Levin, L.A. (2003) Oxygen minimum zone benthos: adaptation and community response to hypoxia. *Oceanography and Marine Biology: An Annual Review* 41, 1–45.

Levin, L.A. (2005) Ecology of cold seep sediments: interactions of fauna with flow, chemistry and microbes. *Oceanography and Marine Biology: An Annual Review* 43, 1–46.

Levin, L.A., Mendoza, G.F., Gonzalez, J., *et al.* (2009a) Diversity of bathyal macrobenthos on the northeastern Pacific margin: the influence of methane seeps and oxygen minimum zones. *Marine Ecology* 31, 94–110.

Levin, L.A., Mendoza, G.F., Konotchick, T., *et al.* (2009b) Community structure and trophic relationships in Pacific hydrothermal sediments. *Deep-Sea Research II* 56, 1632–1648.

Levin, L.A., Ziebis, W., Mendoza, G.F. *et al.* (2006) Recruitment response of methane-seep macrofauna to sulfide and surrounding habitat. *Journal of Experimental Marine Biology and Ecology* 330, 132–150.

Lösekann, T., Robador, A., Niemann, H., *et al.* (2008) Endosymbioses between bacteria and deep-sea siboglinid tubeworms from an Arctic cold seep (Haakon Mosby Mud Volcano, Barents Sea). *Environmental Microbiology* 10, 3237–3254.

Lutz, R.A. (2000) Deep-sea vents: science at the extreme. *National Geographic Magazine* 198(4), 116–127.

Macpherson, E., Jones, W. & Segonzac, M. (2005) A new squat lobster family of Galatheoidea (Crustacea, Decapoda, Anomura) from the hydrothermal vents of the Pacific-Antarctic Ridge. *Zoosystema* 27, 709–723.

McLay, C.L. (2007) New crabs from hydrothermal vents of the Kermadec submarine volcanoes, New Zealand: *Gandalfus* gen. nov. (Bythograeidae) and Xenograpsus (Varunidae) (Decapoda: Brachyura). *Zootaxa* 1524, 1–22.

Marchetti, C. (1998) Notes on the limits to knowledge explored with Darwinian logic. *Complexity* 3, 22–35.

Marienfeld, P. & Marching, V. (1992) Indications of hydrothermal activity at the Chile spreading centre. *Marine Geology* 105, 241–252.

Marsh, A.G., Mullineaux, L.S., Young, C.M. *et al.* (2001) Larval dispersal potential of the tubeworm *Riftia pachyptila* at deep-sea hydrothermal vents. *Nature* 411, 77–80.

Melchert, B., Devey, C.W., German, C.R., *et al.* (2008) First evidence for high-temperature off-axis venting of deep crustal/mantle heat: the Nibelungen Hydrothermal Field, Southern Mid-Atlantic Ridge. *Earth and Planetary Science Letters* 275, 61–66.

Metaxas, A. (2004) Spatial and temporal patterns in larval supply at hydrothermal vents on the northwest Pacific Ocean. *Limnology and Oceanography*, 49, 1949–1956.

Mills, S.W., Beaulieu, S.E. & Mullineaux, L.S. (2009) Photographic identification guide to larvae at hydrothermal vents in the eastern Pacific. Published at: http://www.whoi.edu/science/B/vent-larval-id. Woods Hole Oceanographic Institution Technical Report WHOI-2009-05.

Miura, T. & Kojima, S. (2006) Two new species of vestimentiferan tubeworm (Polychaeta: Siboglinidae a.k.a. Pogonophora) from the Brothers Caldera, Kermadec Arc, South Pacific Ocean. *Species Diversity* 11, 209–224.

Munroe, T.A. & Hashimoto, J. (2008) A new Western Pacific tonguefish (Pleuronectiformes: Cynoglossidae): the first pleuronectiform discovered at active hydrothermal vents. *Zootaxa* 1839, 43–59.

Mullineaux, L., Juniper, S.K. & Desbruyères, D. (1998) Deep-sea sanctuaries at hydrothermal vents: a position paper. *InterRidge News* 7, 15–16.

Mullineaux, L.S., Peterson, C.H., Micheli, F., *et al.* (2003) Successional mechanism varies along a gradient in hydrothermal fluid flux at deep-sea vents. *Ecological Monographs* 73, 523–542.

Mullineaux, L.S., Mills, S.W., Sweetman, A.K., *et al.* (2005) Vertical, lateral and temporal structure in larval distributions at hydrothermal vents. *Marine Ecology Progress Series* 293, 1–16.

Neubert, M.G., Mullineaux, L.S. & Hill, M.F. (2006) A metapopulation approach to interpreting diversity at deep-sea hydrothermal vents. In: *Marine Metapopulations* (eds. J. Kritzer & P. Sale), pp. 321–350, Elsevier.

Niemann, T., Losekann, D., de Beer, M., *et al.* (2006) Novel microbial communities of the Håkon Mosby mud volcano and their role as a methane sink. *Nature* 443, 854–858.

Niemann, H., Fischer, D., Graffe, D., *et al.* (2009) Biogeochemistry of a low-activity cold seep in the Larsen B area, western Weddell Sea, Antarctica. *Biogeosciences Discussion* 6, 5741–5769.

Olu, K., Duperret, A., Sibuet, M., *et al.* (1996) Structure and distribution of cold seep communities along the Peruvian active margin: relationship to geological and fluid patterns. *Marine Ecology Progress Series* 132, 109–125.

Olu-Le Roy, K., Caprais, J., Fifis, A., *et al.* (2007) Cold-seep assemblages on a giant pockmark off West Africa: spatial patterns and environmental control. *Marine Ecology* 28, 115–130.

Ondreas, H., Cannat, M., Cherkashov, G. *et al.* (2007) High resolution mapping of the Ashadze and Logachev hydrothermal fields, Mid Atlantic Ridge 13–15°N. American Geophysical Union, Fall Meeting 2007.

OSPAR 2009. Draft Background Document for Oceanic Ridges with Hydrothermal Vents/Fields. MASH 08/4/1 Add.27-E: 16 pp.

Paull, C.K., Hecker, B., Commeau, R., *et al.* (1984) Biological communities at the Florida Escarpment resemble hydrothermal vent taxa. *Science* 226, 965–967.

Petersen, J.M. & Dubilier, N. (2009) Methanotrophic symbioses in marine invertebrates. *Environmental Microbiology Reports* 1, 319–335.

Petersen, J.M., Ramette, A., Lott, C., *et al.* (2010) Dual symbiosis of the vent shrimp *Rimicaris exoculata* with filamentous gamma- and epsilon-proteobacteria at four Mid-Atlantic Ridge hydrothermal vent fields. *Environmental Microbiology* (in press).

Plouviez, S., Shank, T.M., Faure, B., *et al.* (2009) Comparative phylogeography among hydrothermal vent species along the East Pacific Rise reveals vicariant processes and population expansion in the South. *Molecular Ecology* 18, 3903–3917.

Pradillon, F., Shillito, B., Young, C.M., *et al.* (2001) Developmental arrest in vent worm embryos. *Nature* 413, 698–699.

Pradillon, F., Zbinden, M., Mullineaux, L.S., *et al.* (2005) Colonisation of newly-opened habitat by a pioneer species, *Alvinella pompejana* (Polychaeta: Alvinellidae), at East Pacific Rise vents sites. *Marine Ecology Progress Series* 302, 147–157.

Ramirez-Llodra, E., Blanco, M. & Arcas, A. (2004) ChEssBase: an online information system on biodiversity and biogeography of deep-sea chemosynthetic ecosystems. Version 1. World Wide Web electronic publications, www.noc.soton.ac.uk/chess/db_home.php.

Roberts, H.H., Fisher, C.R., Bernard, B., *et al.* (2007) *ALVIN* explores the deep northern Gulf of Mexico slope. *EOS Transactions, American Geophysical Union* 88, 341–342.

Rouse, G.W., Wilson, N.G., Goffredi, S.K., *et al.* (2009) Spawning and development in *Osedax* boneworms (Siboglinidae, Annelida). *Marine Biology* 156, 395–405.

Sanders, H.L. (1969) Benthic marine diversity and stability-time hypothesis. In *Brookhaven Symposium. Diversity and Stability in Ecological Systems* 22, 71–81.

Santos, R.S., Colaço, A. & Christiansen, S. (2003) Planning the Management of Deep-sea hydrothermal vent fields MPAs in the Azores Triple Junction (workshop proceedings). *Arquipélago–Life and Marine Sciences, Supplement 4: xii + 70pp. Ponta Delgada: University of the Azores.*

Schnabel, K.E. & Bruce, N.L. (2006) New records of *Munidopsis* (Crustacea: Anomura: Galatheidae) from New Zealand with description of two new species from a seamount and underwater canyon. *Zootaxa* 1172, 49–67.

Schander, C., Rapp, H.T., Kongsrud, J.A., *et al.* (2009) The fauna of hydrothermal vents on the Mohn Ridge (North Atlantic). *Marine Biology Research* 6, 155–171.

Sellanes, J., Quiroga, E. & Gallardo, V.A. (2004) First direct evidence of methane seepage and associated chemosynthetic communities in the bathyal zone off Chile. *Journal of the Marine Biological Association of the United Kingdom* 84, 1065–1066.

Sellanes, J., Quiroga, E. & GNeira, C. (2008) Megafauna community structure and trophic relationships at the recently discovered Concepcion Methane Seep Area, Chile, ~36degS. *ICES Journal of Marine Science* 65, 1102–1111.

Shank, T.M. (2004) The evolutionary puzzle of seafloor life. *Oceanus* 42, 19–22.

Shank, T.M. (2006) Preliminary Characterization of Vent Sites and Evolutionary Relationships of Vent Fauna on the Southern Atlantic Ridge. ChEss Biogeography and Diversity of Chemosynthetic Ecosystems Meeting, Barcelona, Spain.

Shank, T.M. & Halanych, K.M. (2007) Toward a mechanistic understanding of larval dispersal: insights from genomic fingerprinting of deep-sea hydrothermal vent populations. *Marine Ecology* 28, 25–35.

Shank, T.M., Fornari, D.J., Von Damm, K.L., *et al.* (1998) Temporal and spatial patterns of biological community development at nascent deep-sea hydrothermal vents along the East Pacific Rise. *Deep-Sea Research* 45, 465–515.

Shank, T.M., Fornari, D.J., Yoerger, D., *et al.* (2003) Deep submergence synergy: Alvin and ABE explore the Galápagos Rift at 86°W. *EOS Transactions* 84(41), 425–440.

Shea, K., Metaxas, A., Young, C.R., *et al.* (2008) Processes and interactions in macrofaunal assemblages at hydrothermal vents: a modelling perspective. In: *Magma to Microbe: Modelling Hydrothermal Processes at Ocean Spreading Centres.* (eds. R.P. Lowell, M.R. Perfit, J. Seewald, A. Metaxas) *Geophysical Monograph* 178, pp. 259–274. Washington, DC: American Geophysical Union.

Sibuet, M. & Olu, K. (1998) Biogeography, biodiversity and fluid dependence of deep-sea cold-seep communities at active and passive margins. *Deep-Sea Research II* 45, 517–567.

Smith C. & Baco, A. (2003) The ecology of whale falls at the deep-sea floor. *Oceanography and Marine Biology: An Annual Review* 41, 311–354.

Sohn, R.A., Willis, C., Humphris, S., *et al.* (2008) Explosive volcanism on the ultraslow-spreading Gakkel ridge, Arctic Ocean. *Nature* 453, 1236–1238.

Southward, A.J. & Jones, D.S. (2003) A revision of stalked barnacles (Cirripedia: Thoracica: Scalpellomorpha: Eolepadidae: Neolepadinae) associated with hydrothermalism, including a description of a new genus and species from a volcanic seamount off Papua New Guinea. *Senckenbergiana Maritima* 32, 77–93.

Speer, K.G., Maltrud, M. & Thurberr, A. (2003) A global view of dispersion on the mid-oceanic ridge. In *Energy and Mass Transfer in Marine Hydrothermal Systems.* (eds. P. Halbach, V. Tunnicliffe & J. Hein), pp. 287–302. Berlin: Dahlem University Press.

Treude, T., Smith, C.R., Wenzhöfer, F., *et al.* (2009) Microbial sulfur and carbon turnover at a deep-sea whale fall, Santa Cruz Basin, northeast Pacific. *Marine Ecology Progress Series* 382, 1–21.

Tunnicliffe, V., McArthur, A.G. & McHugh, D. (1998) A biogeographical perspective of the deep-sea hydrothermal vent fauna. *Advances in Marine Biology* 34, 353–442.

Tunnicliffe, V., Juniper, K.S. & Sibuet, M. (2003) Reducing environments of the deep-sea floor. In: *Ecosystems of the World*, Volume 28, *Ecosystems of the Deep Oceans* (ed P.A. Tyler), pp 81–110. London: Elsevier.

Turnipseed, M., Knick, K.E., Dreyer, J., *et al.* (2003) Diversity in mussel-beds at deep-sea hydrothermal vents and cold seeps. *Ecology Letters* 6, 518–523.

Tyler, P.A., Young, C.M., Dolan, E., *et al.* (2007) Gametogenic periodicity in the chemosynthetic cold-seep mussel, *Bathmodiolus childressi. Marine Biology* 150, 829–840.

Tyler, P.A., German C.R., Ramirez-Llodra E., *et al.* (2003) Understanding the biogeography of chemosynthetic ecosystems. *Oceanologica Acta* 25, 227–241.

Van Dover, C.L. (2000) *The Ecology of Deep-Sea Hydrothermal Vents.* Princeton University Press. 424 pp.

Van Dover, C.L. (2007) The biological environment of polymetallic sulphides deposits, the potential impact of exploration and mining on this environment, and data required to establish environmental baselines in exploration areas. In *Polymetallic Sulphides and Cobalt-Rich Ferromanganese Crusts Deposits: Establishment of Environmental Baselines and an Associated Monitoring Programme During Exploration. Proceedings of the International Seabed Authority's Workshop held in Kingston, Jamaica, 6–10 September 2004.* Kingston, Jamaica: International Seabed Authority.

Van Dover, C.L. & Trask, J. (2001) Biodiversity in mussel beds at a deep-sea hydrothermal vent and a shallow-water intertidal site. *Marine Ecology Progress Series* 195, 169–178.

Van Dover, C.L., German, C.R., Speer, K.G., *et al.* (2002) Evolution and biogeography of deep-sea vent and seep invertebrates. *Science* 295, 1253–1257.

Van Gaest, A.L. (2006) Ecology and early life history of *Bathynerita naticoidea*: evidence for long-distance larval dispersal of a cold seep gastropod. MSc thesis, University of Oregon, USA.

Van Gaever, S., Moodley, L., de Beer, D., *et al.* (2006) Meiobenthos at the Arctic Håkon Mosby Mud Volcano with a parental caring nematode thriving in sulphide-rich sediments. *Marine Ecology Progress Series* 321, 143–155.

Vanreusel, A., Andersen, A.C., Boetius, A., *et al.* (2009) Biodiversity of cold seep ecosystems along the European margins. *Oceanography* 22, 110–127.

Vrijenhoek, R.C. (2009a) Hydrothermal vents as deep-sea habitat islands. In: *Encyclopedia of Islands* (eds. R. Gillesie & D. Clague). Berkeley, California: University of California Press.

Vrijenhoek, R.C. (2009b) Cryptic species, phenotypic plasticity, and complex life histories: Assessing deep-sea faunal diversity with molecular markers. *Deep-Sea Research Part II* 56, 1713–1723.

Warén, A. & Bouchet, P. (2009) New gastropods from deep-sea hydrocarbon seeps off West Africa. *Deep-Sea Research II* 56, 2326–2349.

Webber, W.R. (2004) A new species of *Alvinocaris* (Crustacea: Decapoda: Alvinocarididae) and new records of alvinocarids from hydrothermal vents north of New Zealand. *Zootaxa* 444, 1–26.

Yoerger, D.R., Bradley, A.M., Jakuba, M., *et al.* (2007) Autonomous and remotely operated vehicle technology for hydrothermal vent discovery, exploration and sampling. *Oceanography Magazine* 20, 152–161.

Young, C.M. (2003) Reproduction, development and life-history traits. In: *Ecosystems of the World: Ecosystems of the Deep Oceans* (ed. P.A. Tyler), pp. 381–426. Amsterdam: Elsevier.

Young, C.M., Vázquez, E., Metaxas, A., *et al.* (1996) Embryology of vestimentiferan tube worms from deep-sea methane/sulphide seeps. *Nature* 381, 514–516.

Zekely. J., Sorensen, V. & Bright, M. (2006) Three new nematode species (Monhysteridae) from deep-sea hydrothermal vents. *Meiofauna Mar* 15, 25–42.

Chapter 10

Marine Life in the Arctic

Rolf Gradinger[1], Bodil A. Bluhm[1], Russell R. Hopcroft[1], Andrey V. Gebruk[2], Ksenia Kosobokova[2], Boris Sirenko[3], Jan Marcin Węsławski[4]

[1]School of Fisheries and Ocean Sciences, University of Alaska Fairbanks, Fairbanks, Alaska, USA
[2]P.P. Shirshov Institute of Oceanology, Russian Academy of Sciences, Moscow, Russia
[3]Zoological Institute, Russian Academy of Sciences, St. Petersburg, Russia
[4]Institute of Oceanology, Polish Academy of Sciences, Sopot, Poland

10.1 Introduction

The Arctic Ocean Diversity project (ArcOD), one of the regional field projects of the international Census of Marine Life, is an international collaborative effort to inventory biodiversity in Arctic marine realms on a pan-Arctic scale. Over 100 scientists in a dozen nations have contributed to ArcOD-related efforts, including many conducted during the International Polar Year 2007–9.

The Arctic seas are among the most extreme regions on Earth. Total darkness in winter is paired with low temperatures, strong winds, and heavy snow cover, whereas in summer permanent light produces ice and snow melt with temperatures around the freezing point. Arctic marine biota must deal with extreme seasonality of light, temperature, salinity, and sea ice, and year-round seawater temperatures that are close to freezing. The prevalence of such conditions for millions of years has led to the evolution of truly unique Arctic endemic flora and fauna.

The in- and outflow of water, mainly through Fram Strait and Bering Strait (Fig. 10.1), and cross-Arctic currents plus animal migrations make the Arctic Seas a mixing bowl of different species assemblages that compete for resources like light, substrate, nutrients, and food. Nevertheless, distinct community patterns have arisen within individual Arctic seas, realms, and/or water masses. These biological communities sustain very productive marine food webs regionally and provide subsistence foods around the Arctic.

Historical collections and identification of marine organisms are valuable resources for today's Arctic research. They not only led to the description of many new species, for example by Steller during Bering's expedition (1738–1740), but also to industrial exploitation of the Arctic seas by commercial whalers and quick extinction of the great auk (in 1844) and the Arctic Steller's sea cow (in 1768) shortly after their description. The central Arctic Ocean was the focus of scientific curiosity for decades, including theories of an ice-free central Arctic Ocean in the nineteenth century by German geographer Petermann (Tammiksaar et al. 1999). Although many of the ideas about the central Arctic were wrong, they promoted Arctic exploration. The FRAM drift led by F. Nansen (1893–1896) is particularly noteworthy because of the wealth of physical and biological data collected, including species descriptions of then unknown ice biota.

During the mid-twentieth century, drifting ice stations became long-term research platforms for the USA and the Soviet Union (Kosobokova 1980; Perovich et al. 1999). In 1991, modern non-nuclear research vessels sampled the North Pole area for the first time in a systematic way (see, for example, Gradinger & Nürnberg 1996). Even today, the central Arctic remains the domain of ice camps and ice breakers with access mainly in the summer months. In contrast, the shallow seasonally ice-covered Arctic shelves

Life in the World's Oceans, edited by Alasdair D. McIntyre
© 2010 by Blackwell Publishing Ltd.

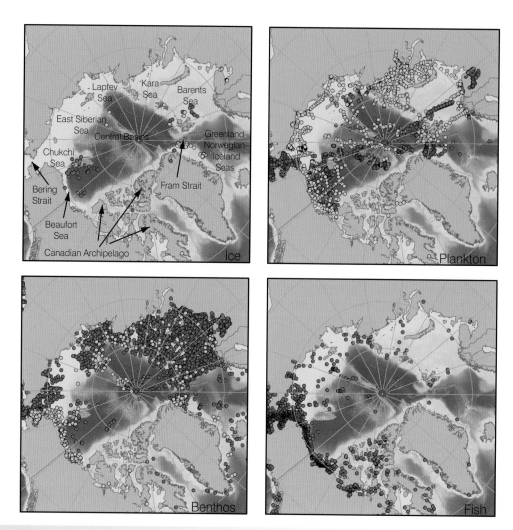

Fig. 10.1

The Arctic data records compiled by ArcOD. Red dots are records already available on OBIS (www.iobis.org). Yellow dots are records prepared for posting online, but not online yet.

have always been more accessible. On the extensive Russian shelves faunistic exploration began over 200 years ago: in the late seventeenth century, the Zoological Museum in St. Petersburg acquired its first collections from the Barents, Kara, and White Seas, with these extensive Russian collections leading to a detailed species list of Arctic invertebrates (Sirenko 2001). On the North American shelf, the onset of oil drilling in the nearshore Beaufort Sea in the late 1970s initiated major research efforts, resulting in a wealth of biological data (see, for example, Horner 1981).

Over recent decades drastic changes have occurred in the Arctic, most notably in the physical settings. Sea ice has decreased in the summer months, reducing not only the substrate for ice-related flora and fauna, but also increasing light levels and temperatures in regions previously covered with ice continuously (Perovich *et al.* 2007). Although

some of the observed changes are related to natural causes, the main driver is thought to be the human footprint, and a completely ice-free Arctic (in summer) is predicted for 2030–2050, or at the latest by 2100 (Walsh 2008).

The predicted total loss of summer ice and the increased human presence will alter Arctic ecosystem functioning (Fig. 10.2) with regional changes in primary production, species distributions (including extinctions and invasions), toxic algal blooms, and indigenous subsistence use (Bluhm & Gradinger 2008). To address these issues scientifically, new research in poorly studied regions is needed with the rescue of historical data on species' distributions. Using recent ArcOD achievements, we discuss some of the urgent issues listed above, and suggest future research and Census activities in the Arctic beyond the end of the first Census.

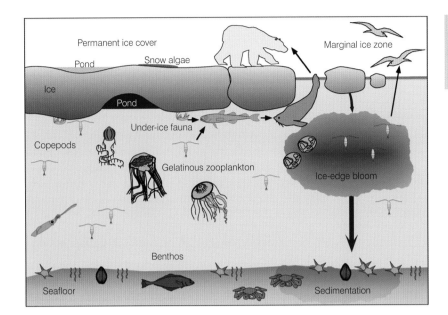

Fig. 10.2

The Arctic's three realms: sea ice, water column, and benthos. The realms are tightly linked through life cycles, vertical migration, and carbon flux.

10.2 The Background

ArcOD's main effort focused on the least explored waters of the Arctic Ocean with its southern boundaries in Bering Strait of the Pacific Sector, and Fram Strait and the Barents Sea of the Atlantic sector, while including the sub-Arctic to some extent. True Arctic boundaries are difficult to define, as currents and ice drift distribute biota within and outside the above boundaries. Definitions vary among countries, agencies, and habitats in focus. Based on water temperature and ice cover, the Arctic extends well south of the Arctic Circle on the western side of the North Atlantic and North Pacific. In contrast, Arctic waters are displaced by communities of more southern fauna along the eastern side of the North Atlantic in the Barents Sea, and by Pacific water in the Bering and Chukchi Seas. Consequently the Arctic Ocean's flora and fauna are a varying mixture of Pacific, Atlantic, and true Arctic endemic species.

10.2.1 The environment

The Arctic Ocean contains 31% of the world ocean's shelves with 53% of the Arctic Ocean shallower than 200 m (Jakobsson *et al.* 2004, Fig. 10.1). Shelf extent varies from very narrow shelves in the Beaufort Sea to the wide Russian shelves. The central Arctic is a deep-sea system divided into abyssal basins by the Gakkel and Lomonosov ridges. The only current deep water connection to the world's ocean is through Fram Strait. The connection to the Pacific has opened and closed several times over the past few million years related to glacial and interglacial periods, with its last deep water connection about 80 million years ago (Bilyard & Carey 1980).

The well-adapted Arctic marine biota comprises viruses, bacteria, protists, and metazoans, including marine mammals. Abiotic forcing factors shape biological patterns and community composition, and cause strong seasonality of biological production and animal migrations. Arctic seas are exposed to winter months of complete darkness followed by intense summer solar irradiance that exceeds daily irradiances measured at the equator. Sea ice and associated snow cover with high albedo and attenuation effectively reduce the available light for phytoplankton growth to a few percent of surface irradiance (Perovich *et al.* 2007) making the timing and extent of sea ice and its melt a major controlling factor throughout the Arctic.

Sea ice covers the entire Arctic during winter with its maximum extent in February (around 14 million km²) (Thomas & Dieckmann 2009) and a minimum summer ice extent in September of historically around 7 million km² (so-called multi-year ice). Recent trends indicate a drastic loss in the extent of the summer multi-year ice by about 8.6% per decade (Serreze *et al.* 2007) and a decrease in sea ice thickness (Rothrock *et al.* 1999). Arctic pack ice drifts with ocean currents in two major drift systems, the anticyclonic Beaufort Gyre and the Transpolar Drift System. Some seasonal coastal sea ice is attached to land and stationary, therefore called fast-ice.

The central Arctic Ocean is permanently stratified owing to the input of fresh water from huge, mostly Russian river systems that reduce the salinity of Arctic surface waters to typically less than 32, whereas deep-water salinities typically exceed 34. River plumes can extend for hundreds of kilometers into the central Arctic. Melting of relatively fresh sea ice causes reduced-salinity lenses that are 5–40 m thick in the marginal ice zones (Perovich *et al.* 1999). Inorganic nutrient concentrations exhibit strong regional gradients from high nutrient regimes, like the Chukchi Sea shelf, to oligotrophic conditions in the Beaufort Gyre (Gradinger 2009a) that are maintained by ocean currents combined with upwelling along shelf slopes and by riverine inputs.

Sea floor sediments are typically muddy on the outer shelves and in the central basins, and coarser with sand and gravel on the inner shelves or at locations with stronger ocean currents (Naidu 1988). Local accumulations of boulders and rocky islands like Svalbard provide hard substrates. Sedimentation is often dominated by terrigenous materials from riverine discharge and coastal erosion or by glacial deposits, while organic content is greatest in areas of high nutrient concentration and productivity.

10.2.2 Knowledge of Arctic marine species before the Census

Before the Census, the only web-based resource containing Arctic marine information was the non-searchable database by the US National Marine Fisheries Service on plankton. Additional information was scattered in reports, publications, and reviews mainly for pelagic and benthic biota. The most complete taxonomic list had been compiled by Sirenko (2001) (Table 10.1) listing 4,784 free-living invertebrate species.

10.2.2.1 Biota in sea ice

Sea ice is a habitat, feeding ground, refuge, breeding and/or nursery ground for several metazoan species (Fig. 10.3), as well as autotrophs, bacteria, and protozoans (Fig. 10.2) including ice-endemic species. The specialized, sympagic (=ice-associated) community lives within a brine filled network of pores and channels or at the ice-water interface. Several hundred diatom species are considered the most important sympagic primary producers (Horner 1985; Quillfeldt et al. 2003), while realizing the significance of flagellated protists (Ikävalko & Gradinger 1997). Ice algal activity exhibits strong regional gradients (Gradinger 2009a) with maximum contributions of approximately 50% of total primary productivity in the central Arctic (Gosselin et al. 1997). Typically ice algal blooms start mid-March and are released during ice melt.

Protozoan and metazoan ice meiofauna, in particular acoels, nematodes, copepods, and rotifers, can be abundant in all ice types, whereas nearshore larvae and juveniles of benthic taxa like polychaetes migrate seasonally into the ice matrix (Gradinger 2002). The variety of under-ice structures provides a wide range of different microhabitats for a partly endemic fauna, mainly gammaridean amphipods (Bluhm et al. 2010b). Amphipod abundances vary from fewer than 1 to several hundred individuals per square meter. They transfer particulate organic matter from the sea ice to the water column through the release of fecal pellets and are a major food source for Arctic cod (*Boreogadus saida*) that occurs with sea ice and acts as the major link from the ice-related food web to seals and whales (Gradinger & Bluhm 2004).

Biodiversity in sea ice habitats was – and still is – poorly known for several groups, but sea ice faunal species richness is low compared with water column and interstitial sediment faunas, with only a few species per higher taxonomic group (Table 10.2), likely because of extreme temperatures (to below $-10\,°C$), high brine salinities (to greater than 100) in the ice interior during winter and early spring, and because of size constraints within the brine channel network (Gradinger 2002).

10.2.2.2 Biota in the water column

Pelagic communities are intricately coupled to the seasonal cycles of pelagic primary production and the seasonal downward flux of ice-algae during breakup (section 10.2.1). Typically phytoplankton production begins with ice melt in April and ends in early September with a growth curve characterized by a single peak in primary production in late June to early July (Sakshaug 2004). Enhanced plankton activity occurs on the Arctic shelf areas, where the seasonal retreat of the sea ice allows for the formation of ice-edge algal blooms with reduced surface salinity increasing vertical stability. The often large herbivorous zooplankton species accumulate substantial lipid reserves for winter survival and early reproduction in the following spring (Pasternak et al. 2001). Predatory zooplankton species rely on continuous availability of their prey, and generalists and scavengers show broad flexible diets (Laakmann et al. 2009). In all cases, the low metabolic rates at cold temperatures allow low rates of annual primary production to support relatively large stocks of zooplankton.

Phytoplankton blooms in spring are mainly dominated by diatoms and *Phaeocystis pouchetii* (Gradinger & Baumann 1991). Arctic estuarine systems harbor defined phytoplankton species assemblages, dominated by freshwater, brackish water, or full marine taxa (Nöthig et al. 2003); however, relatively few studies have closely examined the taxonomic composition of the phytoplankton communities (Booth & Horner 1997). The relevance of bacteria and heterotrophic protist communities and their role in the Arctic ecosystem (Sherr et al. 1997) was largely unknown, causing large uncertainties regarding their contribution to the Arctic carbon cycle (Pomeroy et al. 1990).

Owing to high abundance and ease of capture, the taxonomic composition and life history of the larger more common copepods in the Arctic Ocean was relatively well understood (Smith & Schnack-Schiel 1990). Historically, effort has concentrated on abundant copepods of the genus *Calanus*; however, although smaller copepod taxa are numerically dominant, relatively few studies have used sufficiently fine meshes to assess their contribution fully (Kosobokova 1980). A broad assemblage of other

Table 10.1

Species numbers of free-living invertebrates in the Arctic Seas.

Reference	Total invertebrate species	White Sea	Barents Sea	Kara Sea	Laptev Sea	East Siberian Sea	Chukchi Sea	Canadian Arctic	Central Basins
Zenkevitch 1963	N/A	1,015	1,851	1,432	522	N/A	820		
Sirenko & Piepenburg 1994	3,746	1,100	2,500	1,580	1,337	962	946		
Sirenko 2001	4,784	1,817	3,245	1,671	1,472	1,011	1,168		837
Sirenko 2004[a]; Sirenko & Vassilenko 2009[b]; P. Archambault personal communication[c]; ArcOD[d]	>5,000[d]				1,793[a]		1,469[b]	>1,405[c]	>1125[d]

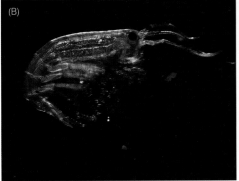

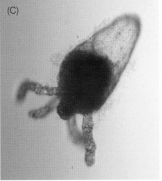

Fig. 10.3

Examples of Arctic sea ice fauna. (**A**) Arctic cod, *Boreogadus saida* (about 10 cm long). (**B**) Under-ice amphipod, *Apherusa glacialis* (approximately 1 cm long). (**C**) Sea ice hydroid, *Sympagohydra tuuli* (approximately 400 μm long), a species new to science. Photographs: A, K. Iken; B, B. Bluhm; C, R. Gradinger; all University of Alaska Fairbanks.

holoplanktonic groups was only occasionally reported in full detail (Mumm 1991). These understudied non-copepod groups held the greatest promise for discovery of new species and trophic importance. Like other oceans, knowledge of deep-water zooplankton was poor because of the time and logistics associated with their collection (Kosobokova & Hirche 2000).

Among the non-copepod groups, larvaceans (=appendicularians) are abundant in Arctic polynyas (Deibel & Daly 2007) and the central Arctic (Kosobokova & Hirche 2000). The basic biodiversity and importance of gelatinous animals were particularly under-appreciated (Stepanjants 1989; Siferd & Conover 1992). Arctic chaetognaths represent considerable biomass (Mumm 1991), and can control

Calanus populations (Falkenhaug & Sakshaug 1991) as can hyperiid amphipods (Auel & Werner 2003).

Sirenko (2001) (Table 10.1) listed about 300 species of multicellular holozooplankton with about half of these copepods, and the arthropods contributing about three-quarters total. Of the remainder, the cnidarians contributed about 50 species, whereas others contributed a dozen species or less each. Sirenko's list also contained about 125 species of planktonic heterotrophic protists, with several important heterotrophic groups still unconsidered. The number of described phytoplankton taxa has increased over time from 115 to more than 300 (Sakshaug 2004).

10.2.2.3 Biota at the sea floor

Benthic communities generally depend on food supplied from the water column, with sediment and water mass characteristics as environmental forcing factors (section 10.2.1). In high latitudes, the quantity of settling food particles rather than temperature per se is restraining the growth and survival of benthic organisms. Faunal densities generally decrease with water depth and sediment thickness in response to the decreasing food supply (Schewe & Soltwedel 2003). On the Arctic shelves, organic particle input is relatively large over the ice-free period, and benthos, therefore, plays a greater role in the marine carbon cycle than at lower latitudes (Grebmeier & Barry 1991). High benthic biomass in some areas provides major feeding grounds for resident and migrating mammals and sea birds (see, for example, Gould *et al.* 1982) in particular at frontal systems, polynyas, and along ice edges (Schewe & Soltwedel 2003). The Arctic shelf macro- and megafauna had received the most attention whereas meiofauna and microbial communities were considerably less studied.

Nematodes and copepods are the most abundant metazoan meiofauna (Schewe & Soltwedel 1999). Less common taxa include kinorhynchs, tardigrades, rotifers, gastrotrichs, and tantulocarids. Foraminifera dominate unicellular meiofauna and can constitute more than 50% of total meiofauna abundance (Schewe & Soltwedel 2003). Macrofaunal abundance and biomass are typically dominated by crustaceans, in particular amphipods, polychaetes, and bivalve mollusks (Grebmeier *et al.* 2006) with massive biomass levels on some Arctic shelves like the northern Bering and southern Chukchi Seas (Sirenko & Gagaev 2007). The most species-rich macrofaunal groups include amphipods and polychaetes (Sirenko 2001). Studies on slope and deep-sea benthos found low infaunal abundances and biomass (Kröncke 1998) dominated by deposit feeding groups (Iken *et al.* 2005), with abundances overlapping with the lower values from the North Atlantic deep sea. Epibenthic megafauna (visible fauna on underwater imagery and caught in trawls) was mostly studied on shelves, where echinoderms, particularly ophiuroids, dominated with up to several hundred individuals per square meter (Piepenburg *et al.* 1996). Other abundant epibenthic faunal taxa include crabs, anemones, sea urchins, and sea cucumbers (Feder *et al.* 2005). For shelf epifauna, bryozoans and gastropods are particularly species rich, followed by sponges and echinoderms (Sirenko 2001; Feder *et al.* 2005).

Over 90% of the Arctic invertebrate species inventory are benthic, and most are macrofaunal (Sirenko 2001) (Table 10.2). By far the highest numbers of species were recorded for the Barents Sea, largely because of its long research history and the occurrence of many boreal-Atlantic species. In other Arctic Seas, numbers ranged from just over 1,000 to almost 3,000, again mostly benthic. Before ArcOD-related research, approximately 350–400 benthic macro- and megafauna species were listed for the deep central Eurasian Arctic.

10.3 ArcOD Activities

ArcOD was from the beginning an international pan-Arctic effort initiated mainly by US and Russian scientists, but including many European and Canadian researchers. In addition to its international character, ArcOD also placed emphasis on rescuing and consolidating historic and new data and making those available through the Census database, the Ocean Biogeographic Information System (OBIS). So far (April 2010), ArcOD has posted 42 datasets to OBIS representing 200,000 records (Fig. 10.1), likely exceeding 250,000 by the end of 2010.

ArcOD scientists collected new samples and generated new observations. Challenges of sampling in ice-covered waters are numerous and they impair the ability to tow collecting gear that collect the most mobile species or reach the area of interest because of ice. In a few cases, failure to generate the interest of professional taxonomists in a less common group has created gaps. ArcOD identified the need for a complete set of taxonomic guides for all Arctic groups that is coming to fruition under the leadership of Zoological Institute of the Russian Academy of Sciences (Vassilenko & Petryashov 2009; Sirenko & Buzhinskaya, personal communication). Online species pages (www.arcodiv.org) provide additional information and imagery useful to the interested public as well as ecologists and taxonomists, and will ultimately become accessible through the Encyclopedia of Life initiative.

Much knowledge has been gained in the field of Arctic biodiversity in the past decade under ArcOD, other programmatic umbrellas, and many individual studies with a significant fraction of this information, including most results gathered during the International Polar Year 2007–9, to be published after this book is printed. Below, we summarize knowledge gained in specific areas with strongest ArcOD participation, including examples of progress based on taxonomic, regional, methodological, and hypothesis-driven efforts.

10.3.1 Improvements in traditional and molecular taxonomic inventories

ArcOD's discovery of over 60 invertebrate species new to science is based on substantial efforts dedicated to new collections and to more complete re-analyses of previously collected materials in different habitats and Arctic regions.

In the sea ice realm, ArcOD-related efforts added to the ice-associated species inventory in all size classes and in a variety of taxa. Sea-ice cores from Bering Sea shelf pack ice are currently being analyzed for bacterial and archaeal diversity using molecular tools (R. Gradinger & G. Herndl, unpublished observations). A comprehensive review of the pan-Arctic literature ice-associated protists (excluding ciliates) resulted in a list of more than 1,000 sympagic species (M. Poulin *et al.*, personal communication). For meiofauna, the first true predator in the brine channel system, the hydroid *Sympagohydra tuuli*, was described (Piraino *et al.* 2008) (Fig. 10.3 and Table 10.2). Juveniles of the polychaete *Scolelepis squamata* were identified as a seasonally common taxon in coastal fast ice in the Chukchi and Beaufort Seas (Bluhm *et al.* 2010a) with other less common polychaete species yet to be identified. Specimens of the groups Acoela, Nematoda, Harpacticoida, and Rotifera from various types of sea ice are currently with European taxonomists for species identifications. Within the macrofauna, we discovered large aggregations of an Arctic euphausiid (*Thysanoessa raschii*) under Bering Sea ice in spring 2008, the first record of winter ice-association for the Arctic (R. Gradinger, B.A. Bluhm, & K. Iken, unpublished observations). We also discovered that sea-ice pressure ridges might be crucial for survival of sympagic fauna during periods of enhanced summer ice melt (Gradinger *et al.* 2010) because sea-ice ridges protrude into the deeper higher-salinity water, and hence may be a less stressful environment than encountered under level ice.

Within the plankton, at least six new species of small primarily epibenthic copepods have recently been discovered and are under description (V. Andronov, personal communication) (Table 10.2). Deepwater expeditions increased the known range of several amphipod species (T.N. Semenova, personal communication), as well as discovered a new pelagic ostracod species (M. Angel, personal communication). As expected, the largest gain in knowledge for the zooplankton has occurred within gelatinous groups. By using a remotely operated vehicles (ROV), more than 50 different "gelatinous" taxa were identified in the Canada Basin (Raskoff *et al.* 2010). Of five new species of ctenophores, only two could be placed within known genera (*Bathyctena, Aulacoctena*) (Table 10.2). Within the cnidarians, a new species of hydromedusae was described within a new genus (Raskoff 2010) (Fig. 10.4) that was surprisingly common at a depth of approximately 1 km. At least four described species of hydromedusae were observed in the Arctic for the first time (Raskoff *et al.* 2010). Within the pelagic tunicates one new species was collected at great depth, several other likely new species were observed by ROVs, and the first records of *Fritillaria polaris* and *Oikopleura gorksyi* were made outside of their type locality (R.R. Hopcroft, unpublished observation). Russian taxonomists continue to go through more recent ArcOD deepwater collections to characterize these communities better.

Most of the new species discovered during ArcOD research were in the benthic realm (Table 10.2), where species richness is generally highest. Most of those were found in the Arctic deep sea, specifically in the polychaetes and crustaceans (Gagaev 2008, 2009), two particularly species-rich groups in soft sediments. More unexpected was the finding of five new bryozoan species around Svalbard (Kuklinski & Taylor 2006, 2008) (Fig. 10.5), because Svalbard's fjords, in particular Kongsfjorden and Hornsund Fjord, are well-studied by the many international field stations located there. Similarly noteworthy are the finds of three new gastropod species in the Bering and Chukchi Seas (Chaban 2008; Sirenko 2009), two of which were actually collected over 70 years ago. All of these and other species, including several amphipods (B. Stransky, unpublished observations), cnidarians (Rodriguez *et al.* 2009), and a sea cucumber (Rogacheva 2007), are in the larger and better studied size fractions. Considerably less taxonomic effort was spent on meiofaunal groups during ArcOD, but new species were recorded among benthic and hyperbenthic copepods (see, for example, Kotwicki & Fiers 2005), Komokiacea (O. Kamenskaya, unpublished observations), and the nematodes (J. Sharma, personal communication). A benthic boundary layer study in the Beaufort Sea (Connelly 2008) discovered six new copepod species.

The compilation of close to 10,000 data records of western Arctic fishes and verification of most of the identifications in museums around the world has resulted in major improvements regarding the taxonomy and distribution of Arctic fishes (Mecklenburg *et al.* 2007, 2008). Some species like Pacific cod, *Gadus macrocephalus*, now present in the western Arctic were historically absent from the area (C.W. Mecklenburg, personal communication). The black snailfish *Paraliparis bathybius* collected from the Canada Basin in 2005 is the first record of this species from the western Arctic. For other species, the known range in the region was extended further north: walleye pollock, *Theragra chalcogramma*, was found 200 km north of its previous northernmost record (Mecklenburg *et al.* 2007). Instances of misidentifications were uncovered, for example virtually all Arctic specimens identified as sturgeon poacher, *Podothecus accipenserinus*, turned out to be the veteran poacher *P. veternus*.

Most of the discoveries of new species were related to (1) the exploration of previously poorly studied areas such as the Canada Basin (Gradinger & Bluhm 2005; Bluhm

Table 10.2

Arctic marine species inventory by taxa and realm. Estimates are primarily based on Sirenko (2001) with estimates for additional taxa per references provided. Updates to Sirenko's estimates are based on contributions by ArcOD researchers (mostly cited in the text) and are to be considered conservative.

Taxon	Species numbers, marine Arctic (Sirenko 2001 and updated)	Arctic sea ice	Arctic plankton	Arctic benthos	Species new to science (range extensions) in ArcOD
Bacteria	4,500–450,000[a]	>115[b]	1,500[c]	?	Many
Archaea	Up to 5,000[a]	?	1,400[d]	?	Many
Cyanobacteria	1		1		
Macrophytes	130–160[e]			130–160	
Bacillariophyta	1,227[f]	731[f]	1059[f]		
Other Protista	1,568[f]	296[f]	815[f]	570	
Porifera	163			163	
Cnidaria	227	1	83	161	5 (7)
Ctenophora	7		7		5
Tentaculata	341			341	6
Sipunculida	12			12	
Plathelminthes	137	>3		134	
Gnathostomulida	1			1	
Nemertini	80		2	78	
Aschelminthes	422	>11	16	403	
Mollusca	487		5	482	3 (7)
Annelida	571	4	6	565	11 (27)
Tardigrada	7			7	
Arthropoda	1,547	>20	214	1,317	31 (12)
Chaetognatha	5		5		
Hemichordata	1			1	
Echinodermata	151			151	1 (1)
Urochordata	60		3	57	1 (2)
Pisces	415	2			(3)
Aves	82				
Mammalia	16	7		3	

[a] Estimates, C. Lovejoy *et al.*, unpublished observations.
[b] Brinkmeyer *et al.* (2003).
[c] Actually found, D. Kirchman *et al.*, unpublished observations.
[d] Actually found, surface and deep waters, Galand *et al.* (2009).
[e] R. Wilce and D. Garbary, personal communication.
[f] M. Poulin *et al.*, unpublished observations, for "Other Protista" combined with Sirenko (2001).

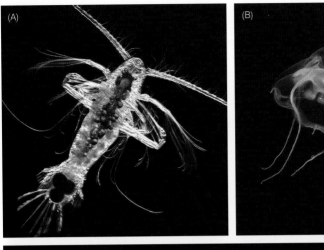

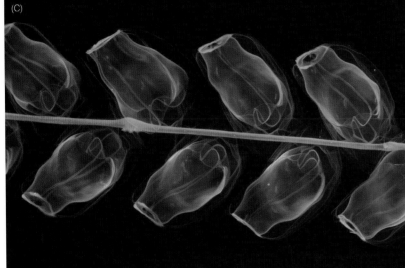

Fig. 10.4

Examples of Arctic zooplankton. (**A**) Copepod, *Euaugaptilus hyperboreus* (about 1 cm long). (**B**) Species of narcomedusa new to science (up to 3 cm). (**C**) Close-up of anterior nectophore region of siphonophore, *Marrus orthocanna* (whole specimen up to 2 m). Photographs: A, R. Hopcroft, University of Alaska Fairbanks; B and C, K. Raskoff, Monterey Peninsula College.

et al. 2010a), (2) study of poorly studied taxonomic groups such as gelatinous zooplankton (Raskoff *et al.* 2005, 2010), (3) little-studied habitats such as the benthic boundary layer (Connelly 2008), or (4) the All-Taxa-Biodiversity-Inventory program in Svalbard. This long-term survey, the first of its kind, part of the European Union's marine biodiversity program BIOMARE, so far assembled over 1,400 marine taxa from an area of approximately 50 km² and depths ranging from 0 to 280 m (http://www.iopan.gda.pl/projects/biodaff/). The estimated number of species, assessed from species accumulation curves, shows near completeness for single taxa like Mollusca (Wlodarska-Kowalczuk 2007), but substantial gaps for other taxa like minute Crustacea. Altogether, more than 2,000 metazoan species are expected to be identified in this small coastal Arctic area. The number of families of Polychaeta, for example, is a good indicator of marine species diversity for soft bottom Arctic benthos (Wlodarska-Kowalczuk & Kedra 2007). This implies that, at least for Hornsund, species richness of a single, well-known taxon might be an indicator for general species richness of the area.

New records of known species are at least as important as new species discoveries. Recent intense taxonomic study in the Chukchi Sea added over 300 species to the Sirenko (2001) inventory, doubling the number of known species since Ushakov (1952) (Sirenko & Vassilenko 2009). The recent additions were primarily in groups such as Foraminifera, Polychaeta, and Mollusca, whereas other groups such as Plathelminthes, Nematelminthes, and Harpacticoida are still poorly studied. New records for the Canada Basin relative to the Sirenko (2001) list include at least 40 benthic species, mainly polychaetes from one expedition, 21 of which were not listed to occur anywhere in the Arctic (MacDonald *et al.* 2010). Reasons for new records may be previous poor sampling or actual range extensions possibly related to climate warming (Mecklenburg *et al.* 2007; Sirenko & Gagaev 2007).

In addition to traditional species identifications and descriptions, ArcOD has contributed to the international Barcoding effort. Molecular "barcoding" uses a short DNA sequence from the cytochrome *c* oxidase mitochondrial

Fig. 10.5

Examples of Arctic benthos. (**A**) Sea star, *Ctenodiscus crispatus* (5 cm across). (**B**) Sea cucumber *Kolga hyalina* (about 2 cm long). (**C**) A new bryozoan species, *Callopora weslawski*. Photographs: A and B, B. Bluhm, University of Alaska Fairbanks; C, P. Kuklinski, Institute of Oceanology Polish Academy of Sciences.

region (MtCOI) as a molecular diagnostic for species-level identification (Hebert *et al.* 2003). Within the microbes, metagenomics and pyrosequencing are additionally applied (Sogin *et al.* 2006). Conservative estimates of the number

of distinct Arctic bacteria are now approximately 1,500 (D. Kirchman *et al.*, unpublished observations) and approximately 700 for the Archaea (Galand *et al.* 2009) in both surface and deep waters. At present, extrapolating these estimates to the various water masses presenting the entire Arctic has large uncertainty, but 4,500–45,000 types of Eubacteria, 500–5,000 types of Archaea, and 450–4,500 eukaryotic protists might exist in the Arctic (C. Lovejoy, personal communication). Viral diversity still remains largely unknown, but first inventories are underway for Svalbard (B. Wrobel, personal communication).

Within the metazoan zooplankton, Bucklin *et al.* (2010) sequenced 41 species, including cnidarians, arthropod crustaceans, chaetognaths, and a nemertean (Table 10.3). Overall, MtCOI barcodes accurately discriminated known species of 10 different taxonomic groups of Arctic Ocean holozooplankton. Work continues on building a comprehensive DNA barcode database for the Arctic holozooplankton in conjunction with the Census of Marine Zooplankton (see Chapter 13).

Within the Arctic benthos, over 300 species from 96 families were barcoded (C. Carr, personal communication; S. Mincks, personal communication), mostly polychaete (116) and amphipod species (63) (Table 10.3). For several morphological species, several unique haplotypes were found that could represent different species based on the molecular evidence (C. Carr, personal communication).

Within the fish, 93 species were barcoded from the North Pacific, the Aleutians, and the northern Bering and Chukchi Seas (Mecklenburg & Mecklenburg 2008) (Table 10.3; more in progress). Results supported the distinction between some species whose validity had been questioned, whereas other accepted species appear to be synonymous (Mecklenburg & Mecklenburg 2008). The method has also linked juvenile stages with the adults of the species, which previously had not been recognized as such.

Ongoing collaboration with the Census of Antarctic Marine Life (see Chapter 11) seeks to determine if bipolar species are truly bipolar based on MtCOI. Sequences for other target regions have also been published to help aid and resolve the separation of sibling species (see, for example, Lane *et al.* 2008), and to resolve haplotype structure within populations (Nelson *et al.* 2009).

10.3.2 Regional inventories: the Chukchi Sea and adjacent Canada Basin

Two expeditions in 2002 and 2005 aimed at improving the biological baseline of the Canada deep-sea Basin, one of the least explored regions in the Arctic Ocean (Gradinger & Bluhm 2005; Bluhm *et al.* 2010a). Although biomass and abundance of the sea ice meiofauna (mainly

Table 10.3

Arctic marine taxa barcoded under the ArcOD umbrella.

Taxon	Number of species barcoded	Number of families barcoded	Investigators/reference
Cnidaria: Hydrozoa	6 (pelagic)	6	Bucklin *et al.* 2010
Cnidaria: Scyphozoa	1	1	Bucklin *et al.* 2010
Cnidaria: Anthozoa	2	2	S. Hardy Mincks, personal communication
Nemertea	1 (pelagic)	1	Bucklin *et al.* 2010
Polychaeta	152 (benthic)	26	C. Carr & A. Smith, personal communication
Mollusca: Bivalvia	7	6	S. Hardy Mincks, personal communication
Mollusca: Gastropoda	12	6	S. Hardy Mincks, personal communication
Mollusca: Polyplacophora	1	1	S. Hardy Mincks, personal communication
Copepoda	28 (pelagic)	12	Bucklin *et al.* 2010
Amphipoda	5 (pelagic) 63 (benthic)	16	Bucklin *et al.* 2010; C. Carr & A. Smith, personal communication; S. Hardy Mincks, personal communication
Decapoda	1 (pelagic) 5 (benthic)	1 3	Bucklin *et al.* 2010 S. Hardy Mincks, personal communication
Euphausiacea	1	1	Bucklin *et al.* 2010
Chaetognatha	2	2	Bucklin *et al.* 2010
Urochordata, Ascidiacea	1	1	S. Hardy Mincks, personal communication
Echinodermata: Asteroidea	13	5	S. Hardy Mincks, personal communication
Echinodermata: Ophiuroidea	2	2	S. Hardy Mincks, personal communication
Echinodermata: Holothuroidea	2	2	S. Hardy Mincks, personal communication
Pisces	92	27	C.W. Mecklenburg and D. Steinke, personal communication
Total	397	121	

Acoela, Nematoda, and Harpacticoida) (Fig. 10.6) and amphipods was similar to records from other Arctic offshore regions (Gradinger *et al.* 2005, 2010), abundances were significantly higher along deep-reaching keels of sea ice ridges (Gradinger *et al.* 2010). Community structure of net-caught zooplankton, mainly copepods (Fig. 10.6), was distinctly depth-stratified, with composition comparable to other Arctic basins, except for the several Pacific expatriates present (Hopcroft *et al.* 2005; Kosobokova & Hopcroft 2010). Assemblages of gelatinous zooplankton were also depth-stratified with shallower stations dominated by siphonophores and ctenophores and deeper stations by medusae (Raskoff *et al.* 2005, 2010). Large

predatory scyphomedusae in the upper 100 m were dominant in the chlorophyll maximum layer, where copepod biomass was also highest. Smaller cnidarian and ctenophore species occurred immediately underneath the sea ice (Purcell *et al.* 2009).

Abundance, biomass, and diversity of benthic macrofauna (approximately 100 taxa) declined with increasing water depth and clustered into groups characterized by depth and location with overall low abundances and biomass similar to findings from the Eurasian Arctic deep sea (Bluhm *et al.* 2005; MacDonald *et al.* 2010). High abundance of the sea cucumber *Kolga hyalina* (Fig. 10.6) characterized a suspected pockmark on the Chukchi Cap

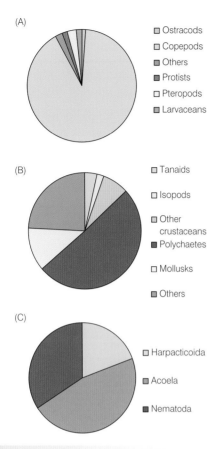

Fig. 10.6

Composition of Arctic net-caught zooplankton **(A)**, benthic macrofauna communities **(B)**, and sea ice meiofauna **(C)**. Example is from the Canada Basin (2005). Data from Gradinger *et al.* (2010), Kosobokova & Hopcroft (2010), and MacDonald *et al.* (2010).

Small demersal fishes (see Mecklenburg *et al.* (2007) for taxonomy) and ichthyoplankton on the Chukchi Shelf also formed distinct regional assemblages related to hydrographical features and sediment type (Norcross *et al.* 2010).

High abundances and biomass of macrobenthos north-west of Bering Strait were dominated by the bivalve *Macoma calcarea* (Sirenko & Gagaev 2007) and were linked to local hydrography retaining the larval pool. Benthic epifaunal biomass was dominated by echinoderms and crustaceans and represented in six distinct assemblages, separated largely based on substrate type and latitude with less influence by indices of food availability (Bluhm *et al.* 2009). Comparisons with previous studies in the region suggest an increase in overall epibenthic biomass since 1976. Regional differences in mean stable isotopic signatures in the benthic food web were mainly driven by the isotopically depleted particulate organic matter source in the Alaska Coastal Water (Iken *et al.* 2010).

10.3.3 Natural variability versus climate warming

Concern over global biodiversity loss is widespread, and Arctic biodiversity is believed to be changed by climate warming (Bluhm & Gradinger 2008). Although Arctic endemic taxa may be endangered, overall species numbers might increase with species-rich warm-water communities thriving in the region. Only a few long-term studies, some with ArcOD involvement, have been performed: several in Svalbard including a 30-year-long study of the rocky sublittoral (Beuchel *et al.* 2006), a 10-year-long zooplankton survey (Hop *et al.* 2006), and a 10-year-long study of the soft bottom (Kedra *et al.* 2009). Repeated sampling efforts included a shelf megafauna survey at Svalbard resurveyed after 50 and 100 years (Dyer *et al.* 1984), occurrence of Decapoda in Isfjorden investigated after 50 and 100 years (Berge *et al.* 2009), a soft bottom survey in VanMijen Fjord after 20 years (Renaud *et al.* 2007), and a Svalbard intertidal survey repeated after 20 years (Wiktor & Węsławski 2008). The results of those surveys demonstrate the high stability of the presence of the species pool in coastal–fjordic waters and very drastic interannual changes in the occurrence of single species. With ongoing warming, warmer-water species replaced cold-water species, but played the same role in the ecosystem. Examples of such species pairs include species in the genera *Sclerocrangon*, *Calanus*, *Themisto*, *Gammarus*, and *Limacina* (Węsławski *et al.* 2008).

Low Arctic biodiversity is usually associated with high population density of a few species. The charismatic icons of the Arctic are huge numbers of seabirds, seals, and walruses, unmatched anywhere on Earth during the feeding season. The underlying reason is the large size of polar marine herbivores (copepods, krill, pteropods) and their high lipid content (Pasternak *et al.* 2001). Exceptionally

(MacDonald *et al.* 2010). Only six putative demersal fish species were observed from ROV imagery (Stein *et al.* 2005). During opportunistic visual surveys, six and seven marine mammal species were encountered in 2002 and 2005, respectively, and 16 bird species were encountered in 2005 with highest sighting numbers related to specific oceanographic features (Harwood *et al.* 2005; Moore *et al.* 2010). A long (stable-isotope-based) food web of four trophic levels points towards low food availability and a high degree of organic matter reworking (Iken *et al.* 2005).

The Russian–American Long-term Census of the Arctic (RUSALCA) conducts long-term research relevant to climate change in Bering Strait and the Chukchi Sea (Bluhm *et al.* 2009). Zooplankton communities were represented by six assemblages coinciding with prevalent thermohaline water mass characteristics (Hopcroft *et al.* 2008). One of the numerically dominant copepod genera, *Pseudocalanus*, was represented by three species with distinct spatial distribution patterns, although their weight-specific egg production rates were similar (Hopcroft & Kosobokova 2010).

efficient and often short polar food chains (diatom–krill–whale) are now under change because of increasing inflow of warm, Atlantic waters that bring smaller species of herbivores. This leads directly to a change in food web structure as primary production is going to be dissipated among several small, fast-growing subarctic species (Węsławski et al. 2007, 2008) and a higher contribution of pelagic versus benthic secondary production (Carroll & Carroll 2003).

To separate this temporal from spatial variability one needs to know the scale of patch sizes of equal biodiversity. Several recent studies, testing this problem in nested sampling approaches (see, for example, Wlodarska-Kowalczuk & Weslawski 2008), demonstrated that on even, flat soft-sediment seabed, the patch size of uniform biodiversity was several hundreds of meters in diameter. The uniformity was lower (diversity is more patchy) in undisturbed shelf areas, and higher (diversity is low and even) at coastal sites under the influence of siltation and strong glacial sedimentation. Similar analyses were performed for Arctic sea ice, which supports mosaic and patchy distribution of organisms, based on local differences in snow cover, ice thickness, undersurface of ice, sediment load, etc. (Wiktor & Szymelfenig 2002; Gradinger et al. 2009, 2010). Low diversity of marine habitats appears to contribute to the overall low species richness in the Arctic.

10.3.4 Biogeography

Global changes of climate, water mass circulation, and geomorphology in the Pliocene and Pleistocene modified the composition and distribution of Arctic benthic fauna over time. In ArcOD we evaluated the different origins of modern Arctic fauna as a whole, as well as of faunas of certain large regions and bathymetric zones in the Arctic Ocean. Faunistic elements recognized in the modern Arctic benthos include (1) faunas originating from the North Pacific and (2) North Atlantic, (3) deep-sea cosmopolitans, and (4) endemic species of autochthonous (local) origin. The opening of Bering Strait approximately 5.3 million years ago resulted in intensive colonization of the Arctic Basin from the North Pacific. Formation of the warm Gulf Stream moved distribution limits of many boreal species northward. Geomorphological and hydrological changes of the Faeroe–Iceland Rise opened dispersal pathways into the Arctic Basin to the North Atlantic deep-sea fauna that in turn has strong links to the Antarctic. At the same time, changes in the Arctic Basin stimulated local species radiation.

Although a comprehensive review will be published by A.N. Mironov and A.V. Gebruk (editors), some example results are given here. For Arctic shallow-water asteroid fauna (A.B. Dilman, unpublished observations) the number of species of Atlantic origin exceeds that of Pacific origin. The only exception is the Chukchi Sea, where more species are in common with the Pacific. At the same time, however,

asteroid species dominating on the Arctic shelves belong to genera that dispersed from the Pacific Ocean. The ratio of species of Pacific origin decreases from the Barents Sea towards the Laptev Sea, but then increases in the East-Siberian Sea and the Chukchi Sea. The East-Siberian Sea acts as a barrier for the dispersal of species, which can be seen in various biogeographical indicators, such as species richness, biogeographical structure of fauna, patterns of vertical distribution, and the ratio of cold-temperature to warm-temperature species (A.N. Mironov & A.B. Dilman, unpublished observations).

10.3.5 Beyond the Arctic and ArcOD: Arctic–Antarctic comparisons

The common textbook notion is that biodiversity in Arctic seas is low compared with the Antarctic and particularly compared with temperate and warm waters. Although this is supported by higher total species numbers in warmer seas, it is less supported when comparing species numbers in specific comparable habitats, or within comparable taxonomic groups. Kendall and Aschan (1993), who analyzed soft-bottom benthos from tropical, temperate, and Arctic sites, found almost identical values for indices of diversity at all sites when including the same type of sediment and water depths. More recently, polychaete diversity was found to be equal at an evolutionary old Antarctic site and evolutionary young Arctic site (Wlodarska-Kowalczuk et al. 2007). This implies that differences in total species richness between areas are driven by habitat diversity. In the Arctic, biogenic reefs, caves, and deep rocky structures are rare, absent or, in some cases, un(der)-sampled such as the deep-sea Arctic benthos. Complete lists on overall species richness for the Arctic and Antarctic are still being compiled and numbers for metazoan species currently range around 8,200 for the Antarctic (www.scarmarbin.be/rams.php?p=stats) and about 6,000 for the Arctic (Table 10.2) (see Chapter 11). Extensive Arctic–Antarctic comparisons are ongoing in collaboration with the Census of Antarctic Marine Life.

10.4 Unknown Aspects

Arctic regions contain a variety of complex habitats that are difficult to access and historically have not been in the focal point of political and scientific interests. Despite a recent increase in overall interest, numerous Arctic research cruises, well-equipped field-stations, drifting stations, and easier access to many areas (because of substantial shrinkage of the ice cover), certain geographic areas, taxa, and habitats still remain poorly sampled. The previous lack of interest is now leading to uncertainties about the extent of ongoing changes. The ecological consequences and implications of ongoing change to biodiversity can never be understood if we do not fully understand the status quo with its regional and temporal variability.

10.4.1   Taxonomic gaps: microbes

Contrary to earlier opinions (Pomeroy *et al.* 1990), recent findings indicate an active microbial contribution in the Arctic (Kirchman *et al.* 2009) (see Chapter 12). We now know that viruses (Le Romancer *et al.* 2007; B. Wrobel, personal communication), Archaea, Eubacteria, and protists (Lovejoy *et al.* 2006; C. Lovejoy *et al.*, unpublished observations) thrive in all Arctic habitats, from nearshore to deep water, with gaps in their regional patterns, biodiversity inventory, and physiological adaptations. Arctic and Antarctic eubacterial communities are distinctly different (Bano *et al.* 2004), and Arctic bacterial diversity in Arctic samples is lower than Antarctica (Junge *et al.* 2002; Fuhrman *et al.* 2008). Microhabitats (Meiners *et al.* 2008) add to bacterial diversity together with unique habitats like naturally occurring methane and oil seeps along slopes and ridges (LaMontagne *et al.* 2004). However, most bacteria have never been cultured to study their physiology (Ducklow *et al.* 2007) and we do not know how microbial diversity connects to the Arctic food web structures.

Although species diversity for protists with hard structures like diatoms has been studied in Arctic waters for decades (Horner 1985; M. Poulin *et al.*, unpublished observations), only limited information on species inventories and abundance is available for other flagellated taxa. Cyanobacteria appear to be less relevant in offshore regions, but more abundant close to shore (Waleron *et al.* 2007) whereas flagellated eukaryotes (for example prasinophytes) occur in high abundances with apparently low diversity (Lovejoy *et al.* 2006).

10.4.2   Regional gaps

Despite considerable sampling efforts in the past decade, the Arctic deep-sea remains undersampled for all realms (Fig. 10.1). Biodiversity inventories of other deep-sea areas have revealed increased species numbers with increasing sampling effort (Brandt *et al.* 2007). Based on species discoveries during two recent deep-sea cruises and a lack of an asymptote in species accumulation curves (B.A. Bluhm *et al.*, unpublished observations), we estimate that possibly hundreds of species (excluding microbes) await discovery in the Arctic deep sea.

Of the shelf seas, the East Siberian Sea is the most understudied in terms of biodiversity (Fig. 10.1) whereas other Russian seas were intensively sampled. Information is also scarce in the Canadian Archipelago and northern Greenland, partly related to the typically heavy ice cover. Biodiversity work there has primarily concentrated around field stations and research institutes, like in Disco Bay, northeast Greenland or Resolute Bay (Conover & Huntley 1991; Michel *et al.* 2006).

10.4.3   Poorly explored habitats

Examples of underexplored habitats include the deep-sea ridge systems that extend thousands of kilometers across the Arctic sea floor. Although biodiversity on both ridge sides may not be that different (Schewe & Soltwedel 1999; K.N. Kosobokova *et al.*, unpublished observations), faunal diversity and densities may vary greatly between ridge tops and sides (Kosobokova & Hirche 2000), possibly related to variability in the overlying nutrient concentrations and primary production. The scarceness of studies, however, precludes general conclusions on the biodiversity at any of the prominent Arctic ridge systems.

Within the realm of the sea ice, sea-ice pressure ridges house major unknowns with respect to their biology. Ridges form when ice piles up under pressure, reaching drafts greater than 20 m even in summer (Eicken *et al.* 2005). Based on ArcOD findings (Gradinger *et al.* 2010), we propose that pressure ridges will be especially relevant for the survival of sea-ice-related invertebrates over the coming decades in areas of dramatic sea ice loss.

Biodiversity is probably better studied in nearshore areas around field stations or logistics centers than anywhere else in the Arctic, but is not studied at all in other nearshore locations, because larger research vessels cannot venture into shallow water. The highly localized geographic coverage relative to transect-oriented station grids typically used elsewhere is problematic because even relatively nearby locations of similar habitats can have rather different biotic inventories (Iken & Konar 2009). Projects like an inventory of many coastal Chukchi Sea sites in 2010 (S. Jewett & D. Dasher, personal communication) will improve the situation.

Arctic seamounts and pockmarks are poorly mapped and inventoried. An ArcOD survey of a suspected pockmark feature on the Chukchi Plateau showed elevated densities of megafauna but no signs of seepage (MacDonald *et al.* 2010), but it is unclear if this observation can be generalized.

10.4.4   Underexplored temporal variability

Environmental conditions in the Arctic are extremely variable and natural variability is overlaid by – and often difficult to separate from – long-term climate change. As the most climatically sensitive region of the Northern hemisphere, the Arctic has experienced changes exceeding the natural variability, including shrinkage of the sea ice cover and thickness, increased precipitations and river run off, seasonal warming and other atmospheric changes, increased ocean mixing, wave generation, and coastal flooding (Walsh

2008). Yet, few time series of biological variables have been collected because of the difficulty associated with re-sampling communities throughout the year and/or over multiple years or even decades. Seasonal changes in algal and faunal abundance and diversity are most dramatic in sea ice and the water column (Story-Manes & Gradinger 2009; Makabe *et al.* 2010), with dampened variability in benthic communities because of their typically slower growth rates and longer lifespans. Interannual and inter-decadal alterations have been detected both in pelagic (Brodeur *et al.* 2002) and benthic habitats (Grebmeier *et al.* 2006).

Opposing trends, the difficulty to identify underlying causes clearly, the overall scarceness of even short time series, and the lack of complete inventories stress the need for the creation of integrated ecosystem biological observing systems as part of the global ocean observing system effort (section 10.5). Different ecosystem components respond to variability in different ways and should not be evaluated independently of each other. Some efforts underway in this direction are outlined in section 10.5.1.

10.5 Into the Future

10.5.1 Time series, networks, monitoring strategies

Long-term time series using a combination of both traditional and recent sampling technologies are essential to understand the changing Arctic. Future observational networks will need to consist of regional nodes, connected in a pan-Arctic scheme, as regional trends may differ. For example, the Chukchi and Beaufort Seas have experienced substantial thinning and reduction of the ice cover (Serreze *et al.* 2007), whereas less change (if any) occurred north of Greenland (Melling 2002).

No comprehensive pan-Arctic observational network is currently in place to assess changes in biodiversity. Several networks, however, are under discussion such as through the International Arctic Council, where ArcOD participates in the marine expert group within the Conservation of Arctic Flora and Fauna (CAFF) program (Vongraven *et al.* 2009). These could be operational as the first Census draws to a close in late 2010 with the caveat that little financial support will be in place at that time. On a national level, the Study on Environmental Arctic Change (SEARCH), an interagency US-based effort to understand system-scale change in the Arctic, has as one of their science-guiding questions "what changes in populations, biodiversity, key species, and living resources are associated with Arctic Change" (yet none of their 70 projects include "biodiversity" in the title). Some successful examples for national programs do exist. Any future network must be interdisciplinary in nature, linking changes in ocean physics and chemistry to biological patterns, in order to create meaningful predictions.

Key components within monitoring networks are the following:

- a series of long-term monitoring sites (biological observatories) situated around the Arctic shelves and in the deep basins (section 10.5.3);
- pan-Arctic surveys using largely autonomous methodologies (section 10.5.2);
- exploratory expeditions completing the Arctic species inventory and conducting new research that augments the observatories.

For biodiversity research, the strategy must be to first identify a set of indices to assess changes in biodiversity, and make the connections between those changes and potential stressors (Vongraven *et al.* 2009). Such indices could, for example, include biomass of key ecosystem components, abundance of dominant or commercially important taxa, traditional biodiversity indices, distribution ranges, or ratios of Arctic to sub-Arctic species. Process measurements should also be included, as organic carbon partitioning between pelagic and benthic food webs is predicted to shift towards pelagic-dominated food webs (Carroll & Carroll 2003; Bluhm & Gradinger 2008). Such food web shifts will cascade through all trophic levels, and could change the key species within the affected regions. Indices should provide the information required to make sound political decisions to mitigate negative impacts to the ecosystems and Arctic human populations.

In addition to continuing the few existing time series, new observatories should be placed in regions where the rates of change appear to be greatest (Chukchi Sea: Hopcroft *et al.* 2010; Barents Sea: Stempniewicz *et al.* 2007). From a logistical perspective, the relative proximity of these regions to major research ports, and their longer ice-free season, greatly simplifies access to them relative to the central Arctic. Areas that have historically been perpetually covered by ice remain worthy of basic exploratory survey because they are so poorly characterized.

10.5.2 Technological developments

A major impediment in understanding the biological impacts of Arctic change is owing to logistical constraints. Few icebreakers are capable of hosting multidisciplinary research teams and expeditions are major investments for funding agencies. International cooperation is needed to connect national efforts, to standardize techniques, and create an operational network with fast data transfer and delivery.

Technology should extend the short-term observations made by traditional expeditions to year-round observations, although the broadest taxonomic coverage will remain achievable only using ships and land-based stations. Moored video systems, or even ice-mounted webcams in nearshore waters, can collect information during seasons when access to the region of interest is difficult. Autonomous underwater vehicles with image capturing capabilities will extend our options to look at the distribution of larger taxa on scales of hundreds of kilometers, even in ice-covered waters (Dowdeswell *et al.* 2008). ROVs (Fig. 10.7) place an observer virtually into the deep-sea environment to survey the larger, rarer, and more mobile animals (Raskoff *et al.* 2005, 2010). Moored instrumentation, including *in situ* imaging flow cytometers, video plankton recorders, or fish-tracking networks, is now commercially available. Autonomous underwater vehicles and gliders equipped with listening devices might extend the range of acoustic curtains, and record the presence of marine mammals in ice-covered waters. The major hurdle appears not to be readiness of scientists or instrumentation, but sufficient logistic and financial commitment by Arctic nations to put such a network in place.

10.5.3　Forward look

ArcOD has contributed to the considerable recent progress regarding biodiversity of the Arctic. The most urgent current need now is to determine how the human footprint is affecting the Arctic Ocean's flora and fauna, and the overall properties of the ecosystem these organisms collectively shape. Current model projections suggest climatic changes may become even more exaggerated over the coming decades (Overland & Wang 2007). Despite the need for continued observation, there is as yet no strong indication that species-level surveys will be expanded in the Arctic. This leads to the situation where our understanding of long-term patterns will be built upon temporally and regionally fragmented observations often focused on specific ecosystem components or by reducing the Arctic's complexity down to biogeochemical units.

As an indirect consequence of ice reduction, the human footprint within the Arctic is already increasing in numerous ways (Gradinger 2009b). Ice-free Arctic seas are considered the future shipping corridor between the industrial centers in the North Pacific and Europe. Reduced summer sea ice will allow intensified resource exploration, as evidenced by the recent oil and gas lease sales on the US Chukchi Sea shelf, and expanding exploration on the Russian and Norwegian shelves. The opening of Arctic seas in summer could allow commercial extraction of living marine resources in areas currently untouched by commercial exploitation. Increased human presence will inevitably leave its trace in the marine environment in

Fig. 10.7

Traditional and modern sampling techniques used in ArcOD. (**A**) Sea ice sampling using an ice corer. (**B**) Water column and benthic sampling using an ROV that can be operated in pack ice. (**C**) Under-ice sampling by SCUBA divers in Arctic pack ice. Photographs: A and B, B. Bluhm; C, S. Harper, both University of Alaska Fairbanks.

the form of industrial contaminants, harmful species, and changes in food web structure through commercial harvests.

Using historical information and estimates of what might happen over the next century (see, for example, Bluhm & Gradinger 2008; Gradinger 2009b) we believe the following patterns are likely. The currently observed changes in species patterns and ecosystem functioning will continue over the next decade(s), likely at an accelerated rate with changes at all trophic levels. These changes will impact the use of Arctic living resources by humans, both commercial and subsistence, and may even lead to long-term biodiversity changes in the Pacific and North Atlantic Oceans as increased species exchange occurs across the Arctic. The Pacific diatom *Neodenticula seminae* has already been observed in the North Atlantic (Reid *et al.* 2007). Arctic endemic biota will likely be most negatively affected, whereas less ice and higher temperatures will allow sub-Arctic species to move northward. The design of observational networks documenting biological change should develop benchmarks against which to follow such change. A discussion and agreement on the biological components of an ocean observing network is long overdue.

Acknowledgments

We thank the many international contributors to ArcOD, listed at http://db.coml.org/community/ for their work in the field of marine biodiversity, compiled at http://db.coml.org/comlrefbase/. We acknowledge the support by a long list of funding bodies listed at www.comlsecretariat.org. An anonymous reviewer and the editor provided comments that improved this article. We particularly thank Jesse Ausubel, Vice President of Programs for the Alfred P. Sloan Foundation, for his sparkling enthusiasm, extraordinary vision, and encouraging support.

References

Auel, H. & Werner, I. (2003) Feeding, respiration and life history of the hyperiid amphipod *Themisto libellula* in the Arctic marginal ice zone of the Greenland Sea. *Journal of Experimental Marine Biology and Ecology* **296**, 183–197.

Bano, N., Ruffin, S., Ransom, B., *et al.* (2004) Phylogenetic composition of Arctic Ocean archaeal assemblages and comparison with Antarctic assemblages. *Applied and Environmental Microbiology* **70**, 781–789.

Berge, J., Renaud, P.E., Eiane, K., *et al.* (2009) Changes in the decapod fauna of an Arctic fjord during the last 100 year (1908–2007). *Polar Biology* **32**, 953–961.

Beuchel, F., Gulliksen, B. & Carroll, M.L. (2006) Long-term patterns of rocky bottom macrobenthic community structure in an Arctic fjord (Kongsfjorden, Svalbard) in relation to climate variability (1980–2003). *Journal of Marine Systems* **63**, 35–48.

Bilyard, G.R. & Carey, A.C. Jr. (1980) Zoogeography of western Beaufort Sea Polychaeta (Annelida). *Sarsia* **65**, 19–26.

Bluhm, B. & Gradinger, R. (2008) Regional variability in food availability for Arctic marine mammals. *Ecological Applications* **18**, S77–S96.

Bluhm, B.A., Iken, K., Mincks Hardy, S., *et al.* (2009) Community structure of epibenthic megafauna in the Chukchi Sea. *Aquatic Biology* **7**, 269–293.

Bluhm, B.A., Iken, K. & Hopcroft, R.R. (2010a) Observations and exploration of the Arctic's Canada Basin and the Chukchi Sea: the Hidden Ocean and RUSALCA expeditions. *Deep Sea Research II* **57**, 1–4.

Bluhm, B., Gradinger, R. & Schnack-Schiel, S. (2010b) Sea ice meio- and macrofauna. In: *Sea Ice* (eds. D. Thomas & G. Dieckmann), pp. 357–393. New York: Wiley-Blackwell.

Bluhm, B.A., MacDonald, I.R., Debenham, C., *et al.* (2005) Macro- and megabenthic communities in the high Arctic Canada Basin: initial findings. *Polar Biology* **28**, 218–231.

Booth, B.C. & Horner, R.A. (1997) Microalgae on the Arctic ocean section, 1994: species abundance and biomass. *Deep-Sea Research II* **44**, 1607–1622.

Brandt, A., Gooday, A.J., Brandao, S.N., *et al.* (2007) First insights into the biodiversity and biogeography of the Southern Ocean deep sea. *Nature* **447**, 307–311.

Brinkmeyer, R., Knittel, K., Juergens, J., Weyland, H., Amann, R. & Helmke, E. (2003) Diversity and structure of bacterial communities in Arctic versus Antarctic pack ice. *Applied and Environmental Microbiology* **69**, 6610–6619.

Brodeur, R.D., Sugisaki, H. & Hunt, G.L., Jr. (2002) Increases in jellyfish in the Bering Sea: implications for the ecosystem. *Marine Ecology Progress Series* **233**, 89–103.

Bucklin, A., Hopcroft, R.R., Kosobokova, K.N., *et al.* (2010) DNA barcoding of Arctic Ocean holozooplankton for species identification and recognition. *Deep-Sea Research II* doi:10.1016/j.dsr2.2009.08.005.

Carroll, M.L. & Carroll, J. (2003) The Arctic Seas. In: *Biogeochemistry of Marine Systems* (eds K.D. Black & G.B. Shimmield). pp. 127–156. Boca Raton, Florida: CRC Press.

Chaban, E.M. (2008) Opistobranchiate Mollusca of the orders Cephalaspidea, Thecosomoata and Gymnosomata (Mollusca, Ophistobranchia) of the Chukchi Sea and Bering Strain. Fauna and zoography of benthos of the Chukchi Sea. *Explorations of the Fauna of the Seas* **61**(69), 149–162. [In Russian.]

Connelly, T.L. (2008) *Biogeochemistry of benthic boundary layer zooplankton and particulate organic matter on the Beaufort Sea shelf.* PhD thesis, Memorial University of Newfoundland.

Conover, R.J. & Huntley, M. (1991) Copepods in ice-covered seas – distribution, adaptations to seasonally limited food, metabolism, growth patterns and life cycle strategies in polar seas. *Journal of Marine Systems* **2**, 1–41.

Deibel, D. & Daly K.L. (2007) Zooplankton processes in Arctic and Antarctic polynyas. In: *Polynyas Windows to the World* (eds. W.O. Smith, Jr. & D.G. Barber), pp. 271–322. Amsterdam: Elsevier.

Dowdeswell, J.A., Evans, J., Mugford, R., *et al.* (2008) Autonomous underwater vehicles (AUVs) and investigations of the ice–ocean interface: deploying the Autosub AUV in Antarctic and Arctic waters. *Journal of Glaciology* **54**, 61–672.

Ducklow, H.W., Baker, K., Martinson, D.G., *et al.* (2007) Marine pelagic ecosystems: the West Antarctic Peninsula. *Philosophical Transactions of the Royal Society of London B* **362**, 67–94.

Dyer, M.F., Cranmer, G.J., Fry, P.D., *et al.* (1984) The distribution of benthic hydrographic indicator species in Svalbard waters, 1978–1981. *Journal of the Marine Biological Association of the United Kingdom* **64**, 667–677.

Eicken, H., Gradinger, R., Gaylord, A., *et al.* (2005) Sediment transport by sea ice in the Chukchi and Beaufort Seas: increasing importance due to changing ice conditions? *Deep-Sea Research II* 52, 3281–3302.

Falkenhaug, I. & Saksaug, E. (1991) Prey composition and feeding rate of *Sagitta elegans* var. *arctica* (Chaetognatha) in the Barents Sea in early summer. *Polar Research* 10, 487–506.

Feder, H.M., Jewett S.C., & Blanchard A. (2005) Southeastern Chukchi Sea (Alaska) epibenthos. *Polar Biology* 28, 402–421.

Fuhrman, J.A., Steele, J.A., Hewson, I., *et al.* (2008) A latitudinal diversity gradient in planktonic marine bacteria. *Proceedings of the National Academy of Sciences of the USA*, 105, 774–7778.

Gagaev, S.Y. (2008) *Sigambra healyae* sp. n., a new species of polychaete (Polychaeta: Pilargidae) from the Canada Basin of the Arctic Ocean. *Russian Journal of Marine Biology* 34, 73–75.

Gagaev, S.Y. (2009) *Terebellides irinae* sp. n. – a new species of the genus *Terebellides* (Polychaeta, Terebellidae) from the Arctic Basin. *Russian Journal of Marine Biology* 35, 474–478.

Galand, P.E., Casamayor, E.O., Kirchman, D.L., *et al.* (2009) Unique archaeal assemblages in the Arctic Ocean unveiled by massively parallel tag sequencing. *International Society of Microbial Ecology Journal* 3, 860–869.

Gosselin, M., Levasseur, M., Wheeler, P.A., *et al.* (1997) New measurements of phytoplankton and ice algal production in the Arctic Ocean. *Deep-Sea Research II* 44, 1623–1644.

Gould, P.J., Forsell, D.J. & Lensink, C.J. (1982) *Pelagic distribution and abundance of seabirds in the Gulf of Alaska and eastern Bering Sea*. US Fish and Wildlife Service Report FWS/OBS-82/48. 294 pp.

Gradinger, R. (2002) Sea ice microorganisms. In: *Encyclopedia of Environmental Microbiology* (ed. G. E. Bitten). pp. 2833–2844. New York: Wiley.

Gradinger, R. (2009a) Sea ice algae: major contributors to primary production and algal biomass in the Chukchi and Beaufort Sea during May/June 2002. *Deep-Sea Research II* 56, 1201–1212.

Gradinger, R. (2009b) The changing Arctic sea ice landscape. In: *The Biology of Polar Seas* (eds. G. Hempel & I. Hempel). pp. 239–246. Bonn: Wirtschaftsverlag.

Gradinger, R.R. & Baumann, M.E.M. (1991) Distribution of phytoplankton communities in relation to the large-scale hydrographical regime in the Fram Strait. *Marine Biology* 111, 311–321.

Gradinger, R.R. & Bluhm, B.A. (2004) *In situ* observations on the distribution and behavior of amphipods and Arctic cod (*Boreogadus saida*) under the sea ice of the high Arctic Canadian Basin. *Polar Biology* 27, 595–603.

Gradinger, R. & Bluhm, B.A. (2005) Arctic Ocean exploration 2002. *Polar Biology* 28, 169–170.

Gradinger, R. & Nürnberg, D. (1996) Snow algal communities on Arctic pack ice floes dominated by *Chlamydomonas nivalis* (Bauer) Wille. *Proceedings of the NIPR Symposium on Polar Biology* 9, 35–43.

Gradinger, R.R., Meiners, K., Plumley, G., *et al.* (2005) Abundance and composition of the sea-ice meiofauna in off-shore pack ice of the Beaufort Gyre in summer 2002 and 2003. *Polar Biology* 28, 171–181.

Gradinger, R., Bluhm, B. & Iken, K. (2010) Arctic sea ice ridges – safe havens for sea ice fauna during periods of extreme ice melt? *Deep-Sea Research II* 57, 86–95.

Gradinger, R., Kaufman, M.R. & Bluhm, B.A (2009) The pivotal role of sea ice sediments for the seasonal development of nearshore Arctic fast ice biota. *Marine Ecology Progress Series* 394, 49–63.

Grebmeier, J.M. & Barry, J.P. (1991) The influence of oceanographic processes on pelagic–benthic coupling in polar regions: a benthic perspective. *Journal of Marine Systems* 2, 495–518.

Grebmeier, J.M., Cooper, L.W., Feder, H.M., *et al.* (2006) Ecosystem dynamics of the Pacific-influenced Northern Bering and Chukchi Seas in the Amerasian Arctic. *Progress in Oceanography* 71, 331–361.

Harwood, L.A., McLaughlin, F., Allen, R.M., *et al.* (2005) First-ever marine mammal and bird observations in the deep Canada Basin and Beaufort/Chukchi Seas: expeditions during 2002. *Polar Biology* 28, 250–253.

Hebert, P.D.N., Cywinska, A., Ball, S.L., *et al.* (2003) Biological identifications through DNA barcodes. *Proceedings of the Royal Society of London B* 270, 313–321.

Hop, H., Falk-Petersen, S., Svendsen, H., *et al.* (2006) Physical and biological characteristics of the pelagic system across Fram Strait to Kongsfjorden. *Progress in Oceanography* 71, 182–231.

Hopcroft, R.R. & Kosobokova, K.N. (2010) Distribution and production of *Pseudocalanus* species in the Chukchi Sea. *Deep-Sea Research II* 57, 49–56.

Hopcroft, R.R., Clarke, C., Nelson, R.J., *et al.* (2005) Zooplankton Communities of the Arctic's Canada Basin: the contribution by smaller taxa. *Polar Biology* 28, 197–206.

Hopcroft, R., Bluhm, B., & Gradinger, R. (2008) *Arctic Ocean synthesis: analysis of climate change impacts in the Chukchi and Beaufort Seas with strategies for future research*. Final Report to North Pacific Research Board, Anchorage, 182 pp.

Hopcroft, R.R., Kosobokova, K.N. & Pinchuk, A.I. (2010) Zooplankton community patterns in the Chukchi Sea during summer 2004. *Deep-Sea Research II* 57, 27–39.

Horner, R. (1981) *Beaufort Sea plankton studies*. NOAA Outer Continental Shelf Environmental Program, Final Report, 13, 65–314.

Horner, R. (1985) *Sea Ice Biota*. Boca Raton, Florida: CRC Press.

Ikävalko, J. & Gradinger, R. (1997) Flagellates and heliozoans in the Greenland Sea ice studied alive using light microscopy. *Polar Biology* 17, 473–481.

Iken, K. & Konar, B. (2009) *Essential Habitats in our Arctic front yard: nearshore benthic community structure*. Final Report to Alaska Sea Grant, Fairbanks, Alaska. 11 pp.

Iken, K., Bluhm, B.A. & Gradinger, R. (2005) Food web structure in the high Arctic Canada Basin: evidence from $\delta^{13}C$ and $\delta^{15}N$ analysis. *Polar Biology* 28, 238–249.

Iken, K., Bluhm, B.A. & Dunton, K. (2010) Benthic food web structure serves as indicator of water mass properties in the southern Chukchi Sea. *Deep-Sea Research II* 57, 71–85.

Jakobsson, M., Grantz, A., Kristoffersen, Y., *et al.* (2004) Physiography and bathymetry of the Arctic Ocean. In: *The Organic Carbon Cycle in the Arctic Ocean* (eds. R. Stein & R. MacDonald). pp. 1–6. Berlin: Springer.

Junge, K., Imhoff, F., Staley, T., *et al.* (2002) Phylogenetic diversity of numerically important Arctic sea-ice bacteria cultured at subzero temperature. *Microbial Ecology* 43, 315–328.

Kedra, M., Włodarska-Kowalczuk, M. & Węsławski, J.M. (2009) Decadal change in macrobenthic soft-bottom community structure in a high Arctic fjord (Kongsfjorden, Svalbard). *Polar Biology* 33, 1–11.

Kendall, M.A. & Aschan, M. (1993) Latitudinal gradients in the structure of macrobenthic communities: a comparison of Arctic, temperate and tropical sites. *Journal of Experimental Marine Biology and Ecology* 172, 157–169.

Kirchman, D.L., Hill, V., Cottrell, M.T., *et al.* (2009) Standing stocks, production, and respiration of phytoplankton and heterotrophic bacteria in the western Arctic Ocean. *Deep Sea Research II* 56, 1237–1248.

Kosobokova, K.N. (1980) Seasonal variations in the vertical distribution and age composition of *Microcalanus pygmaeus, Oithona similis, Oncaea borealis* and *O. notopus* populations in the central Arctic basin. *Biologiya Tsentral'nogo Arkicheskogo Basseyna*. Moscow: Nauka, pp. 167–182

Kosobokova, K. & Hirche, H.J. (2000) Zooplankton distribution across the Lomonosov Ridge, Arctic Ocean: species inventory, biomass and vertical structure. *Deep-Sea Research I* 47, 2029–2060.

Kosobokova, K.N. & Hopcroft, R.R. (2010) Diversity and vertical distribution of mesozooplankton in the Arctic's Canada Basin. *Deep-Sea Research II* 57, 96–110.

Kotwicki, L. & Fiers, F. (2005) *Paracrenhydrosoma oceaniae*, new species, Harpacticoida, Arctica, Fjord, Svalbard. *Annales Zoologici* 55, 467–475.

Kröncke, I. (1998) Macrofauna communities in the Amundsen Basin, at the Morris Jesup Rise and at the Yermak Plateau (Eurasian Arctic Ocean). *Polar Biology* 19, 383–392.

Kuklinski, P. & Taylor, P.D. (2006) A new genus and some cryptic species of Arctic and boreal calloporid cheilostome bryozoans. *Journal of the Marine Biological Association of the United Kingdom* 86, 1035–1046.

Kuklinski, P. & Taylor, P.D. (2008) Arctic species of the cheilostome bryozoan *Microporella*, with a redescription of the type species. *Journal of Natural History* 42, 2893–2906.

Laakmann, S., Kochzius, M. & Auel, H. (2009) Ecological niches of Arctic deep-sea copepods: vertical partitioning, dietary preferences and different trophic levels minimize inter-specific competition. *Deep Sea Research I* 56, 741–756.

Lane, P.V.Z., Llinás, L., Smith, S.L., *et al.* (2008) Zooplankton distribution in the western Arctic during summer 2002: hydrographic habitats and implications for food chain dynamics. *Journal of Marine Research* 70, 97–133.

LaMontagne, M.G., Leifer, I., Bergmann, S., *et al.* (2004) Bacterial diversity in marine hydrocarbon seep sediments. *Environmental Microbiology* 6, 799–808.

Le Romancer, M., Gaillard, M., Geslin, C. *et al.* (2007) Viruses in extreme environments. *Reviews in Environmental Science and Biotechnology* 6, 17–31.

Lovejoy, C., Massana, R. & Pedrós-Alió, C. (2006) Diversity and distribution of marine microbial eukaryotes in the Arctic Ocean and adjacent seas. *Applications in Environmental Microbiology* 72, 3085–3095.

MacDonald, I.R., Bluhm, B., Iken, K., *et al.* (2010). Benthic macro-faunal and megafaunal assemblages in the Arctic deep-sea Canada Basin. *Deep-Sea Research II* 57, 136–152.

Makabe, R., Hattori, H., Sampei, M., *et al.* (2010) Regional and seasonal variability of zooplankton collected using sediment traps in the southeastern Beaufort Sea, Canadian Arctic. *Polar Biology* 33, 257–270.

Mecklenburg, C.W., Stein, D.L., Sheiko, B.A. *et al.* (2007) Russian–American long-term census of the Arctic: benthic fishes trawled in the Chukchi Sea and Bering Strait, August 2004. *Northwestern Naturalist* 88, 168–187.

Mecklenburg, C.W. & Mecklenburg, T.A. (2008) *Arctic marine fish distribution and taxonomy*. Minigrant report to Arctic Ocean Diversity Census of Marine Life Project, Auke Bay, Alaska, 3 pp.

Meiners, K., Krembs, C. & Gradinger, R. (2008) Exopolymer parti-cles: microbial hotspots of enhanced bacterial activity in Arctic fast ice (Chukchi Sea). *Aquatic Microbial Ecology* 52, 195–207.

Melling, H. (2002) Sea ice of the northern Canadian Arctic Archi-pelago, *Journal of Geophysical Research C*, 107, 3181.

Michel, C., Ingram, R.G. & Harris, L.R. (2006) Variability in oceanographic and ecological processes in the Canadian Arctic Archipelago. *Progress In Oceanography* 71, 379–401.

Moore, S.E., Staffort, K.M. & Munger, L.M. (2010) Acoustic and visual surveys for bowhead whales in the western Beaufort and far northeastern Chukchi Seas. *Deep-Sea Research II* 57, 153–157.

Mumm, N. (1991) On the summerly distribution of mesozooplankton in the Nansen Basin, Arctic Ocean. *Reports on Polar Research* 92, 1–173. [In German.]

Naidu, A.S. (1988) *Marine surficial sediments*. Section 1.4 *in* Bering, Chukchi and Beaufort Seas: coastal and ocean zones strategic assess-ment data atlas. Washington, DC: NOAA Strategic Assessment Branch, Ocean Assessment Division.

Nelson, R.J., Carmack, E.C., McLaughlin, F.A., *et al.* (2009) Penetra-tion of Pacific zooplankton into the western Arctic Ocean tracked with molecular population genetics. *Marine Ecology Progress Series* 381, 129–138.

Nöthig, E.M., Okolodkov, Y., Larionov, V.V., *et al.* (2003) Phyto-plankton distribution in the inner Kara Sea: a comparison of three summer investigations. In: *Siberian River Run-Off in the Kara Sea* (eds. R. Stein, K. Fahl, D. K. Fuetterer, *et al.*). pp. 163–184. Amsterdam: Elsevier.

Norcross, B.L., Holladay, B.A., Busby, M.S., *et al.* (2010) Demersal and larval fish assemblages in the Chukchi Sea. *Deep-Sea Research II* 57, 57–70.

Overland, J.E. & Wang, M. (2007) Future regional Arctic Sea ice declines. *Geophysical Research Letters* 34, L17705.

Pasternak, A., Arashkevich, E., Tande, K., *et al.* (2001) Seasonal changes in feeding, gonad development and lipid stores in *Calanus finmarchicus* and *C. hyperboreus* from Malangen, northern Norway. *Marine Biology* 138, 1141–1152.

Perovich, D.K., Andreas, E.L., Fairall, C.W., *et al.* (1999) Year on ice gives climate insights. *EOS Transactions American Geophysical Union* 80, 481–486.

Perovich, D.K., Nghiem, S.V., Markus, T., *et al.* (2007) Seasonal evolution and interannual variability of the local solar energy absorbed by the Arctic sea-ice–ocean system. *Journal of Geophysi-cal Research* 112, C03005, 1–13.

Piepenburg, D., Chernova, N.V., von Dorrien, C.F., *et al.* (1996) Megabenthic communities in the waters around Svalbard. *Polar Biology*, 16, 431–446.

Piraino, S., Bluhm, B.A., Gradinger, R., *et al.* (2008) *Sympagohydra tuuli* gen. nov. et sp. nov. (Cnidaria: Hydrozoa) a cool hydroid from the Arctic sea ice. *Journal of the Marine Biological Association of the United Kingdom* 88, 1637–1641

Pomeroy, L.R., Macko, S.A., Ostrom, P.H., *et al.* (1990) The microbial food web in Arctic seawater: Concentration of dissolved free amino acids and bacterial abundance and activity in the Arctic Ocean and in Resolute Passage. *Marine Ecology Progress Series* 61, 31–40.

Purcell, J.E., Hopcroft, R.R., Kosobokova, K.N., *et al.* (2009) Distri-bution, abundance, and predation effects of epipelagic ctenophores and jellyfish in the western Arctic Ocean. *Deep-Sea Research II* doi:10.1016/j.dsr2.2009.08.011.

Quillfeldt, C.H.v., Ambrose, W.G., Jr. & Clough, L.M. (2003) High number of diatom species in first-year ice from the Chukchi Sea. *Polar Biology* 26, 808–816.

Raskoff, K.A. (2010) *Bathykorus bouilloni*: a new genus and species of deep-sea jellyfish from the Arctic Ocean (Hydrozoa, Nar-comedusae, Aeginidae). *Zootaxa* 2361, 57–67.

Raskoff, K.A., Purcell, J.E. & Hopcroft, R.R. (2005) Gelatinous zoo-plankton of the Arctic Ocean: in situ observations under the ice. *Polar Biology* 28, 207–217.

Raskoff, K.A., Hopcroft R.R., Kosobokova K.N., *et al.* (2010) Jellies under ice: ROV observations from the Arctic 2005 hidden ocean expedition. *Deep-Sea Research II* 57, 111–126.

Reid, P.C., Johns, D.G., Edwards, M. *et al.* (2007) A biological con-sequence of reducing Arctic ice cover: arrival of the Pacific diatom *Neodenticula seminae* in the North Atlantic for the first time in 800 000 years. *Global Change Biology* 13, 1910–1921.

Renaud, P.E., Włodarska-Kowalczuk, M., Trannum, H., *et al.* (2007) Multidecadal stability of benthic community structure in a high-Arctic glacial fjord (van Mijenfjord, Spitsbergen). *Polar Biology* 30, 295–305.

Rodriguez, E., Lopez-Gonzalez, P.J. & Daly, M. (2009) New family of sea anemones (Actiniaria, Acontiaria) from deep polar seas. *Polar Biology* 32, 703–717.

Rogacheva, A.V. (2007) Revision of the Arctic group of species of the family Elpidiidae (Elasipodida, Holothuroidea). *Marine Biology Research* 3, 376–396.

Rothrock, D., Yu, Y. & Maykut, G. (1999) The thinning of the Arctic ice cover. *Geophysical Research Letters* 26, 3469–3472.

Sakshaug, E. (2004) Primary and secondary production in the Arctic Seas. In: *The organic carbon cycle in the Arctic Ocean* (eds. R. Stein & R. MacDonald). pp. 57–82. Berlin: Springer.

Schewe, I. & Soltwedel, T. (1999) Deep-sea meiobenthos of the central Arctic Ocean. *International Review in Hydrobiology* 86, 317–335.

Schewe, I. & Soltwedel, T. (2003) Benthic responses to ice-edge-induced particle flux in the Arctic Ocean. *Polar Biolog* 26, 610–620.

Serreze, M.C., Holland, M.M. & Stroeve, J. (2007) Perspectives on the Arctic's shrinking sea-ice cover. *Science* 315, 1533–1536.

Sherr, E.B., Sherr, B.F. & Fessenden, L. (1997) Heterotrophic protists in the central Arctic Ocean. *Deep-Sea Research II* 48, 1665–1682.

Siferd, T.D. & Conover, R.J. (1992) Natural history of ctenophores in the Resolute Passage area of the Canadian high Arctic with special reference to *Metensia* ovum. *Marine Ecology Progress Series* 86, 133–144.

Sirenko, B.I. (2001) List of species of free-living invertebrates of Eurasian Arctic seas and adjacent deep waters. *Explorations of the Fauna of the Seas* 51, 1–129.

Sirenko, B.I. (2004) Fauna and ecosystems of the Laptev Sea and adjacent deepwaters of the Arctic Basin. *Explorations of the Fauna of the Sea* 54(62): 1–145.

Sirenko, B.I. (2009) The prosobranchs of the gastropods (Mollusca, Gastropoda, Prosobranchia) of the Chukchi Sea and Bering Strait, their species composition and distribution. *Explorations of the Fauna of the Seas*, 63(71), 104–153.

Sirenko, B.I. & Gagaev, S.Y. (2007) Unusual abundance of macrobenthos and biological invasions in the Chukchi Sea. *Russian Journal Marine Biology* 33, 355–364.

Sirenko, B.I. & Piepenburg, D. (1994) Current knowledge on biodiversity and faunal zonation patterns of the shelf of the seas of the Eurasian Arctic, with special reference to the Laptev Sea. *Reports on Polar Research* 144, 69–74.

Sirenko, B.I. & Vassilenko, S.V. (2009) Fauna and zoogeography of benthos of the Chukchi Sea. *Explorations of the Fauna of the Seas* 61(69), 1–230.

Smith, S.L. & Schnack-Schiel, S.B. (1990) Polar zooplankton. In: *Polar Oceanography* (ed. W. O. J. Smith). pp. 527–598. San Diego: Academic Press.

Sogin, M.L., Morrison, H.G., Huber, J.A., *et al.* (2006) Microbial diversity in the deep sea and the underexplored "rare biosphere". *Proceedings of the National Academy of Sciences of the USA* 103, 12115–12120.

Stein, D.L., Felley, J.D. & Vecchione, M. (2005) ROV observations of benthic fishes in the Northwind and Canada Basins, Arctic Ocean. *Polar Biology* 28, 232–237.

Stempniewicz, L., Blachowiak-Samolyk, K. & Weslawski, J.M. (2007) Impact of climate change on zooplankton communities, seabird populations and arctic terrestrial ecosystem – a scenario. *Deep Sea Research II* 54, 2934–2945.

Stepanjants, S.D. (1989) Hydrozoa of the Eurasian Arctic Seas. In: *The Arctic Seas: climatology oceanography geology and biology*

(ed. Y. Herman). pp. 397–430. New York: Van Nostrand Reinhold.

Story-Manes, S. & Gradinger, R. (2009) Small scale vertical gradients of Arctic ice algal photophysiological properties. *Photosynthesis Research* 102, 53–66.

Tammiksaar, E., Sukhova, N.G. & Stone, I.R. (1999) Hypothesis versus fact: August Petermann and polar research. *Arctic* 52, 237–244.

Thomas, D. & Dieckmann, G. (eds.) (2009) *Sea ice: An Introduction to its Physics Chemistry Biology and Geology*. New York: Wiley-Blackwell.

Ushakov, P.V. (1952) *Chukchi Sea and its bottom fauna. Krainii severo-vostok Soyuza SSR. Fauna I flora Chukotskogo moray* (Extreme southeast of the USSR, Fauna and Flora of the Chukchi Sea), vol. 2, pp. 5–83 [In Russian.]

Vassilenko, S.V. & Petryashov, V.V. (eds.) (2009) *Illustrated keys to free-living invertebrates of Eurasian Arctic Seas and adjacent deep waters Vol. 1. Rotifera Pycnogonida Cirripedia Leptostraca Mysidacea Hyperiidae Caprellidea Euphausiacea Dendrobranchiata Pleocyemata Anomura and Brachyura*. Alaska Sea Grant AK-SG-09-02, University of Alaska Fairbanks.

Vongraven, D. & the Marine Expert Monitoring Group (2009) Background paper: Circumpolar marine biodiversity monitoring program. http://cbmp.arcticportal.org/images/stories/memg_background _paper_final.pdf, accessed 2 September 2009.

Waleron, M., Waleron, K., Vincent, W.F., *et al.* (2007) Allochthonous inputs of riverine picocyanobacteria to coastal waters in the Arctic Ocean. *FEMS Microbiology Ecology* 59, 356–365.

Walsh, J.E. (2008) Climate of the Arctic environment. *Ecological Applications* 18, S3–S22.

Węsławski, J.M., Kwasniewski, S., Stempniewicz, L., *et al.* (2007) Biodiversity and energy transfer to top trophic levels in two contrasting Arctic fjords. *Polish Polar Research* 27, 259–278.

Węsławski, J.M., Kwaśniewski, S. & Stempniewicz, L. (2008) Warming in the Arctic may result in the negative effects of increased biodiversity. *Polarforschung* 78, 105–108.

Wiktor, J. & Węsławski, J.M. (2008) Svalbard intertidal 1988 and 20 years later. http://www.iopan.gda.pl/projects/SIP-2008/, accessed 4 October 2009.

Wiktor, J. & Szymelfenig, M. (2002) Patchiness of sympagic algae and meiofauna from the fast ice of NOW polynya, Canadian Arctic. *Polish Polar Research* 23, 175–184.

Wlodarska-Kowalczuk, M. (2007) Molluscs in Kongsfjorden (Spitsbergen, Svalbard): a species list and patterns of distribution and diversity. *Polar Research* 26, 48–63.

Wlodarska-Kowalczuk, M. & Kedra, M. (2007) Surrogacy in natural patterns of benthic distribution and diversity: selected taxa versus lower taxonomic resolution. *Marine Ecology Progress Series* 351, 53–63.

Wlodarska-Kowalczuk, M. & Weslawski, J.M. (2008) Mesoscale spatial structures in soft-bottom macrozoobenthic communities: effects of physical control and impoverishment. *Marine Ecology Progress Series* 356, 215–224.

Wlodarska-Kowalczuk, M., Sicinski, J., Gromisz, S., *et al.* (2007) Similar soft-bottom polychaete diversity in Arctic and Antarctic marine inlets. *Marine Biology* 151, 607–616.

Zenkevitch, V.P. (1963) *Biology of the Seas of the USSR*. Moscow: Interscience.

Chapter 11

Marine Life in the Antarctic

Julian Gutt[1], Graham Hosie[2], Michael Stoddart[3]

[1]*Alfred Wegener Institute, Bremerhaven, Germany*
[2]*Department of the Environment, Water, Heritage and the Arts, Australian Antarctic Division, Hobart, Australia*
[3]*Institute for Marine and Antarctic Studies, University of Tasmania, Hobart, Australia*

11.1 Introduction

The Southern Ocean covers 35 million km² and comprises about 10% of the Earth's oceans. Of the 4.6 million km² of continental shelf, one-third is covered by floating ice shelves (Clarke & Johnston 2003). The sea ice oscillates between a coverage of 60% in winter and 20% in summer and is, together with the sea beneath, the main driver of the Antarctic ecosystem and the Earth's ocean circulation. These conditions have caused a partial isolation of the ecosystem in the past 30 million years, and the unique environment has allowed an evolutionary dispersal of Antarctic species into the adjacent ocean's deep sea and vice versa. Recent ecological conditions in Antarctic waters not only attract the charismatic great whales, but also birds and deep-sea invertebrates from the entire world's ocean. The Census of Marine Life recognized that the Southern Ocean is home of a key component of the Earth's biosphere and launched the Census of Antarctic Marine Life (CAML) in 2005, considered the major marine biodiversity contribution to the International Polar Year 2007–8. It followed international initiatives such as the SCAR projects "BIOMASS" (see BIOMASS Scientific Series), "Ecology in the Antarctic Sea-Ice Zone (EASIZ, Arntz & Clarke 2002; Clarke *et al.* 2006), "Evolution in the Antarctic" (EVOL-ANTA, Eastman *et al.* 2004), and "Evolution and Biodiversity in the Antarctic" (EBA, results of the 10th SCAR-Biology Symposium to be published as a special volume of *Polar*

Science) as well as the projects "European Polarstern Study" (EPOS, Hempel 1993), "Investigación Biológica Marina en Magallanes relacionada con la Antártida" (IBMANT, Arntz & Ríos 1999; Arntz *et al.* 2005), "ANtarctic benthic DEEP-sea biodiversity (ANDEEP, Brandt & Ebbe 2007), and "Latitudinal Gradient Project" (LGP, Balks *et al.* 2006). Consequently, CAML was based on a very active international scientific community and covered a broad spectrum of organisms ranging from microbes to mammals. It cooperated closely with other Census projects, especially the Ocean Biogeographic Information System (OBIS), Census of Marine Zooplankton (CMarZ), Biogeography of Deep-Water Chemosynthetic Ecosystems (ChEss), Arctic Ocean Diversity (ArcOD), and Census of Diversity of Abyssal Marine Life (CeDAMar), because of two aspects. First, by combining all three other oceans by the Antarctic Circumpolar Current (ACC), the Southern Ocean provides a link for most large marine ecosystems. Second, a considerable part of the rich Antarctic fauna is unique and thus contributes significantly to the world's total marine biodiversity.

The scientific aim of CAML was to provide essential knowledge to answer the most challenging question of the future of the Antarctic ecosystem in a changing world. The strategic objective was to create a network of knowledge within the research community and to provide a forum for communication, including the most intensive outreach activities that ever concerned the work of Antarctic marine biologists. Thanks to the CAML two overarching initiatives, the biogeographic data portal SCAR-MarBIN and the barcoding initiative, intensified their efficiency, providing essential tools for scientists to share data. CAML was one of the leading Antarctic projects of the International Polar

Life in the World's Oceans, edited by Alasdair D. McIntyre
© 2010 by Blackwell Publishing Ltd.

Year 2007–8 and was part of the biology program of the Scientific Committee on Antarctic Research (SCAR). Although the Census/CAML was able to support scientific coordination, the field work was funded by the national Antarctic research programs.

This review is compiled at an early stage of CAML's synthesis phase. It provides a preliminary overview and concentrates mainly on results from core projects presented in the Genoa workshop in May 2009, to be published in *Deep-Sea Research II* in 2010 and edited by S. Schiaparelli *et al*. All references cited herein as "submitted" refer to this special CAML volume.

11.2 The Background

11.2.1 Environmental settings

The extreme seasonality in the Antarctic results in a permanently dark winter and a summer with 24 hours sunshine south of 66° 33′ S. The low temperature, and consequently the formation of the sea ice, is due to the low angle of irradiation of the sun, the high albedo of ice, and the zonal atmospheric and oceanographic circulation. The marine habitat is geographically limited to the south by a glaciated coast. The ACC combines all three other ocean basins and in the north it adjoins warmer waters at the Antarctic Convergence (Fig. 11.1). Over evolutionary time the Antarctic ecosystem experienced a permanent advance and retreat of continental glaciation which started with the formation of the ACC 25 million to 30 million years ago and has continued with obvious glacial–interglacial cycles in the past 900,000 years.

11.2.2 History of Antarctic research and exploitation

The era of early naturalists was related to both the discovery of the unknown region and the exploitation of natural resources. One example is the German naturalist Georg Forster, who participated with his father Johann Reinhold in James Cook's second trip around the world (1772–75, Fig. 11.2). Another example is the Weddell seal, which was named after the Scottish sealer James Weddell who in 1823 reached 74°34′ S, the most southerly position ever reached at that time. The famous Adélie penguin was named after the wife of the French explorer Jules Dumont d'Urville, who traveled twice to Antarctica between 1838 and 1840. Milestones of taxonomic surveys (Dater 1975) started with the famous *Challenger* expedition (1872–76) which resulted in 38 volumes of scientific results: 4,714 new species were discovered of which several were from the Antarctic. The *Belgica* undertook the first truly scientific expedition to high-latitude Antarctic waters, during which she advanced farther south than any ship before and overwintered in 1898–99 west of the Antarctic Peninsula. The *Valdivia* expedition of 1898–99 contributed substantially to the understanding of global oceanography and included biological deep-sea sampling in the sub-Antarctic. Highly efficient were also the German Antarctic expedition with the *Gauss* (1901–03), the Swedish South Polar Expedition with the *Antarctica* (1901–04), and the British *Scotia* expedition (1902–04) which conducted trawling and dredging studies of pelagic and benthic organisms. The period 1925–39 was dominated by the *Discovery* expeditions from which publications, including those of recent surveys, are still ongoing.

The exploitation of natural resources started at the beginning of nineteenth century. Populations of Antarctic fur and elephant seals crashed close to extinction by the 1820s. Whaling started at the beginning of the twentieth century. The biomass of the largest species – blue, fin, humpback, southern right, and sei whales – were reduced to between 50% and 0.5% of their original worldwide stock whereas the smallest, the Antarctic minke, became most abundant (Laws 1977). Thus, the natural dominance pattern of whale species was turned upside-down. Interesting calculations have been made about the negative impact of the whaling to deep-sea animals since whale carcasses have no longer been important food sources for marine organisms (Jelmert & Oppen-Berntsen 1995). Bottom trawling in the 1960s reduced the stocks of the marbled rock cod (*Notothenia rossii*) and mackerel ice fish (*Champsocephalus gunnari*) west of the Antarctic Peninsula (Kock 1992) within very short periods, and devastated slow-growing benthic communities. The exploitation of natural resources was the most effective anthropogenic impact that Southern Ocean biodiversity ever experienced. However, the hitherto inviolacy of most high-latitude Antarctic marine habitats is almost unique on Earth, but the ecosystem is increasingly threatened by the new longline fishing and by the impact of climate change.

11.2.3 Modern pre-CAML biodiversity studies

In the 1980s, ecological analyses using bulk parameters (see, for example, http://ijgofs.whoi.edu) tried to solve so-called "process orientated" questions without spending much time determining species diversity. Among the few studies with high taxonomic resolution, outstanding progress was made by the work on the evolutionary radiation of fish (Eastman & Grande 1989). In this phase the macrobenthos became known to be regionally dominated by sessile suspension feeders (Bullivant 1967); their communities later turned out to be more dynamic than previously expected (Dayton 1990; Arntz & Gallardo 1994; Gutt 2000, 2006; Gutt & Piepenburg 2003; Potthoff *et al*. 2006; Barnes & Conlan 2007; Seiler & Gutt 2007; Smale

(A)

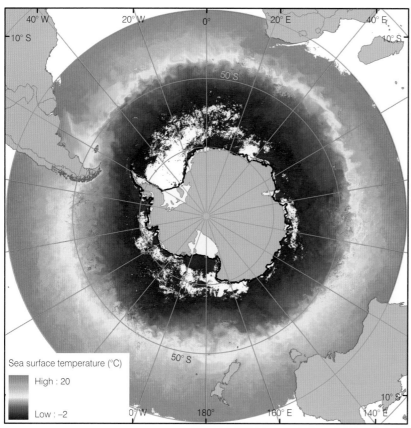

(B)

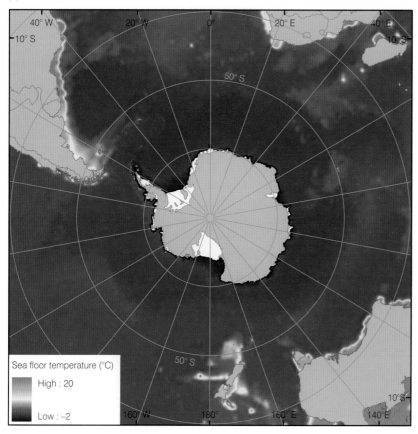

Fig. 11.1

Temperature of the Southern Ocean; at the sea surface **(A)**, where the Antarctic Convergence is clearly indicated by the sharp gradient between warm (red) to cold (blue) temperatures, white areas within Antarctic waters indicate no data due to sea-ice cover; at the sea floor **(B)**. For details of the occurrence of relatively warm water west of the Antarctic Peninsula, see Clarke *et al.* (2009). Graph by H. Griffiths and A. Fleming, British Antarctic Survey; data: NASA.

Fig. 11.2

Original drawing of the chinstrap penguin, *Pygocelis antarctica* (J.R. Forster, 1781) by Georg Forster, *Handschriftenabteilung der Thüringischen Universitäts- und Landesbibliothek Jena*, Germany, MsProv.f 185 (1).

et al. 2008). Plankton studies added substantial information to the traditional view of the simple Antarctic pelagic system consisting only of algae, krill (*Euphausia superba*), and few apex predators. Small organisms became known to contribute to the microbial loop by being relevant for the re-mineralization in a partly iron-limited "high nutrient – low chlorophyll" system. Improved sea-ice research elucidated the diversity not only of unicellular algae but also of metazoans living in and associated with this unique habitat (Thomas & Dieckmann 2009), including the trophic key species of the Antarctic food web, the Antarctic krill (Thomas *et al.* 2008).

11.3 CAML Projects: Advancing Knowledge

11.3.1 The scientific strategy

At first the term "census" had to been interpreted literally: species and specimens were identified and counted. Secondly, CAML researchers raised the question why some of

these species co-exist in specific communities whereas others do not, the answers demanding both evolutionary and ecologically approaches at various spatial scales.

11.3.2 What were the major gaps?

The scientific effort during the pre-CAML phase reflected the good accessibility of the area around the Antarctic Peninsula and historical developments in poorly accessible areas – for example the inner Weddell and Ross Seas – with large gaps in between. The Antarctic deep sea was only known from studies with selective samples with a reduced taxonomic scope. Life in some typical Antarctic habitats was very poorly known, especially from under the ice shelves and the permanent pack ice. The biodiversity not only of microorganisms, but also of rare charismatic species, for example toothed whales, had almost been overlooked and some historic data were hardly accessible. The identification of many invertebrate eggs and larvae to the species level was impossible, and only hypotheses existed in relation to cryptic species. The question about the relation between ecosystem functioning and biodiversity has a long tradition but it is still – at least for the Antarctic – difficult to address. Finally, the pre-CAML era was characterized by the knowledge that climate change would not stop at the Antarctic Circle, but background information and observations on its impact to the ecosystem were scarce.

11.3.3 Approaches to closing gaps

Core strands of CAML were scientific expeditions and the data management allowing overarching analyses. Success has also been reached through the standardization of field methods, for example by using the standard nets, continuous plankton recorders, video-equipped remotely operated vehicles (ROVs), or sleds. The major tool for ensuring information management is the "Marine Biodiversity Information Network of SCAR" (SCAR-MarBIN, www. scarmarbin.be), being the local node of the Census of Marine Life/UNESCO OBIS network. It was initiated by the Royal Belgian Institute of Natural Sciences and CAML became its major research partner. So far, over 1 million geo-referenced records from 156 datasets are available. The Register of Antarctic Marine Species (RAMS) comprises 6,551 primarily benthic and 702 pelagic species (as at May 2010) and is constantly updated by over 70 editors and contributing scientists (De Broyer & Danis submitted). Datasets range from historic information going back to 1781 to recent and genetic data. A barcode manager supported CAML scientists in analyzing over 11,000 sequences (Grant & Linse 2009). Thus, CAML contributes to the Barcode of Life project (BOLD; www.barcodinglife.org) and the "Fish

Barcode of Life Initiative" (FISH-BOL, www.fishbol.org). Spatially explicit ecological models were developed, for example to predict potential fish habitats and to simulate the succession of biodiversity after disturbance (Potthoff *et al.* 2006). A new tool, "GeoPhyloBuilder" (www.nescent.org/wg_EvoViz/GeoPhyloBuilder), and network analyses (Raymond & Hosie 2009) are being used to visualize phylogeographic data.

11.3.4 Evolutionary large-scale patterns and non-circumpolar cryptic species

The question of bipolar species experienced a renaissance under CAML. No doubt exists about the annual pole-to-pole migrations of the blue, humpback and fin whales as well as seabirds such as the Arctic tern. In addition, a bipolar occurrence of a few benthic and pelagic invertebrate species had been controversially discussed. A recent comparison between the Register of Antarctic Marine Species and the ArcOD database revealed approximately 230 species names to which occurrences from both polar regions were attributed. Recent attempts to provide evidence for their existence with genetic methods were successful, for example for the amphipod *Eurythenes gryllus* occurring at the upper slopes of the Canadian Arctic, around Antarctica, and in the deep sea in between (France & Kocher 1996;s De Broyer *et al.* 2007). Such evidence failed for the pteropod *Limacina helicina*, being so far considered as one species but having 32% divergence between both polar regions (J.M. Strugnell, unpublished observations). Another weak example is the sponge *Stylocordyla borealis*, which has two sympatric distinct growth forms even within the Antarctic, one with a thick stalk, the other like a lollipop. For the widespread and well-known deep-sea holothurian *Elpidia glacialis*, which has strong polar emergence, six subspecies are known and, using traditional methods, it is only a matter of interpretation not to consider these as six true species. A morphologic and genetic documentation of the existence of bipolar species among deep-sea komokiaceans and other foraminiferan-like protists was highlighted by Brandt *et al.* (2007a). A high genetic and morphologic similarity was found for the planktonic anthomedusa genus *Pandea* between the north Pacific near Japan and East Antarctica (D. Lindsay *et al.*, unpublished observations). In conclusion, it remains open whether genetic methods will continue to confirm the bipolar occurrence of species and, consequently, gene-flow over extremely long distances or whether true bipolar species will remain rare exceptions.

Before we can understand the role of the Southern Ocean within global biodiversity patterns and underlying evolutionary processes, our knowledge of geographic coverage has to be completed, especially for the deep sea of

the Southern Ocean covering 27.9 million km². Recent investigations, especially those of the ANDEEP expeditions, revealed an extraordinarily high species richness at abyssal depths. More than 1,400 species of invertebrates were identified (from only the taxa investigated) and more than 700 of these were assumed to be new to science (Brandt *et al.* 2007a). For example, within protists, the formaminiferan-like komokiaceans were not known from the Southern Ocean deep sea. Now 50 species are reported from that area of which 35 are undescribed (Godday *et al.* 2007). Within the macrofauna, the isopods were the most diverse taxon with 674 species, of which 87% are putative new species. If we compare these numbers with the more than 4,400 known marine isopod species from the world oceans, the recent Southern Ocean deep-sea expeditions will add approximately 15% to our knowledge on the worldwide zoogeography of that taxon. For the megafauna, the occurrence of new Hexactinellida (glass sponges) and carnivorous demosponges and the first report of Southern Ocean calcareous sponges (Calcarea) were among the most surprising results (Janussen & Reiswig 2009; Rapp *et al.* in press).

Despite the incomplete faunal knowledge, several studies show linkages between the Antarctic fauna and that of the adjacent deep sea. These studies benefited from a new biologically orientated view on Antarctic seawater temperature. Satellite images show that the well-known Southern Ocean hydrodynamic isolation separating warm surface water in the north from cold water in the south along the Antarctic Convergence is superimposed by horizontal gyres (Fig. 11.1). These allow floating material, for example larvae, other pelagic organisms, pieces of algae, or material serving as substratum for benthic species to penetrate this boundary in both directions (Clarke *et al.* 2005; Barnes *et al.* 2006). Thus, it is mainly the temperature difference that allows only very few species to survive at both sides of the Antarctic Convergence, rather than the front acting as a hydrodynamic barrier. The comparison between the surface and near-seabed temperature shows more obviously than ever before how less isolated are the Antarctic bottom-dwelling fauna – including those on the Antarctic shelf – from those in the adjacent deep sea (Fig. 11.1). This has relevance not only for future scenarios under climate change but also major implications for the dispersal of animals at evolutionary and ecological timescales.

Hypotheses have always existed about such large-scale dispersal processes. The colonization of the deep sea by Antarctic organisms seemed to be most likely and most common, after the post-Gondwana breakup and establishment of the ACC. Using genetic techniques, phylogenetic trees can be better linked to plate tectonics, especially the opening of deep-water basins between Antarctica and adjacent continents and the resulting global water mass circulation. Recently, evidence has been provided for an evolutionary dispersal of deep-sea octopods that evolved from common Antarctic ancestors around 30 million years ago into the northerly adjacent deep sea, called tropic submergence (Strugnell *et al.* 2008). Similar development can be reconstructed for isopods (Asellota, Antarcturidae, Acanthaspidiidae, Serolidae, Munnidae, and Paramunnidae; Raupach *et al.* 2004, 2009; Brandt *et al.* 2007b), the amphipod *Liljeborgia*, of which the Antarctic representatives still have eyes whereas their deep-sea relatives are blind (d'Udekem d'Acoz & Vader 2009), and the mollusk *Limopsis* (K. Linse, unpublished observations). In the opposite direction, multiple evolutionary invasions from the deep sea to the Antarctic shelf, called polar emergence, are very likely for some other isopods, for example Munnopsidae, Desmosomatidae, and Macrostylidae because of their lack of eyes (Raupach *et al.* 2004, 2009). Similar interpretations are made for representatives of the deep-sea octopod *Benthoctopus* (Strugnell *et al.* in press). Such examples of long-term evolutionary dispersal have also been described for other taxa such as hexactinellid sponges, pennatularians, stalked crinoids, and elasipod holothurians but have never been studied in detail. Using techniques to decipher the molecular clock, the echinoid *Sterechinus* and the ophiuroid *Astrotoma agassizii* (Hunter & Halanych 2008; Díaz *et al.* in press) were found to be examples of a split between shallow Antarctic and subantarctic species, which occurred not more than 5 million years ago when glacial–interglacial cycles started. This was long after Antarctica disconnected from South America and the Antarctic Convergence formed. Similar results are available for the limpet *Nacella* (González Wevar *et al.* in press) and the bivalve *Limatula* (Page & Linse 2002). Perhaps the most extreme example for cryptic speciation is the sea slug *Doris kerguelensis*, from which approximately 29 lineages are derived (Wilson *et al.* 2009). This puts the development of the Antarctic Convergence 25 million years ago as a main agent of vicariance in question. Surprisingly, this relatively recent split of species within a broad geographical range happened independently of their dispersal potential, because these taxa clearly differ from each other in their early life history traits.

If, despite these few faunistic teleconnections, Antarctica's fauna differs considerably from that of the adjacent slope and the deep sea, for example in the Weddell Sea (Kaiser *et al.* in press) and from that north of the Antarctic Convergence as for deep-sea gastropods (Schwabe *et al.* 2007; Schrödl *et al.* submitted), the reasons must be searched for in polar-, slope-, or deep-sea-specific environmental parameters. At the level of evolution one major mechanism to generate such biogeographical heterogeneity on the Antarctic shelf is the climate diversity pump, being a modified vicariance concept (Clarke & Crame 1989). Until a few years ago this concept was used to explain a relatively high richness of species with a predominantly circumpolar distribution. It was assumed that during glacial periods populations were spatially separated by grounded ice shelves and as a consequence a radiation of species occurred. At the end of a glacial period when the ice retreated, these new species supposedly mixed around the continent but were obviously not able to interbreed anymore. This has resulted in sibling species, for example ten sympatric octopods of the genus *Pareledone* (Allcock 2005; Allcock *et al.* 2007, in press), analogous to approximately eight cryptic species of the isopod *Ceratoserolis* (Raupach & Wägele 2006) and six allopatric species of *Glyptonotus* (Held 2003; Held & Wägele 2005; Leese & Held 2008; C. Held, unpublished observations). Mostly allopatric cryptic species also occur among the dendrochirote and aspidochirote holothurians, for example among *Laetmogone wyvillethomsoni* and *Psolus charcoti* (O'Loughlin *et al.* in press) and the amphipod *Orchomene* sensu lato (Havermans *et al.* submitted). Significant genetic differences have also been found among the pantopod *Nymphon* in the East Antarctic Peninsula and Weddell Sea (Arango *et al.* in press) and the comatulid crinoid *Promachocrinus* west of the Peninsula and in the Weddell Sea (Wilson *et al.* 2007) as well as off East Antarctica (L. Hemery & M. Eléaume, unpublished observations). The narcomedusa *Solmundella bitentaculata* was previously thought to be a single ubiquitous species but molecular studies suggest that it contains at least two cryptic species (D. Lindsay *et al.*, unpublished observations).

Resulting from this, a milestone in evolutionary biodiversity research of the past years might be the paradigm shift from an assumed circumpolar macrobenthos to an obviously long-term patchy occurrence of closely related sibling or cryptic species in many taxa.

If, however, the large-scale pattern of the shelf-inhabiting Antarctic macrobenthos is analyzed, using the current best available dataset (Fig. 11.3), only one single bioregion is found (Griffiths *et al.* 2009). The exception is gastropods following the pattern of a split into the Scotian subregion mainly comprising the Antarctic Peninsula and the High Antarctic Province, as proposed by Hedgpeth (1969), which was already questioned a few years later (Hedgpeth 1977). The difference between the interpretations is that the one-bioregion result is based on fully reproducible presence/absence datasets with an incomplete systematic coverage. Hedgpeth's conclusion of two provinces included impressions of abundances and consequently of dominance patterns referring mainly to higher taxa and life forms. Additional bias can be caused by the fact that traditional results from the Peninsula were mainly from shallow waters whereas the rest of the Antarctic shelf was sampled at greater depth.

The Southern Ocean Continuous Plankton Recorder (CPR) Survey (Hosie *et al.* 2003) was the major contribution of the CAML to the research on the Antarctic pelagic system and provided a close link to the Convention on the Conservation of Antarctic Marine Living Resources (CCAMLR). Use of the CPR has significantly increased our

(A)

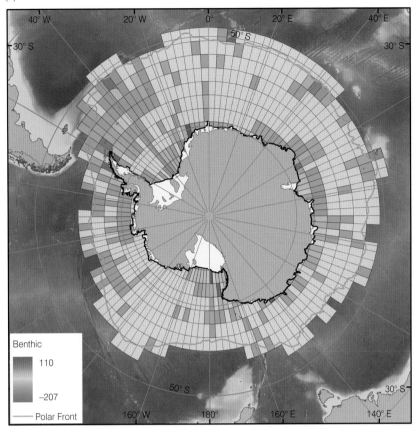

(B)

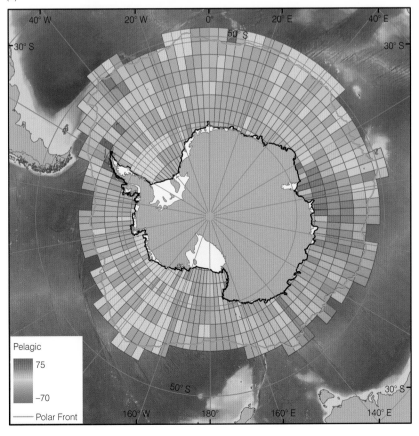

Fig. 11.3

Species richness represented by color-coded residuals. Red implies higher than expected numbers of species (for the number of samples) and green lower than expected. Numbers of species ranged from 1 to 400 benthic **(A)** and from 1 to 52 pelagic **(B)**. The benthic group covers a broad range of invertebrates. Pelagic includes all zooplankton, fish, sea birds, seals, penguins, and whales. Neither group includes plants and microbes as the available data are insufficient. Residuals are calculated from the regression of observed species number on sample number per 3° × 3° grid cell in benthic and pelagic data from the 122 datasets available in SCAR-MarBIN as of May 2009 (www.scarmarbin.be/scarproviders.php; De Broyer & Danis). Sampling effort is eliminated statistically, but intensive sampling by the Continuous Plankton Recorder off East Antarctica remains visible (see also Griffiths *et al.* submitted). Graph and data processing: H. Griffiths and B. Danis.

knowledge of Antarctic plankton communities by extending the time series and increasing the geographic coverage of the Southern Ocean CPR Survey to approximately 70% of the region, with the highest resolution off East Antarctica. In the 2007/2008 CAML-campaign alone, 15 nations were involved using eight ships conducting 88 successful tows and over 23 transects at 10 m water depth. Since 1991, 25,791 samples have been taken with a resolution of 5 nautical miles, covering a total of 128,955 nautical miles (Southern Ocean CPR Data Set; http://data.aad.gov.au/aadc/cpr). In terms of large-scale patterns, previous analyses of the Southern Ocean CPR data have shown latitudinal zonation of zooplankton across the ACC, the Sub-Antarctic Front (SAF) acting as a geographic barrier with different species found north and south of it (Hunt & Hosie 2003, 2005). The copepod *Oithona similis* is not only an example for the large-scale pattern (Fig. 11.4) but also for temporal changes (see below).

South of the SAF and moving toward the continent, distinct assemblages could be identified which were associated with zones within the ACC. Differences between the assemblages were subtle and based primarily on variation in abundances of species relative to each rather than differences in species composition itself. The CAML provided the opportunity to assess circum-Antarctic patterns. Only night data from the period between December and February were used, rare taxa were excluded, adults and juveniles were merged, and unidentified groups removed. The results on the fauna sampled by the CPR showed no clear longitudinal differences between sectors. In other words, the species composition and abundances of zooplankton within any band of the ACC are effectively the same: it is one community. Tows in January 2008 across Drake Passage did show lower abundances and diversity, but no substantial differences from other transects were observed later in February. The Bellingshausen Sea did show very low abundances and fewer plankton species. The large concentrations of krill, especially in the West Atlantic sector (see Atkinson *et al.* 2008), were not sufficiently covered by this survey. Probably because of the method used, a neritic community only became obvious among the semipelagic, cryopelagic (ice preferring), and pelagic fish (Koubbi *et al.*

Fig. 11.4

Predictions for the spatial patterns of relative abundance of the cyclopoid copepod *Oithona similis* in January using boosted regression tree modeling. Data from the Southern Ocean Continuous Plankton Recorder survey were combined with environmental variables such as chlorophyll *a*, bathymetry, ice cover, sea surface temperature, and nutrients, to predict the circum-Antarctic distribution of *O. similis* for bioregionalization. Gray indicates areas with insufficient combined data. From Pinkerton *et al.* (2010); oceanographic fronts according to Orsi *et al.* (1995).

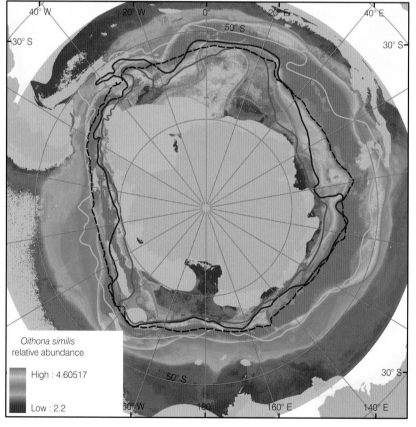

Oithona similis
relative abundance

High : 4.60517

Low : 2.2

——— Polar Front

– – – Northern average limit of sea ice

——— Southern Antarctic Circumpolar Current Front

——— Southern Boundary, Antarctic Circumpolar Current

submitted), which is dominated west of the Antarctic Peninsula by Antarctic rock cod *Notothenia* and at high Antarctic latitudes by *Trematomus*, Channichthyidae (icefish) (Fig. 11.5A), and the pelagic *Pleuragramma antarcticum* (O'Driscoll *et al.* in press). Other planktonic studies embedded in CMarZ (see Chapter 13) used nets with smaller mesh sizes and sampled at greater depth than before. As a consequence, not only were the planktonic fauna more diverse than previously thought, but also many new species were discovered, including the ice-associated fauna.

Microorganisms and the gelatinous plankton likely belonged to the most under-represented groups of organisms in Antarctic surveys. During the CAML phase, the understanding of both the extent and ecological variability of Antarctic marine bacterioplankton diversity was greatly enhanced. In just one study approximately 400,000 sequence tags spanning a short hypervariable region of the SSU rRNA gene were determined for 16 samples collected from four regions (Kerguelen Islands, Antarctic Peninsula, Ross Sea, and Weddell Sea) (Ghiglione & Murray, unpublished observations). This effort revealed over 25,000 different sequence tags, of which 13,000 represented equivalents to new species (at a distance greater than 0.03 from the nearest known sequence in public databases). Samples at a low-activity cold seep in the Larsen B area, west of the Antarctic Peninsula, revealed 29 seep-related operational taxonomic units of bacteria and 10 of Archaea, of which 20–30% have no closely cultivated relatives (Niemann *et al.* 2009). The numbers of gelatinous plankton species increased by a factor of 2–3, especially among hydromedusae, siphonophores, and scyphomedusae, particularly within the neritic assemblage (Lindsay *et al.*, unpublished observations).

Apex predators were also included in the CAML studies. An extensive census in the Atlantic sector of the Southern Ocean, mainly west and east of the Antarctic Peninsula (Scheidat *et al.* 2007a), showed that whale diversity was higher than expected. Four rare toothed whales from the family of the beaked whales (Ziphiidae) were registered: Arnoux's beaked whale (*Berardius arnuxii*), Gray's beaked whale (*Mesoplodon grayi*), strap-toothed whale (*M. layardii*), and southern bottlenose whale (*Hyperoodon planifrons*), the last with occurrences only in waters deeper than 500 m. Some of the sightings were southernmost records (Scheidat *et al.* 2007b).

11.3.5 Ecologically driven community heterogeneity between extremes

One milestone to which CAML researchers contributed is a paradigm shift from a supposed Antarctic circumpolar benthos being rich in species, life forms, and biomass (Figs. 11.5B, C and D) to the general understanding that there is a full range of benthic assemblages from extremely diverse to extremely meager (Fig. 11.6).

Within such a heterogeneous patchwork, poor assemblages were already known decades ago; however, during the CAML phase these were more intensively studied, for example on seamounts (Fig. 11.5E) (Bowden *et al.* submitted) and in areas formerly covered by the ice shelf (Fig. 11.5F) (Gutt *et al.* in press). This extreme variability can also be attributed to the pelagic system, where on the one hand krill swarms are extremely rich in biomass, but on the other hand extremely low biomass and production are known from the winter season, with a deepest-ever recorded Secchi depth of 80 m, measured on October 13, 1986 in the Weddell Sea (Gieskes *et al.* 1987). At the seafloor, extremely low abundances can be found in different habitats; at shallow depths with permanent disturbance, in fresh iceberg scours (Gutt & Piepenburg 2003), and under the ice shelf (Gutt 2007). The question of how extremely low abundances can be explained is especially challenging. Unfavorable environmental conditions can lead to the total absence of specific life forms or ecological guilds, such as filter feeders. If food supply is poor then perhaps no more than a few individuals the size of a tennis ball in an area of a tennis court could exist. However, abundances in the formerly ice shelf covered Larsen B area east of the Antarctic Peninsula remained at obviously even lower levels, observed *in situ* during a *Polarstern* expedition in 2007, five years after the ice shelf disintegrated (Gutt *et al.* in press). Because reduced long-term dispersal capacity, at least of species with a circumpolar distribution, can hardly explain this alone, a hypothesis was developed that a poor temporal predictability of food supply during the early life phase could explain extremely rare abundance of adults (Gutt 2007).

Very low biodiversity is also known from different seamounts. At the Admiralty Seamount (East Antarctic), high local densities of stalked crinoids (Hyocrinidae, Fig. 11.5E), brachiopods, and suspension-feeding ophiuroids (*Ophiocamax*) may reflect ecological conditions such as low predation pressure and low food supply or evolutionary factors (Bowden *et al.* submitted). The sediment here was dominated by crinoid ossicles, indicating a long persistence of these populations. In contrast, the benthos of the Scott Seamount less than 400 km away at the same latitude was characterized by a higher abundance of predators, including lithodid crabs, regular sea urchins, and sea stars. A very similar pattern had previously been found on the Spiess Seamount, with large specimens of sea urchins (*Dermechinus horridus*) as well as lithodid crabs (*Paralomis elongata*) being the most conspicuous species and, like the Admiralty Seamount, the seafloor was almost completely covered by spine debris (J. Gutt, unpublished observations).

These differences of dominant species might not only represent temporal parallel ecological processes leading

Fig. 11.5

(A) Antarctic ice fish (*Pagothenia macropterus*) exhibit the most developed adaptation to low temperatures. Thus they are traditionally a target of evolutionary, physiological, genetic, and ecological studies. Repository reference DOI: 120.1594/PANGAEA.702107, also for Fig. 5F. (Photograph: J. Gutt and W. Dimmler; © AWI/Marum, University of Bremen.) **(B)** Hexactinellid sponges (*Rossella nuda, Scolymastra joubini*) are common on the Antarctic shelf, where they grow to a size of up to 2 m. They indicate areas free of disturbance for long periods owing to their slow growth when they are adult. Eastern Weddell Sea, 233 m water depth. (Photograph: J. Gutt and W. Dimmler; © AWI/Marum, University of Bremen.) **(C)** Concentrations of bryozoans can form together with hydroids and demosponges a microhabitat for other animals (for example holothurians) as seen here north of D'Urville Island, West of the Antarctic Peninsula, at *ca.* 230 m water depth. Owing to their life traits, they can serve as indicator species for Vulnerable Marine Ecosystems for CCAMLR. (Courtesy of S. Lockhart and D. Jones; © US-AMLR program.) **(D)** The concentrations of hydrocorals of the genus *Errina* and other sessile organisms such as sponges (background) at the George V Shelf, 65.7° S 140.5° E, 680 m depth, were the reason for designating this area as a "Vulnerable Marine Ecosystem". (Courtesy of A. Post and M. Riddle; © Australian Antarctic Division.) **(E)** Stalked crinoids (Hyocrinidae) dominate the macro-epibenthos on parts of Admirality Seamount (67° S 171° E) at 550–600 m depth. They are unknown from elsewhere on the Antarctic shelf. (Courtesy of D. Bowden, National Institute of Water and Atmospheric Research; © Land Information New Zealand.) **(F)** Ascidians (*Molgula pedunculata*) can form almost monospecific assemblages in highly dynamic areas owing to iceberg scouring or disintegrating ice shelves. The Larsen B area, east of the Antarctic Peninsula, was covered by ice shelf five years before the photograph was taken, 188 m water depth. (Photograph: J. Gutt and W. Dimmler; © AWI/Marum, University of Bremen.)

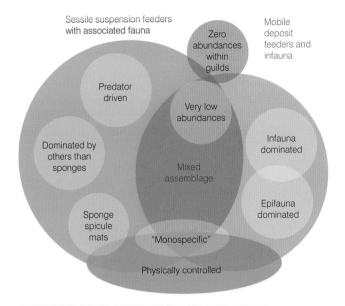

Fig. 11.6
Scheme of Antarctic macro-benthic assemblages. From Turner *et al.*
(2009), modified after Gutt (2007).

to different results: they could also represent differently advanced stages of long-term developments because stalked crinoids resemble ancient Palaeozoic assemblages and predators indicate a more modern benthos. Generally, *in situ* images of crinoids could even be used to sample wide-range information on near-bottom current, which is important in explaining benthic community structures (Eléaume *et al.* in press). Also, early recolonization stages of iceberg scours or areas after ice-shelf disintegration can, but do not necessarily always, consist of almost monospecific assemblages such as bryozoans, cnidarians, or ascidians (Fig. 11.5E). A dominance of one single species due to an assumed competitive displacement seems to be rare on the Antarctic shelf, but was observed, for example for the sponge *Cinachyra barbata* in the Weddell Sea. Favorable environmental conditions can cause a clear dominance on shallow hard and soft substrata, for example of the scallop *Adamussium colbecki*, the limpet *Nacella concinna* (Barnes & Clarke 1995a; Chiantore *et al.* 2001), or the infaunal clam *Laternula elliptica* and the deep-sea holothurian *Elpidia glacialis* (Gruzov 1977; Gutt & Piepenburg 1991). The richness of species of the Southern Ocean deep sea has already been discussed and does not support the hypothesis of a gradient of decreasing richness toward the south (Brandt *et al.* 2007a). Even the opposite was found for gastropods (Schrödl *et al.* submitted), which is in contrast to shallow habitats. Communities on the shelf can reach extremely high values for wet weight biomass, up to 12 kg m^{-2} (Gerdes *et al.* 2003), with relatively high

biodiversity compared with the Arctic shallow water. And they are not always defined by the well-known sponge concentrations: recently an assemblage shaped by the elsewhere rare hydrocoral *Errina* has been discovered (CCAMLR 2008) (Fig. 11.5D).

A high geographical turnover of macrobenthic assemblages within larger regions can generally be explained mainly by sea-ice conditions and proxies for food supply such as current and pigments in the sediment (Gutt 2007; V. Cummings, unpublished observations). Such a regional co-existence of different communities is found almost everywhere, at the Antarctic Peninsula (Lockhart & Jones 2008), at smaller places such as the well-investigated Admiralty Bay including macroalgae (Siciński *et al.* submitted), in the Weddell and the Ross Seas (Gutt 2007; V. Cummings, unpublished observations), or off East Antarctica (Gutt *et al.* 2007). For selected deep-sea polychaetes, the challenging question of how allied species can coexist without out-competing each other was answered by their different food preferences, analyzed by biochemical analyses (Würzberg *et al.* submitted).

Use of the CPR to study plankton patterns has shown that the large-scale zonation around Antarctica (subantarctic, sea ice, neritic) is consistent with latitudinal oceanographic zones as defined by Orsi *et al* (1995) or Sokolov and Rintoul (2002); see also Takahashi *et al.* (2002), Umeda *et al.* (2002), Hunt & Hosie (2003, 2005, 2006a, b), and Takahashi *et al.* (2010). These patterns are superimposed by the sea-ice margin and related melting processes, which directly affect the success of some species and consequently the entire community (Raymond & Hosie 2009). The Bellingshausen Sea, for example, exhibited low diversity and abundances. Temporal changes during the past decade have been observed with a decrease in the dominance of krill in the sea-ice zone of eastern Antarctic and an increase in dominance by smaller zooplankton more typical of the permanent open ocean zone, notably the cyclopoid copepod *Oithona similis*, small calanoid copepods *Calanus simillimus* and *Ctenocalanus citer*, foraminiferans, and larvaceans. Zooplankton abundances in general increased by about 50 times but probably with a lower effect on the total biomass because generally the shift was to small species. Besides the above-described coarse and well-known circumpolar pattern, no latitudinal zonation became obvious within a broad band of the ACC ranging from approximately 52° to 64° S covering a temperature range between a sea surface temperature of 2 and 6 °C. However, within the single surveys differences between predominantly north–south-orientated transects became visible. It is too early to speculate whether the temporal and spatial turnover in community structure is a result of global change, or a shift between two natural events.

The mesopelagic fish fauna, mainly comprising myctophids, were only recently recognized as a key component in the open ocean system because they prey on mesozoo-plankton, especially copepods, and in turn are the major prey of top predators. They also contribute to fast vertical energy flux of organic material due to their vertical migrations (Koubbi *et al.* submitted). Major success has been made in understanding their ecological demands and physical environment, for example in terms of chlorophyll *a*, sea surface temperature, salinity, and nutrients. Based on that, predictions can be made about suitable habitats for their biodiversity, even for areas from which no such biological data exist.

11.3.6 Small-scale heterogeneity, a contribution to large-scale biodiversity

Epibiotic relationships have become more obvious since the first seabed photographs were taken in the late 1950s. However, for a long period, symbiotic associations, which include parasitic relationships, were judged to be rare in Antarctica (AAVV 1977, p. 389). Later, sponges, bryozoans, and cnidarians (Figs. 11.5B, C and D) were recognized as the main substratum for a variety of echinoderms. In total 347 of such interspecific relationships were found using imaging techniques (Gutt & Schickan 1998). More recent and detailed studies revealed that cidaroid sea-urchin spines alone provide the microhabitat for some 156 species, for example bryozoans, sponges, bivalves, and holothurians, of which some even live obligatorily on the spines. So far, 23 especially close associations (encompassing commensalism, associational defense, and parasitism) have been reported for the Antarctic. Hosts are generally echinoderms, whereas the symbionts are mainly mollusks and polychaetes (S. Schiaparelli, unpublished observations). The more such symbioses are searched for, the more that are found, for example between a polynoid polychaete and the holothurian *Bathyplotes bongraini* (Schiaparelli *et al.* in press). Many Antarctic symbioses represent relict interactions, already present before the isolation and cooling of the continent. They might play an important role in explaining an ecological coexistence of species. Such specific relationships are also considered to characterize a mature ecosystem. In the area where the Larsen Ice Shelf recently disintegrated, the composition of epibiotic species did not differ from that living on boulders (Hétérier *et al.* 2008; Linse *et al.* 2008; Hardy *et al.* in press), which indicates a transitional stage of ecological development. A possible hypothesis could be that, in general, symbionts not only linearly contribute to biodiversity as all other species do, but also by providing potential living substrata they might instead accelerate the increase in Antarctic macrobenthic biodiversity by attracting other species.

11.3.7 Applied aspects and biodiversity change

Antarctic waters might be the best protected marine areas on Earth owing to the "Protocol on Environmental Protection to the Antarctic Treaty" (the "Madrid Protocol"). Science managers and politicians have not given high priority to specific marine nature conservation actions for a long time. Recently, interest in such approaches has increased. In 2008 the CCAMLR adopted a proposal by Australia, based on CAML's CEAMARC expedition Collaborative East Antarctic Census of Marine Life (Hosie *et al.* 2007), to declare two areas of the Southern Ocean mentioned above as Vulnerable Marine Ecosystems (VMEs) because of their complex and vast coralline assemblages. The purpose of the classification is to protect the sites from longline fishing impact, a major concern after bottom-trawl fishing was banned in the most profitable area west of the Antarctic Peninsula (CCAMLR 2008). In addition to several VMEs, one of the world's largest Marine Protected Areas (MPAs) has recently been designated in an area south of South Georgia.

CCAMLR initiated a bioregionalization project (Grant *et al.* 2006), which predicts potential habitats for key ecological species and assemblages in order to identify biological hot spots (Koubbi *et al.* submitted). The United Nations Environmental Program developed criteria to define Ecologically and Biologically Significant Areas as determined by the Convention on Biological Diversity in 2008, which is independent of any sustainable use of the ecosystem. The Scientific Committee on Antarctic Research recently compiled a comprehensive Report on the "Antarctic Climate Change and the Environment" (Turner *et al.* 2009), which addresses the necessity for baseline information and long-term observations to monitor the mainly climate-induced affects on the ecosystem. All these initiatives were significantly supported, some even initiated, by leading CAML scientists. The results also provide a valuable basis to keep the Red List of Threatened Species (http://www.iucnredlist.org) updated.

Bioprospection in the Antarctic is still in its infancy. Providing that international law and conventions are respected, CAML can contribute to a further development of this opportunity, and consequently of Antarctic ecosystem services to the benefit of human well-being. Marine biodiversity is protected from large-scale offshore fertilization for CO_2 mitigation by the Convention on Biological Diversity and the Madrid Protocol to the Antarctic Treaty. Fish stocks around the Antarctic Peninsula have been protected from bottom trawling since the 1990–91 season. The development of fish stocks was observed around Elephant Island and the lower South Shetland Islands in December 2006 – January 2007 surveys (K.-H. Kock, unpublished observations). One of the most abundant species before exploitation, the mackerel icefish (*Champsocephalus gunnari*), has not yet recovered. The status of the second target species of the fishery, the marbled notothenia (*Notothenia rossii*), is unclear as no specific surveys for the species have been conducted. Bycatch species, icefish, and nototheniids, appear to have recovered since the area was closed to commercial fishing. Unexplainable so far is the recruitment failure in the past seven or eight years of the yellow notothenia (*Gobionotothen gibberifrons*). The stock currently consists to a very large extent only of adult fish.

11.4 Blueprint for the Future

11.4.1 Hot and cold spots of biodiversity

During the CAML period, knowledge of Antarctic biogeography steadily increased, in some cases exponentially, in others blank spots were filled. We are now able to identify more local biodiversity hot (and cold) spots. However, regionally comparable criteria are still difficult to apply. In this context it is necessary to establish a systematic geographic coverage, as spatially homogenous as possible for as many as possible relevant regions, and not so much to reach detailed results at one single location. Then, the total number of species in the Antarctic must not remain unknown forever (Gutt *et al.* 2004). It can be mathematically extrapolated (Chao 2005), although most algorithms demand information on rare species. Consequently, not only presences but also information on the absence and abundance of species is strongly needed.

Ecological modeling will become increasingly important to fill gaps in our understanding of biodiversity dynamics. The first objective is to understand the relations between environmental parameters, biological traits, and biodiversity patterns. If robust correlations are found, then in a second step the flora and fauna can be deduced from well-known environmental patterns and perhaps vice versa. When ecological interactions are well understood, then the most difficult challenge of predicting the future becomes possible. Some important steps in this direction have already been made. The community approach tries to classify ecologically complex relationships and to identify key species (Gutt 2007; Post *et al.* in press). The habitat suitability or potential habitat modeling approach tries to extrapolate from known environmental parameters using information on the ecological demands of key organisms to fill geographical gaps in their distribution and, thus, also contribute to a general understanding of ecosystem functioning (Koubbi *et al.* submitted). The bioregionalization/ecoregionalization approach reaches a complete geographic coverage in ecologically relevant parameters that have been classified (Grant

et al. 2006; see also Fig. 11.4) and databases provide a comprehensive source for species numbers and their occurrence. A handful of international working groups could reach highly synergistic effects by integrating these approaches, which could result in both a circumpolar mapping of biological processes and structures and, finally, an integrated "Antarctic biodiversity and ecology model".

11.4.2 Biodiversity and ecosystem functioning

Integrated research projects are a fundamental basis to decipher the relation between biodiversity and ecosystem functioning. One complex set of questions centers around the environmental processes generating biodiversity hot and cold spots. What are the main physical and biological drivers of the rich benthic suspension feeder communities? If silicate alone supported the abundant sponges, they would grow everywhere on the high-latitude shelf. Instead, they show complex population patterns. Or is the near-bottom current the main driver providing high amounts of food for the benthos, which is regionally and temporally highly unpredictable (see, for example, Montes-Hugo 2009)? Why are some areas dominated by single species whereas others are highly diverse? We know assemblages shaped by prey-predator relationships (Dayton *et al.* 1994) but why are they so rare? Is the co-occurrence of vivipary and indirect development an evolutionary adaptation to glacial conditions with isolated habitats and interglacials with large areas for colonization but with interspecific competition (Teixidó *et al.* 2006)? Are rare species just a quirk of nature, and only algae, nematodes, and krill matter, or does biodiversity contribute to the resilience of the ecosystem through stability and adaptation? Does iron limitation of the pelagic system support high or low diversity among primary and secondary producers? How can the pelagic system lose large amounts of its key prey Antarctic krill through climate-induced changes in the sea-ice dynamics, but allow whales to recover? Answers to these questions will not only contribute to advances in fundamental research but will also contribute to the needs of politicians and other decision-makers.

11.4.3 Response of marine biodiversity to global change and ecosystem services

When dealing with the response to climate change of any ecosystem there are two major concerns: (1) the reduction in ecosystem services; (2) the irreversible loss of biodiversity. For the latter, CAML has provided a benchmark of current biodiversity in certain habitats, but also changes have already been observed. Compared with the situation in other continents, most effects of climate change are still poorly understood. We do not know whether this is just the beginning of a rapid development similar to that seen in the Arctic or whether the Antarctic will generally remain relatively isolated from anthropogenic processes. To track such process we need long-term, regular observations as proposed for the Southern Ocean Observation System, supported by repeated CAML-type census activities conducted at regular intervals, station-based surveys, as well as the use of moorings. These must be at (1) sensitive habitats presumed to be affected in the near future, for example along the Antarctic Convergence (Cheung *et al.* 2009), (2) areas that can act as refuges, and (3) systems that have already experienced a significant change, for example west of the Antarctic Peninsula (Ducklow *et al.* 2007). Attention should also be paid to the Antarctic deep sea, because virtually nothing is known about its sensitivity to change (Kaiser & Barnes 2008). Several CAML projects have provided valuable information for predictive models on ecosystem developments, which need higher spatial resolution in physical parameters, information on extreme climate events, and more data on life-history traits of representative key ecological species. It might become easy – but currently it is not trivial – to correlate biodiversity information with environmental parameters, for example changes in the ice-loving biota related to changes in sea-ice dynamics. The biggest challenge, however, is to simulate synergistic effects through the trophic system, which amplify or buffer the effects of climate changes. In the case of acidification, future research should consider not only pteropods and the famous coccolithophorid *Emiliana huxleyi* but a cascade of associations including the planktonic anthomedusa *Pandea rubra*, pycnogonids, amphipods, and baby cuninid medusae (Lindsay *et al.* 2008), and benthic calcifying organisms, e.g. the coral *Errina*. It should examine the effects of increased ultraviolet radiation on the composition of food for benthic and pelagic consumers, and the consequences of increased particulate matter resulting from the retreat of the coastal glaciation, which affects primary producers and filter feeders.

11.4.4   New technology and gaps

Antarctica's biodiversity has two clear characteristics: there are only a few large and charismatic species such as penguins and whales, but tens of thousands of invertebrates. There are also legions of microbes, almost all of which are unknown. For the last two groups, the quality of next-generation biodiversity studies will depend upon how well we can identify species. Traditional methods of species identification may be supplanted by genetic methods, which may contribute to larger ecological and evolutionary concepts. Barcoding must be as applicable in the future as traditional methods are today (including the accessibility of information about the species described so far). The new genetic technology has a bright future if existing knowledge is not wasted and if it does not remain an elitist tool for geneticists. It must be a complementary method for use by all biologists as they would use computers or microscopes today. Because ecological studies depend on the biological species concept of interbreeding populations, it must also be agreed that genetically defined species serve as good proxies for the biologically defined species bringing the same degree of confidence as morphologically defined species did in the past. In addition, reconstruction of phylogeny demands the application and further development of modern genetic techniques, for example a better understanding of the "molecular clock". Until these aims are reached, traditional taxonomic work must continue to be supported including the development of new strategies to speed up the publication of hundreds of putative new species.

One of the biggest challenges to discovering Antarctica's life is to survey the large areas underneath the large floating ice shelves, some being up to several hundreds of meters thick. Only ROVs, autonomous underwater vehicles, gliders, and crawlers are suited to operate in this kind of habitat. Let us imagine such vessels are equipped with autonomously working gene sequence analyzers. That would be the "key" to surveying the biodiversity of this extreme habitat and answering major evolutionary and ecological questions. The same could be applied to permanent pack-ice areas and 60% of the Southern Ocean when it is ice-covered in winter. It has recently been discovered that benthic and pelagic life does not necessarily slow down during winter as formerly suggested, for example, by Gruzov (1977). Thus, it is very important for a general understanding of the Antarctic ecosystem to continue with studies on the adaptation of key ecological organisms to the extreme winter conditions (see, for example, Barnes & Clarke 1995b; Schnack-Schiel 2001).

At the ecological level, a promising strategy to discover unknown processes might be studies in well-known hot spots in one ecological subsystem but with poor knowledge of the rest of the ecosystem. These could be pelagic and benthic studies near feeding grounds of vertebrates or benthic deep-sea areas with and without krill concentrations. In addition, areas with intensive downwelling can be of specific scientific interest. They are rare, but such sites have been found at the slope of East Antarctica and are assumed to exist in the western Weddell Sea. It is not only interesting how organic material as food for pelagic and benthic life is rapidly transported from shallow to deeper waters but also what the role is of such a current for the evolutionary and short-term dispersal of shelf animals into the deep sea. The relevance of ecological interfaces for Antarctic biodiversity is frequently discussed in the scientific community. Results for interactions between ice and water column, euphotic zone and underlying water masses, or the pelagic–benthic coupling at all water depths should not remain tantalizingly out of reach.

11.5 Conclusions

CAML has demonstrated that life in the coldest marine ecosystem on Earth is rich, unique, and worthy of high-priority study. We have taken a significant step forward toward the long-term aim of a complete documentation of its biodiversity and a comprehensive understanding of its sculpting forces. The major findings at the evolutionary scale are of a large systematic coverage of cryptic species with distinct non-circumpolar occurrence. Extreme ecological heterogeneity exists at various spatial, biological, and temporal scales. The key to convincing decision-makers to finance a progressive continuation of this work is to assess the contribution of Antarctica's biodiversity to human well-being and ecosystem services. This can include sustainable exploitation of genetic information and the recognition of the role of the ecosystem as a natural CO_2 sink. Antarctica's life is part of the global biodiversity in which causes of and biological response to anthropogenic impact are spatially separated. It is recognized as "the canary in the coal mine", able to provide early warning of dire environmental effects of global warming (IPCC 2007). In this respect, future marine biodiversity surveys have an important role to fill.

Acknowledgments

We are very grateful to the Alfred P. Sloan Foundation and the national programs for financial support, and Angelika Brandt, Claude De Broyer, Russell Hopcroft, Alison Murray, Lucia de Siqueira Campos, Victoria Wadley (CAML secretariat), Stefano Schiaparelli, Sigrid Schiel, and Huw Griffiths for integration efforts and ideas. We also thank the CAML research community for allowing us to use preliminary results presented at the CAML Genoa symposium. Thanks are also due to the providers of data from SCAR-MarBIN and which are used in Fig. 11.3.

References

AAVV (1977) The structure and function of marine benthic ecosystems. Discussion of part II: In: *Adaptations within the Antarctic Marine Benthic Ecosystem. Proceedings of the third SCAR symposium on Antarctic Biology.* (ed. G. A. Llano), pp. 1277–1230. Washington, DC: Smithsonian Institution.

Allcock, A.L. (2005) On the confusion surrounding *Pareledone charcoti*: endemic radiation in the Southern Ocean. *Zoological Journal of the Linnean Society* 143, 75–108.

Allcock, A.L., Strugnell, J.M., Prodöhl, P. *et al.* (2007) A new species of *Pareledone* from Antarctic Peninsula waters. *Polar Biology* 30, 883–893.

Allcock, A.L., Barratt, I., Eléaume M., *et al.* (in press) Cryptic speciation and the circumpolarity debate: a case study on endemic Southern Ocean octopuses using the *coxI* barcode of life. *Deep-Sea Research II.*

Arango, C.P., Soler Membrives, A. & Miller, K.J. (in press) Genetic differentiation in the circum-Antarctic sea spider *Nymphon australe* (Pycnogonida: Nymphonidae). *Deep-Sea Research II.*

Arntz, W.E. & Gallardo, V.A. (1994) Antarctic benthos: present positions and future prospects. In: *Antarctic Science* (ed. G. Hempel), pp. 243–277. Berlin: Springer.

Arntz, W.E. & Rios, C. (eds.) (1999) Magellan – Antarctic ecosystems that drifted apart. *Scientia Marina* 63 (Suppl. 1). 518 pp.

Arntz, W.E. & Clarke, A. (2002) *Ecological Studies in the Antarctic Sea Ice Zone.* Berlin: Springer.

Arntz, W.E., Lovrich, G.A. & Thatje, S. (eds.) (2005) The Magellan–Antarctic connection: links and frontiers at southern high latitudes. *Scientia Marina* 69 (Suppl. 2). 373 pp.

Atkinson, A., Siegel, V., Pakhomov, E.A., *et al.* (2008) Oceanic circumpolar habitats of Antarctic krill. *Marine Ecology Progress Series* 362, 1–23.

Balks, M.R., Cummings, V., Green, T.G.A., *et al.* (eds.) (2006) The Latitudinal Gradient Project (LGP). *Antarctic Science*, 18 (Special issue).

Barnes, D.K.A. & Clarke, A. (1995a) Epibiotic communities on sublittoral macroinvertebrates at Signy Island, Antarctica. *Journal of the Marine Association of the United Kingdom*, 75, 689–703.

Barnes, D.K.A. & Clarke, A. (1995b) Seasonality of feeding activity in Antarctic suspension feeders. *Polar Biology* 15, 335–340.

Barnes, D.K., Hodgson, D.A., Convey, P. *et al.* (2006) Incursion and excursion of Antarctic biota: past, present and future. *Global Ecology and Biogeography* 15, 121–142.

Barnes, D.K.A. & Conlan, C. (2007) Disturbance, colonization and development of Antarctic benthic community. *Philosophical Transactions of the Royal Society of London B* 362, 11–38.

Bowden, D., Schiaparelli, D.A., Clark, M.R., *et al.* (submitted) A lost world? Archaic crinoid-dominated assemblages on an Antarctic seamount. *Deep-Sea Research II.*

Brandt, A. & Ebbe, B. (2007) Antarctic benthic DEEP-sea biodiversity: colonization, history and recent community patterns (ANDEEP-III). *Deep-Sea Research II* 54, 1645–1904.

Brandt, A., Gooday, A.J., Brandão, S.N., *et al.* (2007a) First insights into the biodiversity and biogeography of the Southern Ocean deep sea. *Nature*, 447, 307–311.

Brandt, A., De Broyer, C., De Mesel, I., *et al.* (2007b) The biodiversity of the deep Southern Ocean benthos. *Philosophical Transactions of the Royal Society of London B* 362, 39–66.

Bullivant, J.S. (1967) Ecology of the Ross Sea benthos. *Bulletin – New Zealand Department of Scientific and Industrial Research* 176, 49–78.

CCAMLR (2008) *Report of the Twenty-Seventh Meeting of the Commission.* Hobart: Commission for the Conservation of Antarctic Marine Living Resources.

Chao, A. (2005) Species estimation and applications. In: *Encyclopedia of Statistical Sciences* (eds. N. Balakrishnan, C.B. Read & B. Vidakovic), pp. 7907–7916. New York: Wiley.

Cheung, W.W.L., Lam, V.W.Y., Sarmiento, J.L., *et al.* (2009) Projecting global marine biodiversity impacts under climate change scenarios. *Fish and Fisheries* 10, 235–251.

Chiantore, M., Cattaneo-Vietti, R., Berkman, P.A., *et al.* (2001) Antarctic scallop (*Adamussium colbecki*) spatial population variability along the Victoria Land Coast, Antarctica. *Polar Biology* 24, 139–143.

Clarke, A. & Crame, J.A. (1989) The origin of the Southern Ocean marine fauna. In: *Origins and Evolution of the Antarctic Biota* (ed. J.A. Crame), pp. 253–268. Geological Society Special Publications, vol. 7. London: The Geological Society.

Clarke, A. & Johnston, N.M. (2003) Antarctic marine benthic diversity. *Oceanography and Marine Biology* 41, 47–114.

Clarke, A., Barnes, D.K.A. & Hodgson, D.A. (2005) How isolated is Antarctica? *Trends in Ecology and Evolution* 20, 1–3.

Clarke, A., Arntz, W.E. & Smith, C.R. (2006) EASIZ: ecology of the Antarctic Sea Ice Zone. *Deep-Sea Research II*, **53** (Special volume), 803–1140.

Clarke A., Griffiths, H.J., Barnes D.K.A., *et al.* (2009) Spatial variation in sea-bed temperatures in the Southern Ocean: Implications for benthic ecology and biogeography. *Journal of Geophysical Research* **114**, G03003.

Dater, H.M. (1975) History of Antarctic exploration and scientific investigation. In: *Antarctic Map Folio Series*, Folio 19, (ed. V. C. Bushnell), pp. 1–6. New York: American Geographical Society.

Dayton, P.K., Mordida, B.J. & Bacon, F. (1994) Polar marine communities. *American Zoologist* **34**, 90–99.

Dayton, P.K. (1990) Polar benthos. In: *Polar Oceanography Part B: Chemistry Biology and Geology* (ed. W. O. Smith), pp. 631–685. London: Academic Press.

De Broyer C., Lowry, J.K., Jazdzewski, K., *et al.* (2007) Catalogue of the gammaridean and corophiidean Amphipoda (Crustacea) of the Southern Ocean with distribution and ecological data. In: *Census of Antarctic Marine Life. Synopsis of the Amphipoda of the Southern Ocean*, Vol. 1. (ed. C. De Broyer), Bulletin de l'Institute Royal des Sciences Naturelles de Belgique Series Biologie, **77** (Suppl.), 1–325.

De Broyer, C. & Danis, B. *SCAR-MarBIN: The Antarctic Marine Biodiversity Information Network*. http://www.scarmarbin.be/

De Broyer, C. & Danis, B. (submitted) How many species in the Southern Ocean? Towards a dynamic inventory of Antarctic marine species. *Deep-Sea Research II*.

Díaz, A., Féral, J.-P., David B., *et al.* (in press) Evolutionary pathways among shallow and deep sea echinoids of the genus *Sterechinus* in the Southern Ocean. *Deep-Sea Research II*.

Ducklow, H.W., Baker, K., Martinson, D.G., *et al.* (2007) Marine pelagic ecosystems: the West Antarctic Peninsula. *Philosophical Transactions of the Royal Society of London B* **362**, 67–94.

d'Udekem d'Acoz, C. & Vader, W. (2009) On *Liljeborgia fissicornis* (M. Sars, 1858) and three related new species from Scandinavia, with a hypothesis on the origin of the group *fissicornis*. *Journal of Natural History* **43**, 2087–2139.

Eastman, J.T. & Grande, L. (1989) Evolution of the Antarctic fish fauna with emphasis on the recent notothenioids. *Geological Society of London Special Publications* **47**, 241–252.

Eastman, J., Gutt, J. & di Prisco, G. (eds.) (2004) *Adaptive evolution of Antarctic marine organisms*. *Antarctic Science* **16** (1) (Special issue). 93 pp.

Eléaume, M., Beaman, R.J., Griffiths, H.J., *et al.* (in press) Near bottom current direction inferred from comatulid crinoid feeding postures on the Terre Adélie and George V Shelf, East Antarctica. *Deep-Sea Research II*.

France, S.C. & Kocher, T.D. (1996) Geographic and bathymetric patterns of mitochondrial 16S rRNA sequence divergence among deep-sea amphipods, *Eurythenes gryllus*. *Marine Biology* **126**, 633–643.

Gerdes, D., Hilbig, B. & Montiel, A. (2003) Impact of iceberg scouring on macrobenthic communities in the high-Antarctic Weddell Sea. *Polar Biology* **26**, 295–301.

Gieskes, W.W.C., Veth, C., Woehrmann, A., *et al.* (1987) Secchi disc visibility world record shattered. *EOS* **68**, 123.

Godday, A.J., Cedhagen, T., Kamenskaya, O.E., *et al.* (2007) The biodiversity and biogeography of komokiaceans and other enigmatic foraminiferan-like protists in the deep Southern Ocean. *Deep-Sea Research II* **54**, 1691–1719.

González Wevar D., David, B. & Poulin, E. (in press) Phylogeography and demographic inference in *Nacella (Patinigera) concinna* (Strebel, 1908) in the western Antarctic Peninsula. *Deep-Sea Research II*.

Grant, R.A. & Linse, K. (2009) Barcoding Antarctic biodiversity: current status and the CAML initiative, a case study of marine invertebrates. *Polar Biology* **32** 1629–1637.

Grant, S., Constable, A., Raymond, B., *et al.* (2006) Bioregionalisation of the Southern Ocean: Report of Experts Workshop, Hobart, September 2006. WWF-Australia and ACE CRC.

Griffiths, H.J., Barnes, D.K.A. & Linse, K. (2009) Towards a generalized biogeography of the Southern Ocean. *Journal of Biogeography* **36**, 162–177.

Griffiths H.J., Danis, B. & Clarke, A. (submitted) Quantifying Antarctic marine biodiversity: the SCAR-MarBIN data portal. *Deep-Sea Research II*.

Gruzov, E.N. (1977) Seasonal alternations in coastal communities in the Davis Sea. In: *Adaptations within Antarctic Ecosystems* (ed. G. A. Llano), pp. 263–278. Washington, DC: Smithsonian Institution.

Gutt, J., & Piepenburg, D. (1991) Dense aggregations of deep-sea holothurians in the southern Weddell Sea, Antarctica. *Marine Ecology Progress Series* **253**, 77–83.

Gutt, J. & Schickan, T. (1998) Epibiotic relationships in the Antarctic benthos. *Antarctic Science* **10**, 398–405.

Gutt, J. (2000) Some "driving forces" structuring communities of the sublittoral antarctic macrobenthos. *Antarctic Science* **12**, 297–313.

Gutt, J. & Piepenburg, D. (2003) Scale-dependent impact on diversity of Antarctic benthos caused by grounding of icebergs. *Marine Ecology Progress Series* **253**, 77–83.

Gutt, J., Sirenko, B.I., Smirnov, I.S., *et al.* (2004) How many macro-zoobenthic species might inhabit the Antarctic shelf? *Antarctic Science* **16**, 11–16.

Gutt, J. (2006). Coexistence of macro-zoobenthic species on the Antarctic shelf: an attempt to link ecological theory and results. *Deep-Sea Research II* **53**, 1009–1028.

Gutt, J. (2007) Antarctic macro-zoobenthic communities: a review and a classification. *Antarctic Science* **19**, 165–182.

Gutt, J., Koubbi, P. & Eléaume, M. (2007) Mega-epibenthic diversity off Terre Adélie (Antarctica) in relation to disturbance. *Polar Biology* **30**, 1323–1329.

Gutt J., Barratt I., Domack, E., *et al.* (in press) Biodiversity change after climate-induced ice-shelf collapse in the Antarctic. *Deep-Sea Research II*.

Hardy, C., David, B., Rigaud, T., *et al.* (in press) Ectosymbiosis associated with cidaroids (Echinodermata: Echinoidea) promoted benthic colonization of the seafloor in the Larsen embayments, western Antarctica. *Deep-Sea Research II*.

Havermans, C., Nagy, Z.T., Sonet, G., *et al.* (submitted) DNA barcoding reveals cryptic diversity in Antarctic species of *Orchomene* sensu lato (Crustacea: Amphipoda: Lysianassoidea). *Deep-Sea Research II*.

Hedgpeth, J.W. (1969) Introduction to Antarctic Zoogeography: Distribution of selected groups of marine invertebrates in waters south of 35°S latitude. In: *Antarctic Map Folio Series*, Folio 11, (eds. V.C. Bushnell & J.W. Hedgpeth), pp. 1–29. New York: American Geographical Society.

Hedgpeth, J.W. (1977) The Antarctic marine ecosystem. In: *Adaptations within Antarctic Ecosystems. Proceedings of the Third SCAR Symposium on Antarctic Biology* (ed. G.A. Llano), pp. 3–10. Washington, DC: Smithsonian Institution

Held, C. (2003). Molecular evidence for cryptic speciation within the widespread Antarctic crustacean *Ceratoserolis trilobitoides* (Crustacea, Isopoda). In: *Antarctic Biology in a Global Context* (eds. A.H. Huiskes, W.W. Gieskes, J. Rozema, *et al.*), pp. 135–139. Leiden: Backhuys.

Held, C. & Wägele, J.-W. (2005). Cryptic speciation in the giant Antarctic isopod *Glyptonotus antarcticus* (Isopoda: Valvifera: Chaetiliidae). *Scientia Marina* **69**, 175–181.

Hempel, G. (1993) *Weddell Sea Ecology*. Berlin: Springer.

Hétérier, V., David, B., De Ridder, C., *et al.* (2008) Ectosymbiosis is a critical factor in the local benthic biodiversity of the Antarctic deep sea. *Marine Ecology Progress Series* **364**, 67–76.

Hosie, G.W., Fukuchi, M. & Kawaguchi, S. (2003) Development of the Southern Ocean Continuous Plankton Recorder Survey. *Progress in Oceanography* **58**, 263–283.

Hosie, G.W., Stoddart, D.M., Wadley, V., *et al.* (2007) The Census of Antarctic Marine Life and the Australian-French-Japanese CEAMARC (Collaborative East Antarctic Marine Census) contribution. In: *Proceedings of the International Symposium Asian Collaboration in IPY 2007–2008*, pp. 47–50. Tokyo: National Institute of Polar Research, Tokyo.

Hunt, B.P.V. & Hosie, G.W. (2003) The Continuous Plankton Recorder in the Southern Ocean: a comparative analysis of zooplankton communities sampled by the CPR and vertical net hauls along 140°E. *Journal of Plankton Research* **25**(12), 1–19.

Hunt, B.P.V. & Hosie, G.W. (2005) Zonal structure of zooplankton communities in the Southern Ocean south of Australia: results from a 2150 kilometre Continuous Plankton Recorder transect. *Deep-Sea Research I* **52**, 1241–1271.

Hunt, B.P.V. & Hosie, G.W. (2006a) Seasonal zooplankton community succession in the Southern Ocean south of Australia, part I: the seasonal ice zone. *Deep-Sea Research I* **53**, 1182–1202.

Hunt, B.P.V. & Hosie, G.W. (2006b) Seasonal zooplankton community succession in the Southern Ocean south of Australia, part II: the sub-Antarctic to polar frontal zones. *Deep-Sea Research I* **53**, 1203–1223.

Hunter, R.L. & Halanych, K.M. (2008) Evaluating connectivity in the brooding brittle star *Astrotoma agassizii* across the Drake Passage in the Southern Ocean. *Journal of Heredity* **99**, 137–148.

IPCC (2007) *Climate Change 2007*, IPCC Fourth Assessment Report. New York: WMO and UNEP.

Janussen, D. & Reiswig, H.M. (2009): Hexactinellida (Porifera) from the ANDEEP III expedition to the Weddell Sea, Antarctica. *Zootaxa* **2136**, 1–20.

Jelmert, A. & Oppen-Berntsen, D.O. (1995) Whaling and deep-sea biodiversity. *Conservation Biology* **10**, 653–654.

Kaiser, S., Griffiths, H.J., Barnes D.K.A., *et al.* (in press) Is there a distinct continental slope fauna in the Antarctic? *Deep-Sea Research II*.

Kaiser, S. & Barnes, D.K.A. (2008) Southern Ocean deep-sea biodiversity, sampling and predicting responses to climate change. *Climate Research* **37**, 165–179.

Kock, K-H. (1992). *Antarctic Fish and Fisheries*. Cambridge, UK: Cambridge University Press.

Koubbi, P., Moteki, M., Duhamel, G. *et al.* (submitted) Ecological importance of micronektonic fish for the ecoregionalisation of the Indo-Pacific sector of the Southern Ocean: role of myctophids. *Deep-Sea Research II*.

Laws, R.M. (1977) The significance of vertebrates in the Antarctic marine ecosystem. In: *Adaptations within Antarctic Ecosystem* (ed. G. A. Llano), pp. 411–438, Houston: Gulf Publishing.

Leese, F. & Held, C. (2008). Identification and characterization of microsatellites from the Antarctic isopod *Ceratoserolis trilobitoides*: nuclear evidence for cryptic species. *Conservation Genetics* **9**, 1369–1372.

Lindsay, D.J., Pagès, F., Corbera, J., *et al.* (2008) The anthomedusan fauna of the Japan Trench: preliminary results from in situ surveys with manned and unmanned vehicles. *Journal of the Marine Biological Association of the United Kingdom* **88**, 1519–1539.

Linse, K., Walker, L.J. & Barnes, D.K.A. (2008) Biodiversity of echinoids and their epibionts around the Scotia Arc, Antarctica. *Antarctic Science* **20**, 227–244.

Lockhart, S.J. & Jones, C.D. (2008) Biogeographic patterns of benthic invertebrate megafauna on shelf areas within the Southern Ocean Atlantic sector. *CCAMLR Science* **15**, 167–192.

Montes-Hugo, M., Doney, S.C., Ducklow, H.W., *et al.* (2009) Recent changes in phytoplankton communities associated with rapid regional climate change along the Western Antarctic Peninula. *Science* **323**, 1470–1473.

Niemann, H., Fischer, D., Graffe, D., *et al.* (2009) Biogeochemistry of a low-activity cold seep in the Larsen B area, western Weddell Sea, Antarctica. *Biogeosciences* **6**, 5741–5769.

O'Driscoll, R.L., Macaulay, G.J., Gauthier, S., *et al.* (in press) Distribution, abundance and acoustic properties of Antarctic silverfish (*Pleuragramma antarcticum*) in the Ross Sea. *Deep-Sea Research II*.

O'Loughlin, P.M., Paulay, G., Davey, N., *et al.* (in press) The Antarctic Region as a marine biodiversity hotspot for echinoderms: diversity and diversification of sea cucumbers. *Deep-Sea Research II*.

Orsi, A.H., Whitworth III, T. & Nowlin Jr., W.D. (1995) On the meridional extent and fronts of the Antarctic Circumpolar Current. *Deep-Sea Research I* **42**, 641–673.

Page, T.J. & Linse, K. (2002) More evidence of speciation and dispersal across the Antarctic Polar Front through molecular systematics of Southern Ocean *Limatula* (Bivalvia: Limidae). *Polar Biology* **25**, 818–826.

Pinkerton, M.H., Smith, A.N.H., Raymond, B., *et al.* (2010) Spatial and seasonal distribution of adult *Oithona similis* in the Southern Ocean: predictions using boosted regression trees. *Deep-Sea Research I* **57**, 469–485.

Post, L.A., Beaman, R.J., O'Brien, P.E., *et al.* (in press) Community structure and benthic habitats across the George V Shelf, East Antarctica: trends through space and time. *Deep-Sea Research II*.

Potthoff, M., Johst, K. & Gutt, J. (2006) How to survive as a pioneer species in the Antarctic benthos: minimum dispersal distance as a function of lifetime and disturbance. *Polar Biology* **29**, 543–551.

Rapp, H.T., Janussen, D. & Tendal, O.S. (in press) Calcareous sponges from abyssal and bathyal depths in the Weddell Sea, Antarctica. *Deep-Sea Research II*.

Raupach, M.J., Held, C. & Wägele, J.-W. (2004) Multiple colonization of the deep sea by the Asellota (Crustacea: Peracarida: Isopoda). *Deep-Sea Research II* **51**, 1787–1795.

Raupach, M. & Wägele, J.-W. (2006) Distinguishing cryptic species in Antarctic Asellota (Crustacea: Isopoda) – a preliminary study of mitochondrial DNA in *Acanthaspidia drygalskii*. *Antarctic Science* **18**, 191–198.

Raupach M.J., Mayer C., Malyutina M., *et al.* (2009) Multiple origins of deep-sea Asellota (Crustacea: Isopoda) from shallow waters revealed by molecular data. *Proceedings of the Royal Society of London B* **276**, 799–808.

Raymond, B. & Hosie, G. (2009) Network-based exploration and visualisation of ecological data. *Ecological Modelling* **220**, 673–683.

Scheidat, M., Kock, K.-H., Friedlaender, A., *et al.* (2007a) *Using helicopters to survey Antarctic minke whale abundance in the ice SC/59/IA20. Working paper presented at the International Whaling Commission 2007*. 10 pp.

Scheidat, M., Kock, K.-H., Friedlaender, A., *et al.* (2007b) *Preliminary results of aerial surveys around Elephant Island and the South Shetland Islands SC/59/IA21. Working paper presented at the International Whaling Commission 2007*. 7 pp.

Schiaparelli S., Alvaro M.C., Bohn J., *et al.* (in press) 'Hitchhiker' polynoids (Phyllodocida: Polynoidae) in cold waters: an overlooked case of parasitic association in Antarctica. *Antarctic Science*.

Schnack-Schiel, S.B. (2001) Aspects of the study of the life cycles of Antarctic copepods. *Hydrobiologia* **453–454**, 9–24.

Schrödl, M., Bohn J.M., Brenke, N., *et al.* (submitted) Abundance, diversity and latitudinal gradients of south-eastern Atlantic and Antarctic abyssal gastropods. *Deep-Sea Research II*.

Schwabe, E., Bohn, J.M., Engl, W., *et al.* (2007) Rich and rare – first insights into species diversity and abundance of Antarctic abyssal Gastropoda (Mollusca). *Deep-Sea Research II* **54**, 1831–1847.

Seiler, J. & Gutt, J. (2007) Can dead sponges still talk? *Antarctic Science* 19(3), 337–338.

Siciński, J., Jażdżewski, K., De Broyer, C., *et al.* (submitted) Admiralty Bay benthos diversity: a long-term census. *Deep-Sea Research II*.

Smale, D.A., Brown, K.M., Barnes, D.K.A., *et al.* (2008) Ice scour disturbance in Antarctic waters. *Science* 321, 371.

Sokolov, S. & Rintoul, S.R. (2002) Structure of Southern Ocean fronts at 140°E. *Journal of Marine Systems* 37, 151–184.

Strugnell, J.M., Rogers, A.D., Prodöhl, P.A., *et al.* (2008) The thermohaline expressway: the Southern Ocean as a centre of origin for deep-sea octopuses. *Cladistics* 24, 853–860.

Strugnell, J., Cherel, Y., Cooke, I.R., *et al.* (in press) The Southern Ocean: source and sink? *Deep-Sea Research II*.

Takahashi, K., Kawaguchi, S., Kobayashi, M., *et al.* (2002) Zooplankton distribution patterns in relation to the Antarctic Polar Front Zones recorded by Continuous Plankton Recorder (CPR) during 1999/2000 Kaiyo Maru cruise. *Polar Bioscience* 15, 97–107.

Takahashi, K.T., Kawaguchi, S., Hosie, G.W., *et al.* (2010) Surface zooplankton distribution in the Drake Passage recorded by Continuous Plankton Recorder (CPR) in late austral summer of 2000. *Polar Science* 3, 235–245.

Teixidó, N., Gili, J.-M., Uriz, M.-J., *et al.* (2006) Observations of asexual reproductive strategies in Antarctic hexactinellid sponges from ROV video records. *Deep-Sea Research II* 53, 972–984.

Thomas, D.N., Fogg, G.E., Convey, P. *et al.* (2008) *The Biology of Polar Regions*. Oxford: Oxford University Press.

Thomas, D.N. & Dieckmann, G.S. (2009) *Sea Ice*, 2nd edn. Oxford: Blackwell Publishing Ltd.

Turner, J., Convey, P., di Prisco, G. *et al.* (2009) *Antarctic Climate Change and the Environment*. Cambridge, UK: Scientific Committee on Antarctic Research & Scott Polar Research Institute.

Umeda, H., Hosie, G.W., Odate, T., *et al.* (2002) Surface zooplankton communities in the Indian Ocean Sector of the Antarctic Ocean in early summer 1999/2000 observed with a Continuous Plankton Recorder. *Antarctic Record* 46, 287–299.

Wilson, N.G., Belcher, R.L., Lockhart, S.J., *et al.* (2007) Multiple lineages and an absence of panmixia in the 'circumpolar' crinoid *Promachocrinus kerguelensis* in the Atlantic sector of Antarctica. *Marine Biology* 152, 895–904.

Wilson, N.G., Schrödl, M., Halanych, K.M. (2009) Ocean barriers and glaciation: evidence for explosive radiation of mitochondrial lineages in the Antarctic sea slug *Doris kerguelensis* (Mollusca, Nudibranchia). *Molecular Ecology* 18, 965–984.

Würzberg, L., Peters, J., Schüller, M., *et al.* (submitted) Different lifestyles of deep-sea polychaetes: first insights from fatty acid analyses. *Deep-Sea Research II*.

PART III
Oceans Present – Global Distributions

Chapter 12

A Global Census of Marine Microbes

Linda Amaral-Zettler[1], Luis Felipe Artigas[2], John Baross[3], Loka Bharathi P.A.[4], Antje Boetius[5], Dorairajasingam Chandramohan[6], Gerhard Herndl[7], Kazuhiro Kogure[8], Phillip Neal[1], Carlos Pedrós-Alió[9], Alban Ramette[5], Stefan Schouten[7], Lucas Stal[10], Anne Thessen[1], Jan de Leeuw[7], Mitchell Sogin[1]

[1]*Marine Biological Laboratory, Woods Hole, Massachusetts, USA*
[2]*Université Lille Nord de France, Université du Littoral Côte d'Opale, CNRS UMR 8187 LOG, Wimereux, France*
[3]*University of Washington, Seattle, Washington, USA*
[4]*Microbiology Laboratory, National Institute of Oceanography, Dona Paula, Goa, India*
[5]*Max Planck Institute for Marine Microbiology, Bremen, Germany*
[6]*Ratnapuri Colony, J.N. Salai, Koyambedu, Chennai, India*
[7]*The Royal Netherlands Institute for Sea Research, Den Burg, Texel, The Netherlands*
[8]*Ocean Research Institute, University of Tokyo, Tokyo, Japan*
[9]*Institut de Ciències del Mar, Barcelona, Spain*
[10]*Netherlands Institute of Ecology, Yerseke, The Netherlands*

12.1 Introduction

12.1.1 Importance

The oceans abound with single cells that are invisible to the unaided eye, encompassing all three domains of life – Bacteria, Archaea, and Eukarya – in a single drop of water or a gram of sediment (Figs. 12.1A, B, C, and D). The microbial world accounted for all known forms of life for more than 80% of Earth's history. Today, microbes continue to dominate every corner of our biosphere, especially in the ocean where they might account for as much as 90% of the total biomass (Fuhrman *et al.* 1989; Whitman *et al.* 1998). Even the most seemingly inhospitable marine environments

host a rich diversity of microbial life (Figs. 12.1E and H). For the past six years, microbial oceanographers from around the world have joined the effort of the International Census of Marine Microbes (ICoMM, Box 12.1) to explore this vast diversity. In this chapter we provide a brief history of what is known about marine microbial diversity, summarize our achievements in performing a global census of marine microbes, and reflect on the questions and priorities for the future of the marine microbial census.

From the time of their origins, single-cell organisms – initially anaerobic and later aerobic – have served as essential catalysts for all of the chemical reactions within biogeochemical cycles that shape planetary change and habitability. Marine microbes carry out half of the primary production on the planet (Field *et al.* 1998). Microbial carbon re-mineralization, with and without oxygen, maintains the carbon cycle. Microbes account for more than 95% of the respiration in the oceans (Del Giorgio & Duarte 2002). They control global utilization of nitrogen through

Fig. 12.1

Microbial life spans all three domains of life inclusive of Bacteria, Archaea, and Eukarya and their associated viruses. This collage shows examples of the types of marine microbes and diverse habitats included in the microbial census. Photograph credits are given in parentheses. From the leftmost panel, **(A)** a *Synechococcus* phage (John Waterbury), **(B)** filaments of the marine cyanobacterium *Lyngbya* (David Patterson, used under license), **(C)** the hyperthermophilic archaeon "GR1" (Melanie Holland), and **(D)** a single-celled eukaryote called an acantharian (Linda Amaral-Zettler, used under license). Examples of diverse environments sampled as part of the microbial census include the following: **(E)** the Lost City Hydrothermal Vent flange actively venting heated hydrogen and methane rich fluids, (IFE, URI-IAO, UW, Lost City science party, and NOAA); **(F)** the sandy coastline from the North Sea island Sylt (Angélique Gobet); **(G)** the open ocean waters of the South Pacific Ocean (Katsumi Tsukamoto), and **(H)** the waters off the Antarctic Peninsula (Hugh Ducklow).

N_2 fixation, nitrification, nitrate reduction, and denitrification, and drive the bulk of sulfur, iron, and manganese biogeochemical cycles (Kirchman 2008; Whitman *et al.* 1998). Marine microbes regulate the composition of the atmosphere, influence climate, recycle nutrients, and decompose pollutants. Without microbes, multicellular animals on Earth would not have evolved or persisted over the past 500 million years.

Measuring microbial diversity in a broad range of marine ecosystems (see, for example, Figs. 12.1E, F, G, and H) will facilitate quantification of the magnitude and dynamics of the microbial world and its stability through space and

Box 12.1

A Brief History of ICoMM

ICoMM is one of 14 Census of Marine Life ocean realm projects that explores the diversity, distribution, and abundance of microbial life in the oceans. ICoMM's leadership represents a collaborative effort between the Royal Netherlands Institute for Sea Research (NIOZ), in Texel, The Netherlands, and the Marine Biological Laboratory (MBL) in Woods Hole, MA, USA. Collectively ICoMM has provided a means to galvanize the microbial oceanographic community in conducting a global census of marine microorganisms. The goal of ICoMM is to determine the range of genetic diversity and relative numbers of different microbial organisms at sampling sites throughout the world's oceans.

Since 2004, ICoMM has provided support for training workshops and meetings including five primary working groups (Benthic, Open Ocean and Coastal Systems, Technology, Informatics and Data Management, and Microbial Eukaryotes), and its Scientific Advisory Council that engage the international community of marine microbiologists. In 2006, ICoMM served as the coordinating body that helped to secure funding from the W. M. Keck Foundation for a 454 DNA pyrosequencing system dedicated to DNA tag sequencing projects. Additional information about ICoMM's membership, scope and activities can be found on the ICoMM website: icomm.mbl.edu.

time. The phylogenetic and physiological diversities of microbes are considerably greater than those of animals and plants, and microbial interactions with other life-forms are correspondingly more complex (Pace 1997). Measuring marine microbial diversity and determining corresponding associated functions will thus provide a wealth of information about specific microbial processes of great significance such as wastewater treatment, industrial chemical production, pharmaceutical production, bioremediation, and global warming. Examining the relationships between microbial populations and whole communities within their dynamic environment will allow us to formulate better the definition of what constitutes an ecologically relevant species in the microbial world. Molecular methods rely upon measures of genetic similarity to describe operational taxonomic units (OTUs). Statistical treatments can use the relative number of distinct OTUs to estimate diversity, but these inferences do not translate directly into numbers of microbial species. Microbiologists have not reached consensus on the definition of microbial species using either molecular or phenotypic approaches. However, ecological concepts of microbial species based upon molecular data will inform theoretical applications and guide solutions to major challenges facing science and human society.

12.1.2 Microbial diversity and abundance

The reliance upon traditional cultivation and staining techniques led to gross underestimates of microbial abundance and species richness in both oceanic and terrestrial

environments (Jannasch & Jones 1959; Zimmermann & Meyer-Reil 1974; Hobbie *et al.* 1977) (Fig. 12.2). The application of fluorescence-based microscopy coupled with DNA staining methods revealed the great "plate count anomaly", which posits that microbiologists have underestimated microbial abundances by at least three orders of magnitude. Instead of a mere 100 cells per milliliter of seawater, nucleic-acid staining technology showed the number of bacteria in the open ocean exceeds 10^{29} cells, with average cell concentrations of 10^6 per milliliter of seawater (Whitman *et al.* 1998). In marine surface sediments, cell abundances are 10^8–10^9 per gram, and even in the greatest depths of the subsurface seabed, more than 10^5 cells per gram are encountered (Jørgensen & Boetius 2007). The ocean also hosts the densest accumulations of microbes known on Earth, reaching 10^{12} cells per milliliter, like the photosynthetic mats thriving in hypersaline environments, and the methanotrophic mats of anoxic seas, resembling ancient microbial assemblages before the advent of eukaryote grazers (Knittel & Boetius 2009). Archaeal cell abundances rival those of bacteria in certain parts of the ocean and the seabed, and microbial eukaryotic (protistan) densities vary widely from tens of cells per liter to bloom conditions that can surpass 10^6 cells per milliliter of seawater.

As of 2010, cultivation-based studies have described more than 10,000 bacterial and archaeal species (http://www.bacterio.cict.fr/number.html) and an estimated 200,000 protistan species (Corliss 1984; Lee *et al.* 1985; Patterson 1999; Andersen *et al.* 2006). Cultivation-independent studies that rely upon molecular methods such

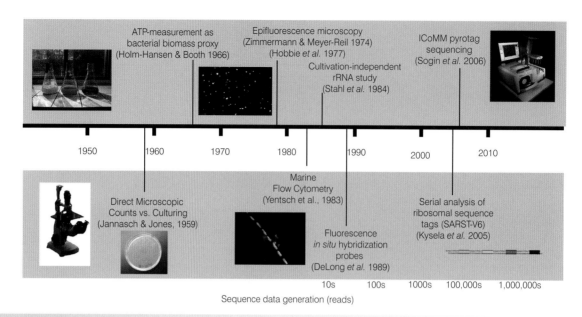

Fig. 12.2

A timeline showing milestones in advances in technology that have enabled the microbial census (Jannasch & Jones 1959; Holm-Hansen & Booth 1966; Zimmermann & Meyer-Reil 1974; Hobbie *et al.* 1977; Yentsch *et al.* 1983; Stahl *et al.* 1984; DeLong *et al.* 1989; Kysela *et al.* 2005; Sogin *et al.* 2006). Upper right photograph by Tom Kleindinst, Woods Hole Oceanographic Institution.

as the sequencing of 16S ribosomal RNA (rRNA) genes show microbial diversity to be approximately 100 times greater (Pace 1997). With each new molecular survey, this window on the microbial world has increased in size.

12.2 Challenges of a Microbial Census

The ocean covers 70% of the Earth's surface (an estimated volume of about 2×10^{18} m³) and has an average depth of 3,800 m. Strategies for conducting a census must consider the enormous geographical area to be surveyed, an almost unimaginable number of cells, and the impact of spatial gradients and temporal shifts on microbial assemblages. In fact, before ICoMM, little was known about global patterns in microbial communities. Basic questions such as "is there a distinct difference between pelagic and benthic microbial communities?" or "what is the temporal turnover in microbial cells between two sampling dates?" profoundly influenced our sampling strategies.

Contemporary molecular approaches typically use rRNA sequences as proxies for the occurrence of different microbial genomes in an environmental DNA sample (coding regions for functional genes can also provide information about microbial population structures). However, the expense of conventional DNA sequencing has constrained the number of homologous sequences that microbial

ecologists typically collect to describe community composition. Relative to the number of microbes in most samples, these surveys superficially describe microbial community structures. There are more than 10^8 microorganisms in a liter of seawater or a gram of soil (Whitman *et al.* 1998). Few studies collect more than 10^4 sequences, which correspond to 0.01% of the cells in a liter of seawater or a gram of soil. The detection of organisms that correspond to the most abundant OTUs or species equivalents requires minimal molecular sampling, whereas the recovery of sequences from rare taxa that constitute the "long tail" of low abundance organisms in taxon rank–abundance curves demands surveys that are orders of magnitude larger.

As an alternative to analyzing nearly full-length polymerase chain reaction (PCR) amplicons of rRNA genes from environmental DNA samples, short sequence tags from hypervariable regions in rRNAs (pyrotags) can provide measures of diversity (species or OTU richness) and relative abundance (evenness) of OTUs in microbial communities. When combined with the massively parallel capacity of "next generation" DNA sequencing technology that allows for the simultaneous sequencing of hundreds of thousands of templates (Margulies *et al.* 2005), it becomes possible to increase the number of sampled gene sequences in an environmental survey by orders of magnitude (Huber *et al.* 2007; Sogin *et al.* 2006). Enumerating the number of different rRNA pyrotags provides a first-order description of the relative occurrence of specific microbes in a population. The highly variable nature of the tag sequences and paucity of positions do not allow direct inference of phylogenetic frameworks. However, when tag sequences are queried against a comprehensive reference database of hypervariable regions within the context of full-length sequences, it is possible to extract information about taxonomic identity and microbial diversity. Initial tests of this innovative technology examined the microbial population structures of samples from the meso- and bathypelagic realm of the North Atlantic Deep Water Flow and two diffuse flow samples from Axial Seamount on the Juan de Fuca ridge off the west coast of the United States (Sogin *et al.* 2006). These initial data sets led ICoMM investigators to the discovery of the "rare biosphere", a rich diversity of novel, low-abundance populations and dormant or slow growing microbes. A single liter of seawater, on average containing 10^8–10^9 bacteria, represents about 20,000 "species" of bacteria, a number that is one or two orders of magnitude higher than estimated earlier (Venter *et al.* 2004). When plotted on a two dimensional x–y microbial rank distribution diagram, this species-richness shows an extraordinarily long tail, the long tail including low-abundance taxa, many of which represent types of microbes that have never been seen before. Huber *et al.* (2007) extended this approach to the Archaea, also targeting the V6 16S rRNA hypervariable region and reported species richness estimates to be on the order of 3,000 "species" per liter of seawater. Amaral-Zettler *et al.* (2009) developed a tag sequencing strategy

for the V9 hypervariable region of the 18S rRNA gene in eukaryotes and determined that estimates of microbial eukaryotic (protist) species richness can be on the order of magnitude seen in the archaeal domain but may be an order of magnitude lower in more extreme environments such as Antarctic waters.

The International Census of Marine Microbes subsequently adopted this pyrotag strategy in a coordinated microbial census of samples from globally distributed marine environments. A study of lipid molecular structures from marine microbes complements the pyrotag survey. The database MICROBIS (http://icomm.mbl.edu/microbis) serves information to ICoMM, and its website provides access to this information including the capacity to retrieve contextual data information for all samples (Fig. 12.3).

The database VAMPS (Visualization Analysis of Microbial Population Structures, http://vamps.mbl.edu) and its links to MICROBIS provide full access to the pyrotag sequences, the contextual data, analytical and graphical tools for comparing microbial population structures for different sites, search tools for locating sequences in each of our samples, descriptions of community composition at taxonomic ranks of phyla, class, order, family, or, when possible, genus for all samples, and rarefaction and diversity analyses for all of ICoMM's data. Figure 12.4 depicts the geospatial breadth of pyrotag and lipid data for this global study of microbes in the world's oceans. It includes

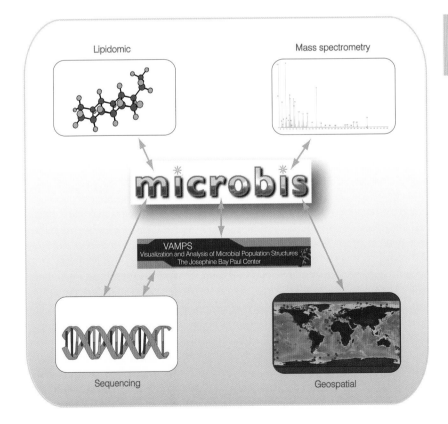

Fig. 12.3

An overview of MICROBIS and its relationship to VAMPS and the microbial lipid database.

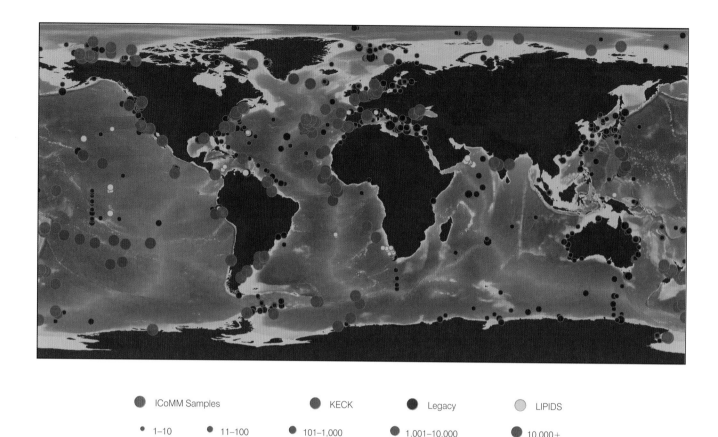

ICoMM Samples   KECK   Legacy   LIPIDS

1–10   11–100   101–1,000   1,001–10,000   10,000+

Fig. 12.4

The global distribution of observations gathered/recorded in the MICROBIS database. These include 454-pyrosequenced DNA pyrotag data generated during the ICoMM community sequencing project (red) and the Keck core sequencing projects (green), as well as legacy molecular data observations compiled from the literature (blue) and lipid-based analyses (yellow). The diameter of the circle represents the $\log_{10}$ of the sample size.

a subset of more than 18 million DNA sequence reads distributed among 583 bacterial, 120 archaeal, and 59 eukaryotic datasets from a larger dataset of >25 million sequences from >1,200 samples. The samples represent all major oceanic systems including the Atlantic, Pacific, Arctic, Southern, and Indian Oceans, and sediment and water samples from estuaries to deep-water environments including vents and seeps, seamounts, corals, sponges, microbial mats and biofilms, and polar regimes. Table 12.1 describes the origin of samples, targeted domains, project descriptions, and relevance to other Census ocean-realm projects. Here we present a broad-brush synthesis of our data emphasizing the most abundant pyrotags recovered from our surveys. Although a comprehensive synthesis of these data lies beyond the scope of this chapter, Figures 12.5, 12.6, 12.7 and 12.8 and the highlights that follow offer a glimpse into novel insights that will soon emerge from this international study of microbial community structures of the world's oceans. More detailed meta-analyses will frame the bulk of ICoMM's working groups during 2010.

12.3 Highlights of ICoMM Investigations

12.3.1 The "Abundant Biosphere"

The pie charts in Figures 12.5, 12.6, and 12.7 summarize the most abundant tags in our bacterial, archaeal, and micro-eukaryotic datasets respectively. As expected from the work of S.J. Giovannoni in the Sargasso Sea (Giovannoni *et al.* 1990), pyrotags corresponding to α-Proteobacteria and specifically SAR11 represented the most abundant organisms (primarily in planktonic samples) in our global survey. This heterotrophic α-Proteobacterial lineage plays a critical role in the cycling of carbon, nitrogen, and sulfur and accounts for approximately 25% of the biomass and 50% of the cell abundance in the ocean. More recently, researchers discovered that members of this group of bacteria contain proteorhodopsin, which potentially enables the harvesting of

Table 12.1

ICoMM microbial population structures of the world's oceans projects

Primary Investigator	PI First Name	Code	Project description	Domain	Examples of relevant projects
Aguiar	Paula	ASV	Azorean Shallow Vents	B	ChEss/MAR-ECO
Amaral-Zettler	Linda	MHB	Mount Hope Bay	BE	
Andersen	Robert	SAB	Surreptitious Algal Bacteria	B	
Artigas	Felipe	LCR	LaCAR Cooperative Run	BAE	NaGISA/CMarZ
Bharathi	Loka	ICR	Indian Ocean Cooperative Run	B	COMARGE/CeDAMar
Bertilsson	Stefan	BSP	Baltic Sea Proper	B	
Bolhuis	Henk	CMM	Coastal Microbial Mats	BA	NaGISA
Brazelton	William	LCY	Lost City	BA	ChEss
Caron	David	GPS	Global Protist Survey	E	CAML/CMarZ
Chistoserdov	Andrei	CAR	Cariaco Basin	B	COMARGE
Chistoserdov/Artigas	Andrei/Felipe	AGW	Amazon-Guianas Water	B	NaGISA
Coolen	Marco	WBS	Black Sea	E	HMAP/CMarZ
Dennett	Mark	DOF	Deep Ocean Flux	E	HMAP/CMarZ
D'Hondt	Steven	KNX	Station KNOX	BA	
Epstein	Slava	SSD	Spatial Scaling Diversity	B	NaGISA
Franklin	Rima	AOT	Atlantic Ocean Transect	B	MAR-ECO
Gaidos	Eric	CRS	Coral Reef Sediment	BA	CReefs
Gallardo	Victor	VAG	Humboldt Marine Ecosystem	B	COMARGE/ChEss
Gerdts	Gunnar	MPI	Helgoland	B	
Gilbert	Jack	PML	English Channel	B	
Hamasaki	Koji	ABR	Active but Rare	BA	CAML
Herndl	Gerhard	NADW	North Atlantic Deep Water	BA	
Huber	Julie	EEL	Eel Pond Winter Pilot Study	B	
Huber	Julie	SMT	Seamounts	BA	ChEss, CenSeam
Kirchman	David	ACB	Arctic Chukchi Beaufort	BA	ArcOD
Lovejoy	Connie	DAO	Deep Arctic Ocean	BA	ArcOD
Maas	Els	NZS	New Zealand Sediment	B	
Martins	Ana	AWP	Azores Waters Project	BAE	MAR-ECO/CMarZ
Murray	Alison	CAM	Census Antarctic Marine	B	CAML
Pawlowski	Jan	DSE	Deep Sea Eukarya	E	CeDAMar/CMarZ

Primary Investigator	PI First Name	Code	Project description	Domain	Examples of relevant projects
Polz	Martin	CNE	Coastal New England	B	NaGISA
Pommier	Thomas	BMO	Blanes Bay Microbial Observatory	B	
Post	Anton	GOA	Gulf of Aqaba	B	
Prosser	James	SMS	Station M Sediments	B	CeDAMar
Ramette	Alban	FIS	Frisian Island Sylt	B	NaGISA
Rappé	Michael	HOT	Hawaii Ocean Time Series	B	
Reysenbach	Anna-Louise	ALR	Lau Hydrothermal Vent	BA	ChEss
Rocap	Gabrielle	HCW	Hood Canal Washington	B	NaGISA
Rooney-Varga	Juliette	JRV	Gulf of Maine	E	GoMA/CMarZ
Sogin	Mitchell	LSM	Little Sippewissett Salt Marsh	B	NaGISA
Staley	James	BSR	Black Sea Redox	B	
Stoeck	Thorsten	APP	Anaerobic Protist Project	E	COMARGE/CMarZ
Sunagawa	Shinichi	CCB	Caribbean Coral Bacteria	BA	CReefs
Teske	Andreas	GMS	Guaymas Methane Seeps	BA	ChEss
Teske	Andreas	ODP	Ocean Drilling Project	BA	CeDAMar
Wagner	Michael	SPO	Sponges	B	CReefs
Webster	Gordon	CFU	Deep Subseafloor Sediment	BA	CeDAMar
Yager	Patricia	ASA	Amundsen Sea Antarctica	B	CAML

B, Bacteria; A, Archaea; E, Eukarya.

energy from light (Fuhrman *et al.* 2008; Giovannoni *et al.* 2005). The presence of this clade in different habitats (Fig. 12.8) including coastal waters, seamounts, polar waters, and the open ocean (not shown) reflects its ubiquity in the marine pelagic environment. The 20 most abundant tags in our bacterial analyses also include members of the Rhodobacteraceae. One member of this group, *Roseobacter* sp., is cultivable by adding extracts of algal secreted organic matter to the medium (Mayali *et al.* 2008). The worldwide association of *Roseobacter* with algal blooms suggests it has a role in controlling bloom outbreaks.

The most abundant tag sequence derived from a photosynthetic bacterium belonged to a member of the Prochlorales (Cyanobacteria) and shares 100% V6 rRNA region sequence identity with the cultivar *Prochlorococcus marinus*.

The picocyanobacteria (smaller than 2 μm) *Prochlorococcus* spp. along with *Synechococcus* spp. dominate the oceans with cell numbers of up to 10^5–10^6 per milliliter (Heywood *et al.* 2006; Scanlan *et al.* 2009). Collectively they contribute up to 50% of oceanic primary production (Li 1994). Cyanobacteria represent an ancient group of organisms. These inventors of oxygenic photosynthesis drove the oxygenation of the Earth's atmosphere 2.5 billion years ago. The evolution of aerobic Bacteria and Archaea made possible the origins of plants and animals about 0.5 billion years ago when the oxygen concentration in the atmosphere reached its present-day level. Today, Cyanobacteria produce about 50% of the oxygen on Earth. Most Cyanobacteria occur in marine communities (Garcia-Pichel *et al.* 2003).

Top 20 most abundant bacterial tags

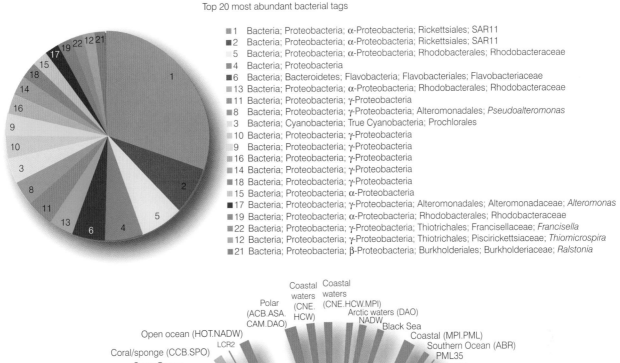

- ■ 1 Bacteria; Proteobacteria; α-Proteobacteria; Rickettsiales; SAR11
- ■ 2 Bacteria; Proteobacteria; α-Proteobacteria; Rickettsiales; SAR11
- □ 5 Bacteria; Proteobacteria; α-Proteobacteria; Rhodobacterales; Rhodobacteraceae
- ■ 4 Bacteria; Proteobacteria
- ■ 6 Bacteria; Bacteroidetes; Flavobacteria; Flavobacteriales; Flavobacteriaceae
- ■ 13 Bacteria; Proteobacteria; α-Proteobacteria; Rhodobacterales; Rhodobacteraceae
- ■ 11 Bacteria; Proteobacteria; γ-Proteobacteria
- ■ 8 Bacteria; Proteobacteria; γ-Proteobacteria; Alteromonadales; *Pseudoalteromonas*
- □ 3 Bacteria; Cyanobacteria; True Cyanobacteria; Prochlorales
- □ 10 Bacteria; Proteobacteria; γ-Proteobacteria
- □ 9 Bacteria; Proteobacteria; γ-Proteobacteria
- ■ 16 Bacteria; Proteobacteria; γ-Proteobacteria
- ■ 14 Bacteria; Proteobacteria; γ-Proteobacteria
- ■ 18 Bacteria; Proteobacteria; γ-Proteobacteria
- □ 15 Bacteria; Proteobacteria; α-Proteobacteria
- ■ 17 Bacteria; Proteobacteria; γ-Proteobacteria; Alteromonadales; Alteromonadaceae; *Alteromonas*
- ■ 19 Bacteria; Proteobacteria; α-Proteobacteria; Rhodobacterales; Rhodobacteraceae
- ■ 22 Bacteria; Proteobacteria; γ-Proteobacteria; Thiotrichales; Francisellaceae; *Francisella*
- ■ 12 Bacteria; Proteobacteria; γ-Proteobacteria; Thiotrichales; Piscirickettsiaceae; *Thiomicrospira*
- ■ 21 Bacteria; Proteobacteria; β-Proteobacteria; Burkholderiales; Burkholderiaceae; *Ralstonia*

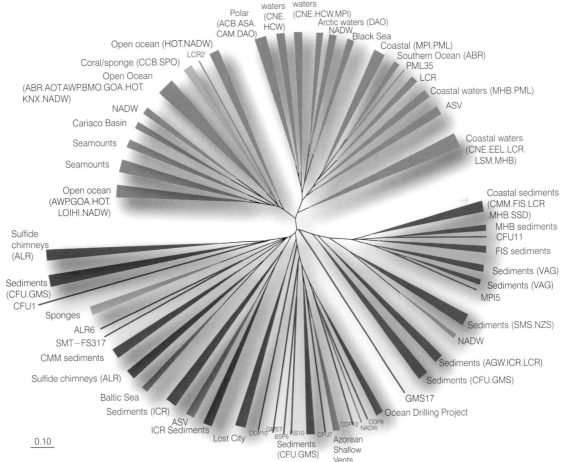

Fig. 12.5

A summary of results from pyrotag bacterial projects. Top, the taxonomic breakdown of the top 20 most abundant bacterial sequences found across 583 bacterial datasets. The rankings are based on the sum of the relative abundances of individual sequences from each sample. Taxonomies are based on the Global Alignment for Sequence Taxonomy (GAST) procedure (Huse *et al.* 2008). The numbering has been adjusted to match the tag sequence numbering in Figure 12.8 and is ordered in descending order of abundance. Bottom, a radial dendrogram of clustered bacterial datasets. Clusters are based on similarity calculations of presence/absence data of the most abundant pyrotag sequences. Brown, benthic samples; blue, water-column samples; orange, sponge- or coral-associated samples. See Table 12.1 for descriptions of project abbreviations.

Top 20 most abundant archaeal tags

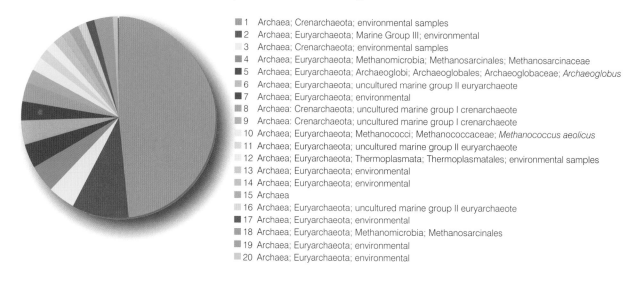

■ 1 Archaea; Crenarchaeota; environmental samples
■ 2 Archaea; Euryarchaeota; Marine Group III; environmental
□ 3 Archaea; Crenarchaeota; environmental samples
■ 4 Archaea; Euryarchaeota; Methanomicrobia; Methanosarcinales; Methanosarcinaceae
■ 5 Archaea; Euryarchaeota; Archaeoglobi; Archaeoglobales; Archaeoglobaceae; *Archaeoglobus*
■ 6 Archaea; Euryarchaeota; uncultured marine group II euryarchaeote
■ 7 Archaea; Euryarchaeota; environmental
■ 8 Archaea; Crenarchaeota; uncultured marine group I crenarchaeote
■ 9 Archaea; Crenarchaeota; uncultured marine group I crenarchaeote
 10 Archaea; Euryarchaeota; Methanococci; Methanococcaceae; *Methanococcus aeolicus*
■ 11 Archaea; Euryarchaeota; uncultured marine group II euryarchaeote
□ 12 Archaea; Euryarchaeota; Thermoplasmata; Thermoplasmatales; environmental samples
■ 13 Archaea; Euryarchaeota; environmental
□ 14 Archaea; Euryarchaeota; environmental
■ 15 Archaea
■ 16 Archaea; Euryarchaeota; uncultured marine group II euryarchaeote
■ 17 Archaea; Euryarchaeota; environmental
■ 18 Archaea; Euryarchaeota; Methanomicrobia; Methanosarcinales
■ 19 Archaea; Euryarchaeota; environmental
■ 20 Archaea; Euryarchaeota; environmental

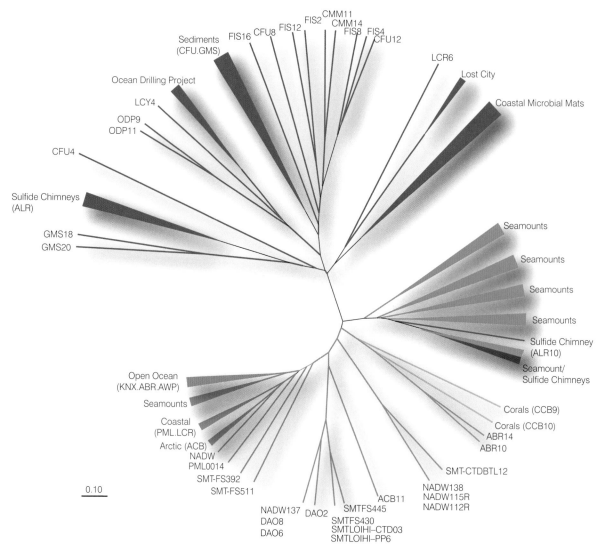

Fig. 12.6

A summary of results from pyrotag archaeal projects. Top, the taxonomic breakdown of the top 20 most abundant archaeal tag sequences found in 120 archaeal datasets. The rankings are based on the sum of the relative abundances of individual sequences from each sample. Taxonomies are based on the GAST procedure. Bottom, a radial dendrogram of clustered archaeal datasets. Clusters are based on similarity calculations of presence/absence data of the most abundant tag sequences. Brown, benthic samples; blue, water column samples; orange, coral-associated samples. See Table 12.1 for descriptions of project abbreviations.

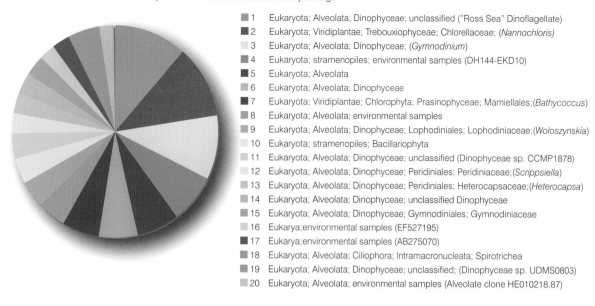

Top 20 most abundant microbial eukaryotic tags

1 Eukaryota; Alveolata; Dinophyceae; unclassified ("Ross Sea" Dinoflagellate)
2 Eukaryota; Viridiplantae; Trebouxiophyceae; Chlorellaceae; (*Nannochloris*)
3 Eukaryota; Alveolata; Dinophyceae; (*Gymnodinium*)
4 Eukaryota; stramenopiles; environmental samples (DH144-EKD10)
5 Eukaryota; Alveolata
6 Eukaryota; Alveolata; Dinophyceae
7 Eukaryota; Viridiplantae; Chlorophyta; Prasinophyceae; Mamiellales;(*Bathycoccus*)
8 Eukaryota; Alveolata; environmental samples
9 Eukaryota; Alveolata; Dinophyceae; Lophodiniales; Lophodiniaceae;(*Woloszynskia*)
10 Eukaryota; stramenopiles; Bacillariophyta
11 Eukaryota; Alveolata; Dinophyceae; unclassified (Dinophyceae sp. CCMP1878)
12 Eukaryota; Alveolata; Dinophyceae; Peridiniales; Peridiniaceae;(*Scrippsiella*)
13 Eukaryota; Alveolata; Dinophyceae; Peridiniales; Heterocapsaceae;(*Heterocapsa*)
14 Eukaryota; Alveolata; Dinophyceae; unclassified Dinophyceae
15 Eukaryota; Alveolata; Dinophyceae; Gymnodiniales; Gymnodiniaceae
16 Eukarya;environmental samples (EF527195)
17 Eukarya;environmental samples (AB275070)
18 Eukaryota; Alveolata; Ciliophora; Intramacronucleata; Spirotrichea
19 Eukaryota; Alveolata; Dinophyceae; unclassified; (Dinophyceae sp. UDMS0803)
20 Eukaryota; Alveolata; environmental samples (Alveolate clone HE010218.87)

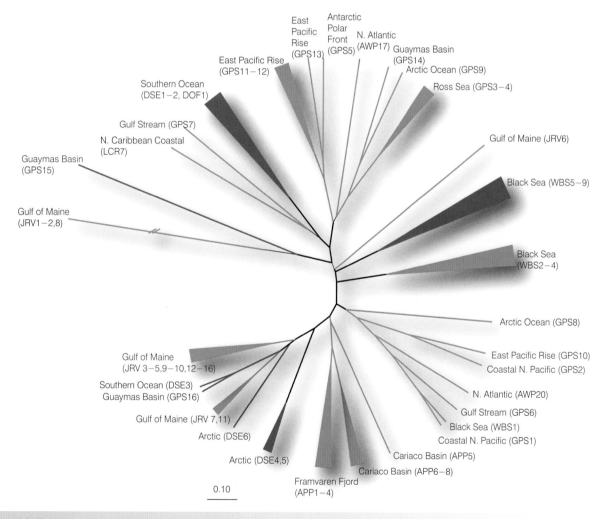

Fig. 12.7

A summary of eukaryotic pyrotag projects. Top, the taxonomic breakdown of the top 20 most abundant microbial eukaryotic tag sequences found in 59 eukaryotic datasets. All tags with metazoan-associated taxonomy have been removed from the analysis. The rankings are based on the sum of the relative abundances of individual sequences from each sample. Taxonomies are based on a combination of the GAST procedure and BLAST. Tags with a 100% match to a representative sequence/taxon in GenBank have that representative affiliation in parentheses. Bottom, a radial dendrogram of clustered eukaryotic datasets. Clusters are based on similarity calculations of presence/absence data of the most abundant tag sequences. Brown, benthic samples; blue, water column samples. See Table 12.1 for descriptions of project abbreviations.

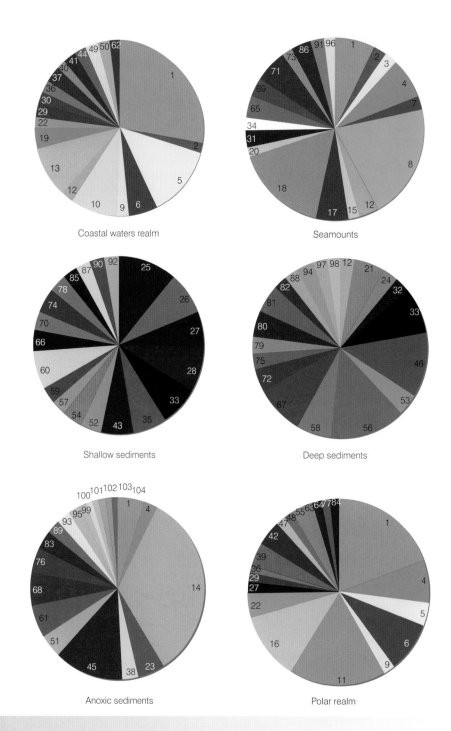

Coastal waters realm

Seamounts

Shallow sediments

Deep sediments

Anoxic sediments

Polar realm

Fig. 12.8

A summary of the major ocean realms sampled showing the top 20 most abundant bacterial tag sequences for each habitat. Realms (abbreviations from Table 12.1) shown include Coastal Waters (CNE, HCW, LCR, MPI, PML, EEL, LSM, MHB), Seamounts (SMT), Shallow Sediments (AGW, CMM, CRS, FIS, GMS, ICR, LCR, SSD, VAG, MHB), Deep Sediments (CFU, ICR, NZS, SMS, ODP), Anoxic Sediments (BSR, CAR), and Polar Regions (ABR, ACB, ASA, CAM, DAO). Numbers facilitate comparisons between samples. The lowest possible taxonomic rank assigned for each tag follows the number designation: (1) SAR11; (2) SAR11; (3) Prochlorales; (4) Proteobacteria; (5) Rhodobacteraceae; (6) Flavobacteriaceae; (7) *Sulfurovum*; (8) *Pseudoalteromonas*; (9) γ-Proteobacteria; (10) γ-Proteobacteria; (11) γ-Proteobacteria; (12) *Thiomicrospira*; (13) Rhodobacteraceae; (14) γ-Proteobacteria; (15) α-Proteobacteria; (16) γ-Proteobacteria; (17) *Alteromonas*; (18) γ-Proteobacteria; (19) Rhodobacteraceae; (20) α-Proteobacteria; (21) *Ralstonia*; (22) *Francisella*; (23) Actinobacteria; (24) γ-Proteobacteria; (25) *Bacillus*; (26) *Clostridium*; (27) Flavobacteriaceae; (28) γ-Proteobacteria; (29) Flavobacteriaceae; (30) α-Proteobacteria; (31) *Alteromonas*; (32) Ectothiorhodospiraceae; (33) γ-Proteobacteria; (34) *Marinobacter*; (35) *Clostridium*; (36) *Methylophilus*; (37) *Roseovarius*; (38) *Pseudomonas*; (39) γ-Proteobacteria; (40) Flavobacteriaceae; (41) Rhodobacteraceae; (42) γ-Proteobacteria; (43) *Bacillus*; (44) *Glaciecola*; (45) Bacteria; (46) *Erythrobacter*; (47) Flavobacteriaceae; (48) α-Proteobacteria; (49) γ-Proteobacteria; (50) Rhodospirillales; (51) Proteobacteria; (52) *Paenibacillus*; (53) *Diaphorobacter*; (54) *Bacillus*; (55) γ-Proteobacteria; (56) JS1; (57) *Clostridium*; (58) γ-Proteobacteria; (59) *Tepidibacter*; (60) *Bacillus*; (61) Bacteria; (62) Flavobacteriaceae; (63) Bacteria; (64) *Sulfitobacter*; (65) *Pseudomonas*; (66) *Methylophaga*; (67) Actinobacteria; (68) Bacteria; (69) *Thiomicrospira*; (70) *Bacillus*; (71) *Burkholderia*; (72) Gemmatimonadetes; (73) Comamonadaceae; (74) γ-Proteobacteria; (75) γ-Proteobacteria; (76) Bacteria; (77) Flavobacteriaceae; (78) γ-Proteobacteria; (79) γ-Proteobacteria; (80) *Hyphomicrobium*; (81) Clostridia; (82) γ-Proteobacteria; (83) γ-Proteobacteria; (84) *Methylophaga*; (85) *Desulfosarcina*; (86) *Caminibacter*; (87) Flavobacteriales; (88) γ-Proteobacteria; (89) Bacteria; (90) Chromatiales; (91) *Thioreductor*; (92) Desulfobulbaceae; (93) γ-Proteobacteria; (94) Dehalococcoidetes; (95) Desulfobacterium; (96) *Thioreductor micantisoli*; (97) γ-Proteobacteria; (98) γ-Proteobacteria; (99) Proteobacteria; (100) γ-Proteobacteria; (101) Deferribacteres; (102) γ-Proteobacteria; (103) Bacteria; (104) Deferribacteres.

Members of the phylum Crenarchaeota dominated the Archaeal pyrotag surveys. Microbial ecologists originally thought that all Crenarchaea represented extremophiles, until the discovery of their ubiquity in everyday marine and terrestrial environments (DeLong 1992; Fuhrman *et al.* 1992; Simon *et al.* 2000). Crenarchaeotal abundances can exceed bacterial abundances below 100 m depth in the ocean where they are metabolically active and can contribute to the oceanic carbon cycle (Herndl *et al.* 2005). In the Arctic, Marine Group III Euryarchaeota can dominate the deep water masses such as the deep Atlantic Layer in the central Arctic Ocean (Galand *et al.* 2009b). Sequences related to this group represented the second most commonly encountered pyrotags in our global archaeal dataset. In many cases, we only detected other abundant archaeal pyrotags in specific samples. For example, the methanogens Methanosarcinales occurred primarily in Lost City Hydrothermal Vents (see below), whereas *Archaeglobus*- and *Methanococcus*-related tags specifically associated with sulfide chimneys.

The pyrotag studies showed that dinoflagellates dominated most eukaryotic microbial communities. Members of the dinoflagellates include phototropic and heterotrophic representatives and many co-occur with or may be responsible for harmful algal blooms, making them commercially and ecologically important. The high frequency of dinoflagellate tags likely reflects bias introduced by the very large copy number of rRNA genes in the genomes of most dinoflagellates (as many as 12,000 copies in species such as *Akashiwo sanguinea* (Zhu *et al.* 2005)). Indeed, many of the tags recovered among the top 20 most abundant microbial eukaryotes included members of the picoeukaryotes (0.2–2 μm in size) that numerically dominate, but tend to have lower copy numbers of their 18S rRNA genes. The diversity of these Lilliputians of the protist world was only first recognized at the beginning of the twenty-first century (Díez *et al.* 2001; López-García *et al.* 2001; Moon-van de Staay *et al.* 2001).

The top most abundant tag among our eukaryotic datasets displayed 100% identity with the sequence of an unclassified dinoflagellate within the *Karenia/Karlodinium* group that Gast *et al.* (2006) first identified in the Ross Sea, Antarctica. In some cases these cells occur at densities up to 29,000 cells per liter. Equally intriguing, our global analyses of eukaryotic tags revealed this pyrotag also occurs in the Arctic, Pacific, and Atlantic Oceans (from the Caribbean to the Gulf of Maine), the Framvaren Fjord in Norway and the Black Sea. Whether this tag represents the same cosmopolitan species or closely related ecotypes that extend over the globe remains unknown.

Because of differences in their gene copy number in different taxa, the relative abundance of eukaryotic pyrotags does not reflect the number of cells in a sampled environment. However, these data provide important taxonomic information at the species level for many morphologically rich eukaryotic microbes including dinoflagellates (for example *Scrippsiella*, *Heterocapsa*, *Woloszynskia*) and for members of the picoeukaryotes such as *Bathycoccus* that are harder to distinguish morphologically.

12.3.2 The "Rare Biosphere"

In the microscopic realm, ICoMM's sampling of many diverse marine ecosystems has reinforced the concept of a ubiquitous rare biosphere (Pedrós-Alió 2006; Sogin *et al.* 2006). This forces us to reconsider the potential feedback mechanisms between shifts in extremely complex microbial communities and global change, as well as how microbial communities and the genomes of their constituents change over evolutionary timescales. Minor population members may serve as functional keystone species in microbial consortia, or they might be the products of historical ecological change with the potential to become dominant in response to shifts in environmental conditions, for example when local or global change favors their growth. The absence of information about the global distribution of members of the rare biosphere makes it impossible to ascertain if they represent specific biogeographical distributions of bacterial taxa, functional selection by particular environments, or cosmopolitan distribution of all microbial taxa – the "everything is everywhere" hypothesis. Data from recent pyrosequencing efforts, however, are beginning to shed light on this topic. Galand *et al.* (2009a), for example, compared pyrotags from Arctic Ocean samples. These samples included both surface and deep waters, as well as winter and summer samples from different locations. When they clustered the samples, they observed nearly identical patterns for all abundant sequences (more than 1% of all tags), or only rare sequences (less than 0.01% of all tags). This indicates that in this system, the rare OTUs have the same biogeography as the abundant OTUs. This opens up two possibilities: either the pyrosequencing approach is only targeting the most abundant of the rare biosphere and the actual tails are even longer; or the rare OTUs must be a dynamic lot, able to grow and experience losses due to predation and viral attack in some intriguing and unknown way.

To explore whether high-abundance taxa represent physiologically active populations whereas low-abundance pyrotags represent less active or dormant microbes, Hamasaki *et al.* (2007) applied the bromodeoxyuridine (BrdU, thymidine analogs for detecting *de novo* DNA synthesis) magnetic bead immunocapture method to examine both the abundant and rare members of the microbial community. They applied this technique to surface seawater samples from four stations along a north–south transect in the South Pacific taken from November 2004 to March 2005 during the KH-04-5 cruise of the R/V *Hakuho-maru* (Ocean Research Institute, the University of Tokyo and

JAMSTEC) (Fig. 12.1G). They incubated subsamples on board after adding BrdU and then compared the pyrotag microbial community structures in treated and untreated samples. Their results indicated that some high-abundance taxa incorporated BrdU whereas others did not, and some rare taxa represented only by singletons in the resulting tag dataset also took up BrdU. More importantly, some BrdU-labelled taxa were detected only in the BrdU-labelled fraction but were absent in the untreated fractions. These results suggested that the rare biosphere is not restricted to physiologically inactive populations but can include taxa involved in biogeochemical cycles. This study further illustrates that there may be dynamic exchange between abundant and rare microbial populations.

12.3.3 High archaeal microdiversity in the Lost City Hydrothermal Field

The Lost City Hydrothermal Field on the Mid-Atlantic Ridge is the first deep-sea environment discovered where exothermic water-rock reactions in the sub-seafloor and not magmatic sources of heat drive hydrothermal fluid flow. These reactions create a combination of extreme conditions never before seen in the marine environment: the venting of high-pH (from 9 to 11), warm (40–91 °C) hydrothermal fluids with high concentrations of hydrogen, methane, and other hydrocarbons of low molecular mass. The Lost City Hydrothermal Field may thus represent a new type of life-supporting system in the deep sea. Mixing of the warm, high-pH fluids with seawater precipitates carbonate that drives the growth of chimneys that tower up to 60 m above the seafloor (Fig. 12.1E). These carbonate towers, some of which remain active for thousands of years, house extensive microbial biofilms that are dominated by a single group of Archaea, the *Methanosarcinales*. However, multiple lines of evidence including morphology, and evidence for diversity of specific genes such as the genes involved in N_2 fixation and methane production and anaerobic oxidation, indicate physiological diversity within this *Methanosarcinales*-dominated biofilm.

Brazelton and colleagues (Brazelton *et al.* 2009, 2010) correlated archaeal and bacterial V6 tag distributions with isotopic ages of carbonate chimneys collected from the Lost City Hydrothermal Field spanning a 1,200 year period. Clear shifts in the archaeal and bacterial communities were evident over time, and many of the shifts featured rare sequences that dramatically increased in relative abundance to become dominant in older chimneys. These results indicate that some organisms can remain rare at a location for many years before "blooming" and becoming dominant when the environmental conditions allow. Furthermore, the very low overall diversity of the Lost City chimneys revealed that each of the dominant archaeal and bacterial

sequences represented one member of a large pool of similar but much rarer sequences. For example, the most abundant archaeal sequence was more than 90% similar to 1,771 different sequences clustering into 517 operational taxonomic units at 97% sequence similarity, all of which were too rare to be detected by clone library sequencing. Further work in the Baross laboratory has shown that this microdiversity in the V6 region is correlated with microdiversity in a more variable marker, the intergenic transcribed spacer (ITS) region, indicating that it is not generated by pyrosequencing error and that the archaeal population contains even more microdiversity than represented by the 1,771 variants detected in the V6 region. These results confirm that there are many rare species of *Methanosarcinales* that had not been previously identified from 16S rRNA gene clone libraries that likely represent multiple "ecotypes" within the biofilm and that the ecotype composition changes depending on the age of the carbonate structure.

12.3.4 Rapid temporal turnover in sands of the North Sea island of Sylt

Permeable sandy sediments play a critical role in the recycling of carbon and nitrogen, and act as natural filters that may concentrate microorganisms, nutrients, and organic matter on the extensive continental shelves. Despite the importance of such ecosystems, the extent of microbial diversity and how microbes respond to environmental, spatial, and temporal changes (such as global warming, ocean acidification, and various anthropogenic effects) are still mostly unknown.

Using a 454 massively parallel pyrotag sequencing strategy to describe microbial diversity in temperate sandy sediments from the North Sea island of Sylt (Fig. 12.1F), Ramette and colleagues obtained between 5,000 to 19,000 unique types of bacteria in each gram of sand (A. Gobet, S.I. Böer, J.E.E. van Beusekom, A. Boetius and A. Ramette, unpublished observations). Rarefaction analyses suggest that the OTU richness of sand-associated bacterial communities significantly exceeds the diversity of water column communities from the same environment. The OTU richness also changed dramatically over a few centimeters of sediment depth or between any two consecutive sampling times, with up to 70–80% community turnover. Those remarkable, non-random shifts in community composition may reflect responses to variation of many environmental/biogeochemical parameters (for example temperature, nutrients, pigments, production of extracellular enzymes) at the study site. The reservoir of highly diverse low abundant bacterial types might include taxa that become abundant in response to environmental differences in this system. The comparison of diversity patterns at different taxonomic levels indicated that community shifts occurred at broad taxonomic levels, but that fine-scale patterns in community composition were mostly responsible for the large community turnover observed over sediment depth and sampling time. This study demonstrates the dynamic nature of coastal sandy sediments in terms of microbial diversity, allowing for the formulation of strong ecological hypotheses to explain this phenomenon: strong vertical shifts in nutrient, organic matter, and oxygen availability create a large range of microbial niches, which may support a high turnover of bacterial types in sandy sediments.

12.3.5 Diversity varying with oxygen availability in the Black Sea

The Black Sea, a permanently anoxic basin connected to the ocean through the Bosphorus Sea, has well defined redox gradients and known microhabitats for different metabolic groups of bacteria. C.A. Fuchsman and colleagues (unpublished observations) obtained bacterial pyrotags from four low-oxygen Black Sea water samples: a low-oxygen sample ($30\,\mu M$ oxygen), a sample from the middle of the suboxic zone ($2\,\mu M$ oxygen), a sample from the bottom of the suboxic zone with no detectable oxygen or sulfide, and a sinking particle obtained from the middle of the suboxic zone. The bottom-of-the-suboxic-zone sample ($0\,\mu M$ oxygen) and the particle-attached bacterial sample were more diverse than the $30\,\mu M$ oxygen and $2\,\mu M$ oxygen samples. Although all three samples contained low oxygen and no measurable sulfide, only the microbial community structures for the 0 or $2\,\mu M$ oxygen samples had similar community structures (51% similarity by Bray Curtis) whereas neither resembled the $30\,\mu M$ oxygen samples (11%). Micro-aerophilic heterotrophs and nitrate reducers dominated the $30\,\mu M$ and $2\,\mu M$ oxygen samples. In contrast, the $0\,\mu M$ oxygen sample and the particle-attached bacterial samples were more diverse and contained strikingly different taxonomic groups of bacteria. Enriched populations of *Deferribacter*, δ-Proteobacteria, *Lentisphaera*, ε-Proteobacteria and Planctomycetes occurred in the particle-attached fraction. These taxonomic groups of bacteria are not normally identified as part of the particle-attached community from oxic waters. Pyrotags for the particle-associated ε-Proteobacteria resemble rRNA sequences from epsilon species that oxidize sulfide. The *Deferribacter* species are known to reduce metals including manganese and iron oxides, or nitrate and elemental sulfur. *Lentisphaera* occur in anaerobic environments but little is known about their metabolism. The δ-Proteobacteria, which include known sulfate reducers, were found in the $0\,\mu M$ oxygen sample and in the particle-attached fraction. However, the OTUs of δ-Proteobacteria differed between the samples, with *Desulfobacteraceae* dominating the $0\,\mu M$ oxygen sample whereas the particle-attached group could not be assigned to a cultured species.

These results showed a clear correlation between the fluxes and depth of the chemical species such as O_2, NO_3^-, NH_4^+, CH_4, MnO_2, H_2S and the inferred metabolisms of the bacterial OTUs. Manganese and sulfate reducers and sulfide oxidizers dominated metabolic groups associated with sinking particles whereas microaerophilic and nitrate reducers dominated the water column. This study provides insights into the importance of the particle-attached bacterial communities and points to the potential biases on bacterial diversity estimation when researchers pre-filter samples for microbial diversity studies.

12.3.6 Community signatures of the North Atlantic Deep Water masses

Small size suggests that microbes have high dispersal and high immigration rates, leading to a ubiquitous distribution in the marine environment. However, recent studies demonstrate that microbes can have biogeographic distributions corresponding to individual water masses. Distinct salinity, temperature, and nutrient characteristics differentiate several deep water masses separated by thousands of kilometers of thermohaline ocean circulation. Using bacterial pyrotag sequencing, Herndl and colleagues (unpublished observations) tested this hypothesis by determining the biogeography of bacterioplankton communities following the flow of the North Atlantic Deep Water over a stretch of 8,000 km in the North Atlantic. They focused on the distribution of the abundant versus rare phylotypes to decipher whether rare phylotypes exhibit a similar distribution pattern as the abundant phylotypes or whether they occur ubiquitously. If the rare phylotypes represent a seed bank for the few abundant phylotypes, then the community structure of the rare phylotypes should be fairly uniform across water masses.

Cluster analysis (Fig. 12.9) showed that abundant bacterial phylotypes clustered according to the water masses (Fig. 12.9A). The samples partitioned into one cluster containing bacterial communities from the subsurface zone, two clusters from the mesopelagic waters, three deep water clusters, and one cluster of bacterial communities from the deep Labrador Seawater. Bacterial community composition of deep waters was less similar than samples from subsurface and intermediate waters. Bacterial communities from the same water mass but separated by thousands of kilometers resembled each other more than communities separated by a few hundred meters at individual sites but originating from different water masses.

Fig. 12.9

Non-metric multi-dimensional scaling analysis based on relative abundance of **(A)** abundant tags (frequency greater than 1% within a sample) and **(B)** rare tags (frequency less than 0.01% within a sample). Discrimination among samples by water mass. Superimposed circles represent clusters of samples at similarity values of 60% and 80% **(A)** and 20% **(B)** (Bray-Curtis similarity). LDW, Lower Deep Water; NEADW, Northeast Atlantic Deep Water, AAIW, Antarctic Intermediate Water, tCW- transitional Central Water, SACW, South Atlantic Central Water; LSW, Labrador Sea Water.

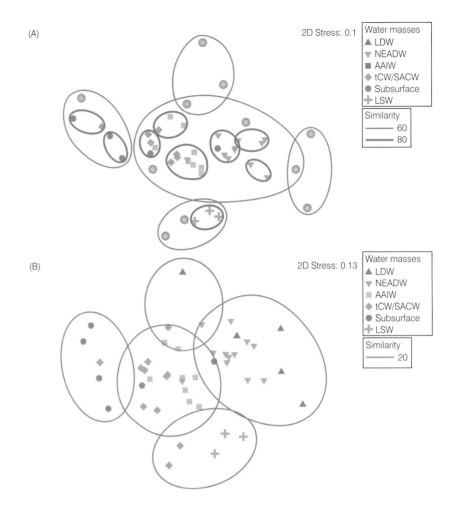

The clustering of the rare sequences (frequency less than 0.01% within a sample, including the singletons; Fig. 12.9B) was similar to the clustering of the abundant sequences (frequency greater than 1% within a sample; Fig. 12.9A), albeit with a generally lower percentage of similarity. Proteobacteria constituted more than 85% of the 1,000 most abundant tags and of these, 52% were α-Proteobacteria, mostly composed of the SAR 11 cluster. The bathypelagic zone had the highest proportion of unassigned bacteria, but showed the highest tag richness and evenness compared to overlying waters. γ-Proteobacteria increased with depth, with higher proportions of Chromatiales and Alteromon- adales (8.7%) in the bathypelagic zone than in the subsur- face zone. The distinct clusters of bacterial communities in specific water masses reflected the presence of unique phy- lotypes specific to distinct water masses. The variability in the abundance of tags increased with decreasing overall abundance. The most abundant pyrotags exhibited a ubiq- uitous distribution pattern whereas the representation of low abundance pyrotags seemed to be specific to different water masses.

In summary, the bacterial rare biosphere in the North Atlantic is water-mass-specific and hence not ubiquitously

distributed as previously suggested. Thus, in this example, it is likely that the rare biosphere originates from the more abundant and/or more active bacterial community through mutation. The biogeochemical role of the high richness of the rare microbial biosphere remains enigmatic and deserves further investigation.

12.3.7   A bipolar distribution of the most abundant bacterial pyrotags

Among the globally distributed samples sequenced as part of the ICoMM Community Sequencing Project, Polar Regions contributed 56 datasets divided among five projects (ABR, ACB, ASA, CAM, and DAO; Table 12.1). The Polar Realm pie chart in Figure 12.8 shows the taxo- nomic affiliation for the top 20 most abundant tags from these samples. One of the striking results from our com- parative study is that the most abundant SAR 11 pyrotag corresponds to the most abundant polar pyrotag. When we map the distribution of the top 20 most abundant tags, we find that all 20 occur in both the Arctic and Southern

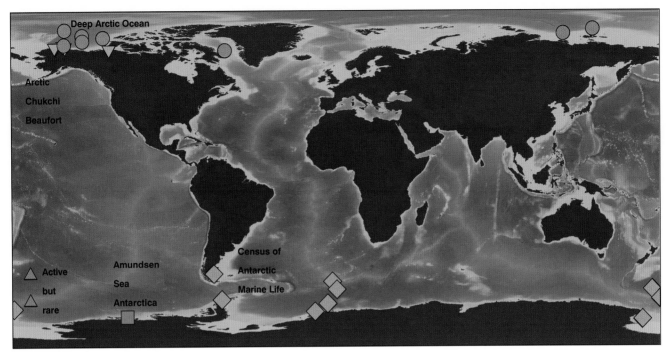

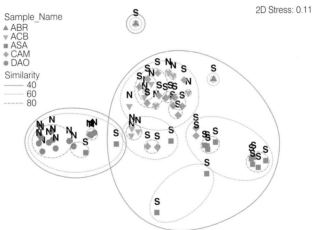

Fig. 12.10

Top, a map detailing the location of all the Arctic and Antarctic datasets examined in this synthesis. Bottom, a non-metric multidimensional scaling plot of 56 Arctic (N) and Antarctic (S) samples based on standardization by total and square-root transformed pyrotag abundances and Bray-Curtis similarities (Clarke & Warwick 2001).

Oceans. Our non-metric multidimensional scaling analysis (Fig. 12.10) confirms the similarity of certain Arctic and Antarctic samples by way of similarity envelopes that encircle datasets that cluster at 80% similarity levels. Although these tags show a "bipolar" distribution using the V6 region as a metric for comparison, we do not know how these populations compare when looking across the entire genome. This question, as well as the source and persistence of bipolar species, awaits the next decade of Census research.

12.4 A Census of Microbial Lipids

12.4.1 MICROBIS and lipid maps

ICoMM has also considered phenotypic characters ranging from microbial physiology to metabolic capability in its global marine microbial census. Intact polar lipids provide a case in point. Lipids can be a powerful tool for deciphering microbial communities in present and past environments. However, in contrast to genetic data, lipids lack a large database linking identification, chemical structure, and biological or environmental sources. Such a linked database would substantially increase the potential for lipids to be used as a tool for relating environmental microbial communities to cultivated microbes and provide support for phylogenetic relationships. Our approach to constructing a lipid database follows MICRO-BIS (http://icomm.mbl.edu/lipids) in which a central database/program/website links several databases including a database of microbial organisms, lipid structures (Lipid Maps), and mass spectra.

The Lipid Maps database (www.lipidmaps.org; Fahy *et al.* 2005) represents the most extensive lipid library, but it currently targets biomedical applications and lacks a comprehensive collection of marine microbial lipids. Collaboration with Lipid Maps to generate a universal lipid collection that includes marine microbial lipids enabled the deposition of more than 200 marine microbial lipid structures in the Lipid Maps database. The Lipid Maps database uses a systems biology approach for the categorization, nomenclature, and chemical representation of lipids (Fahy *et al.* 2005). The classification scheme of Lipid Maps distributes lipids into eight defined categories that are divided into classes and subclasses. Each lipid in the Lipid Maps library can be retrieved using different search criteria including its lipid identification, classification, systematic name, synonym, and chemical name and structure. Furthermore, Lipid Maps assigns unique numbers to lipid structures comparable to the unique accession numbers within GenBank for genomic sequences. New lipid structures can be continuously submitted to Lipid Maps with a proposed lipid identification and thus can evolve continuously if support within the biogeochemical community is strong.

Typically, mass spectrometry identifies lipids in the laboratory. The mass spectra of lipids are generally very comparable between laboratories and thus well suited for database purposes. Unfortunately, at present there are no publicly available mass spectral databases and only commercial libraries such as those from the National Institute of Standards and Technology exist but do not contain a large number of typical marine microbial lipids. Within ICoMM's database MICROBIS we therefore developed a mass spectrometry library that contains lipid data derived from microbes from modern and ancient environments. This library can run under National Institute of Standards and Technology SEARCH software, which is the most commonly used software to search with mass spectra in mass spectral libraries. At this point we have assembled more than 200 mass spectra of the most common and diagnostic lipids of marine microbes.

MICROBIS has laid the foundation for an integrated database for searching and relating lipid data with other molecular and geospatial data. Once further developed, the MICROBIS website will cross-reference lipidomic, taxonomic, DNA sequence, and geospatial data (Fig. 12.3). Currently, ICoMM is designing a search-engine-supported mass spectrometry library wherein cross-referencing between Lipid Maps and the mass spectrometry database will proceed by the LIPID identifications that will also be linked to geospatial data. In the future, MICROBIS will provide a user-friendly interface that will allow searching for taxonomic, DNA sequence, and phylogenetic information, enabling biogeochemists to link lipid data to genomic and geospatial data.

12.4.2 Looking back in time with lipids

ICoMM has also attempted to explore the unknowable: a glimpse at microbial diversity in the past. Dating of evolutionary events within phylogenetic clusters is problematic as it mostly relies on the morphological identification of fossilized remains of microbes, which are usually limited to microbes having inorganic skeletons such as diatoms and coccolithophorids. A new approach is to use fossilized organic molecules. Indeed, fossil DNA occurs in several selected cases (Fish *et al.* 2002; Coolen *et al.* 2004), but findings such as these are rare and controversial. In contrast, fossilized lipids commonly occur in sediments of up to 2 billion years old and thus may be, as long as they are diagnostic for certain microbial phylogenetic clusters, suitable to trace the evolutionary history of microbes.

The usefulness of this approach lies in a detailed study of 18S rRNA genes and lipid biomarkers of more than 100 representative marine diatoms (Sinninghe Damsté *et al.* 2004). This study revealed that several lipid biomarkers are quite specific for phylogenetic clusters within the diatoms. For example, the biosynthesis of so-called highly branched isoprenoid alkenes is restricted to two specific phylogenetic clusters, which independently evolved in the centric and pennate diatoms (Sinninghe Damsté *et al.* 2004; Fig. 12.11). The molecular record of C_{25} highly branched isoprenoid chemical fossils in a large suite of well-dated marine sediments and petroleum reveals that the older cluster, comprising rhizosolenoid diatoms, evolved 91.5 ± 1.5 million years ago (Upper Turonian), enabling an unprecedented accurate dating of diatom evolution.

12.5 Viewing Microbial Diversity through a Community Lens

ICoMM's systematic and high-throughput analyses of the sequence variation of rRNA genes have provided a wealth of microbial community sequence data for numerous,

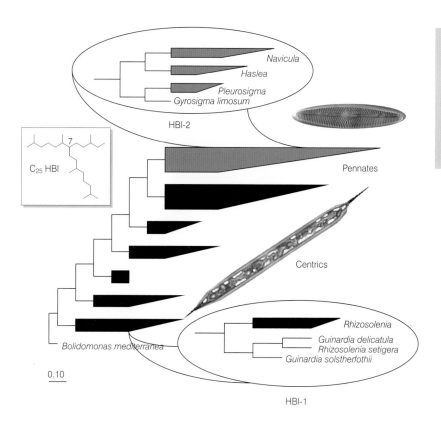

Fig. 12.11

A phylogenetic tree showing how highly branched isoprenoid (HBI) alkenes are restricted to two specific phylogenetic clusters of diatoms (HBI-1 and HBI-2), which independently evolved in the group of centric and pennate diatoms, respectively. Photomicrographs used under license, courtesy of Robert Andersen and David Patterson (http://microscope.mbl.edu).

poorly understood marine environments. When contextual parameters are recorded together with diversity data, it is now possible to assess the impact of space, time, and complex environmental gradients on microbial communities, and to quantify interactions among factors. The integration of laboratory-developed microbiological sensors into observing platforms that track changes at high temporal and spatial resolution will enable autonomous observation of changes in marine microbial diversity in the field (Paul *et al.* 2006). Here we can find answers to as different questions as the following: Why do specific communities flourish in one environment and not in another? Which microbial populations are more successful than others in the competition for energy or space? Which environments host those seed populations that only temporarily dominate communities? The same tools that have been used in classical community ecology are available for the analysis of changes in microbial community patterns, because ultimately standard sample-by-species matrices can be obtained with any high-throughput method. Our next challenge will be the generation of microbial diversity theories that will allow further comparisons with established ecological theories for macroorganisms or that can be tested across various ecosystems. For instance, recent developments in the study of microbial biogeography may be seen as a prelude to a more dramatic revolution in better understanding microbial communities in their complex environments.

12.6 Marine Microbes and Their Roles in a Changing Ocean

The importance of marine microbes to our biosphere cannot be overstated (Box 12.2). Since the microbial census began, several major scientific breakthroughs in microbial diversity and microbial ecology have occurred. Owing to the rapid developments in high-throughput and relatively cost-effective sequencing technologies like massively parallel DNA sequencing, it has become possible to deeply explore microbial (that is, bacterial, archaeal, and eukaryotic) genetic diversity of environmental samples in both qualitative and quantitative ways. Over the past five to ten years, spectacular findings have highlighted new and unexpected roles of microbes in biogeochemical cycling of carbon, nitrogen, sulfur, iron, and many other (trace) elements owing to interdisciplinary research based on the integration of sequencing, membrane lipid research, and isotope techniques. Fascinating examples of new and important microbial shunts in biogeochemical cycles include the following: the existence of anaerobic oxidation of methane by bacterial–archaeal consortia oxidizing methane with sulfate operating in (sub)oxic environments in the ocean and on land (Knittel & Boetius 2009 and the literature cited therein); the anaerobic oxidation of methane by

Box 12.2

Marine Microbial Diversity and Abundance Highlights

- The number of bacteria in the open ocean exceeds 10^{29} cells and microbes in total contribute as much as 90% of the biomass in the ocean.
- Microbes may be more than 100 times more diverse than plants and animals.
- A single liter of seawater can represent approximately 20,000 "species" of bacteria.
- A gram of sand can contain between 5,000 and 19,000 "species" of bacteria.
- Archaeal cell numbers can rival those of bacteria in the ocean but their diversity is 10% that of bacteria.
- Protist diversity rivals that of Archaea in some parts of the ocean.

- Most of marine microbial diversity is represented by low abundance populations.
- Each metazoan may have its own unique microbiome population structure.
- Different water masses possess signature microbial community structures.
- Some microbes are everywhere: ICoMM's most abundant type of bacterial pyrotag matched sequences from the SAR11 marine bacterial group which accounts for 25% of the biomass and 50% of the cell abundance in the pelagic ocean.
- Lipids allow us to look back in time at ancient microbial populations and delve into the unknown.

bacteria oxidizing methane with nitrate (Raghoebarsing *et al.* 2006); anaerobic ammonium oxidation (Anammox), whereby anammox bacteria use specific membrane components to oxidize ammonia with nitrite to form nitrogen gas that escapes from the oceans (Strous *et al.* 1999; Sinninghe Damste *et al.* 2002; Kuypers *et al.* 2003); the discovery that some Crenarchaea use ammonia as an energy source (Konneke *et al.* 2005); crenarchaeotal CO_2 fixation (Wuchter *et al.* 2003); and the incredible phenotypic adaptation of the largest known bacterial cells on Earth, the giant sulfide-oxidizing bacteria, to their environment (Gallardo 1977; Teske & Nelson 2006).

These and many other examples (Giovannoni & Stingl 2005; Karl 2007; Azam & Malfatti 2007; Bowler *et al.* 2009; DeLong 2009; Fuhrman 2009) clearly indicate that marine microbial biogeochemical cycling of elements is even more important than traditionally thought and that microbes dominate these cycles in many known and recently discovered ways. This increased recognition of microbial importance in biogeochemical cycling of elements combined with the discovery of vast microbial diversity, the discovery of the rare biosphere, and the dominance of just a few microbial taxa in environmental settings, makes us aware of the crucial importance of microbes in climate and climate change. In ICoMM we are aware that ongoing human-induced global climate change will and probably already has impacted microbial diversity. Owing to rising seawater temperatures, ocean acidification, and salinity changes, dominant marine microbial taxa may become

dormant and completely unknown taxa present in the environment but extremely rare, may become dominant. Because we have little idea about which marine microbial taxa will become dormant and which will become dominant, predictions of changes in biogeochemical cycles are very difficult to make. Thereby predictions of not whether but rather how the climate will change and how to ameliorate such change present even greater challenges to scientists and policy-makers.

The above can be illustrated by the following. The far greater part of N_2 fixation in the marine environment is currently performed by just a very few bacteria, *Trichodesmium* and an uncultivated 'Group A' putative unicellular cyanobacterium lacking oxygenic photosystem II, whereas the endosymbiotic cyanobacterium *Richelia intracellularis*, as well as other endo- and ectosymbiotic N_2-fixing cyanobacteria, are less important globally (Arrigo 2005). Very preliminary laboratory experiments with artificially acidified seawater indicate that N_2-fixing bacteria react very strongly to lowered pH values, thereby changing the rate and nature of nitrogen and carbon fixation (Hutchins *et al.* 2007). This will impact the complete marine nitrogen and carbon cycles and thereby other biogeochemical cycles (for example, Crenarchaeota use ammonium to fix carbon; phosphate may become the omnipresent limiting nutrient). Similar experiments with the major carbon-fixing organisms in oligotrophic water, *Prochlorococcus* and *Synechococcus*, also indicated that acidification and temperature rise will severely affect their metabolism, leading to changes in the rate of CO_2 fixation (Fu *et al.* 2007).

There are many factors that add to the uncertainty in the estimation of the pathways and the scale of the interaction of marine microbe communities with anthropogenically driven global climate change. Second only to human impacts on the environment, the interplay between environmental parameters and shifts in marine microbial diversity will dominate the course of climate change. The paucity of research on underlying mechanisms severely constrains the ability of policy-makers to make informed decisions about mitigating strategies. Trying to understand microbial diversity and functioning in biogeochemical and nutrient cycling is of major importance for future research on a worldwide scale.

12.7   New Questions

Despite the great diversity we see in microbial communities, there is a high degree of structure and non-random patterns in temporal and spatial scaling, and biological associations. Hence, new questions have emerged from ICoMM studies:

- What is the turnover of microbial populations and communities across various scales in space and time?
- Why are some groups dominating marine habitats globally?
- Why is there such a division between the community structure of pelagic and benthic habitats?
- Are the most diverse taxa also the most numerically abundant and why?
- What kinds of taxa are associated with plants and animals, and to what extent are they unique to each species?
- Why are there so many rare populations?
- Is the "rare biosphere" the result of 3.5 billion years of evolution, of massive horizontal gene transfer, of dispersion in the oceans, of life strategies, or combinations of the above?

To unravel the origin of these phenomena and to address these questions are challenges for the future.

12.8   Outlook

Looking forward to the next decade of microbial census research, we see many opportunities and challenges. As massively parallel DNA sequencing brings an unprecedented volume of data, it also brings challenges in analyzing these data. In this chapter we have chosen to highlight the general findings of our efforts so far, but the necessary computer algorithms and models required to bring us closer

to more robust estimates of microbial diversity are still being developed and the required computational power still being sought. Improving the taxonomy attached to pyrotags is another area that will need attention in the future. Much of this will likely come though improved annotations of the vast amount of full-length or nearly full-length sequences presently housed in public databases, as well as next-generation sequencing providing much longer reads that will enable better taxonomic assignment. Yet, we find that even when definitive taxonomic assignment is not possible, the approach is still very powerful for comparisons of assemblage composition and diversity. The technique reveals substantial diversity undetected by previously used techniques. Even ICoMM formalin-preserved samples belonging to the Joint Global Ocean Flux Study have been successfully sequenced, unveiling the potential of pyrosequencing to become a powerful tool for paleobiological and paleoenvironmental studies. The massive amount of data from pyrosequencing also enables predictions of functions for many OTUs. In short, large-scale patterns could emerge not only of phylogenetic diversity but also functional diversity along geochemical gradients. As the 454-pyrotag-based technology does not distinguish between dead and living microbial taxa, it could be complemented by other techniques for assessing the level of viability within a sample such as starting with RNA samples and reverse transcribing it to reveal the most active populations. Deciphering ecological signals linked to the definition of rare or abundant OTUs at different taxonomic levels is crucial. This definition will influence how diversity patterns are measured. Furthermore, this new paradigm will aid researchers to better understand marine microbial communities and their impact on planetary biogeochemical cycles.

Future endeavors must pay closer attention to the temporal dimension of changes in microbial community structures. As sequencing costs decline, it will be possible to monitor microbial populations and their changes over time in appropriate marine environments on different timescales from minutes to centuries. Developing such monitoring strategies through existing observing systems, time-series stations (for example Bermuda Atlantic Time-Series, Hawaii Ocean Time-Series), and Long Term Ecological Research Sites, we might be better able to predict changes in microbial populations as a consequence of natural and anthropogenic climate change. The ICoMM team recommends that this kind of monitoring approach may thus help substantially to improve our ability to predict climate change, harmful algal blooms, and ultimately our own impact on biodiversity in the ocean.

Acknowledgments

We thank all the project principal investigators who participated in the ICoMM 454 community pyrosequencing effort, as well as all the scientists, technicians, and students who participated in sample collection, processing, and analysis of data in each regional laboratory and especially at the Marine Biological Laboratory, Woods Hole. ICoMM is a project of the Census of Marine Life program funded by the Sloan Foundation. A grant from the W. M. Keck Foundation supported the 454 pyrotag sequencing conducted at the Marine Biological Laboratory, Woods Hole. Local and international funding facilitated project participation in this global initiative. GenBank sequence read archive data include SRP000903, SRP000912, SRP001108, SRP001172, SRP001206-31, SRP001242-45, SRP001259, SRP001265-66, SRP001268-70, SRP001273, SRP001309, and SRP001542-47.

References

Amaral-Zettler, L.A., McCliment, E.A., Ducklow, H.W. & Huse, S.M. (2009) A method for studying protistan diversity using massively parallel sequencing of V9 hypervariable regions of small-subunit ribosomal RNA enes. *PLoS ONE*, 4(7), e6372.

Andersen, R.A., Beakes, G., Blackburn, S., *et al.* (2006) *Report on the Alfred P. Sloan Foundation Workshop on Protistan Barcoding, Reference Material and Cultures, Portland.*

Arrigo, K.R. (2005) Marine microorganisms and global nutrient cycles. *Nature* 437, 349–355.

Azam, F. & Malfatti, F. (2007) Microbial structuring of marine ecosystems. *Nature Reviews Microbiology* 5, 782–791.

Bowler, C., Karl, D.M. & Colwell, R.R. (2009) Microbial oceanography in a sea of opportunity. *Nature* 459, 180–184.

Brazelton, W.J., Sogin, M.L. & Baross, J.A. (2009) Multiple scales of diversification within natural populations of Archaea in hydrothermal chimney biofilms. *Environmental Microbiology Reports* 2, 236–242.

Brazelton, W.J., Ludwig, K.A., Sogin, M.L., Andreishcheva, E.N., Kelley, D.S., Shen, C.-C., Edwards, R.L. & Baross, J.A. (2010) Archaea and bacteria with surprising microdiversity show shifts in dominance over 1,000-year time scales in hydrothermal chimneys. *Proceedings of the National Academy of Sciences of the USA* 107, 1612–1617.

Clarke, K.R. & Warwick, R.M. (2001) *Change in Marine Communities: An Approach to Statistical Analysis and Interpretation*, 2nd edn. Plymouth Marine Laboratory, UK: PRIMER-E.

Coolen, M.J.L., Muyzer, G., Rijpstra, W.C., *et al.* (2004) Combined DNA and lipid analyses of sediments reveal changes in Holocene phytoplankton populations in an Antarctic lake. *Earth and Planetary Science Letters* 223, 225–239.

Corliss, J.O. (1984) The kingdom Protista and its 45 phyla. *BioSystems*, 17, 87–126.

Del Giorgio, P.A. & Duarte, C.M. (2002) Respiration in the open ocean. *Nature* 420, 379–384.

DeLong, E.F. (1992) Archaea in coastal marine environments. *Proceedings of the National Academy of Sciences of the USA* 89, 5685–5689.

DeLong, E.F. (2009) The microbial ocean from genomes to biomes. *Nature* 459, 200–206.

DeLong, E.F., Wickham, G.S. & Pace, N.R. (1989) Phylogenetic stains: ribosomal RNA-based probes for the identification of single cells. *Science* 243, 1360–1363.

Díez, B., Pedrós-Alió, C. & Massana, R. (2001) Study of genetic diversity of eukaryotic picoplankton in different oceanic regions by small-subunit rRNA gene cloning and sequencing. *Applied and Environmental Microbiology* 67, no. 7, 2932–2941.

Fahy, E., Subramaniam, S., Brown, H.A., Glass, C.K., *et al.* (2005) A comprehensive classification system for lipids. *Journal of Lipid Research* 46, 839–861.

Field, C.B., Behrenfeld, M.J., Randerson, J.T. & Falkowski, P. (1998) Primary production of the biosphere: integrating terrestrial and oceanic components. *Science* 281, 237–240.

Fish, S.A., Shepherd, T.J., McGenity, T.J. & Grant, W.D. (2002) Recovery of 16S ribosomal RNA gene fragments from ancient halite. *Nature* 417, 432–436.

Fu, F.-X., Warner, M.E., Zhang, Y., *et al.* (2007) Effects of increased temperature and CO_2 on photosynthesis, growth, and elemental ratios in marine *Synechococcus* and *Prochlorococcus* (Cyanobacteria). *Journal of Phycology* 43, 485–496.

Fuhrman, J.A. (2009) Microbial community structure and its functional implications. *Nature*, 459, 193–199.

Fuhrman, J.A., McCallum, K. & Davis, A.A. (1992) Novel major archaebacterial group from marine plankton. *Nature* 356, 148–149.

Fuhrman, J.A., Schwalbach, M.S. & Stingl, U. (2008) Proteorhodopsins: an array of physiological roles? *Nature Reviews Microbiology* 6, 488–494.

Fuhrman, J.A., Sleeter, T.D., Carlson, C.A. & Proctor, L.M. (1989) Dominance of bacterial biomass in the Sargasso Sea and its ecological implications. *Marine Ecology Progress Series* 57, 207–217.

Galand, P.E., Casamayor, E.O., Kirchman, D.L. & Lovejoy, C. (2009a) Ecology of the rare microbial biosphere of the Arctic Ocean. *Proceedings of the National Academy of Sciences of the USA* 106, 22427–22432.

Galand, P.E., Casamayor, E.O., Kirchman, D.L., *et al.* (2009b) Unique archaeal assemblages in the arctic ocean unveiled by massively parallel tag sequencing. *The International Society of Microbial Ecology Journal* 3, 860–869.

Gallardo, V.A. (1977) Large benthic microbial communities in sulphide biota under Peru–Chile Subsurface Countercurrent. *Nature* 268, 331–332.

Garcia-Pichel, F., Belnap, J., Neuer, S. & Schanz, F. (2003) Estimates of global cyanobacterial biomass and its distribution. *Archive für Hydrobiologie/Algological Studies* 109, 213–228.

Gast, R.J., Moran, D.M., Beaudoin, D.J., *et al.* (2006) Abundance of a novel dinoflagellate phylotype in the Ross Sea, Antarctica. *Journal of Phycology* 42, 233–242.

Giovannoni, S.J., Britschgi, T.B., Moyer, C.L. & Field, K.G. (1990) Genetic diversity in Sargasso Sea bacterioplankton. *Nature* 345, 60–63.

Giovannoni, S.J. & Stingl, U. (2005) Molecular diversity and ecology of microbial plankton. *Nature*, 437, 343–348.

Giovannoni, S.J., Tripp, H.J., Givan, S., *et al.* (2005) Genome streamlining in a cosmopolitan oceanic bacterium. *Science* 309, 1242–1245.

Hamasaki, K., Taniguchi, A., Tada, Y., *et al.* (2007) Actively growing bacteria in the inland Sea of Japan, identified by combined bromodeoxyuridine immunocapture and denaturing gradient gel electrophoresis. *Applied and Environmental Microbiology* 73, 2787–2798.

Herndl, G.J., Reinthaler, T., Teira, E., *et al.* (2005) Contribution of Archaea to total prokaryotic production in the deep Atlantic Ocean. *Applied and Environmental Microbiology* 71, 2303–2309.

Heywood, J.L., Zubkov, M.V., Tarran, G.A., *et al.* (2006) Prokaryoplankton standing stocks in oligotrophic gyre and equatorial provinces of the Atlantic Ocean: evaluation of inter-annual variability. *Deep-Sea Research II* 53, 1530–1547.

Hobbie, J.E., Daley, R.J. & Jasper, S. (1977) Use of nuclepore filters for counting bacteria by fluorescence microscopy. *Applied and Environmental Microbiology* 33, 1225–1228.

Holm-Hansen, O. & Booth, C.R. (1966) The measurement of adenosine triphosphate in the ocean and its ecological significance. *Limnology and Oceanography* **11**, 510–519.

Huber, J.A., Mark Welch, D.B., Morrison, H.G., *et al.* (2007) Microbial population structures in the deep marine biosphere. *Science* **318**, 97–100.

Huse, S.M., Dethlefsen, L., Huber, J.A., *et al.* (2008) Exploring microbial diversity and taxonomy using SSU rRNA hypervariable tag sequencing. *PLoS Genetics* **4**(11), e1000255.

Hutchins, D.A., Fu, F.-X., Zhang, Y., *et al.* (2007) CO$_2$ Control of *Trichodesmium* N$_2$ fixation, photosynthesis, growth rates, and elemental rations: implications for past, present and future biogeochemistry. *Limnology and Oceanography* **52**, 1293–1304.

Jannasch, H.W. & Jones, G.E. (1959) Bacterial populations in sea water as determined by different methods of enumeration. *Limnology and Oceanography* **4**, 128–139.

Jørgensen, B.B. & Boetius, A. (2007) Feast and famine – microbial life in the deep-sea bed. *Nature Reviews Microbiology* **5**, 770–781.

Karl, D.M. (2007) Microbial oceanography: paradigms, processes and promise. *Nature Reviews Microbiology* **5**, 759–769.

Kirchman, D.L. (ed.) (2008) *Microbial Ecology of the Oceans.* Hoboken: John Wiley.

Knittel, K. & Boetius, A. (2009) Anaerobic oxidation of methane: progress with an unknown process. *Annual Review of Microbiology* **63**, 311–334.

Konneke, M., Bernhard, A.E., de la Torre, J.R., *et al.* (2005) Isolation of an autotrophic ammonia-oxidizing marine archaeon. *Nature* **437**, 543–546.

Kuypers, M.M.M., Sliekers, A.O., Lavik, G., *et al.* (2003) Anaerobic ammonium oxidation by anammox bacteria in the Black Sea. *Nature* **422**, 608–611.

Kysela, D.T., Palacios, C. & Sogin, M.L. (2005) Serial analysis of V6 ribosomal sequence tags (SARST-V6): a method for efficient, high-throughput analysis of microbial community composition. *Environmental Microbiology* **7**, 356–364.

Lee, J.J., Hutner, S.H. & Bovee, E.C. (eds.) (1985) *An Illustrated Guide to the Protozoa.* Lawrence, Kansas: Society of Protozoologists.

Li, W.K.W. (1994) Primary production of prochlorophytes, cyanobacteria, and eukaryotic ultraphytoplankton: measurements from flow cytometric sorting. *Limnology and Oceanography* **39**, 169–175.

López-García, P., Rodríguez-Valera, F., Pedrós-Alió, C. & Moreira, D. (2001) Unexpected diversity of small eukaryotes in deep-sea Antarctic plankton. *Nature* **409**, 603–607.

Margulies, M., Egholm, M., Altman, W.E. *et al.* (2005) Genome sequencing in microfabricated high-density picolitre reactors. *Nature* **437**, 376–380.

Mayali, X., Franks, P.J.S. & Azam, F. (2008) Cultivation and ecosystem role of a marine *Roseobacter* clade-affiliated cluster bacterium. *Applied and Environmental Microbiology* **74**, 2595–2603.

Moon-van der Staay, S.Y., De Wachter, R. & Vaulot, D. (2001) Oceanic 18S rDNA sequences from picoplankton reveal unsuspected eukaryotic diversity. *Nature* **409**, 607–610.

Pace, N.R. (1997) A molecular view of microbial diversity and the biosphere. *Science* **276**, 734–740.

Patterson, D.J. (1999) The diversity and diversification of protists. *American Naturalist* **154**, S96–S124.

Paul, J., Scholin, C., Van den Engh, G. & Perry, M.J. (2006) *In situ* instrumentation. *Oceanography* **20**, 70–78.

Pedrós-Alió, C. (2006) Marine microbial diversity: can it be determined? *Trends in Microbiology* **14**, 257–263.

Raghoebarsing, A.A., Pol, A., van de Pas-Schoonen, K.T., *et al.* (2006) A microbial consortium couples anaerobic methane oxidation to denitrification. *Nature* **440**, 918–921.

Scanlan, D.J., Ostrowski, M., Mazard, S., *et al.* (2009) Ecological genomics of marine Picocyanobacteria. *Microbiology Molecular Biology Reviews* **73**, 249–299.

Simon, H.M., Dodsworth, J.A. & Goodman, R.M. (2000) Crenarchaeota colonize terrestrial plant roots. *Environmental Microbiology* **2**, 495–505.

Sinninghe Damsté, J.S., Muyzer, G., Abbas, B., *et al.* (2004) The rise of the rhizosolenid diatoms. *Science* **304**, 584–587.

Sinninghe Damste, J.S., Strous, M., Rijpstra, W.I.C., *et al.* (2002) Linearly concatenated cyclobutane lipids form a dense bacterial membrane. *Nature* **419**, 708–712.

Sogin, M.L., Morrison, H.G., Huber, J.A., *et al.* (2006) Microbial diversity in the deep sea and the underexplored "rare biosphere". *Proceedings of the National Academy of Sciences of the USA* **103**, 12115–12120.

Stahl, D.A., Lane, D.J., Olsen, G.J. & Pace, N.R. (1984) Analysis of hydrothermal vent-associated symbionts by ribosomal RNA sequences. *Science* **224**, 409–411.

Strous, M., Fuerst, J.A., Kramer, E.H.M., *et al.* (1999) Missing lithotroph identified as new planctomycete. *Nature* **400**, 446–449.

Teske, A. & Nelson, D.C. (2006) The genera *Beggiatoa* and *Thioploca*. In: *The Prokaryote* (eds. M. Dworkin, M. Falkow, K.H. Schleifer. & E. Stackebrandt), pp.784–810. New York: Springer.

Venter, J.C., Remington, K., Heidelberg, J.F., *et al.*(2004) Environmental genome shotgun sequencing of the Sargasso Sea. *Science* **304**, 66–74.

Whitman, W.B., Coleman, D.C. & Wiebe, W.J. (1998) Prokaryotes: the unseen majority. *Proceedings of the National Academy of Sciences of the USA* **95**, 6578–6583.

Wuchter, C., Schouten, S., Boschker, H.T.S. & Sinninghe Damsté, J.S. (2003) Bicarbonate uptake by marine Crenarchaeota. *Federation of European Microbiological Societies Microbiology Letters* **219**, 203–207.

Yentsch, C.M., Horan, P.K., Muirhead, K., Dortch, Q., *et al.* (1983) Flow cytometry and cell sorting: a technique for analysis and sorting of aquatic particles. *Limnology and Oceanography* **28**, 1275–1280.

Zhu, F., Massana, R., Not, F., *et al.* (2005) Mapping of picoeukaryotes in marine ecosystems with quantitative PCR of the 18S rRNA gene. *Federation of European Microbiological Societies Microbiology Letters* **52**, 79–92.

Zimmermann, R. & Meyer-Reil, L. (1974) A new method for fluorescence staining of bacterial populations on membrane filters. *Aus dem Institut fur Meereskunde an der Uniiversitat Kiel* **30**, 24–27.

Chapter 13

A Census of Zooplankton of the Global Ocean

Ann Bucklin[1], Shuhei Nishida[2], Sigrid Schnack-Schiel[3], Peter H. Wiebe[4], Dhugal Lindsay[5], Ryuji J. Machida[2], Nancy J. Copley[4]

[1]Department of Marine Sciences, University of Connecticut, Groton, Connecticut, USA
[2]Ocean Research Institute, University of Tokyo, Tokyo, Japan
[3]Alfred Wegener Institute for Polar and Marine Research, Bremerhaven, Germany
[4]Biology Department, Woods Hole Oceanographic Institution, Woods Hole, Massachusetts, USA
[5]Japan Agency for Marine-Earth Science and Technology, Yokosuka City, Japan

13.1 Introduction

The animals that drift with ocean currents throughout their lives (that is, the holozooplankton) include approximately 7,000 described species in 15 phyla. The holozooplankton assemblage is the focus of the Census of Marine Zooplankton (CMarZ; www.CMarZ.org), which has produced comprehensive new information on species diversity, distribution, abundance, biomass, and genetic diversity. Our realm among Census of Marine Life projects is the open ocean; we have sampled biodiversity hot spots throughout the world's oceans: little-known seas of Southeast Asia, deep-sea zones below 5,000 m, and polar seas. We have used traditional plankton nets and newer sensing systems deployed from ships and submersibles. Our analysis has included traditional microscopic and morphological examination, as well as molecular genetic analysis of zooplankton populations and species. CMarZ has contributed to Census legacies in data and information for the Ocean Biogeographic Information System (see Chapter 17) and proven technologies of DNA barcoding. Our photograph galleries of living plankton have captured public interest,

and our training workshops have enhanced taxonomic expertise in many countries. The knowledge gained will provide a new baseline for detection of impacts of climate change, and will contribute to our fundamental understanding of biogeochemical transports, fluxes and sinks, productivity of living marine resources, and marine ecosystem health.

13.2 Historical Perspective

Despite more than a century of sampling the oceans, comprehensive understanding of zooplankton biodiversity has eluded oceanographers because of the fragility, rarity, small size, and/or systematic complexity of many taxa. For many zooplankton groups, there are long-standing and unresolved questions of species identification, systematic relationships, genetic diversity and structure, and biogeography.

There has never been a taxonomically comprehensive, global-scale summary of the current status of our knowledge of biodiversity of marine zooplankton. Although studies of the taxonomy, distribution, and abundance of zooplankton date back as far as the middle of the nineteenth century, worldwide distribution patterns have not been mapped for all described species. The cosmopolitan or circumglobal distributions characteristic of holozooplankton species of many groups have created special

Life in the World's Oceans, edited by Alasdair D. McIntyre
© 2010 by Blackwell Publishing Ltd.

difficulties for accurate biodiversity assessment. The snapshots from different parts of the world ocean have rarely been merged together, in part because the complicated and time-consuming task of compiling the information from numerous individual publications is undervalued (but see Irigoien *et al.* 2004).

For most zooplankton groups, significant numbers of species remain to be discovered. This is especially true for fragile (for example gelatinous) forms that are difficult to sample properly and for forms living in unique and isolated habitats, such as the water surrounding hydrothermal vents and seeps (Ramirez-Llodra *et al.* 2007; Chapter 9). All regions of the deep sea are certain to continue to yield many new species in multiple taxonomic groups. The practical difficulties of exploring these regions are gradually being overcome, and they are likely to continue to yield new species discoveries for many years.

Our perception of zooplankton biodiversity has almost certainly been affected by their small size, resulting in a marked under-description of species and morphological types. Until recently, some pelagic taxa (for example foraminifers, copepods, euphausiids, and chaetognaths) have been thought to be well known taxonomically, but the advent of molecular genetics has altered this perspective. Morphologically cryptic, but genetically distinctive, species of zooplankton are being found with increasing frequency (see, for example, Bucklin *et al.* 1996, 2003; de Vargas *et al.* 1999; Dawson & Jacobs 2001; Goetze 2003) and will probably prove to be the norm across a broad range of taxa. Many putative cosmopolitan species may comprise morphologically similar, genetically distinct sibling species, with discrete biogeographical distributions. This issue is especially relevant for widely distributed species and/or for species with disjoint distributional ranges, including those occupying coastal environments (Conway *et al.* 2003). It is likely that many morphologically defined zooplankton species will be found to consist of complexes of genetically distinct populations, but how many cryptic species are present is currently unknown, even for well-known zooplankton groups.

Marine zooplankton are important indicators of environmental change associated with global warming and acidification of the oceans. A global-scale baseline assessment of marine zooplankton biodiversity, including long-term monitoring and retrospective analysis, is critically needed to provide a contemporary benchmark against which future changes can be measured. Knowledge of previous and existing patterns of zooplankton distribution and diversity is useful for management of marine ecosystems and assessment of their status and health (Link *et al.* 2002). Marine zooplankton are also significant mediators of fluxes of carbon, nitrogen, and other critical elements in ocean biogeochemical cycles (Buitenhuis *et al.* 2006). Species composition of zooplankton assemblages may have strong impacts on rates of recycling and vertical export (see, for example, Gorsky & Fenaux 1998); long-term changes in fluxes into the deep sea (Smith *et al.* 2001) may be related to zooplankton species composition in overlying waters (Roemmich & McGowan 1995; Lavaniegos & Ohman 2003).

Compared with the dimensions of the known – in terms of numbers of species and regions of the world's oceans – the unknown is thought to be many times larger. Introducing his monograph on the biogeography of the Pacific Ocean, McGowan (1971) posed several questions that help frame the unknown territory of zooplankton biodiversity. "What species are present? What are the main patterns of species distribution and abundance? What maintains the shape of these patterns? How and why did the patterns develop?" Nearly 40 years later, the answers to these questions remain poorly known for many ocean regions and most zooplankton groups.

13.3 Approaches to the Study of Marine Zooplankton

13.3.1 Zooplankton sampling

Zooplankton samples for CMarZ have been collected by nets, buckets, water bottles, sediment traps, light traps, remotely operated vehicles (ROVs), submersibles, and divers. Sampling strategies have trade-offs for each type of sampling gear: some may obtain numerous specimens, but under-sample fragile taxa, whereas others may be suited for collecting fragile organisms for taxonomic analysis, but may be unable to sample at spatial resolutions and scales appropriate for accurate characterization of patterns of distribution and abundance.

During CMarZ dedicated cruises in the Atlantic Ocean, zooplankton and micronekton were quantitatively sampled throughout the water column using MOCNESS (Multiple Opening/Closing Net and Environmental Sensing System; Wiebe *et al.* 1985; Wiebe & Benfield 2003). In addition to collecting depth-stratified plankton samples, the MOCNESS transmits environmental data (depth, temperature, salinity, horizontal speed, and volume filtered) to the ship throughout the tow; the data are recorded for subsequent analysis. A uniquely equipped 10-meter MOCNESS allowed CMarZ to sample to 5,000 m in the Atlantic Ocean and rapidly filter large volumes (tens of thousands of cubic meters) to capture rare deep-sea zooplankton (Wiebe *et al.* 2010). The collections included first-ever observation of living specimens of rare deep-sea species (see, for example, Johnson *et al.* 2009; Bradford-Grieve 2010), and offered remarkable opportunities for photographing living specimens (Fig. 13.1) and barcoding novel species.

CMarZ has used modern *in situ* survey technologies, including crewed submersibles, ROVs, towed camera arrays, and visual/video plankton recorders (VPR; Davis

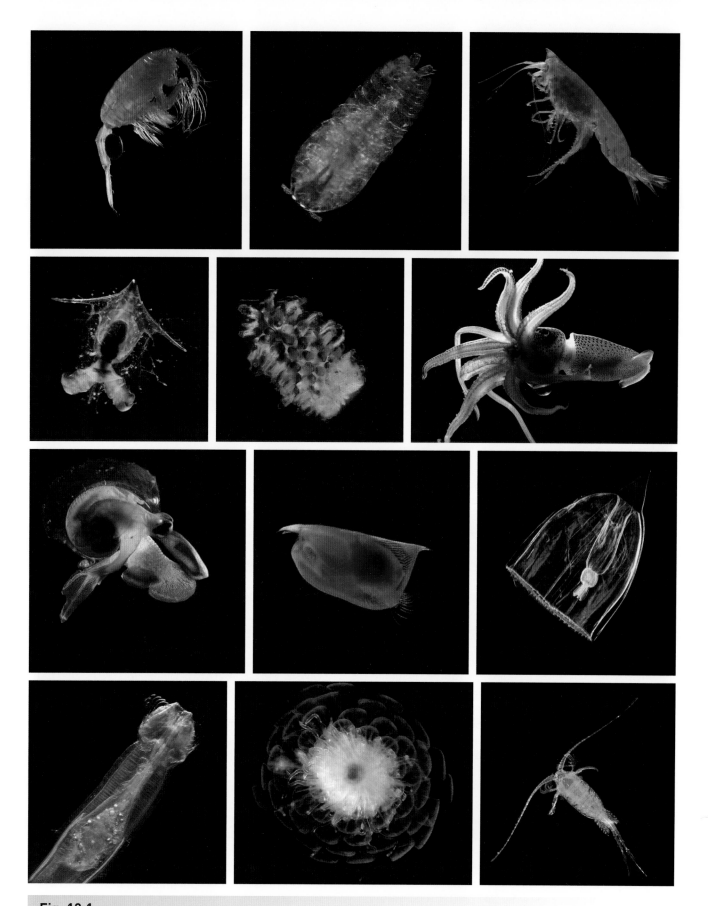

Fig. 13.1

Photograph gallery of living marine zooplankton. Row 1 (left to right): *Valdiviella* sp. and *Sapphirina metalina* (Copepoda); *Cyphlocaris* sp. (Amphipoda); row 2: *Clio cuspidate* (Pteropoda); *Pyrosoma* sp. (Thaliacea); *Histioteuthis* sp. (Cephalopoda); row 3: *Oxygyrus keraudreni* (Heteropoda); *Conchoecissa plinthina* (Ostracoda), *Aglantha* sp. (Hydrozoa); row 4: unidentified Chaetognatha with a copepod; *Athorybia rosacea* (Siphonophora); *Lucicutia* sp. (Copepoda). Photograph credits R.R. Hopcroft and C. Clarke (University of Alaska – Fairbanks) and L.P. Madin (Woods Hole Oceanographic Institution).

et al. 1992) to observe and collect zooplankton, especially fragile gelatinous forms, in many areas of the ocean. These sampling approaches have led to new species discoveries (Haddock *et al.* 2005; Lindsay & Miyake 2007), and rapid advances in our understanding of deep sea biology and ecology (Pagès *et al.* 2006; Ates *et al.* 2007; Fujioka & Lindsay 2007; Kitamura *et al.* 2008a, b; Lindsay *et al.* 2008). In 2006, Dhugal Lindsay (Japan Agency for Marine-Earth Science and Technology) led a pilot study to census gelatinous and hard-bodied zooplankton in Sagami Bay (Japan) using diverse sampling technologies, including an autonomous video plankton recorder (AVPR) with a high-definition video camera for color imagery. The study yielded images and samples of zooplankton and marine snow that are being analyzed to model and predict effects of climate change on carbon cycling and sequestration.

Blue-water SCUBA diving for observing and collecting fragile zooplankton was developed during the past 30 years (Hamner 1975), and has been used to advantage by CMarZ. A group of divers work from an inflatable boat launched from a research vessel; they are connected to a central line by a 10-meter tether line and overseen by a safety-diver. This technique has proven ideal to locate, observe, photograph, and collect live and undamaged specimens of free-swimming gelatinous animals.

A variety of remote plankton-sensing platforms (that is, those deployed from ships that return data – but not necessarily samples) has been developed for the study of zooplankton diversity, distribution, and abundance. CMarZ has used several among the many instruments developed for this purpose, including the video plankton recorder (VPR; Davis *et al.* 1992); underwater video profiler (UVP; Gorsky *et al.* 1992, 2000); optical plankton counter (OPC; Herman 1988); and continuous plankton recorder (CPR; Hardy 1926; Glover 1962). In general, these systems provide higher spatial resolution than nets and more accurate depiction of the animal in its environment (Mori & Lindsay 2008). For those species that can be remotely identified, these instruments are valuable tools in describing the geographical and temporal changes in zooplankton populations in relation to behavior and the environment.

To census the world's oceans, CMarZ has used ships of opportunity to sample zooplankton in open-ocean waters and areas not regularly frequented by large research vessels. Ships of opportunity have deployed ROVs and crewed submersibles, which usually require large ocean-going vessels for their deployment, in studies in Monterey Bay, California (Matsumoto *et al.* 2003; Raskoff & Matsumoto 2004) and off the coast of Japan (Lindsay *et al.* 2004, 2008; Kitamura *et al.* 2005; Lindsay & Hunt 2005; Lindsay & Miyake 2009). In particular, the Plankton Investigatory Collaborative Autonomous Survey System Operon (PICASSO) ROV system was designed for deployment from ships of opportunity to study gelatinous plankton as deep as 1,000 m (Yoshida *et al.* 2007a, b; Yoshida & Lindsay 2007).

13.3.2 Sample preservation

Zooplankton samples for CMarZ have been processed as bulk unsorted samples, especially during cruises of opportunity, and as individual expertly identified specimens, usually during dedicated CMarZ surveys. No single sampling-handling approach can preserve the appearance and morphological, molecular, and biochemical properties of zooplankton specimens. CMarZ developed and has used a sample-splitting protocol that entails immediate bulk processing of a portion of the sample (partly in formalin for morphological analysis and partly in alcohol for molecular analysis), with another portion retained alive for photography, observation, and identification of living specimens, some of which may not be suitable for eventual preservation. Splitting is not recommended for samples with few individuals or rare species, but may in other cases optimize sample use among scientists. Samples for molecular analysis were preserved in 95% non-denatured ethanol or buffer solution (for example RNAlater) and then stored at low temperatures ($-20\,°C$) to slow degradation. Identified specimens were flash-frozen in individual vials in liquid nitrogen. Overall, best results were obtained when DNA extractions were done very soon after collection.

13.3.3 Sample analysis

An essential element of CMarZ has been traditional morphological examination of samples by taxonomic experts, who are essential to validate species identifications for uncertain and possible new species, examine and confirm putative new or cryptic species, and describe new species. Such skills are the domain of a very few specialists worldwide and are a diminishing resource. The lack of manpower – both expert and technical – has been a bottleneck for CMarZ in our progress toward our goal of a global, taxonomically comprehensive biodiversity census.

Consequently, CMarZ has championed integrated morphological and molecular genetic approaches to analysis of zooplankton species' diversity. A revolution in the analysis of global patterns of species diversity has been driven by the widespread use of DNA barcodes (that is, short DNA sequence used for species recognition and discrimination; Hebert *et al.* 2003). The usual barcode gene region for metazoan animals is a 708 base-pair region of mitochondrial cytochrome oxidase I, mtCOI (Schindel & Miller 2005). CMarZ barcoding efforts have included analysis of both targeted taxonomic groups and particular ocean regions or domains. Five CMarZ barcoding centers (at the University of Connecticut, USA; Ocean Research Institute, Japan; Institute of Oceanology, China; Alfred Wegener Institute, Germany; and National Institute of Oceanography, India) have worked together toward a shared goal of determining DNA barcodes for the approximately 7,000 described species of zooplankton. CMarZ has also uniquely demonstrated the use of off-the-shelf automated DNA sequencers in ship-board molecular laboratories, allowing a continuous at-sea analytical "assembly line" from collection, identification, and DNA barcoding.

Environmental DNA surveys (that is, determination of sequences for 16S or 16S-like rRNA coding regions from mixed environmental samples) have transformed our understanding of microbial diversity in the oceans (Pace 1997; Sogin *et al.* 2006). CMarZ has applied this revolutionary approach to the analysis of zooplankton species diversity based upon COI barcodes, using an approach dubbed environmental barcoding (that is, DNA sequencing of the COI barcode region from unsorted bulk samples). This approach has the marked advantage of not requiring morphologically based species identification. For zooplankton, environmental barcoding entails comparison of the resultant DNA sequences with "gold standard" DNA barcode data to identify species and characterize species diversity (Machida *et al.* 2009).

13.3.4 Data and information management

CMarZ uses a centralized distributed data and information management system, an outgrowth of the US GLOBEC Data and Information Management System (Groman & Wiebe 1998; Groman *et al.* 2008), which integrates among three primary data centers: Woods Hole Oceanographic Institution (Woods Hole, USA), Ocean Research Institute (Tokyo, Japan), and Alfred Wegener Institute (Bremerhaven, Germany). The ready and open exchange of information helps ensure that CMarZ project participants can coordinate and avoid duplication of effort, and thus speed progress toward the goal of a comprehensive and complete DNA barcode database for zooplankton.

13.4 Results from CMarZ

13.4.1 Toward a global view of pelagic biodiversity

Compared with the approximately 1 million described terrestrial insects and more than 1 million benthic marine organisms, the diversity of marine zooplankton, with about 7,000 species, is by no means rich. A unique attribute of this assemblage is the relative magnitude of local diversity to global diversity (Angel *et al.* 1997). As an example, the Copepoda – the most species-rich group of marine zooplankton – are very common and species are frequently very abundant. One net sample from oceanic waters may contain hundreds of copepod species or about 10% of the global total of approximately 2,200 species. This ratio is nearly unique among animal groups and habitats. Low

global diversity has been attributed to the homogenous and unstructured pelagic environment compared with terrestrial, intertidal, or benthic habitats. High local diversity has been attributed to the coexistence of many species, through vertical or other modes of niche partitioning, but the exact mechanism for their co-existence is still poorly understood (Lindsay & Hunt 2005; Kuriyama & Nishida 2006). Recently, the contribution of biological associations toward the enhancement of species diversity has been attracting much attention (Pagès *et al.* 2007; Lindsay & Takeuchi 2008; Ohtsuka *et al.* 2009).

Since 2004, CMarZ has completed more than 90 cruises, and samples for CMarZ have been collected at more than 12,000 stations; an additional 6,500 archived samples have been available for analysis. CMarZ has sampled from every ocean basin (Fig. 13.2). For selected groups of zooplankton, CMarZ has made excellent progress toward a new global view of biodiversity. Although zooplankton are not as prevalent as microbes, for which an "everything is everywhere" debate continues (see, for example, Patterson 2009), species with circumglobal distributions are found in every phylum of the zooplankton assemblage from Protista to Chordata. Such broadly distributed species have been a focus of particular attention for CMarZ. The global biogeography of planktonic Foraminifera has been mapped by Colomban de Vargas (CNRS, France), based upon integrated morphological and molecular systematic analysis (de Vargas *et al.* 2002; Moranda *et al.* 2009). Demetrio Boltovskoy (University of Buenos Aires, Argentina) has produced an atlas of Radiolaria (Polycystina) distributions based upon 6,719 samples that reveals relations between radiolarian distributions and worldwide water mass and circulation patterns (Boltovskoy *et al.* 2003, 2005). CMarZ contributed to production of a monograph on the known genera of Hydrozoa in the world ocean (Bouillon *et al.* 2006). Analysis of global patterns of copepod diversity and abundance is being performed by Sigrid Schnack-Schiel (Alfred Wegener Institute, Germany), who is comparing regional patterns in tropical, temperate, and polar seas; in all regions, more than 50% of all species occur in low abundances (not more than 10 individuals per 100 cubic meters); in Antarctic waters, more than 80% of species are rare (Fig. 13.3).

CMarZ has also contributed to new understanding of ocean-basin scale patterns of species diversity through monographic treatments of selected zooplankton groups. Notable among these are analyses of planktonic Ostracoda of the Atlantic Ocean (Angel *et al.* 2007; Angel 2008; Angel & Blachowiak-Samolyk 2009; Angel 2010). Also, Vijayalakshmi Nair (National Institute of Oceanography, India) has advanced understanding of species diversity of the Chaetognatha, a taxonomically challenging group, in the Indian Ocean (Nair *et al.* 2008) and, working with Annelies Pierrot-Bults (University of Amsterdam, The Netherlands), in the Atlantic Ocean. Gelatinous zooplankton

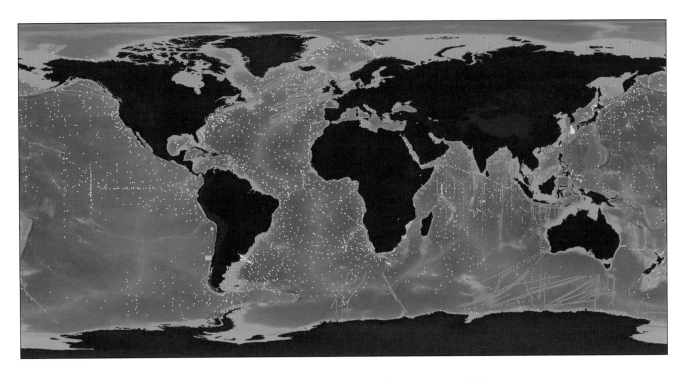

● V. Nair (NIO, India) *

● D. Boltovskoy (UBA, Argentina) *

● P.C. Reid (SAHFOS, UK)

● S. Sun (IOCAS, China)

● S. Schnack-Schiel (AWI, Germany)

● P. Wiebe/L. Madin (WHOI, USA)

● S. Nishida (ORI, Japan)

● Others

* Historical collections

Fig. 13.2

Global map showing collection locations of new zooplankton samples for analysis by CMarZ during 2004–2009. Also shown are two large historical collections that have been analyzed by CMarZ scientists. Colors indicate the various CMarZ participating institutions and individuals. NIO, National Institute of Oceanography, India; UBA, University of Buenos Aires, Argentina; SAHFOS, Sir Alister Hardy Foundation for Ocean Science, United Kingdom; IOCAS, Institute of Oceanology, Chinese Academy of Sciences, China; AWI, Alfred Wegener Institute for Polar and Marine Research, Germany; WHOI, Woods Hole Oceanographic Institution, USA; ORI, Ocean Research Institute, University of Tokyo, Japan.

diversity patterns have been found to differ between the Pacific Ocean and Japan Sea sides of Japan (Lindsay & Hunt 2005), including unique investigations of ctenophores and other fragile gelatinous zooplankton using submersibles below 2,000 m (Lindsay 2006; Lindsay & Miyake 2007). An in-depth study on the gelatinous fauna of the Gulf of Maine was published by Pagès *et al.* (2006). Also, checklists and field guides have been produced to aid in species identification of gelatinous plankton for Japanese waters (Lindsay 2006; Kitamura *et al.* 2008a, b; Lindsay & Miyake 2009); for waters off California (Mills *et al.* 2007; Mills & Haddock 2007); and for the Mediterranean (Bouillon *et al.* 2004).

13.4.2 *Biodiversity hot spots*

Sampling within regions and/or for taxa that have historically been ignored or understudied has been a key objective of CMarZ. Our efforts have been focused on biodiversity hot spots (that is, geographic or taxonomic domains for which there is greatest scope for improved knowledge of species richness), which may be specific areas of the ocean, taxonomic groups, or ecological guilds. Marine ecologists and oceanographers must identify and prioritize such regions, similar to terrestrial ecologists, who have identified 18 biodiversity hot spots based primarily on degree of endemism and impacts of human activities (Wilson 1999).

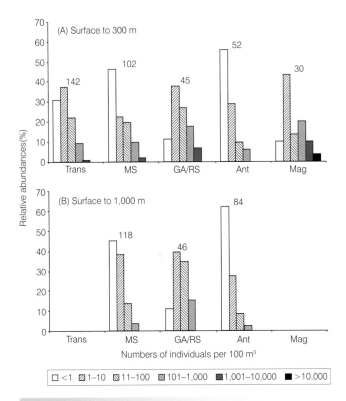

Fig. 13.3

Comparisons for tropical, temperate, and polar regions of patterns of Copepoda species diversity and abundance for two ocean depth strata: **(A)** surface to 300 m and **(B)** surface to 1,000 m. Regions shown top and bottom are as follows: Trans, *Polarstern* Transect 2002 (temperate Atlantic); MS, Meteor Seamount 1998 (subtropical North Atlantic); GA/RS, Gulf of Aqaba, Red Sea 1999 (tropical); Ant, Weddell Sea and Bellingshausen Sea, Antarctica (polar); Mag, Magellan Strait (sub-polar South Atlantic).

Table 13.1

Numbers of new zooplankton species, genera, and families discovered during CMarZ. Species that are not yet formally described are listed separately below.

Taxonomic group	New family (genus)	New species
Described new species		
Phylum Ctenophora	1(0)	
Phylum Cnidaria		
Hydromedusae	1(2)	2
Siphonophora		1
Scyphozoa		2
Phylum Arthropoda		
Copepoda	1(3)	21
Mysidae	0(1)	23
Amphipoda	1(0)	1
Phylum Chaetognatha		2
Total	4(6)	52
Species to be described		
Phylum Ctenophora		1
Phylum Arthropoda		
Copepoda		20
Ostracoda	0(1)	15
Phylum Annelida		
Polychaeta		1
Total	0(1)	37

Among the numerous acknowledged biodiversity hot spots for marine zooplankton, CMarZ has focused on diversity in the deep sea, polar seas, and coastal regions and marginal seas of Southeast Asia. Our taxonomic targets have included gelatinous groups and other taxonomically challenging and under-appreciated groups throughout the zooplankton assemblage. The CMarZ focus on geographic and taxonomic areas with high potential for species discovery has resulted in discoveries of 89 new species, of which 52 have been formally described (Table 13.1).

13.4.2.1 Southeast Asian coastal waters and marginal seas

Comprehensive research has been conducted in the embayed waters, coastal areas, and marginal seas of Southeast Asia. This is a major biodiversity hot spot in the world and has a very complicated geography and geological history. New species discoveries here have been dominated by copepods (including *Pseudodiaptomus, Tortanus (Atortus)*, and species of the families Pontellidae and Pseudocyclopidae) and mysids collected using sledge nets from coastal near-bottom habitats and by night-time or SCUBA sampling in coral reefs, indicating that the high diversity of these habitats has been overlooked by conventional daytime net sampling (see, for example, Nishida & Cho 2005; Murano & Fukuoka 2008). The Sulu Sea, a semi-enclosed marginal sea in the tropical Western Pacific Ocean, has been a particular focus for CMarZ studies (Nishikawa *et al.* 2007) and has yielded several discoveries of new species and genera, including copepods (Ohtsuka *et al.* 2005).

Species discoveries by CMarZ within the Copepoda have added another 8% to the total number of copepod species in Southeast Asia (another 2% to the global total), and new species discoveries of Mysidacea in Southeast Asia have added 15% to the global total for that group. Understanding the significance of these numbers must also take into account the ecological importance of the species and their role in the ecosystem. Regardless, CMarZ has made exceptional progress in improving our knowledge of zooplankton biodiversity in Southeast Asia by building effective teams of expert taxonomists who collaborate with CMarZ scientists.

13.4.2.2 The deep sea

By volume, 88% of the ocean environment is deeper than 1 km and 76% is between a depth of 3 and 6 km (Table 13.2; Menard & Smith 1966; Hering 2002). The deep sea is thus the largest habitat on earth – and also the one least known. Previous studies have yielded several general characteristics of pattern of zooplankton diversity, distribution, and abundance in the deep sea. A primary finding is that numbers of species and their abundances tend to decrease with depth (Longhurst 1995). The decrease in number of species is not linear; there is a peak in mid-water layers and a decrease with greater depth (Fig. 13.4). However, better sampling in

the deep sea and discovery of new species at depth may alter this trend. Latitude affects this general trend, with higher numbers of species at all depths in lower latitudes than at higher latitudes (Angel 2003). Other general trends are that deeper-dwelling species are less likely to be endemic (that is, native and restricted to a particular region) and more likely to be geographically widespread. Usual feeding mode varies through the depth strata, with filter-feeding herbivorous species occurring in the upper water layers, and detritivores and carnivores most abundant below the light-filled surface waters.

Exploration and discovery in the deep sea have been slowed by the inherent difficulties of sampling at great depths. Deeper than 1,500 m in most ocean areas, the very low abundances of most species requires that huge volumes of water be filtered, with sampling over many hours using huge sampling systems deployed from large ships, to collect significant numbers of individuals. Despite these challenges, many new deep-sea species have been discovered in the past decade, strongly indicating that deep-sea biodiversity has so far been markedly underestimated.

CMarZ' unique approach to sampling deep-sea zooplankton using a 10-meter MOCNESS with fine mesh nets has yielded many discoveries and first-time observations of

Table 13.2

Pelagic habitat volumes of the Atlantic, Pacific, and Indian Oceans based on hypsometry presented by Menard & Smith (1966). The ocean pelagic habitat has been divided vertically into five zones: epipelagic, mesopelagic, bathypelagic, abyssopelagic, and hadopelagic (Hedgepeth 1957). The last zone occupies a small fraction of the ocean volume and is present in the ocean's deep-sea trenches.

Habitat zone	Atlantic Ocean volume (10^6 km³)	Volume (%)	Pacific Ocean volume (10^6 km³)	Volume (%)	Indian Ocean volume (10^6 km³)	Volume (%)
Epipelagic (0–200 m)	17311.6	4.76	33248.0	4.28	14685.2	4.59
Mesopelagic (200–1,000 m)	64382.4	17.70	130822.4	16.83	56643.2	17.71
Bathypelagic (1,000–4,000 m)	213140.0	58.60	455499.0	58.560	193879.0	60.62
Abyssopelagic (4,000–7,000 m)	68859.0	18.93	157588.0	20.27	54594.0	17.07
Hadopelagic (>7,000 m)	10.0	0.003	175.0	0.023	0	0
Total	363703.0	100	777332.4	100	319801.4	100

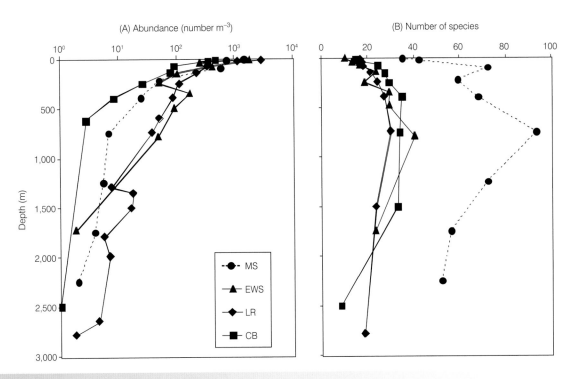

(A) Abundance (number m⁻³)

(B) Number of species

Fig. 13.4

Vertical profiles of abundance **(A)** and numbers **(B)** of calanoid Copepoda species in different geographical regions during summer (excluding the benthopelagic zone). Abbreviations are as follows: Meteor Seamount (MS), Eastern Weddell Sea (EWS), Lomonosov Ridge (LR), Canada Basin (CB). Arctic data from Kosobokova (1989) and Kosobokova & Hirche (2000); Antarctic data from S. Schnack-Schiel (unpublished data).

living specimens. During two CMarZ cruises using this gear to explore the deep tropical/subtropical Atlantic Ocean regions (that is, the Sargasso Sea on the R/V *RH Brown* in 2006, and the eastern Atlantic on the FS *Polarstern* in 2007), zooplankton were collected from the entire water column with a focus on describing species composition and richness and discovering new species in the poorly known meso- and bathypelagic zones. The Sargasso Sea cruise yielded a treasure-trove of specimens, including Ctenophora (22 species), Cnidaria (110 species), Ostracoda (58 of 140 known Atlantic species), Copepoda (134 species), euthecosome pteropod Mollusca (20 of approximately 33 species), heteropod Mollusca (17 of 29 species), Cephalopoda (13 species), and Appendicularia (13 of approximately 70 species). In addition, 3,965 fish specimens were collected, including 127 species of 84 genera from 42 families. Below 1,000 m depth, the MOCNESS-10 collected several little-known species, including the siphonophores *Nectadamas richardi* (Pugh 1992) and *Lensia quadriculata* (Pagès *et al.* 2006).

During the eastern Atlantic cruise, more than 1,000,000 cubic meters of seawater was filtered and approximately 60,000 specimens were identified. In some cases, collec-

tions represented a significant fraction of the species known from the South Atlantic; 104 copepod species were identified of an estimated total of 500 species known (Bradford-Grieve *et al.* 1999). A sample from the bathypelagic captured a putative new copepod species, the third to be described from the family Hyperbionychidae (Bradford-Grieve 2010). From the two CMarZ deep-sea Atlantic cruises, at least 15 novel ostracod species were discovered and are in process of description (Martin Angel, unpublished data).

In recent years, CMarZ' use of *in situ* sampling and observation from submersibles and ROVs has dramatically improved our understanding of deep-sea biodiversity, biology, and ecology. Laurence P. Madin (Woods Hole Oceanographic Institution, USA) led a CMarZ exploration to the Celebes Sea, a tropical sea and biodiversity hot spot in the Indonesia/New Guinea/Philippine triangle between the Pacific and Indian Oceans. Sampling was performed by blue-water diving and net systems; deep-sea observations to 3,000 m used a Global Explorer ROV with high-definition television and benthic-baited video "Ropecams". The team discovered that the overall biomass of the water column was high, with exceptional abundance of the nitrogen fixing, blue-green bacteria *Trichodesmium*. Sperm whales and spinner dolphins were observed at the surface, squid were seen from the ROV, and myctophid fishes were collected in the trawl. Ten of 23 known worldwide species of Salpidae, a group of gelatinous zooplankton, were collected by blue-water divers. Two species thought to be new to science were observed: a black, benthopelagic lobate ctenophore and a large pelagic polychaete worm with ten long cephalic tentacles.

Further CMarZ deep-sea exploration using ROVs and submersibles uncovered a cascade of biological associations of the deep-sea hydromedusan, *Pandea rubra*, which is dependent upon a pteropod mollusk for the polyp stage of its life cycle (Lindsay *et al.* 2008). Ocean acidification is thought to be detrimental to calcareous shell-bearing Mollusca, and the newly discovered linkage between these species may represent a threat to the medusa. *Pandea rubra* was found to host many other species during its deep-sea medusa stage, including Pycnogonida (sea spiders) (Pagès *et al.* 2007), hyperiid Amphipoda, and larval stages of other hydromedusae (Lindsay *et al.* 2008). Invaluable archived video from the 11,000 m ROV *Kaiko* revealed what appeared to be a new order of Ctenophora in the Ryukyu Trench (Japan); a comb jelly was observed floating above and attached by "strings" to the sea floor at a depth of 7,217 m (Lindsay & Miyake 2007).

13.4.2.3 Polar seas

As a general rule across pelagic groups, species diversity is lower at high latitudes than at low latitudes (see Chapters 10 and 11). Although the explanation for this remains unclear, low temperature and dramatic seasonal shifts in light levels and sea ice cover – and thus primary production – surely represent significant challenges to survival. Although the most characteristic feature of polar seas is sea ice, early studies of polar zooplankton were largely restricted to ice-free areas and summer months. This has severely limited our understanding of polar ecosystems, because the sea ice environment is a unique environment harboring a diverse fauna (Bluhm *et al.* 2010) and plays a vital role in ecosystem dynamics of both polar oceans (see, for example, Schnack-Schiel 2001; Arndt & Swadling 2006; Kiko *et al.* 2008; Schnack-Schiel *et al.* 2008).

In the Antarctic, where sea ice is predominantly seasonal, the Southern Ocean krill (*Euphausia superba*) is the keystone species and inhabits the seasonal pack-ice zone of Antarctic Coastal Current (Atkinson *et al.* 2004; Siegel 2005). Copepoda are dominant in many Antarctic regions in terms of both biomass and abundance, with few large species (for example *Calanus propinquus*, *Calanoides acutus*) making up more than 40% of total copepod biomass, and frequently neglected smaller species (for example *Oithona*, *Oncaea*, *Microcalanus*, *Ctenocalanus*, and others) accounting for more than 80% of total copepod abundance (Kosobokova & Hirche 2000; Hopcroft &

Robison 2005; Schnack-Schiel *et al.* 2008). Park & Ferrari (2008) reported a total of 205 calanoid copepod species from the Southern Ocean: 184 species (of which 50 are endemic) were restricted to deep waters, 13 species (8 endemic) were epipelagic, and 8 species (all endemic) were neritic.

The Arctic Ocean is unique owing to its permanent and seasonal ice cover, and restricted exchange of deep-water biota with the Pacific and Atlantic Oceans (see, for example, Carmack & Wassmann 2006). Extreme environmental conditions and limited exchange with the adjacent ocean regions have resulted in a zooplankton assemblage comprising species endemic to the Arctic Ocean and uniquely adapted to cold temperatures (Smith & Schnack-Schiel 1990; Kosobokova & Hirche 2000; Deibel & Daly 2007). Approximately 300 species of holozooplankton have been recorded for the Arctic (Sirenko 2001). The greatest diversity occurs within the Copepoda (approximately 150 species), which dominate the zooplankton community in both abundance and biomass (Kosobokova & Hopcroft 2009). Four large calanoid species (*Calanus glacialis*, *C. hyperboreus*, *C. finmarchicus*, and *Metridia longa*) are by far the most dominant species, contributing 60–70% of total zooplankton biomass (see, for example, Kosobokova *et al.* 1998; Kosobokova & Hirche 2000). Cnidaria are represented by approximately 50 species, mostly hydromedusae; mysids contribute approximately 30 species, most of which are epibenthic. Other groups are each represented by fewer than a dozen described species.

13.4.2.4 Gelatinous zooplankton

Special attention has been paid to the biodiversity of gelatinous plankton as a hot spot for species discovery. Discoveries of novel Cnidaria and Ctenophora species have resulted (Kitamura *et al.* 2005; Fuentes & Pagès 2006; Pagès *et al.* 2006; Hosia & Pagès 2007), some requiring the establishment of new higher taxonomic groups (Lindsay & Miyake 2007). This work has also allowed comparisons among regional faunas in light of geological history and environmental conditions, and revealed novel relationships among gelatinous plankton and other organisms (Ates *et al.* 2007; Pagès *et al.* 2007; Lindsay & Takeuchi 2008; Ohtsuka *et al.* 2009).

13.4.3 DNA barcoding

CMarZ has championed integrated morphological and molecular genetic approaches to analysis of zooplankton species' diversity (see, for example, Lindeque *et al.* 2006; Ueda & Bucklin 2006; Bucklin *et al.* 2007; Bucklin & Frost 2009; Goetze & Ohman 2010; Jennings *et al.* 2010a). Importantly, CMarZ has placed a high priority on "gold-standard" barcoding (that is, determination of a 500+ base-pair DNA sequence for mtCOI for an identified vouchered specimen, with specified metadata for protocols and collections) for described species of zooplankton. The growing CMarZ barcode database – now approaching 2,000 species or about 30% of the approximately 7,000 described species – will serve as a "Rosetta Stone" for species identification of marine zooplankton, linking species names, morphology, and DNA sequence variation. DNA barcodes will thus facilitate rapid characterization of patterns of species diversity and distribution in the pelagic realm.

Taxon-specific barcoding efforts by CMarZ researchers have included analysis of every phylum and taxonomic group within the zooplankton assemblage, including Protista (Morarda *et al.* 2009); Cnidaria (Ortman 2008; Ortman *et al.* 2010), calanoid Copepoda (Machida *et al.* 2006), Euphausiacea (Bucklin *et al.* 2007), Ostracoda (Angel *et al.* 2008), pteropod Mollusca (Jennings *et al.* 2010a), Chaetognatha (Jennings *et al.* 2010b), among other groups. These studies have demonstrated the usefulness of DNA barcodes for identification of known species, discovery of new species, and recognition of cryptic species within widespread or poorly known taxa. Alternatively, CMarZ barcoding campaigns have had a regional focus, with barcoding of all identified specimens collected from a particular ocean region or during a survey cruise. CMarZ has regionally focused barcoding efforts either completed or ongoing in the Arctic Ocean (Bucklin *et al.* 2010a), Sargasso Sea (Bucklin *et al.* 2010b), Eastern Atlantic and the Bay of Biscay, and South China Sea.

DNA barcodes have been used to characterize large-scale patterns of population genetic diversity and structure (that is, the amount and distribution of genetic variance within and among natural populations of organisms). Global-scale sampling and molecular analysis of zooplankton has revealed geographically distinct and genetically differentiated populations of Copepoda (Blanco-Bercial *et al.* 2009; Machida & Nishida 2010); Euphausiacea (Bucklin *et al.* 2007); and Chaetognatha (Peijnenburg *et al.* 2004; Miyamoto *et al.* 2010). Although geographic populations of zooplankton are not reproductively isolated, they are important units of evolution; studies of population genetic diversity and structure, using the barcoding gene region of choice, have been a critical aspect of the CMarZ global biodiversity assessment.

CMarZ has uniquely demonstrated the feasibility and value of performing DNA barcoding at sea during oceanographic research cruises. During two CMarZ biodiversity surveys to the Sargasso Sea and the Eastern Atlantic, DNA extraction, polymerase chain reaction (PCR), and DNA sequencing were performed in shipboard barcoding laboratories. Hundreds of species were barcoded, based on specimens identified by the taxonomic experts also participating in the cruise. During the Sargasso Sea cruise, 329 DNA barcodes were determined for 191 holozooplankton species, including hydrozoans, crustaceans, chaetognaths, and mollusks; barcodes were determined for an additional 35 fish species (Bucklin *et al.* 2010b).

CMarZ has pioneered the use of environmental barcoding for zooplankton communities: Machida *et al.* (2009) sequenced the COI barcode from an unsorted bulk sample collected in the western equatorial Pacific Ocean, detected 189 species of zooplankton based on COI sequences, and demonstrated the usefulness of this powerful approach to estimating species diversity of metazoan animals (Fig. 13.5).

As the DNA barcode database has grown to include described species collected from diverse ocean regions, CMarZ has explored new approaches for computationally efficient analysis and useful presentation of DNA sequence data and results. A novel approach is vector analysis (Sirovich *et al.* 2009), scalable analysis for very large datasets that produces heuristic displays of barcode similarity called heat maps or – because of their resemblance to modern art – "Klee diagrams" (Fig. 13.6).

13.4.4 Patterns of historical change

Time-series observations and monitoring programs that span many years are needed to document long-term changes in zooplankton diversity, distribution, abundance, and biomass in ocean ecosystems (Perry *et al.* 2004). CMarZ has sought to embed our field activities in the context of such valuable programs in order to provide benchmark biodiversity information for the analysis of temporal changes associated with global climate change.

13.4.4.1 Northeast Atlantic Ocean, UK

Since 1946, the Continuous Plankton Recorder (CPR) Survey Program (Sir Alister Hardy Foundation for Ocean Science, UK) has sought to characterize and understand changes in the species composition of North Atlantic zooplankton. Approximately 2.5 million non-zero records have indicated that zooplankton biomass has declined below the long-term average. In the North Sea, present biomass levels are one-half those in 1960 and warm-water species (for example the copepod *Calanus helgolandicus*) are displacing cold-water species (for example *C. finmarchicus*). This change affects the survival of fish larvae that depend on *C. finmarchicus* and has far-reaching consequences for the ecosystem (Beaugrand *et al.* 2002; Edwards *et al.* 2007). Another finding from analysis of CPR data is evidence of seasonal shifts in spawning and other life-history processes related to climate warming, creating mismatch between fish larvae and their food (Edwards & Richardson 2004; Edwards *et al.* 2008) and resulting in low fish recruitment. CPR surveys are beginning to document trans-Arctic migrations, with unknown consequences for pelagic communities (Edwards *et al.* 2008). CPR data have revealed what are called "regime shifts" (that is, markedly increased diversity and decreased productivity of zooplankton) in the North Sea (Edwards *et al.* 2007), as well

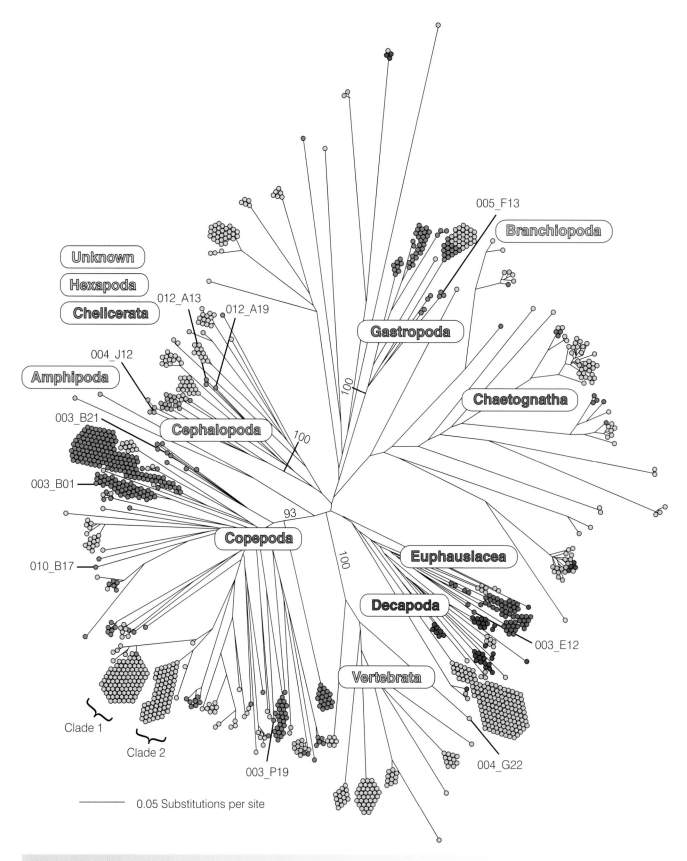

Fig. 13.5

Results of environmental barcoding of a zooplankton sample collected near Pohnpei Island, Micronesia, in the Equatorial Pacific Ocean. The tree includes 828 mitochondrial COI sequences (each 1,500 base pairs in length) analyzed by the Neighbor Joining algorithm using Kimura 2-Parameter distances. A total of 189 species were calculated based on DNA sequence variation: sequences differing by less than 9% were considered to represent the same species, whereas those differing by more than 9% were considered distinct species. Identification of species groups was based on NCBI BLAST searches of the GenBank database. Figure modified from Machida *et al.* (2009).

as the Northwest Atlantic, Northeast Pacific, and Northwest Pacific Oceans (Kane & Green 1990; Pershing *et al.* 2005; Kane 2007; Mackas *et al.* 2007; Tian *et al.* 2008).

13.4.4.2 Northwest Atlantic Ocean, USA

A 40-year survey by the US National Marine Fisheries Service (NMFS) has used the CPR (Jossi & Goulet 1993) to correlate effects on zooplankton populations with decadal-scale forcing by the North Atlantic Oscillation (Greene & Pershing 2000; Conversi *et al.* 2001; Piontkovski *et al.* 2006; Turner *et al.* 2006). Analysis of data and samples collected during 1977–2004 by NMFS' Marine Resources Monitoring, Assessment and Prediction (MARMAP) program (Jossi & Kane 2000; Kane 2007), demonstrated that total zooplankton counts over Georges Bank were at or above long-term average levels between 1989 and 2004 (Kane & Green 1990; Kane 2007). This regime shift was also observed in CPR records from the Gulf of Maine, which showed an increase in smaller-sized species (Pershing *et al.* 2005; Greene & Pershing 2007). The US GLOBEC Northwest Atlantic Program/Georges Bank Study examined zooplankton dynamics during 1994–1999, and provided one of the most comprehensive species datasets available for an historical fishing ground (Wiebe *et al.* 2002). Since 2004, NMFS has provided samples to CMarZ from quarterly ecosystem monitoring surveys over the Northwest Atlantic continental shelf; these samples are being used to barcode 200 of the most abundant zooplankton species, with a goal of allowing rapid DNA-based biodiversity assessments for fisheries management in this area.

13.4.4.3 Benguela Current, South Africa

Zooplankton diversity and biomass have been recorded for the Benguela Current, west of South Africa, since 1951. In contrast to most Eastern Boundary Currents, numerical abundances have increased 100-fold over recent decades. Species composition has shifted, with smaller-bodied Copepoda and Cladocera now dominating in the region, perhaps because of the combined effects of changes in climate, circulation patterns, and predator dynamics (Verheye & Richardson 1998; Verheye 2000).

13.4.4.4 California Current, USA

The California Cooperative Oceanic Fisheries Investigations (CalCoFI) is a 60-year time-series of quarterly surveys off southern and central California. No long-term trend was detectable in total zooplankton carbon biomass, although zooplankton displacement volume has declined in both regions, likely due to declines in salp abundances (which are mostly water and contribute little to carbon biomass; Lavaniegos & Ohman 2007). Zooplankton biomass varied among areas along the coast (Fernandez-Alamo & Färber-Lorda 2006). The time-series showed a marked shift from a "warm" regime with low zooplankton biomass and high sardine populations to a "cool" regime with high zooplankton biomass and high anchovy populations.

13.4.4.5 Western Pacific, Japan

The Odate collection at the Tohoku National Fisheries Research Institute (Japan) contains 20,000 formalin-preserved zooplankton samples collected throughout the western North Pacific from 1950 to 1990. Analysis of this extensive collection revealed decadal oscillations in copepod diversity and abundance. Compared with all other decades, the late 1980s showed a climatic regime shift: the copepod *Metridia pacifica* was dominant, whereas the abundance of *Neocalanus plumchrus* was low (Sugisaki 2006; Tian *et al.* 2008).

13.4.4.6 Indian Ocean, India

The International Indian Ocean Expedition (IIOE), carried out during 1962–1965, was the most intensive sampling program to characterize zooplankton species diversity and abundance in the region. Digitization and analysis of the IIOE data by CMarZ scientists have characterized seasonal variability associated with monsoon conditions and revealed long-term trends in species abundances and biogeographical distributions (Nair *et al.* 2008; Nair & Gireesh 2010). The data and results from IIOE represent an invaluable resource for detecting seasonal, interannual, and long-term shifts in the zooplankton assemblage (Baars 1999).

13.4.4.7 Marginal seas

In the Caspian Sea, the invasive comb jelly *Mnemiopsis leidyi* has severely impacted the entire ecosystem since its introduction in the late 1990s (Kideys *et al.* 2005, 2008). In the Black Sea, 130 years of plankton records (1870–2000) documented a long-term increase in the diversity of tintinnid Ciliata until 1960, followed by a decline into the 1990s. Ctenophore blooms in the 1990s may have impacted the anchovy fishery by causing replacements of copepod species (Gavrilova & Dolan 2007) and generating a trophic cascade with disastrous consequences for the ecosystem (Kideys *et al.* 2005).

13.4.4.8 Arctic Ocean

Changes in zooplankton species diversity, distribution, and abundance can be expected to occur in the Arctic Ocean, as climate change alters water temperature and thus timing and magnitude of productivity cycles (Grebmeier *et al.* 2006; Bluhm & Gradinger 2008). The introduction of exotic species through increased commercial traffic and the establishment of expatriate species from adjoining regions with warming are likely for the Arctic Ocean in the near future (ACIA 2004). Comparisons of zooplankton abundances in recent years with the early 1950s suggest that some taxa were more abundant in the Chukchi and

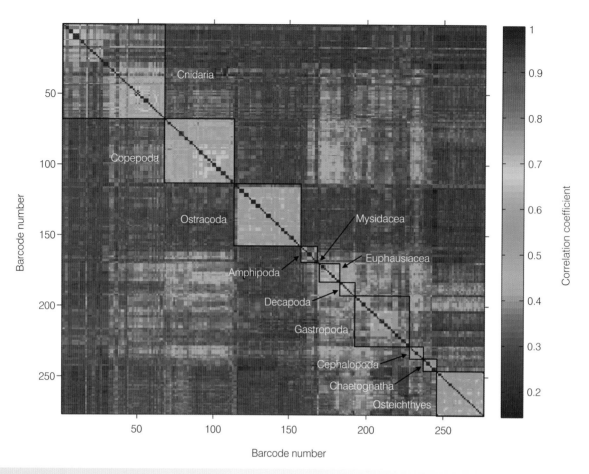

Fig. 13.6

Results of vector analysis shown as a heat map or "Klee diagram" for 329 DNA barcodes for 191 zooplankton species collected from the Sargasso Sea, Northwest Atlantic Ocean, during a CMarZ cruise on the R/V *RH Brown* in April 2006. The vector analysis method is from Sirovich *et al.* (2009). Analysis includes species of diverse zooplankton groups, including Cnidaria (Hydrozoa, 54 species); Crustacea (Amphipoda, 8; Copepoda, 38; Euphausiacea, 10; Ostracoda, 33; other Crustacea, 11); Chaetognatha, 5); and Mollusca (Cephalopoda, 12; Gastropoda, 21), as well as fish (Osteichthyes, 35).

Beaufort shelf and slope regions and the Canada Basin in 2002 than about 50 years ago (Grebmeier *et al.* 2006).

13.5 Significance and Impacts

13.5.1 Taxonomic training

CMarZ has enhanced research capacity in marine biodiversity and zooplankton ecology through the training of graduate students by CMarZ scientists and through international exchanges of students, staff, and researchers among CMarZ laboratories. CMarZ has placed a high priority on training new zooplankton taxonomists, with a total of 252 participants for 27 Taxonomic Training Workshops during 2004–2009. Many students have joined the CMarZ Network, which now includes more than 150 members, to

seek information and access to expertise to facilitate their research. CMarZ has sought to enhance capacity for taxonomic analysis of zooplankton through both traditional morphological and molecular systematic analysis.

13.5.2 Applications of CMarZ results

CMarZ' efforts toward a global assessment of marine zooplankton biodiversity, focusing on geographic and taxonomic hot spots, will provide a benchmark baseline biodiversity assessment for measurement of future changes resulting from climate change or other anthropogenic or natural variation. Zooplankton diversity can also be used as a measure of the health of marine ecosystems, and knowledge of prior and existing patterns of zooplankton distribution and diversity is needed for the management of coastal marine ecosystems (Link *et al.* 2002). Zooplankton are pivotal players in the dynamics of marine ecosystems, and new knowledge is needed of their roles in biogeochemical cycles (Buitenhuis *et al.* 2006).

CMarZ' barcode database, protocols for barcoding diverse marine phyla, and techniques for environmental sequencing of zooplankton will be useful in accelerating analysis of zooplankton diversity and distribution for a variety of applications in ocean research, management, and conservation. DNA barcodes will be used to produce DNA microarray "chips" for automated and/or remote identification and quantification of zooplankton. In the not-too-distant future, ocean-observing stations may include moored instruments with DNA-based detection systems for *in situ* species identification. These same approaches may allow rapid and accurate species identification for ecosystem monitoring and fisheries management, detection of invasive species in ballast water, and other possibilities central to ocean observing, management, and regulation.

13.6 Challenges and Opportunities for the Future

The question of how many species are present still remains, and future studies will continue to challenge our understanding of diversity. Large-scale studies of zooplankton are needed to evaluate patterns of biodiversity at scales appropriate to dispersal ability in ocean currents. Shifts in geographic ranges may underlie apparent temporal changes observed during spatially limited studies. In the case of species introductions, clear definition of potential source populations and likely colonization pathways require an understanding of global-scale distributions. For some cosmopolitan species, there may be little genuine endemism, whereas others may consist of complexes of genetically distinct entities, representing geographically isolated populations or cryptic species.

There remain large gaps in the sampling coverage done by CMarZ, especially in the deep sea and under-sampled ocean regions. Despite the increase of deep-sea studies, these investigations still represent rare snapshots of this huge habitat and vast areas of the deep sea remain unexplored. Despite new capabilities of ice-breaking research vessels, our understanding of polar ecosystems during the dark season remains fragmentary. Despite our efforts, CMarZ has only performed limited sampling in the Central Pacific Ocean, obtaining some samples from ships of opportunity (for example sailing ships associated with the Sea Education Association, Woods Hole, USA). Comprehensive sampling in the Indian and tropical Pacific Oceans has not yet extended below a few hundred meters. The richly diverse benthopelagic communities have received almost no attention because of sampling difficulties.

The urgency and/or vulnerability of regions or taxa to anthropogenic or natural threats must be weighed in determining future research priorities. These include regions where rates and impacts of climate change are most likely to be amplified, and poorly studied areas threatened by anthropogenic inputs (such as near population centers in emerging nations). The availability of baseline data is a critical issue, because evaluation of biodiversity patterns and hot spots requires improved knowledge of existing data and trends. Sites where time-series collections or long-term monitoring studies have been done are of high priority for continued assessment.

An essential feature for continued progress toward a global zooplankton census will be an international partnership, ideally coordinated through a network of regional centers that can identify opportunities for cooperative field work, arrange sampling from ships of opportunity, and lead efforts to secure funding for dedicated cruises. Interdisciplinary collaborations – among oceanographers, ecologists, taxonomists, geneticists, geochemists, and others – will continue to be needed to answer increasingly complex questions and address increasingly critical issues about the future and health of the global ocean. Knowledge of the species diversity, distribution, and abundance of the zooplankton assemblage will continue to be a critical element for monitoring, understanding, and predicting the complex global system.

Acknowledgments

The effort, enthusiasm, and expertise of the CMarZ Steering Group members are the basis of the results reported here. We acknowledge the many contributions by the CMarZ Steering Group members who are not listed as authors herein. They are the following: Martin Angel (National Oceanography Centre, UK); Demetrio Boltovskoy (Universidad de Buenos Aires, Argentina); Janet M. Bradford-Grieve (NIWA, New Zealand); Rubén Escribano (Universidad de Concepción, Chile); Erica Goetze (University of Hawaii, USA); Steven Haddock (Monterey Bay Aquarium Research Institute, USA); Steve Hay (Fisheries Research Services Marine Laboratory, UK); Russell R. Hopcroft (University of Alaska – Fairbanks, USA); Ahmet Kideys (Institute of Marine Sciences, Turkey); Laurence P. Madin (Woods Hole Oceanographic Institution, USA); Webjørn Melle (Institute of Marine Research, Norway); Vijayalakshmi R. Nair (National Institute of Oceanography, India); Mark Ohman (Scripps Institution of Oceanography, USA); Francesc Pagés (deceased) (Institute of Marine Sciences, Barcelona, Spain); Annelies C. Pierrot-Bults (University of Amsterdam, The Netherlands); Philip C. Reid (Sir Alister Hardy Foundation for Ocean Science, UK); Song Sun (Institute of Oceanology, Chinese Academy of Sciences, China); Erik V. Thuesen (The Evergreen State

College, USA); Colomban de Vargas (Roscoff Marine Station, France); Hans M. Verheye (Marine & Coastal Management, South Africa). We acknowledge the support of the Alfred P. Sloan Foundation. This study is a contribution from the Census of Marine Zooplankton (CMarZ; www.CMarZ.org), an ocean realm field project of the Census of Marine Life.

References

ACIA (2004) Impacts of a warming Arctic. In: *Arctic Climate Impact Assessment*. Cambridge University Press. 139 pp.

Angel, M.V. (2008) *Atlas of Atlantic Planktonic Ostracods*. London: Natural History Museum, http://www.nhm.ac.uk/research-curation/research/projects/atlantic-ostracods/index.html.

Angel, M.V., Ormond, R.F.G. & Gage, J.D. (1997) Pelagic biodiversity. In: *Marine Biodiversity: Patterns and Processes*, pp. 35–68. New York: Cambridge University Press.

Angel, M.V. (2003) The pelagic environment of the open ocean. In: *Ecosystems of the Deep Ocean* (ed. P.A. Tyler), pp. 39–79. Amsterdam: Elsevier.

Angel, M.V., Blachowiak-Samolyk, K., Drapun, I., *et al.* (2007) Changes in the composition of planktonic ostracod populations across a range of latitudes in the North-east Atlantic. *Progress in Oceanography* 73, 60–78.

Angel, M.V., Nigro, L. & Bucklin, A. (2008) DNA barcoding of oceanic planktonic ostracoda: species recognition and discovery (abstract). World Conference on Marine Biodiversity, Valencia, Spain, 11–15 November 2008.

Angel, M.V. & Blachowiak-Samolyk, K. (2009) Ostracods. *Reports on Polar and Marine Research* 592, 29–31.

Angel, M.V. (2010) Towards a full inventory of planktonic Ostracoda (Crustacea) for the subtropical Northwestern Atlantic Ocean. *Deep-Sea Research II* (in press).

Arndt, C.E. & Swadling, K.M. (2006) Crustacea in Arctic and Antarctic Sea Ice: distribution, diet and life history strategies. *Advances in Marine Biology* 51, 197–315.

Ates, R., Lindsay, D.-J. & Sekiguchi, H. (2007) First record of an association between a phyllosoma larva and a prayid siphonophore. *Plankton and Benthos Research* 2, 67–69.

Atkinson, A., Siegel, V., Pakhomov, E., *et al.* (2004) Long-term decline in krill stock and increase in salps within the Southern Ocean. *Nature* 432, 100–103.

Baars, M.A. (1999) On the paradox of high mesozooplankton biomass, throughout the year in the western Arabian Sea: Re-analysis of IIOE data and comparison with newer data. *Indian Journal of Marine Sciences* 28, 125–137.

Beaugrand, G., Ibanez, F., Lindley, J.A., *et al.* (2002) Diversity of calanoid copepods in the North Atlantic and adjacent seas: species associations and biogeography. *Marine Ecology Progress Series* 232, 179–195.

Blanco-Bercial, L., Álvarez-Marqués, F. & Bucklin, A. (2009) Global phylogeographies of the planktonic copepod *Clausocalanus* based on DNA barcodes. (abstract). Third International Conference for the Barcode of Life, Mexico City, Mexico, 10–13 November 2009.

Bluhm, B.A. & Gradinger, R. (2008) Regional variability in food availability for Arctic marine mammals. *Ecological Applications* 18 (Suppl.), 77–96.

Bluhm, B., Gradinger, R. & Schnack-Schiel, S. (2010) Sea ice meio- and macrofauna. In: *Sea Ice: An Introduction to its Physics Chemistry Biology and Geology* (eds. D. Thomas, and G. Dieckmann), pp. 357–394. Oxford: Blackwell Publishing Ltd.

Boltovskoy, D., Correa, N. & Boltovskoy, A. (2003) Marine zooplanktonic diversity: a view from the South Atlantic. *Oceanologica Acta* 25, 271–278.

Boltovskoy, D., Correa, N. & Boltovskoy, A. (2005) Diversity and endemism in cold waters of the South Atlantic: contrasting patterns in the plankton and the benthos. *Scientia Marina* 69 (Suppl. 2), 17–26.

Bouillon, J., Medel, M.D., Pagès, F., *et al.* (2004) Fauna of the Mediterranean Hydrozoa. *Scientia Marina* 68, 1–438.

Bouillon, J., Gravili, C., Pagès, F., *et al.* (2006) An introduction to Hydrozoa. *Memoires du Museum d'Histoire Naturelle Paris* 194, 1–591.

Bradford-Grieve, J.M., Markhaseva, E.L., Rocha, C.E.F., *et al.* (1999) Copepoda. In: *South Atlantic Zooplankton* (ed. D. Boltovskoy). Leiden: Backhuys.

Bradford-Grieve, J.M. (2010) *Hyperbionyx athesphatos* n.sp. (Calanoida: Hyperbionychidae), a rare deep-sea benthopelagic species taken from the tropical North Atlantic. *Deep-Sea Research II*, Special Volume: Species Diversity of Zooplankton in the Global Ocean.

Bucklin, A., LaJeunesse, T.C., Curry, E., *et al.* (1996) Molecular genetic diversity of the copepod, *Nannocalanus minor*: genetic evidence of species and population structure in the North Atlantic Ocean. *Journal of Marine Research* 54, 285–310.

Bucklin, A., Frost, B.W., Bradford-Grieve, J., *et al.* (2003) Molecular systematic and phylogenetic assessment of 34 calanoid copepod species of the Calanidae and Clausocalanidae. *Marine Biology* 142, 333–343.

Bucklin, A., Wiebe, P.H., Smolenack, S.B., *et al.* (2007) DNA barcodes for species identification of euphausiids (Euphausiacea, Crustacea). *Journal of Plankton Research* 29, 483–493.

Bucklin, A. & Frost, B.W. (2009) Morphological and molecular phylogenetic analysis of evolutionary lineages within *Clausocalanus* (Crustacea, Copepoda, Calanoida). *Journal of Crustacean Biology* 29, 111–120.

Bucklin, A., Hopcroft, R.R., Kosobokova, K.N., *et al.* (2010a) DNA barcoding of Arctic Ocean holozooplankton for species identification and recognition. *Deep-Sea Research II* 57, 40–48.

Bucklin, A., Ortman, B.D., Jennings, R.M., *et al.* (2010b) A "Rosetta Stone" for zooplankton: DNA barcode analysis of holozooplankton diversity of the Sargasso Sea (NW Atlantic Ocean). *Deep-Sea Research II* (in press).

Buitenhuis, E., Le Quere, C., Aumont, O., *et al.* (2006) Biogeochemical fluxes through mesozooplankton. *Global Biogeochemical Cycles* 20, 1–18.

Carmack, E. & Wassmann, P. (2006) Food webs and physical–biological coupling on pan-Arctic shelves: Unifying concepts and comprehensive perspectives. *Progress in Oceanography* 71, 446–477.

Conversi, A., Piontkovski, S. & Hameed, S.N. (2001) Seasonal and interannual dynamics of *Calanus finmarchicus* in the Gulf of Maine (Northeastern US shelf) with reference to the North Atlantic Oscillation. *Deep-Sea Research II* 48, 519–530.

Conway, D.V.P., White, R.G., Hugues-Dit-Ciles, J., *et al.* (2003) *Guide to the Coastal and Surface Zooplankton of the Southwestern Indian Ocean*, Vol. 15. Plymouth, UK: Marine Biological Association of the United Kingdom.

Davis, C.S., Gallager, S.M., Berman, M.S., *et al.* (1992) The video plankton recorder (VPR): design and initial results. *Archiv für Hydrobiologie Beiheft: Ergebnisse der Limnologie* 36, 67–81.

Dawson, M.N. & Jacobs, D.K. (2001) Molecular evidence for cryptic species of *Aurelia aurita* (Cnidaria, Scyphozoa). *The Biological Bulletin* 200, 92–96.

de Vargas, C., Norris, R., Zaninetti, L. *et al.* (1999) Molecular evidence of cryptic speciation in planktonic foraminifers and their relation to oceanic provinces. *Proceedings of the National Academy of Sciences of the USA* 96, 2864–2868.

de Vargas, C., Bonzon, M., Rees, N.W., *et al.* (2002) A molecular approach to diversity and biogeography in the planktonic foraminifer *Globigerinella siphonifera* (d'Orbigny). *Marine Micropaleontology* 870, 1–16.

Deibel, D. & Daly, K.L. (2007) Zooplankton processes in Arctic and Antarctic polynyas. In: *Arctic and Antarctic Polynyas* (eds. W.O. Smith, Jr. & D.G. Barber), pp. 271–322. Elsevier.

Edwards, M. & Richardson, A.J. (2004) Impact of climate change on marine pelagic phenology and trophic mismatch. *Nature* 430, 881–884.

Edwards, M., Johns, D.G., Licandro, P., *et al.* (2007) Ecological Status Report: results from the CPR survey 2005/2006. *Sir Alister Hardy Foundation for Ocean Science Report* 4, 1–8.

Edwards, M., Johns, D.G., Beaugrand, G., *et al.* (2008) Ecological status report: results from the CPR survey 2006/2007. *Sir Alister Hardy Foundation for Ocean Science Report* 5, 1–8.

Fernandez-Alamo, M.A. & Färber-Lorda, J. (2006) Zooplankton and the oceanography of the eastern tropical Pacific: a review. *Progress in Oceanography* 69, 318–359.

Fuentes, V. & Pagès, F. (2006) Description of *Jubanyella plemmyris* gen. nov. et sp. nov. (Cnidaria: Hydrozoa: Narcomedusae) from a specimen stranded off Jubany Antarctic station and a new diagnosis for the family Aeginidae. *Journal of Plankton Research* 28, 959–963.

Fujioka, K. & Lindsay, D.J. (2007) Deep trenches: the ultimate abysses. In: *The Deep: The Extraordinary Creatures of the Abyss*, pp. 256. Chicago: University of Chicago Press.

Gavrilova, N. & Dolan, J.R. (2007) A note on species lists and ecosystem shifts: Black Sea tintinnids, ciliates of the microzooplankton. *Acta Protozoologica* 46, 279–288.

Glover, R.S. (1962) The continuous plankton recorder. *Rapports et Procés-verbaux des Réunions Conseil Permanent International pour l'Exploration de la Mer* 153, 8–15.

Goetze, E. (2003) Cryptic speciation on the high seas; global phylogenetics of the copepod family Eucalanidae. *Proceedings of the Royal Society of London B* 270, 2321–2331.

Goetze, E. & Ohman, M.D. (2010) Integrated molecular and morphological biogeography of the calanoid copepod family Eucalanidae. *Deep-Sea Research II* (in press).

Gorsky, G., Aldorf, C., Kage, M., *et al.* (1992) Vertical distribution of suspended aggregates determined by a new underwater video profiler. *Annales de l'Institut Océanographique*, 68, 275–280.

Gorsky, G. & R. Fenaux (1998) The role of Appendicularia in marine food chains. In: *The Biology of Pelagic Tunicates* (ed. Q. Bone), pp. 161–169. New York: Oxford University Press.

Gorsky, G., Picheral, M. & Stemmann, L. (2000) Use of the underwater video profiler for the study of aggregate dynamics in the North Mediterranean. *Estuarine Coastal and Shelf Science* 50, 121–128.

Grebmeier, J.M., Cooper, L.W., Feder, H.M., *et al.* (2006) Ecosystem dynamics of the Pacific-influenced Northern Bering and Chukchi Seas in the Amerasian Arctic. *Progress in Oceanography* 71, 331–361.

Greene, C.H. & Pershing, A.J. (2000) The response of *Calanus finmarchicus* populations to climate variability in the Northwest Atlantic: basin-scale forcing associated with the North Atlantic Oscillation. *ICES Journal of Marine Science* 57, 1536–1544.

Greene, C.H. & Pershing, A.J. (2007) Climate drives sea change. *Science* 315, 1084–1085.

Groman, R.C. & Wiebe, P.H. (1998) Data management in the U.S. GLOBEC Georges Bank Program In: Ocean Community Conference'98 Proceedings, pp. 807–812. Marine Technology Society, Baltimore, MD.

Groman, R.C., Chandler, C.L., Allison, M.D., *et al.* (2008) Discovery, access, interoperability, and visualization features of a web interface to oceanographic data. In: Ocean Community Conference '98 Proceedings, 8 pp. ICES CM 2008/R:02.

Haddock, S.H.D., Dunn, C.W. & Pugh, P.R. (2005) A re-examination of siphonophore terminology and morphology, applied to the description of two new prayine species with remarkable bio-optical properties. *Journal of the Marine Biological Association of the United Kingdom* 85, 695–707.

Hamner, W.M. (1975) Underwater observations of blue-water plankton: logistics, techniques, and safety procedures for divers at sea. *Limnology and Oceanography* 20, 1045–1051.

Hardy, A.C. (1926) The herring in relation to its animate environment. Part II: report on trials with the plankton indicator. *Ministry of Agriculture Fisheries and Food Investigative Series II* 8, 1–13.

Hebert, P.D.N., Cywinska, A., Ball, S.L., *et al.* (2003) Biological identifications through DNA barcodes. *Proceedings of the Royal Society of London B* 270, 313–321.

Hedgepeth, J.W. (1957) Classification of Marine Environments. *The Geological Society of America Memoir* 67, 17–27.

Hering, P. (2002) *The Biology of the Deep Ocean.* Oxford University Press.

Herman, A.W. (1988) Simultaneous measurement of zooplankton and light attenuation with a new optical plankton counter. *Continental Shelf Research* 8, 205–221.

Hopcroft, R.R. & Robison, B.H. (2005) New mesopelagic larvaceans in the genus *Fritillaria* from Monterey Bay, California. *Journal of the Marine Biological Association of the United Kingdom* 85, 665–678.

Hosia, A. & Pagès, F. (2007) Unexpected new species of deep-water Hydroidomedusae from Korsfjorden, Norway. *Marine Biology* 151, 177–184.

Irigoien, X., Huisman, J. & Harris, R.P. (2004) Global biodiversity patterns of marine phytoplankton and zooplankton. *Nature* 429, 863–867.

Jennings, R.M., Bucklin, A., Ossenbrügger, H., *et al.* (2010a) Analysis of genetic diversity of planktonic gastropods from several ocean regions using DNA barcodes. *Deep-Sea Research II* (in press).

Jennings, R.M., Bucklin, A. & Pierrot-Bults, A. (2010b) Barcoding of arrow worms (phylum Chaetognatha) from three oceans: genetic diversity and evolution within an enigmatic phylum. *PLoS ONE* 5, e9949. doi:10.1371/journal.pone.0009949.

Johnson, G.D., Paxton, J.R., Sutton, T.T., *et al.* (2009) Deep-sea mystery solved: astonishing larval transformations and extreme sexual dimorphism unite three fish families. *Biology Letters* 5, 235–239.

Jossi, J.W. & Goulet, J.R. (1993) Zooplankton trends: US north-east shelf ecosystem and adjacent regions differ from north-east Atlantic and North Sea. *ICES Journal of Marine Science* 50, 303–313.

Jossi, J.W. & Kane, J. (2000) An atlas of seasonal mean abundances of the common zooplankton of the United States northeast continental shelf ecosystem. *Bulletin of the Sea Fisheries Institute Gdynia*, 67–87.

Kane, J. (2007) Zooplankton abundance trends on Georges Bank, 1977–2004. *ICES Journal of Marine Science* 64, 909–919.

Kane, J. & Green, J. (1990) Zooplankton biomass on Georges Bank 1977–86. Council Meeting of the International Council for the Exploration of the Sea, Copenhagen (Denmark), 11 pp. ICES-CM-1990/L: 22.

Kideys, A.E., Roohi, A., Bagheri, S., *et al.* (2005) Impacts of Invasive Ctenophores on the Fisheries of the Black Sea and Caspian Sea. *Oceanography* – 18 (Black Sea Special Issue), 76–85.

Kideys, A.E., Roohi, A., Eker-Develi, E., *et al.* (2008) Increased chlorophyll levels in the Southern Caspian Sea following an invasion of jellyfish. *Research Letters in Ecology* 2008, 4 pp. Article 185642.

Kiko, R., Michels, J., Mizdalski, E., *et al.* (2008) Living conditions and abundance of surface and sub-ice layer fauna in pack-ice of the western Weddell Sea during early summer. *Deep-Sea Research II* 55, 1000–1014.

Kitamura, M., Lindsay, D.J. & Miyake, H. (2005) Description of a new midwater medusa, *Tiaropsidium shinkai* n. sp. (Leptomedusae, Tiaropsidae). *Plankton Biology and Ecology* **52**, 100–106.

Kitamura, M., Lindsay, D.J., Miyake, H., *et al.* (2008a) Ctenophora. In: *Deep-Sea Life – Biological Observations Using Research Submersibles* (eds. Fujikura, K., Okutani, T. & Maruyama, T.), pp. 321–328. Kanagawa: Tokai University Press.

Kitamura, M., Miyake, H. & Lindsay, D.J. (2008b) Cnidaria. In: *Deep-Sea Life – Biological Observations Using Research Submersibles* (eds. Fujikura, K., Okutani, T. & Maruyama, T.), pp. 295–320. Kanagawa: Tokai University Press.

Kosobokova, K.N. (1989) Vertical distribution of plankton animals in the eastern part of the central Arctic Basin. *Explorations of the Fauna of the Seas, Marine Plankton* **41**, 24–31.

Kosobokova, K.N., Hanssen, H., Hirche, H.J., *et al.* (1998) Composition and distribution of zooplankton in the Laptev Sea and adjacent Nansen Basin during summer, 1993. *Polar Biology* **19**, 63–76.

Kosobokova, K.N. & Hirche, H.J. (2000) Zooplankton distribution across the Lomonosov Ridge, Arctic Ocean: species inventory, biomass and vertical structure. *Deep-Sea Research I* **47**, 2029–2060.

Kosobokova, K.N. & Hopcroft, R.R. (2009) Diversity and vertical distribution of mesozooplankton in the Arctic's Canada Basin. *Deep-Sea Research II* **57**, 96–110.

Kuriyama, M. & Nishida, S. (2006) Species diversity and niche-partitioning in the pelagic copepods of the family Scolecitrichidae (Calanoida). *Crustaceana* **79**, 293–317.

Lavaniegos, B.E. & Ohman, M.D. (2003) Long term changes in pelagic tunicates of the California Current. *Deep-Sea Research I* **50**, 2493–2518.

Lavaniegos, B. & Ohman, M. (2007) Coherence of long-term variations of zooplankton in two sectors of the California Current System. *Progress in Oceanography* **75**, 42–69.

Lindeque, P.K., Hay, S.J., Heath, M.R., *et al.* (2006) Integrating conventional microscopy and molecular analysis to analyse the abundance and distribution of four *Calanus* congeners in the North Atlantic. *Journal of Plankton Research* **28**, 221–238.

Lindsay, D.J. (2006) A checklist of midwater cnidarians and ctenophores from Sagami Bay – species sampled during submersible surveys from 1993–2004. *Bulletin of the Plankton Society of Japan* **53**, 104–110.

Lindsay, D.J., Furushima, Y., Miyake, H., *et al.* (2004) The scyphomedusan fauna of the Japan Trench: preliminary results from a remotely-operated vehicle. *Hydrobiologia* **530/531**, 537–547.

Lindsay, D.J. and Hunt, J.C. (2005) Biodiversity in midwater cnidarians and ctenophores: submersible-based results from deep-water bays in the Japan Sea and North-western Pacific. *Journal of the Marine Biological Association of the United Kingdom* **85**, 503–517.

Lindsay, D.J. & Miyake, H. (2007) A novel benthopelagic ctenophore from 7217 m depth in the Ryukyu Trench, Japan, with notes on the taxonomy of deep sea cydippids. *Plankton and Benthos Research* **2**, 98–102.

Lindsay, D.J. & Miyake, H. (2009) A checklist of midwater cnidarians and ctenophores from Japanese waters – species sampled during submersible surveys from 1993–2008 with notes on their taxonomy. *Kaiyo Monthly* **41**, 417–438.

Lindsay, D.J., Pagès, F., Corbera, J., *et al.* (2008) The anthomedusan fauna of the Japan Trench: preliminary results from *in situ* surveys with manned and unmanned vehicles. *Journal of the Marine Biological Association of the United Kingdom* **88**, 1519–1539.

Lindsay, D.J. & Takeuchi, I. (2008) Associations in the benthopelagic zone: the amphipod crustacean *Caprella subtilis* (Amphipoda: Caprellidae) and the holothurian *Ellipinion kumai* (Elasipodida: Family: Elpidiidae). *Scientia Marina* **72**, 519–526.

Link, J.S., Brodziak, J.K.T., Edwards, S.F., *et al.* (2002) Marine ecosystem assessment in a fisheries management context. *Canadian Journal of Fisheries and Aquatic Sciences* **59**, 1429–1440.

Longhurst, A. (1995) Seasonal cycles of pelagic production and consumption. *Progress in Oceanography* **36**, 77–167.

Machida, R.J., Miya, M.U., Nishida, M., *et al.* (2006) Molecular phylogeny and evolution of the pelagic copepod genus *Neocalanus* (Crustacea: Copepoda). *Marine Biology* **148**, 1071–1079.

Machida, R.J., Hashiguchi, Y., Nishida, M., *et al.* (2009) Zooplankton diversity analysis through single-gene sequencing of a community samples. *BMC Genomics* **10**, 438.

Machida, R.J. & Nishida, S. (2010) Amplified fragment length polymorphism analysis of the mesopelagic copepods *Disseta palumbii* in the equatorial western Pacific and adjacent waters: role of marginal seas for genetic isolation of mesopelagic animals. *Deep-Sea Research II* (in press).

Mackas, D., Batten, S. & Trudel, M. (2007) Effects on zooplankton of a warmer ocean: recent evidence from the Northeast Pacific. *Progress in Oceanography* **75**, 223–252.

Matsumoto, G., Raskoff, K. & Lindsay, D.J. (2003) *Tiburonia granrojo*, a new mesopelagic scyphomedusa from the Pacific Ocean representing the type of a new subfamily (class Scyphozoa, order Semaeostomae, family Ulmaridae, subfamily Tiburoniiae subfam. nov.). *Marine Biology* **143**, 73–77.

McGowan, J.A. (1971) Oceanic biogeography of the Pacific. In: *The Micropaleontology of Oceans* (eds. B.M. Funnell, & W.R. Riedel), pp. 3–74. Cambridge, UK: Cambridge University Press.

Menard, H.W. and Smith, S.M. (1966) Hypsometry of ocean basin provinces. *Journal of Geophysical Research* **71**, 4305–4325.

Mills, C.E. & Haddock, S.H.D. (2007) Key to the Ctenophora. In: *Light and Smith's Manual: Intertidal Invertebrates of the Central California Coast* (ed. J.T. Carlton), pp. 189–199. University of California Press.

Mills, C.E., Haddock, S.H.D., Dunn, C.W., *et al.* (2007) Key to the Siphonophora. In: *Light and Smith's Manual: Intertidal Invertebrates of the Central California Coast* (ed. J.T. Carlton), pp. 150–166. University of California Press.

Miyamoto, H., Machida, R.J. & Nishida, S. (2010) Genetic diversity and cryptic speciation of the deep sea chaetognath *Caecosagitta macrocephala* (Fowler, 1904). *Deep-Sea Research II* (in press).

Morarda, R., Quillévéré, F., Escarguel, G., *et al.* (2009) Morphological recognition of cryptic species in the planktonic foraminifer *Orbulina universa*. *Marine Micropaleontology* **71**, 148–165.

Mori, M. & Lindsay, D.J. (2008) Body pigmentation changes in the planktonic crustacean *Vibilia stebbingi* (Amphipoda: Hyperiidea) under different light regimes, with notes on implications for the development of automated plankton identification systems. *JAMSTEC Report on Research and Development* **8**, 37–45.

Murano, M. & Fukuoka, K. (2008) A systematic study of the genus *Siriella* (Crustacea: Mysida) from the Pacific and Indian Oceans, with description of fifteen new species. *National Museum of Nature and Science Monographs* **36**, 173 pp.

Nair, V.R. & Gireesh, R. (2010) Biodiversity of chaetognaths of the Andaman Sea, Indian Ocean. *Deep-Sea Research II* (in press).

Nair, V.R., Panampunnayil, S.U., Pillai, H.U.K., *et al.* (2008) Two new species of Chaetognatha from the Andaman Sea, Indian Ocean. *Marine Biology Research* **4**, 208–214.

Nishida, S. and Cho, N. (2005) A new species of *Tortanus (Atortus)* (Copepoda: Calanoida: Tortanidae) from the coastal water of Nha Trang, Vietnam. *Crustaceana* **78**, 223–235.

Nishikawa, J., Matsuura, H., Castillo, L.V., *et al.* (2007) Biomass, vertical distribution and community structure of mesozooplankton in the Sulu Sea and its adjacent waters. *Deep-Sea Research II* **54**, 114–130.

Ohtsuka, S., Nishida, S. & Machida, R. (2005) Systematics and zoogeography of deep-sea hyperbenthic Arietellidae (Copepoda: Calanoida) collected from the Sulu Sea. *Journal of Natural History* 39, 2483–2514.

Ohtsuka, S., Tanimura, A., Machida, R.J., *et al.* (2009) Bipolar and antitropical distributions of planktonic copepods. *Fossils* 85, 6–13.

Ortman, B.D. (2008) *DNA barcoding the medusozoa and ctenophora.* Ph.D. thesis, University of Connecticut.

Ortman, B.D., Bucklin, A., Pages, F. *et al.* (2010) DNA barcoding of the Medusozoa. *Deep-Sea Research II* (in press).

Pace, N.R. (1997) A molecular view of microbial diversity and the biosphere. *Science* 276, 734–740.

Pagès, F., Corbera, J. & Lindsay, D.J. (2007) Piggybacking pycnogonids and parasitic narcomedusae on *Pandea rubra* (Anthomedusae, Pandeidae). *Plankton Benthos Research* 2, 83–90.

Pagès, F., Flood, P. & Youngbluth, M. (2006) Gelatinous zooplankton net-collected in the Gulf of Maine and adjacent submarine canyons: new species, new family (Jeanbouilloniidae), taxonomic remarks and some parasites. *Scientia Marina* 70, 363–379.

Park, E.T. & Ferrari, F.D. (2008) Species diversity and distributions of pelagic calanoid copepods from the Southern Ocean. In: *Smithsonian at the Poles: Contributions to the International Polar Year Science* (eds. I. Krupnik, M.A. Lang & S.E. Miller), pp. 143–180. Washington, DC: Smithsonian Institution Scholarly Press.

Patterson, D.J. (2009) Seeing the big picture on microbe distribution. *Science* 325, 1506–1507.

Perry, R.I., Batchelder, H.P., Mackas, D.L., *et al.* (2004) Identifying global synchronies in marine zooplankton populations: issues and opportunities. *ICES Journal of Marine Science* 61, 445–456.

Pershing, A.J., Greene, C.H., Jossi, J.W., *et al.* (2005) Interdecadal variability in the Gulf of Maine zooplankton community, with potential impacts on fish recruitment. *ICES Journal of Marine Science* 62, 1511.

Peijnenburg, K.T.C.A., Breeuwer, J.A.J., Pierrot-Bults, A.C., *et al.* (2004) Phylogeography of the planktonic chaetognath *Sagitta setosa* reveals isolation in European seas. *Evolution* 58, 1472–1487.

Piontkovski, S.A., O'Brien, T.D., Umani, S.F., *et al.* (2006) Zooplankton and the North Atlantic Oscillation: a basin-scale analysis. *Journal of Plankton Research* 28, 1039–1046.

Pugh, P.R. (1992) The status of the genus *Prayoides* (Siphonophora: Prayidae). *Journal of the Marine Biological Association of the United Kingdom* 72, 895–909.

Ramirez-Llodra, E., Shank, T.M. & German, C.R. (2007) Biodiversity and biogeography of hydrothermal vent species: thirty years of discovery and investigations. *Oceanography* 20, 30.

Raskoff, K.A. & Matsumoto, G.I. (2004) *Stellamedusa ventana*, a new mesopelagic scyphomedusa from the eastern Pacific representing a new subfamily, the Stellamedusinae. *Journal of the Marine Biological Association of the United Kingdom* 84, 37–42.

Roemmich, D. & McGowan, J.A. (1995) Climatic warming and the decline of zooplankton in the California Current. *Science* 267, 1324–1326.

Schindel, D.E. & Miller, S.E. (2005) DNA barcoding a useful tool for taxonomists. *Nature* 435, 17.

Schnack-Schiel, S.B. (2001) Aspects of the study of the life cycles of Antarctic copepods. *Hydrobiologia* 453, 9–24.

Schnack-Schiel, S.B., Haas, C., Michels, J., *et al.* (2008) Copepods in sea ice of the western Weddell Sea during austral spring 2004. *Deep-Sea Research II* 55, 1056–1067.

Siegel, V. (2005) Distribution and population dynamics of *Euphausia superba*: summary of recent findings. *Polar Biology* 29, 1–22.

Sirenko, B.I. (2001) *List of Species of Free-living Invertebrates of Eurasian Arctic Seas and Adjacent Deep Waters.* St. Petersburg: Russian Academy of Sciences.

Sirovich, L., Stoeckle, M.Y. & Zhang, Y. (2009) A scalable method for analysis and display of DNA sequences. *PLoS ONE* 4, e7051.

Smith, S.L. & Schnack-Schiel, S.B. (1990) Polar Zooplankton. In: *Polar Oceanography Part B: Chemistry Biology and Geology* (ed. W.O. Smith Jr.), pp. 527–598. San Diego: Academic Press.

Smith, K., Jr, Kaufmann, R., Baldwin, R. & Carlucci, A. (2001) Pelagic–benthic coupling in the abyssal eastern North Pacific: an 8-year time-series study of food supply and demand, *Limnology and Oceanography* 46, 543–556.

Sogin, M.L., Morrison, H.G., Huber, J.A., *et al.* (2006) Microbial diversity in the deep sea and the underexplored "rare biosphere". *Proceedings of the National Academy of Sciences of the USA* 103, 12115.

Sugisaki, H. (2006) Studies on long-term variation of ocean ecosystem/climate interactions based on the Odate collection: outline of the Odate Project. *PICES Press* 14, 12–15.

Tian, Y., Kidokoro, H., Watanabe, T., *et al.* (2008) The late 1980s regime shift in the ecosystem of Tsushima warm current in the Japan/East Sea: Evidence from historical data and possible mechanisms. *Progress in Oceanography* 77, 127–145.

Turner, J.T., Borkman, D.G. & Hunt, C.D. (2006) Zooplankton of Massachusetts Bay, USA, 1992–2003: relationships between the copepod *Calanus finmarchicus* and the North Atlantic Oscillation. *Marine Ecology Progress Series* 311, 115–124.

Ueda, H. & Bucklin, A. (2006) *Acartia (Odontacartia) ohtsukai*, a new brackish-water calanoid copepod from Ariake Bay, Japan, with a redescription of the closely related *A. pacifica* from the Seto Inland Sea. *Hydrobiology* 500, 77–91.

Verheye, H.M. (2000) Decadal-scale trends across several marine trophic levels in the southern Benguela upwelling system off South Africa. *Ambio* 29, 30–34.

Verheye, H.M. & Richardson, A.J. (1998) Long-term increase in crustacean zooplankton abundance in the southern Benguela upwelling region (1951–1996): bottom-up or top-down control? *ICES Journal of Marine Science* 55, 803–807.

Wiebe, P.H., Morton, A.W., Bradley, A.M., *et al.* (1985) New developments in the MOCNESS, an apparatus for sampling zooplankton and micronekton. *Marine Biology Heidelberg* 87, 313–323.

Wiebe, P.H., Beardsley, R., Mountain, D., *et al.* (2002) US GLOBEC Northwest Atlantic/Georges Bank Program. *Oceanography* 15, 13–29.

Wiebe, P.H. & Benfield, M.C. (2003) From the Hensen net toward four-dimensional biological oceanography. *Progress in Oceanography* 56, 7–136.

Wiebe, P.H., Bucklin, A., Madin, L.P., *et al.* (2010) Deep-sea holozooplankton species diversity in the Sargasso Sea, Northwestern Atlantic Ocean. *Deep-Sea Research II* (in press).

Wilson, E.O. (1999) *The Diversity of Life.* New York: W.W. Norton.

Yoshida, H. & Lindsay, D.J. (2007) Development of the PICASSO (Plankton Investigatory Collaborating Autonomous Survey System Operon) System at the Japan Agency for Marine-Earth Science and Technology. *Japan Deep Sea Technology Society Report* 54, 5–10.

Yoshida, H., Aoki, T., Osawa, H., *et al.* (2007a) Newly-developed devices for two types of underwater vehicles. In: Oceans 2007 Conference Proceedings, pp. 1–6.

Yoshida, H., Lindsay, D.J., Yamamoto, H., *et al.* (2007b) Small hybrid vehicles for jellyfish surveys in midwater. In: *Proceedings of the 17th International Offshore and Polar Engineering Conference*, pp. 127.

PART IV
Oceans Present –
Animal Movements

Chapter 14

Tracking Fish Movements and Survival on the Northeast Pacific Shelf

John Payne[1], Kelly Andrews[2], Cedar Chittenden[3], Glenn Crossin[4], Fred Goetz[5], Scott Hinch[6], Phil Levin[2], Steve Lindley[7], Scott McKinley[8], Michael Melnychuk[9], Troy Nelson[10], Erin Rechisky[9], David Welch[11]

[1]Pacific Ocean Shelf Tracking Project, Vancouver Aquarium, Vancouver, British Columbia, Canada
[2]Northwest Fisheries Science Center, National Marine Fisheries Service, National Oceanic and Atmospheric Administration, Seattle, Washington, USA
[3]Faculty of Biosciences, Fisheries and Economics, University of Tromsø, Tromsø, Norway
[4]Centre for Applied Conservation Research, University of British Columbia, Vancouver, British Columbia, Canada
[5]School of Aquatic and Fishery Sciences, University of Washington, Seattle, Washington, USA
[6]Department of Forest Sciences, Centre for Applied Conservation Research, University of British Columbia, Vancouver, British Columbia, Canada
[7]Southwest Fisheries Science Center, National Marine Fisheries Service, National Oceanic and Atmospheric Administration, Santa Cruz, California, USA
[8]West Vancouver Laboratory – Animal Science, University of British Columbia, West Vancouver, British Columbia, Canada
[9]Department of Zoology and Fisheries Centre, University of British Columbia, Vancouver, British Columbia, Canada
[10]Fraser River Sturgeon Conservation Society, Vancouver, British Columbia, Canada
[11]Kintama Research Corporation, Nanaimo, British Columbia, Canada

14.1 Introduction

The Pacific Ocean Shelf Tracking Project (POST) is one of the 14 field projects of the Census of Marine Life. POST began in 2001 (see Box 14.1) as an ambitious experiment to study the movements and survival of salmon in the ocean using a large seabed network of acoustic receivers to track individual acoustically tagged fish. The successful proof-of-concept, and the fact that compatible receivers and tags were in use by other researchers on the West Coast, helped POST mature and diversify into a complex infrastructure that is now regarded as an indispensable tool for under-

standing the behavior of many marine species that move along the continental shelves. Operationally, POST is a non-profit program run by an independent board, and hosted by the Vancouver Aquarium. POST's mission is to facilitate the development of a large-scale acoustic telemetry network along the entire length of the West Coast of North America, working through contractors and partners who deploy the array, and through collaborative relationships with independent principle investigators who conduct their own research projects using the array. POST maintains a public database where currently over 6.2 million detections of over 12,000 tags and 18 species are securely stored, and may be searched and shared by anyone.

POST is distinguished by three attributes:

1) A reliance on acoustic tags and a large network of strategically located receivers (Fig. 14.1).

Life in the World's Oceans, edited by Alasdair D. McIntyre
© 2010 by Blackwell Publishing Ltd.

Box 14.1

POST Technology and History

Acoustic tags have been in use for 50 years (Johnson 1960), but in 2001 a fisheries biologist, David Welch, and his colleagues proposed to the Alfred P. Sloan and Gordon and Betty Moore Foundations to design and build a very large network of listening lines to track salmon in the ocean. They reasoned that satellite tags were too big to use on salmon, archival tags were too unlikely to be recovered (and their light-based geo-location estimates were too inaccurate at the time), and radio tags, although useful for tracking salmon in rivers, were useless in the ocean because of rapid attenuation of electromagnetic signals in seawater.

POST was built around acoustic tags and receivers manufactured by a Canadian company, Vemco (www.vemco.com). Vemco's tags could be implanted in small fish, detected at relatively long distances, programmed to have relatively long tag lifespans and, most importantly, the system generated few false-positive signals. Several studies have assessed the effects of the tags on the survival and behavior of fish that carry them (Lacroix et al. 2004; Zale et al. 2005; Welch et al. 2007; Chittenden et al. 2009a; Rechisky & Welch 2009), and helped to define fish size limits for tagging. Early on, the number of available unique tag identification numbers was small so they were re-used, which quickly became very confusing on the large scale of the POST array. POST helped to motivate the development of a system with many identification numbers that are unique worldwide.

Early Vemco receivers had short battery lives and could not be used in deep water, but by the time POST was scaling up, several thousand second-generation VR-2 receivers had been sold on the West Coast. These receivers were tough, reliable, had batteries that lasted one year, and, with a maximum depth of about 500 m, could be deployed almost anywhere along the continental shelf. Most importantly, all of the tags and receivers were compatible. However, the early receivers had to be physically retrieved to download the data. Most of the original POST network has now been replaced with a newer generation of VR-3 receivers equipped with long-lived batteries (four to seven years) and acoustic modems by which a boat can download data from the surface without physically recovering the receiver. This has generated significant cost savings over the life of the array and made it easier to keep receivers in position full-time, year-round.

Welch's research and development company, Kintama Research Corporation, tackled the problems of deploying large-scale arrays and developed the architecture and tag programming for the original demonstration array forming the core of POST. They designed specialized protective flotation collars and anchors (Fig. 14.3), improved moorings to reduce losses to trawling and storms, and built portable surgery stations and data-recording systems for large-scale tagging. They are currently modeling optimal array geometries for specific research projects, which depend on a host of factors including the research objectives, noise level in the area of the line, behavior of the tagged animal, tag parameters (loudness and programming), and position of the receiver relative to features such as the surface, the bottom, thermoclines, and haloclines. Where measurable, POST lines have obtained high enough detection efficiencies to produce useful survival estimates for juvenile salmon (Melnychuk 2009).

With support from US and Canadian government agencies and foundations, the POST array is maturing into a network of highly engineered, long listening lines that now spans 3,000 km from California to Alaska and is maintained year-round for use by any researcher. POST shares data with independent researchers who maintain their own, smaller receiver networks (some in grids or other geometries). We have begun the process of integrating POST data into large-scale ocean-observing systems including OBIS (see Chapter 17), the Ocean Tracking Network, and the Global Ocean Observing System (GOOS) system.

2] A focus on studying the behaviors of marine species, including long-distance migrations.

3] A focus on estimating survival by deploying acoustic receivers in long, relatively straight lines that stretch from the coastline to the edge of the continental shelf, or across straits between land bodies (Figs. 14.1 and 14.2). The lines are designed to have a high probability of detecting animals that cross them, and the effect is to compartmentalize large areas so that survival can be estimated within each area.

Although not an exclusive focus, POST fills a technological gap as a method to study the movements of small-bodied marine animals (10 cm – 1 m in length, including the juveniles of larger species), which are abundant, important in oceanic food chains, and generally difficult to study.

Fig. 14.1
The POST array as of 2009. Lines of acoustic receivers are shown in red. Permission of POST 2009.

Continental Shelf (<200m deep)　　Acoustic receiver lines

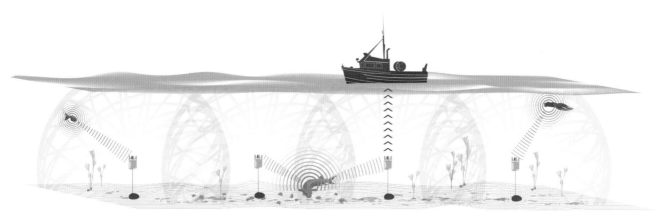

Fig. 14.2
A conceptual diagram of a POST receiver line. The spheres show hypothetical detection limits for the receivers. The tagged animals are being detected by different receivers (yellow objects tethered to the sea floor), and the ship is downloading information from a receiver, via an acoustic modem. Permission of POST 2009.

14.1.1 The questions that motivated POST

POST was originally designed to answer two questions about salmon. Salmon are among the most culturally and economically important species on the Pacific Northwest coast of North America, and have sustained human popula-tions there since prehistoric times. The freshwater portion of salmon life cycles may be observed with relative ease, yet in the early 1990s, the renowned salmon biologist William Pearcy noted that the ocean life-history of Pacific salmon was a "black box", constrained by the enormous difficulty of studying salmon on the high seas (Pearcy 1992).

The first of the questions was, "When and where do juvenile salmon die in the ocean?" It was known that very few of the juvenile salmon that left a river mouth would return as adults, and there were many theories about what happened to those fish. One theory was that the out-migrating cohorts suffered high mortality immediately upon entry to the ocean, and that the magnitude of that mortality would determine how many adult salmon returned from each cohort. Receiver lines were also designed to help clarify the details of movement patterns, which were known in a general way from recaptures of passive tags.

The second question was, "Are salmon as specific in their use of the ocean as they are in their use of rivers?" This was nicknamed the "two zip code" theory, the idea being that salmon could have one address in freshwater and a second address in the ocean. Even in mid-ocean, the marine environment is much less homogeneous than it appears to the human eye. At different scales, natural change encompasses variation such as decadal oscillations that affect sea surface temperatures and upwelling over vast areas, seasonally fluctuating currents that bring warm water into cold seas and spin off rings of warmer water which persist for months, and short-term turbulence from surface winds and currents that can mix thermally stratified water layers or concentrate surface debris, creating feeding grounds for marine animals. The POST idea was that some of the puzzling variation in survival that we observe in neighboring salmon populations might be due to their occupation of different parts of the ocean, or the timing of those uses, as some earlier studies suggested (Mckinnell *et al.* 1997). This question was distinguished from the first by the need to track juvenile salmon longer and farther and the need to understand how they used deeper water beyond the edge of the continental shelf.

POST's founders knew the network might prove useful for other species as well. The continental shelves are the most productive oceanic regions, and many commercially important species spend much of their life cycles on the shelf (Pauly & Christensen 1995). Seasonal upwelling of cold, nutrient-rich water supports high species diversity, commercially important fisheries, and large populations of marine mammals, seabirds, and fish in the Pacific Northwest (Keiper *et al.* 2005). The shelves also suffer the heaviest anthropogenic impacts from fishing, shipping, oil exploration, marine aquaculture, and land-based activities that export sediment, fertilizers, and pollutants.

14.1.2 How successful has POST been?

The signature of the early POST results is just how surprising they have been. The effort to study salmon survival in the ocean, while still in its early stages, has led to an explosion of applications to other species and a great expansion

of the questions being asked: in a sense, POST has highlighted how little we know about the behavior of marine animals (Table 14.1). POST has now been used to study 18 species including salmon, salmon sharks, two species of squid, rockfish, ling, white sturgeon, and English sole.

However, there is still a long way to go toward answering the original questions. Measuring the survival of salmonids along the continental shelf with high accuracy and precision is still a goal that will require further methodological refinement and testing, as well as an expansion beyond the large, hatchery-raised fish that have been the focus of many salmon studies, to smaller wild fish and a larger range of life histories. Making sense of survival estimates in the context of natural and anthropogenic variation in ocean conditions will require long-term studies for which POST has begun to lay a baseline. Results from salmonid studies so far have highlighted enormous diversity among populations. Some appear to suffer high mortality in-river, others in inland marine waters and still others in deeper water beyond the reach of the current array. Although some populations appear to have experienced high mortality immediately upon ocean entry in some years, enough populations have shown alternative mortality patterns that it is safe to conclude there is no single answer to where and when salmon die in the ocean. From the perspective of conservation and management, this means that it may be difficult to identify the causes of weak or strong returns of adult salmonids without detailed, population-specific data.

The question of whether salmon have "two zip codes" was technologically the more difficult of the two original questions that motivated POST. Evidence from shelf waters suggests that there is inter-annual variation in migration patterns, for example in the proportion of Georgia Strait fish that migrate north rather than south in a given year. POST has been unable to test the "two-zip code" theory in deeper waters, but it is a goal that seems closer than ever to being answerable. In the chapter summary we discuss some new directions in technology that will help to move tracking off the continental shelf.

14.1.3 The value of movement and survival data

Nearly every marine species moves in order to take advantage of the ocean's physical and biological diversity. There are large differences in temperature, pressure, salinity, oxygen levels, and other parameters from the surface to the depths, from the equator to the poles, and from the shoreline to the mid-ocean. Productivity varies over many orders of magnitude from rich coastal upwelling zones to the barren mid-ocean waters. Where an animal goes determines what it experiences. We attempt to understand where animals go, the conditions they experience, and the internal

Table 14.1

Examples of technological and research questions currently being addressed by projects that use the POST array, with a brief summary of results to date.

Technological question	Application	Results so far
Can receiver lines be engineered so that it is possible to estimate the survival rates of migrating fish?	Juvenile salmonids in rivers Juvenile and adult salmonids on the marine continental shelf	Yes. Detection efficiency can vary with conditions such as flow rates Yes, but work remains to calibrate estimates, esp. on outer coast. High detection efficiency (85–95%) appears routinely achievable.
Do acoustic tags (which are larger than some other tag types) cause little enough additional mortality to be useful for survival studies of small animals?	Salmon smolts as small as 130 mm length, herring, small squid, other forage fish	Positive evidence so far; more studies are underway. Tag-to-body-size ratios do influence survival, and smaller tags will enable a wider variety of species, life stages, and stocks to be tagged.

Research goal/question	Example	Results so far
Describe residency, coast-wide movements, and interchange between major river basins of anadromous species	Green sturgeon and white sturgeon (California to British Columbia)	For green sturgeon, unusual northwards winter migration was discovered, plus complex, previously unknown marine behavior including substantial interchange between rivers.
Characterize movements of an apex predator at nested scales	Sixgill sharks in Puget Sound	Sharks are relatively sedentary on short time scales, move much more on longer scales. Puget Sound may be a nursery.
Characterize speed of migration	Salmonids	Speed is more variable in freshwater than in the marine environment, and migration of some species is faster and more directed than others. Juveniles of some species (for example, steelhead) cover long distances very fast.
Partition mortality between life-history phases (downriver migration, estuary, and early ocean)	Steelhead (British Columbia and Puget Sound)	The baseline of survival rates thus far suggests a diversity of patterns for different stocks, and finds differences between hatchery and wild fish.
Locate areas of high mortality for endangered stocks	Coho salmon (Strait of Georgia)	High-mortality areas seem to be stock-specific. Evidence for fall migration out of the strait; possible mortality in summer.
Investigate the impact of freshwater mitigation efforts on anadromous species	Chinook salmon (Columbia and Fraser Rivers)	Surprising and still controversial preliminary results find no evidence that passage through dams causes delayed mortality, and find similar survival rates in a dammed and an undammed river.
Search for physiological explanations of a major mortality event	Later-run adult sockeye salmon (Fraser River) that do not delay river entry and die in-river	High water temperatures probably exceed salmon physiological limits; hormonal changes while fish are at sea may cause them to enter the river early.

mechanisms that drive their behaviors, because for marine species, being in the right place at the right time may mean surviving instead of dying, being able to grow instead of going hungry, or reproducing instead of having no offspring. Even not moving has consequences, because the ocean is dynamic and change can visit a sedentary animal.

Movement and survival data are useful in many aspects of fisheries management. These data can help us to understand immigration and emigration from fish populations and to parse out some of the complexity observed in natural mortality rates; both are among the factors that limit the accuracy of stock size predictions. Some species are vulnerable during a brief window when they congregate at some

life-history stage, or are sedentary, and movement data are useful for describing how species use habitat. Spatially resolved survival data are critical for restoration efforts, which benefit from a detailed understanding of a species' full life cycle to identify survival bottlenecks. Marine protected areas are an understudied new management tool, and there are many questions about how – and even whether – they work. Dispersal patterns may determine when a protected area will be a source of juveniles that can recolonize nearby exploited areas and boost depleted populations, or when will it be too small or unconnected to do so (Botsford *et al.* 2009). Finally, there is growing acknowledgement in fisheries management that it is important to preserve natural genetic and life-history biodiversity, and movement data suggest that there may be variation we are not aware of in the life histories of many species.

Indirect human impacts on the ocean – including climate change, acidification, and a host of effects that are produced by a growing population (increased shipping, increased light and noise, increased mineral exploitation, and others) – now threaten to exceed, perhaps greatly, our direct impact through fisheries. The accumulation of carbon dioxide in the Earth's atmosphere is changing global atmospheric temperatures, oceanic currents, ocean chemistry, and weather (IPCC 2007; Fabry *et al.* 2008). If we are to have any ability to respond to coming changes, we must understand a great deal more about the extent to which movement behavior is evolutionarily flexible, and how it may be changed by intense natural selection.

14.1.4 Acoustic tracking in the context of other technologies

Tagging can be used to understand where an animal goes, the conditions it experiences and, to a limited degree, its internal state. The smaller an animal is, the more difficult it is to track and the less sophisticated the tag can be. Tags can be categorized by how much information they provide about the time between capture and recovery, and by how data are retrieved. It is relatively easy to attach a sophisticated tag to a marine animal. The challenge is to retrieve the data, and there are only two solutions: to recover the tag physically, or to transmit the data to a receiver (which may be on land, in space, or underwater on a variety of platforms including fixed moorings, gliders, or other animals).

14.1.4.1 Chemical, biological, and genetic tracking

Many animals have natural markings that can be useful in movement studies (Payne *et al.* 1983). In addition, our environment leaves "natural tags" in our bodies that can be deciphered for information about where we have been, including ratios of stable isotopes, chemical signatures in otoliths (Barnett-Johnson *et al.* 2008), and even parasites (Timia 2007). Genetics are widely used for information about origins and mixing of stocks (Habicht *et al.* 2007; Seeb *et al.* 2004), and fatty acid ratios in the tissues of predators can help to identify the prey species and proportions eaten (Iverson *et al.* 2004). These methods are very useful because (1) the marking is already done for us by nature, and (2) even the smallest larvae retain readable signatures. However, calibration of the methods is complex and none yields detailed movement information. In addition, some require lethal sampling. The only way to study the movements of tiny larvae has been to take regular samples at grid points (see www.calcofi.org) but only very limited inferences about movement can be made from such data.

14.1.4.2 Non-electronic tags

The simplest tags are passive physical devices or marks such as freeze brands, fin clips, and spaghetti tags used to identify individual fish. Since 1968, US and Canadian institutions have released over 600 million salmon batch-marked with coded-wire tags in the largest tagging program on Earth. The main drawback to non-transmitting tags is that the vast majority are never seen again. Therefore, large sample sizes are needed, the species of interest must be the target of a substantial fishery, and any tagging data must be interpreted cautiously because the results are influenced by the movements, techniques, gear, and reporting behavior of the fishermen themselves. Information gathered from a physical tag usually can be summarized as two data points (release and recapture locations), plus associated dates and measurements. These studies leave many questions open. What route did the animal take? How did it respond to the conditions it encountered? What happened to most of the animals that were not seen again? Is the behavior observed representative of the population?

14.1.4.3 Electronic tags

A passive integrated transponder (PIT) tag is a semi-passive radio-frequency device that transmits a unique identification number when excited by a signal from a scanner. The scanner must be close to the tag (usually 45 cm or less). PIT tags have been used to tag 1 million to 2 million salmon per year since the 1980s, and fish are recorded as they pass through dams where expensive infrastructure has been used to channel and separate tagged juveniles and adults from other fish. In these situations, PIT tags are powerful tools, although the information each tag provides is limited.

14.1.4.4 Archival tags

Archival tags store data from one or more sensors on a computer memory chip. Sensors may record internal or external conditions, such as light levels and temperature that can be analyzed to provide rough estimates of latitude and longitude. Non-transmitting archival tags must be physically retrieved. Larger archival tags transmit data by radio signals to satellite or cell phone networks, and this capability to obtain fisheries-independent, detailed tracks is unmatched. With archival tags, it is possible to begin to understand the complex questions of why animals go where they go, how they navigate, and to observe detailed behavior in the wild. Unfortunately, most species are too small to carry satellite tags, and archival tags are not particularly useful for measuring survival, except in studies of hooking mortality.

14.1.4.5 Acoustic tags

The POST system is based on tags that transmit data to a network of submerged receivers. Each tag transmits an

Fig. 14.3
Acoustic receiver deployed in a tank at the Vancouver Aquarium. The receiver (soda-bottle-sized black object) is protected by the yellow flotation collar, which suspends it off the bottom and protects it from trawl nets and other disturbances, without compromising its ability to listen for tags. In an ocean deployment, the anchor would be much larger. Photograph: John Healy, Vancouver Aquarium.

14.2 Contributions from the POST Array to Marine Science

This section reviews species-specific accounts of some of the most important results so far from research that has taken advantage of the POST array.

14.2.1 A brief introduction to salmon

The fact that POST began in the North American Pacific Northwest made it almost inevitable that it would focus on anadromous species like salmon which use both fresh and saltwater. Salmon (*Oncorhynchus* spp.) spawn in freshwater rivers and streams, their eggs develop and hatch, and after a few days to two years in freshwater the juveniles swim to the ocean. In the ocean, they mature and grow to full size and sexual maturity before returning to spawn in their natal stream. Anadromous species come into more intimate contact with humans than purely marine species. In the prosperous Northwest, the human population now affects every part of the salmon freshwater life-stage. Many Northwest salmon populations declined dramatically during the twentieth century, and contributing factors include habitat destruction by agriculture, logging, and development, pollution, overfishing, negative impacts of large hatchery programs, hydropower and water storage dams, and changing climate patterns.

Luckily, salmon are extremely adaptable. The evolutionary history of salmon is one of repeated recolonization of rivers as the Northwest was covered and uncovered by glaciers (Waples *et al.* 2008), and Pacific salmon fisheries are still economically, socially, and environmentally important. Even today, many insects, birds, and mammals living as far inland as 1,500 km from the ocean owe much of their growth to ocean-derived nutrients carried in the bodies of salmon, although the total weight of Pacific salmon runs in this southern portion of their range may now be less than 10% of historical levels (Gresh *et al.* 2000).

Salmon management in the Pacific Northwest is a complex pastiche of efforts to regulate and mitigate human impacts. Some of the more complicated controversies involve efforts to reduce mortality of juveniles and adults at hydropower dams, to supplement natural populations with hatchery-raised fish, and to farm non-native Atlantic salmon along the Pacific coast.

14.2.2 Studies of survival

The next three sections on steelhead, coho salmon, and chinook salmon highlight the original purpose of the POST array, which was to study when and where salmon die in

individual identification code as a train of acoustic pulses at 69 kHz, approximately once a minute for 4–12 months, depending on battery and programming. The smallest tags (7 mm diameter × 18 mm length) can be surgically implanted in fish as small as 130 mm in length, and a new generation of higher-frequency tags may be used in fish as small as 95 mm. The two tag models most commonly used in POST have an approximate functional range of 200–400 m, depending on a host of conditions including ambient noise. Receivers record tag identification numbers plus associated detection times. Vemco's acoustic communications (see Box 14.1) are engineered to have a low rate of false-positive signals. The receivers wait for echoes to die down between pulses received from the tag, so the amount of information that can be transmitted is small (Grothues 2009). However, acoustic tags cost less than one-tenth as much as satellite tags, and may be manufactured with very long battery lives (more than 10 years), so in addition to being the method of choice for small species, they are also useful for long deployments on larger species.

the ocean. Conservation strategies rely on knowledge of when and how much mortality occurs in the population of interest, and why. Most marine species are observed only when they are caught, usually as adults. An estimate of ocean survival for a salmon cohort may be a single number that integrates a 15,000 kilometer voyage from Oregon to Japan via Alaska and back over as long as five years, but it has been impossible to partition such survival estimates by location and time period. Despite sometimes being able to measure freshwater survival fairly accurately, we observe large, unexplained year-to-year variation in overall survival rates, as well as surprisingly large differences between closely neighboring stocks, both of which indicate the importance of ocean survival.

Technically, it is more difficult to study survival than movement, because when an animal is not detected, we must determine what happened. As long as the animal cannot swim around the end of a line, there are only four possibilities: (1) the animal died, (2) it slipped through the line undetected, (3) its tag stopped functioning or was lost, and (4) the animal took up residence between lines. The problem of tag loss can be solved by double-tagging studies (Wetherall 1982), and the residence problem can be addressed by adding additional lines or by using active tracking to locate missing tags. The most difficult problem is estimating the probability of detecting an animal that passes through a line (Fig. 14.4), and it is traditionally solved by jointly estimating detection probabilities and survival with mark-recapture models (Amstrup 2005). Producing survival estimates with narrow confidence intervals requires lines that are highly likely to detect passing fish. One of the real triumphs of POST has been to demonstrate that it is possible to maintain high-efficiency marine receiver lines and to produce survival estimates for migrating juvenile salmon in rivers and on the continental shelf. However, the system works best when animals are migrating in one direction and pass through lines that completely cross rivers or inland waterways, and work remains to understand better methodological issues such as tag effects, and to expand the range of salmonid stocks, sizes, and life histories tagged to a more representative sample of the natural range.

14.2.2.1 Steelhead

Background

Steelhead are renowned for their long-distance migrations in the North Pacific. They are the anadromous form of rainbow trout (*Oncorhynchus mykiss*), but are often managed alongside the five species of Pacific salmon based on their behavior. Steelhead juveniles rear in swift streams and creeks, then migrate as smolts to the ocean for their adult life. Unlike other Pacific salmonids, steelhead are iteroparous, meaning adults can spawn in freshwater, return to the ocean for one or more years, and then spawn again. Widely distributed from California to Alaska, steel-

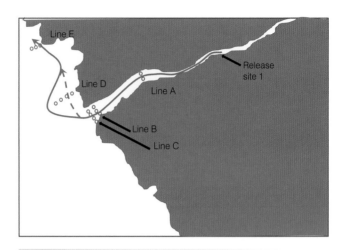

Fig. 14.4

The probability of detecting a fish that swims through a line of receivers can be estimated when a group of tagged fish is observed again after passing through the line, and the number detected can only unambiguously be related to detection probability if they *must* have passed through it. In the figure, if fish are migrating in the direction indicated by the blue arrow, detection efficiency can be measured for the array lines A, B, and C. However, detection efficiency cannot be estimated at line E (the terminal line, given the direction of the fish movement) without additional assumptions. Because fish may go around the end of the line at D, the probability that a fish passes undetected through line D (dashed line) is confounded with the probability that it swims around the end of the line.

head are highly sought after by anglers in recreational fisheries, but most southern populations are currently much smaller than they were historically.

Much like coho salmon, the downstream and early marine migration period is thought to be critical in determining recruitment of steelhead (Pearcy 1992). Smolt-to-adult survival rates generally declined throughout much of their southern range beginning in the 1990s, mirroring the declines in abundance. Low smolt-to-adult survival rates (less than 5%) have generally persisted to the present. There is regional variation in smolt-to-adult survival even at relatively small scales, however; populations from Washington State's inshore Puget Sound typically have had lower survival than those from the outer west coast, and similarly, populations from British Columbia's east coast of Vancouver Island bordering on the Strait of Georgia have typically had lower survival than those from the west (outer) coast of Vancouver Island despite geographic proximity. Variation in survival rates among years and watersheds raises a number of questions. Where and when do mortality periods predominantly occur? Does variation in migration rate or behavior contribute to variation in mortality?

Findings

Several steelhead populations were studied under POST to quantify their mortality rates during this critical period, and

Table 14.2

Survival rates of steelhead, from studies that tagged a total of 21 experimental groups of wild and hatchery fish.

Watershed	Survival (%) to marine entry		Survival (%) to exit from inshore waters	
	Wild	Hatchery	Wild	Hatchery
Cheakamus River (British Columbia)[a]	64–84	33–43	18–39	3
Hood Canal (Washington)[b]	78–96	88	22–40	15
Puget Sound rivers (Washington)[c]	74–87	74–76		
Keogh River (British Columbia)[d]			55	17–47

[a] Estimates from Melnychuk *et al.* (2009), four wild and three hatchery groups.
[b] Estimates from Moore *et al.* (2010), four wild and one hatchery groups.
[c] Estimates from Goetz *et al.* (2010), four wild (Skagit, Green, Puyallup, Nisqually Rivers) and two hatchery (Green, Puyallup) groups.
[d] Estimates from Welch *et al.* (2004), one wild and two hatchery groups.

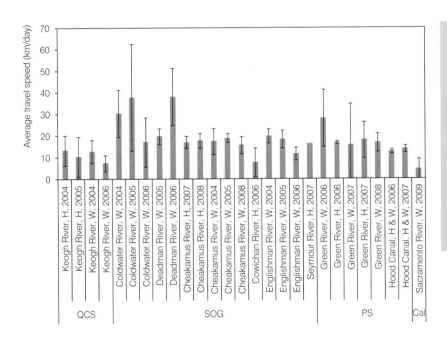

Fig. 14.5

Average travel speeds of steelhead smolt populations during the early ocean migration. Travel rates are calculated as shortest in-water migration distances from river mouth receiver stations to stations at exit points from inshore waters, divided by the average travel time to complete this distance. Error bars show ±1 standard deviation. Populations are grouped by area: Queen Charlotte Strait (QCS), British Columbia; Strait of Georgia (SOG), British Columbia; Puget Sound (PS), Washington State; and California (Cal). Wild (W) and hatchery-reared (H) populations are distinguished. Travel speeds are calculated from POST data (QCS, SOG, PS-Green River) or provided by Moore *et al.* (2010) and B. McFarlane (personal communication).

to compare survival among wild and hatchery-reared populations (Table 14.2). Survival from release to the river mouth and from release to exit from inshore areas was assessed for populations from British Columbia and Washington State. Populations differed in the distances they migrated, but within each watershed, survival of wild populations was generally greater than that of their hatchery-reared counterparts during the smolt migration. The high mortality incurred during the downstream and early ocean migrations is surprising, considering how little time steelhead spend in these areas. Steelhead smolts generally exhibited rapid movements downstream, through estuaries, and out of the inshore areas of the Strait of Georgia or Puget Sound within a few weeks of being released. Travel speeds downstream varied widely, depending mostly on river flow. Those from Fraser River populations varied from 53 to

81 km per day on average, whereas those from smaller rivers varied from 0.2 to 17 km per day. After ocean entry, however, average travel speeds were remarkably similar among populations, varying little with mean body length (which ranged from 161 to 203 mm; travel speeds were estimated over distances ranging from less than 20 to more than 400 km (Fig. 14.5).

It appears that mortality rates (on a per-day or per-kilometer basis) are considerably higher during the downstream migration than during the early ocean migration, but the agents of this freshwater mortality are not well known. Much mortality occurs soon after release, especially in hatchery fish that generally have little exposure to predators or natural selection pressures before release. Considerable mortality also occurs beyond the areas of study of the current POST system over the remaining year(s) of ocean life, as smolt-to-adult survival of many populations is typically less than 5%.

One exception to the fast, directed migrations of steelhead smolts is that typically around 5–10% of fish within a cohort will residualize, or fail to migrate downstream (Melnychuk *et al.* 2009). Residual steelhead may either delay their migration for a year or take up permanent freshwater residence. This sort of flexibility in life-history strategies is yet another example of the variability that makes it challenging to estimate survival of salmonids. There is still much to be learned about steelhead ecology and life history from tagging studies, and additional technologies will likely be required to extend investigations further along the ocean life-history trajectory of steelhead.

14.2.2.2 Coho salmon

Background

Coho salmon (coho; *Oncorhynchus kisutch*) in the Strait of Georgia have exhibited unusual behavior and survival patterns during recent decades. Once the target of a major year-round fishery, the Strait of Georgia coho all but disappeared during the mid-1990s (Beamish *et al.* 1999). Their marine survival had dropped from 10% (during the 1980s) to 2% (Beamish *et al.* 2000), and marked coho that would normally have spent their entire lives within the strait were being observed off the outer coast of Vancouver Island (Weitkamp *et al.* 1995). The general opinion at the time was that overfishing was to blame, but when the fishery was closed in 1998, no noticeable effect on coho marine survival followed (Bradford & Irvine 2000).

Some investigators hypothesized that climate was playing a key role (Beamish *et al.* 1999). Correlations were found between the sudden changes in coho survival and climate regime shifts (Hare *et al.* 1999). As average sea surface temperatures in the Pacific increased, northern coho populations grew, whereas many southern stocks faced extinction (Coronado & Hilborn 1998). The advancing onset of spring plankton blooms was identified as a possible negative influence on the later-migrating species such as coho (Beamish *et al.* 2008). Further potential causes emerged, including the long-term effects of hatchery production, increasing predator populations, pollutant levels, and other side effects of human development (Araki *et al.* 2007; Bradford & Irvine 2000).

Research on the marine portion of the coho life cycle has been limited by technology. Trawl surveys were initiated in 1997 to examine the distribution and growth of juvenile coho in the Strait of Georgia (Beamish *et al.* 2008). Catch surveys and mark–recapture methods have been used to examine the distribution and growth of juvenile coho in the ocean (Beamish *et al.* 2008). However, these methods are less effective as population sizes and catch rates decline. The POST array has provided researchers with another

opportunity to investigate the behavior and potentially the survival patterns of these coho stocks.

Findings

The migratory behavior and survival of coho smolts from various river systems were examined to identify key freshwater and marine mortality areas. The Thompson River (a tributary of the Fraser River) coho population is extremely endangered and concern has been raised about poor habitat quality in the watershed and a lack of research being done (Irvine & Bradford 2000). For these reasons, 190 hatchery-reared coho smolts were implanted with acoustic tags over three consecutive years and tracked using the POST array. Survival to the mouth of the Fraser River was found to be extremely low during 2004 and 2005 (0–6% and 7%, respectively). The freshwater survival of other Thompson River salmon species was higher (Welch *et al.* 2008, Chittenden *et al.* 2010), as were the survival rates of coho during 2006 and in other river systems (Chittenden *et al.* 2008). The low freshwater survival of Thompson River coho may be a key reason for the endangered status of this stock. Further work needs to pinpoint high mortality areas in the Thompson/Fraser watershed and possible causes for the low survival.

Marine survival was evaluated by tracking 173 tagged juvenile coho in the Strait of Georgia during 2006 and 2007. The fish left the Strait of Georgia through the Juan de Fuca Strait primarily from October to December, and the remaining coho either died or took up residence in the strait (Fig. 14.6) (Chittenden *et al.* 2009b). The proportion of fish surviving and migrating from the strait was smaller in a group of fish tagged in July (19%) than in a group tagged in September (52%), suggesting that coho may have suffered high mortality during the summer (Chittenden *et al.* 2009b). A small proportion of the acoustically tagged

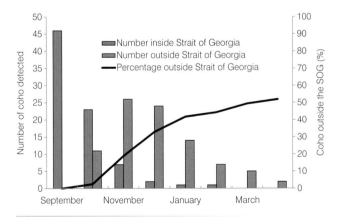

Fig. 14.6
Evidence of unexpected autumn migration timing for juvenile coho from protected waters to the open ocean: the location of tagged juvenile coho salmon by month, from September 2006 to April 2007, expressed as the number of fish detected within or outside of the Strait of Georgia.

Strait of Georgia coho (4%) was also detected on the outer coast POST array as far south as Oregon (750 km from the release site) (Chittenden *et al.* 2009b). These studies demonstrate that coho survival is stock-specific and probably dependent on ecosystem dynamics in both fresh and saltwater.

The long-term effects of hatchery releases on wild coho populations may be part of the reason for lower survival rates. A study comparing hatchery and wild smolts in Campbell River, British Columbia, found differences in physiology, travel time, survival, and migratory behavior, with wild fish spending less time in the river and estuary, arriving at POST arrays sooner than hatchery fish (Chittenden *et al.* 2008). Scientists are working to understand what effects these differences may have on ecosystem dynamics in the Strait of Georgia as well as the relative roles that genetics and rearing environment play on the phenotypic expression of coho young.

The story of the disappearance of coho salmon from the Strait of Georgia is complex. Some stocks have been found to suffer high mortality before they leave freshwater, whereas others have high mortality in the early marine phase. Long-term telemetry studies will help to understand the effects of climate, and other ecosystem dynamics, on the inter-annual variability of coho distributions, behavior, and survival.

14.2.2.3 Chinook salmon

Background

Within the Columbia River basin in the Northwestern United States, the completion of the Snake River dams in the late 1970s coincided with a "regime shift" to warmer ocean conditions that was generally considered to be deleterious to salmon survival in the southern portions of their range (Hare *et al.* 1999). The subsequent rapid decline of spring-type chinook (*Oncorhynchus tshawytscha*) in the Snake River (an upper tributary of the Columbia) resulted in this population being classified as threatened under the US Endangered Species Act. The listing had major economic consequences for the fishing industry and for the operators of the dams, who have spilled water to accommodate salmon migration, resulting in lost potential revenue ever since. Over the past several decades, improvements in fish passage at the dams and the implementation of a fish transportation program to bypass the impounded section of the river basin have improved the survival of seaward migrating juveniles (smolts) (Muir *et al.* 2001), but adult return rates have not substantially improved (Williams 2008; Schaller *et al.* 2007).

Because high in-river smolt survival has not improved adult returns from the ocean, it has been hypothesized that dam passage or transportation around the dams may reduce the fitness of Snake River smolts by impairing their survival after they migrate from the hydrosystem (the impounded

section of the river basin) into the coastal ocean. This is referred to as the "delayed mortality" hypothesis. POST brings a new dimension to our ability to test the hypothesis. Survival estimates were previously available only for the entire life-history period from smolt out-migration until adult return to the river, some two to three years later, resulting in a significant knowledge gap about ocean survival and especially whether the very poor adult survival still observed was caused by the operation of the hydro dams (see, for example, Schaller *et al.* 1999; Budy *et al.* 2002; Budy and Schaller 2007).

Findings

From 2006 to 2009, 4,000 hatchery-reared juvenile stream-type chinook salmon (130–200 mm fork length) originating from the Columbia River basin were tagged and tracked down the river and along the Pacific shelf as they swam north, in order to address several significant management issues. Five of the smolts were detected in Alaska, 2,500 km from their release point. A summary of the findings includes the following.

- A comparison of acoustic tags with much smaller PIT tags demonstrated that in-river survival estimates from the two tag types were statistically indistinguishable (Fig. 14.7), which suggests that there was not enough additional mortality caused by the larger acoustic tags greatly to bias the survival estimates relative to estimates made with PIT tags.

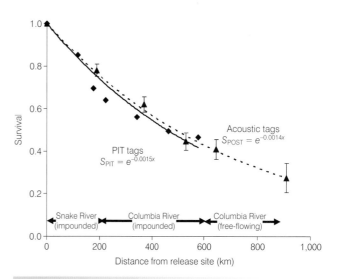

Fig. 14.7

Survivorship of spring chinook smolts released in the Snake River basin in 2006 (first published in Welch *et al.* (2008)). Diamonds represent NOAA PIT tag survival estimates within impounded sections of the river. Triangles represent POST survival estimates within the river and onto the continental shelf of Washington State, USA. Error bars show one standard error.

● A large-scale experimental comparison of survival in two tributaries of the Columbia, the Snake and Yakima Rivers (which join the Columbia mainstream at approximately the same location), to test the delayed mortality hypothesis also found similar survival rates. Juvenile salmon from the Snake River must pass through eight dams on their way to the sea, whereas juveniles from the Yakima River pass through only the four lowest dams. As survival of Yakima fish to adulthood is four times greater than the survival of the Snake River fish, these contrasting adult survival rates contributed to the development of the delayed mortality hypothesis.

Initial results from the 2006 (Rechisky et al. 2009) and 2008 (Porter et al. 2009) studies indicated that lower-river survival (from below the final dam to the first ocean detection line, a distance of 274 km) was similar for the Snake and Yakima River populations. Survivorship from release to the first ocean detection site (911 km for the Snake populations, 655 km for the Yakima populations) was also similar, indicating that the fourfold greater survival observed in adult return rates in favor of the Yakima population may develop later in the marine life-history phase and could possibly be attributed to different ocean life-history strategies (perhaps owing to migration to different ocean regions). Because of the current rather coarse spacing of the marine sub-arrays, finer-scale arrays are needed to better assess the behavior and survival of the two populations.

● A costly mitigation effort was evaluated: many out-migrating Snake River juvenile salmon have been transported via barge around the entire dam system for several decades in hopes that eliminating dam-caused mortality would improve adult return rates (Williams 2008) as over half the in-river migrants die before reaching the final hydropower dam. However, despite near-perfect survival of the barged fish during transport, transported smolts do not return at twice the rate of the smolts that migrated downstream through the dams before reaching the ocean (Schaller et al. 2007; Muir et al. 2001).

The POST system was used to compare survival of transported and in-river fish as they swam through the estuary and onto the continental shelf. Smolts migrating the entire length of the river and smolts first transported by barge were observed to have similar survival through the estuary to the first ocean detection site in both 2006 and 2008. Overall, the number of transported hatchery fish surviving to the first ocean sub-array was greater than that of the in-river migrants, as expected based on the initial survival benefits from transportation. However, the data also indicate that survival rates in the ocean and river are very similar. The solution to the transportation paradox may therefore have a very simple explanation: survival per unit time may be quite similar in the ocean and the river. If true, transporting smolts around the dams may provide little benefit simply because the smolts are transported between two environments that currently have approximately equal survival prospects; because the total lifespan is unchanged, smolts that would otherwise die in-river during migration may simply die in the ocean – at about the same rate – if they are transported.

● A large-scale acoustic telemetry comparison was made of the dammed Columbia and the un-dammed Fraser Rivers (Welch et al. 2008), which found that in-river survival rates of steelhead and chinook hatchery smolts were similar in both rivers. The reasons for the similarity are not yet clear.

The ability to estimate survival over thousands of kilometers is now in the near future. The POST array will contribute to testing numerous hypotheses about the effect of dam passage on ocean survival, migration timing, and survival by size, as well as how ocean climate change will affect the future sustainability of salmon populations, all of which is essential knowledge for improving decision-making regarding the sustainability of Pacific salmon.

14.2.3 Studies of movements, life history, and habitat use

The biggest surprise of the POST project occurred when green and white sturgeon were detected by chance far beyond areas they were thought to use. The very large scale of the POST array and its integrated data management system will make such serendipitous discoveries ever more likely as the number of tagged species increases. Such discoveries emphasize how little we know about the movements of marine animals. In this section, we report on three studies of sharks and sturgeon, which all show how the POST array can be combined with smaller, local receiver networks to study movements. Although lines of receivers are not strictly necessary for studying movement, they have proven useful.

14.2.3.1 Green sturgeon

Background

Green sturgeon (*Acipenser medirostris*) are a rare and poorly understood species. They are anadromous, and known to use just three rivers for spawning. Aggregations of green sturgeon have been noted in summer months in various Pacific coast estuaries, especially Willapa Bay and Grays Harbor, Washington, the lower Columbia River in Washington and Oregon, and several smaller estuaries on the Oregon coast (Adams et al. 2007). Much of the lifetime of green sturgeon is spent in marine waters, and although they have been observed in coastal waters between Baja

California (Mexico) and the Bering Sea (Moyle 2002), nothing is known of their migratory behavior.

In 2007, the US listed the southern distinct population segment of green sturgeon as a threatened species under the US Endangered Species Act (Adams *et al.* 2007). Several critical questions were unresolved in the status review that led to this listing, however, including the following:

- Where do green sturgeon in summer aggregations originate?
- What proportion of the population occurs in these aggregations?
- Do green sturgeon migrate frequently among different areas and habitat types or do they reside in more restricted areas?

The deployment of the POST array provided a backbone of receiver arrays in the coastal ocean throughout the purported range of green sturgeon from southeast Alaska to Washington starting in 2003, which provided an opportunity to address these critical questions. The POST array was augmented by receivers in the coastal ocean off central Oregon and in Monterey Bay, California, in estuaries and bays in Washington, Oregon, and California, and in known (Rogue River, Klamath-Trinity River, Sacramento River) or suspected (Umpqua River, Oregon, and Eel River, California) spawning rivers for green sturgeon. Green sturgeon

were captured and tagged in spawning rivers and in estuaries where summer aggregations occur.

Findings

The augmented POST array showed that green sturgeon made surprisingly long and rapid seasonal migrations along the west coast of North America (Lindley *et al.* 2008). Many tagged green sturgeon moved northward past the northern end of Vancouver Island in the late fall, overwintering somewhere north of Vancouver Island, and then made rapid southward migrations in the spring to return to spawning rivers or summering grounds in various estuaries in Washington, Oregon, and California. This sort of poleward migration in winter is highly unusual in the animal kingdom. Many green sturgeon were observed making these migrations annually, while a subset appeared to overwinter off Oregon without making extensive migrations (Fig. 14.8).

Closer examination of the migratory behavior of individual green sturgeon showed that there is substantial diversity among fish that is not explained simply by the location of capture and release. Most notably, many green sturgeon tagged in the Rogue River, but also some from the Sacramento and Klamath Rivers, spend summers almost exclusively in the Umpqua River estuary, whereas other groups of fish exhibited characteristic patterns of use of the

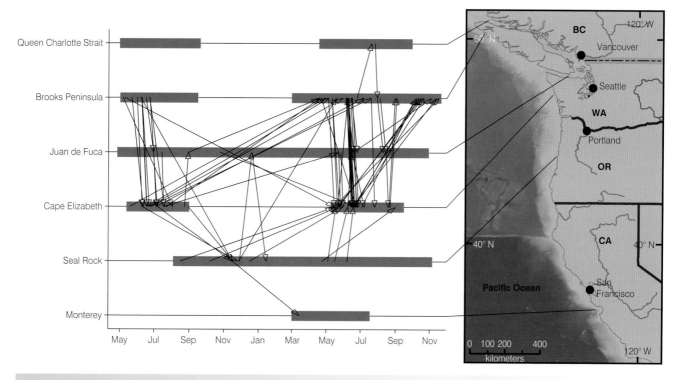

Fig. 14.8

Sturgeon movements along the Pacific coast, by date (modified from Lindley *et al.* 2008). The blue bars show periods during which receivers were deployed, and the arrows between them link detections of individual sturgeon. The map shows the locations of the receivers.

lower Columbia River, Grays Harbor, Willapa Bay, and San Francisco Bay.

Although the appearance of green sturgeon as bycatch in specific coastal fisheries had provided some indication of the species' marine distribution, the telemetry results have greatly expanded our insight into the extent and diversity of their marine movements. The data have opened and sharpened several questions. How stable are these behavior patterns and aggregations over years, and to what extent do they represent behaviors and habitat use peculiar to discrete sub-populations? What mechanism(s) underlies the surprising northward migration in the fall and winter? More generally, an important but difficult question is the basis for behavioral diversity in green sturgeon: is migratory behavior determined genetically or is it learned (Culum & Kevin 2003)?

Acoustic telemetry has spurred a quantum advance in our understanding of the migratory behavior of adult green sturgeon and has generated valuable demographic information (Erickson & Webb 2007). It would be extremely fruitful to apply the same methods to sub-adult green sturgeon, whose behavior is completely unknown, as well as to other migratory acipenserids, including the Atlantic sturgeon (*A. oxyrinchus*) and the shortnose sturgeon (*A. brevirostrum*).

14.2.3.2 White sturgeon

Background

White sturgeon are the largest (by length and weight) freshwater fish in North America, and one of the longest freshwater fishes in the world. In the Fraser River of British Columbia, white sturgeon can attain confirmed lengths to 6.1 m and weights to 629 kg (Scott & Crossman 1973). They are also late to mature (15–20 years for males, 20–30 years for females) and long lived (150+ years). They are truly living fossils, exhibiting little change in morphology from fossil records dating back 65 million years. Current research suggests that green sturgeon and white sturgeon use different sets of watersheds for spawning (Lindley *et al.* 2008). Distribution of white sturgeon is limited to western North America, with strong evidence that spawning occurs in only three major watersheds: the Sacramento River (California), the Columbia River (Oregon/Washington/British Columbia), and the Fraser River (British Columbia). Overfishing and habitat loss has decreased the populations of white sturgeon significantly, and in Canada all of the Fraser River stocks have been classified as "endangered" by the Committee on the Status of Endangered Wildlife in Canada and most have been listed for protection under the Species at Risk Act.

Findings

Current stock monitoring and assessment of lower Fraser River white sturgeon by the Fraser River Sturgeon Conservation Society suggests that this population is declining, and has suffered a 27% loss in total abundance since 2003 (Nelson *et al.* 2008). Interestingly, the greatest decline in this population has occurred in the lower age groups (fish younger than 10 years old), which suggests that there is either a recruitment issue or perhaps young sturgeon are leaving the lower Fraser River (that is, migrating to marine environments). To address the question of residency and migration of lower Fraser River white sturgeon, 110 specimens of various sizes and ages were captured and acoustically tagged in the lower Fraser River over three seasons (summer and fall 2008, spring 2009). Twelve acoustic receiver stations have been established at strategic locations in the lower Fraser River and estuary to augment existing POST receiver stations in the same area. All acoustic tags have been detected since release, and the majority of tagged sturgeon have moved considerable distances within the lower Fraser River study area. This three-year project, conducted in partnership with POST and local Aboriginal communities, is providing novel and useful information regarding inter- and intra-annual movements and habitat preferences of endangered lower Fraser River white sturgeon.

14.2.3.3 Sixgill sharks

Background

The sixgill shark, *Hexanchus griseus*, is one of the largest predatory sharks and is found in nearly all temperate and tropical seas of the world, preying on a wide variety of resources. Their basic life history (slow growth, late maturity, low fecundity) suggests they would be susceptible to exploitation and other environmental perturbations. Understanding how large predators use habitat through time and space is critical to successful management. The spatial distributions of populations are created by movement behavior (Turchin 1991); the impact of predators on local prey populations is relative to how much time they spend moving through different habitats (Fortin *et al.* 2005); and the susceptibility of populations to exploitation or environmental perturbations is dependent on their movement patterns (Kritzer & Sale 2006).

Findings

The project reported here took advantage of an array of acoustic receivers at more than 200 sites throughout Puget Sound established by a consortium of researchers, in addition to active acoustic tracking methods. From September 2004 to October 2008, Vemco acoustic transmitters were implanted into 59 sub-adult sixgill sharks ranging in size from 109 cm to 293 cm total length (6–173 kg). Movements of individuals were monitored for up to four years.

The spatial patterns of movement for sixgill sharks differ greatly depending on the temporal scale measured. Daily, individuals moved very little, averaging 0.2–3.1 km per day, with smaller individuals moving four times as far as larger

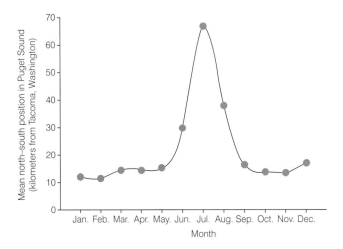

Fig. 14.9
An example of seasonal movements, which occur on a scale midway between smaller daily movements, and larger dispersal over the scale of years. Mean location of sixgill sharks in Puget Sound in 2006 as measured along a latitudinal gradient from the southern main basin (Tacoma, Washington) northward.

individuals (Andrews *et al.* 2007). Moreover, passive receivers detected sharks at the same location as the day before 76% of the time. Seasonally, tagged sixgill sharks occupied a narrow region (8 km latitudinal range) in the southern main basin of Puget Sound during the autumn and winter and then dispersed northward during the late spring and summer to all parts of the main basin (120 km latitudinal range) (Fig. 14.9). At the yearly scale, most tagged sixgill sharks were exclusive residents of Puget Sound from 2004 to 2007. In 2006, 10% of tagged individuals left Puget Sound, whereas in the summer of 2008, 46% of tagged individuals had left Puget Sound. The sharks that left were detected most frequently in the Strait of Georgia (approximately 350 km) or Strait of Juan de Fuca (200 km), with the farthest detected individual at Point Reyes, CA (approximately 1400 km). Four individuals that have left and come back to Puget Sound were absent from 1 month to 1.5 years.

Sixgill sharks show consistent patterns of movement, both vertically (Andrews *et al.* 2009) and horizontally, across multiple temporal scales, with most sub-adults residing in Puget Sound yearlong. The most plausible hypothesis for these stable patterns is that sharks are following prey populations at various temporal scales until they reach a certain size or age, when they make large movements out of Puget Sound. It is possible that the sub-adults that left Puget Sound in 2008 were members of a large recruiting cohort that is now leaving its rearing area. Pregnant and other large individuals (longer than 3 meters) have been seen in Puget Sound, but most adults are seen on the outer coast as bycatch from longline fisheries.

14.2.3.4 Other species

The POST array has also been used to study habitat use by ling cod in Prince William Sound, Alaska, and rockfish in the Strait of Georgia, and the movements of bull trout, market squid, and English sole in Puget Sound. The study of English sole was particularly interesting because it showed that sole, which are used as indicators of sediment contamination, spend less time at contaminated sites than had been presumed. This suggests that contaminated sediments may have larger impacts on fish health than has been previously calculated (Johnson *et al.* 2002; Moser *et al.* 2010).

14.3 Interdisciplinary Studies as a Model for Future Research: Sockeye Salmon

This section reviews several linked studies that have taken a multi-disciplinary approach to investigating why a valuable population of salmon has suffered very high mortality. Some of the studies were done in laboratories or outside the POST array. The array was used to identify the travel rates, migration timing, and fates of individual fish in a series of manipulative field experiments. Acoustic tracking thus acted as a bridge between laboratory studies of physiological limits and field studies of actual performance of migrating fish, and helped to integrate those results with physiological measures of fish condition and genetic stock identification. The section demonstrates how experimental approaches can be combined with observation and auxiliary data to derive the maximum value from acoustic telemetry. As such, it is a model for future studies.

14.3.1 Background

Recent advances in physiological telemetry are now making possible the study of marine species at spatiotemporal scales that encompass the complete ocean life history of the individual animal, thus enabling investigators to explore questions about "why" and "how" animals behave or survive the way they do. The behavioral physiology of Pacific salmon spawning migrations has been most thoroughly studied in sockeye salmon (*O. nerka*) hailing from the Fraser River in British Columbia (reviewed in Hinch *et al.* 2006). The Fraser River is Canada's most productive salmon system, and sockeye are the most abundant species after pink salmon (*O. gorbuscha*). Like all Pacific salmon, sockeye exhibit a high degree of genetic and phenotypic diversity, and in the Fraser River there are upwards of 150 distinct stocks (that is collections of distinct populations) which spawn throughout the watershed. Some of these travel as little as 100 km upriver to reach natal spawning areas whereas others travel over 1,200 km. At the start of the freshwater spawning migration, sockeye stocks can differ greatly in characteristics like morphology, fecundity, and energetics, which provides ample opportunity to explore the intraspecific bases of diversity.

One of the most distinctive behavioral traits in Fraser River sockeye is the highly predictable within-stock timing of their spawning migrations. The migrations of sockeye into the Fraser River are classified into four broad run-timing groups, each composed of several distinct stocks and populations. Fraser sockeye begin departing the high seas of the North Pacific Ocean in early June, with the early Stuart stock complex entering the river in early July, the early summer stocks in mid-July, the summer-run stocks in early August, and the late-run stocks in September and October. Understanding the mechanisms that control this variation in migration timing is interesting from an evolutionary point of view, but this understanding is critical for those tasked with the management of Fraser River sockeye salmon fisheries, particularly when so much depends on pre- and in-season predictions of coastal and river migration timing of these stocks.

14.3.2 Conservation crisis and research approach

Late-run sockeye will characteristically delay their migration from the ocean to the river for four to six weeks by "holding" in the Strait of Georgia within the Fraser River Estuary (Fig. 14.1) (Cooke *et al.* 2004). This scale of holding behavior is unique among sockeye anywhere, and has likely evolved to avoid peak summertime water temperatures in the Fraser River and prolonged exposure to various pathogens (Hinch 2009). Since 1995, segments of all late-run stocks have entered the river with little or no delay in the Strait of Georgia, and these early timed migrants suffer extraordinarily high mortality during the migration (Fig. 14.10). Depending on the year, 50–95% of late-run sockeye have perished during the river migration (Cooke *et al.* 2004) representing more than 4,000,000 individuals over that time period. The problem of early migration is a clear and ongoing economic and conservation crisis, and the long-term sustainability of key sockeye stocks in the Fraser is in jeopardy (Cooke *et al.* 2004; Crossin *et al.* 2008).

A team of investigators determined that potential causes for this seemingly maladaptive change in migration behavior involved changes in how maturing salmon use endogenous or exogenous cues to time their migration into freshwater (Cooke *et al.* 2004). However, it was clear that uncovering the specific factors responsible would be challenging as there was surprisingly little basic information

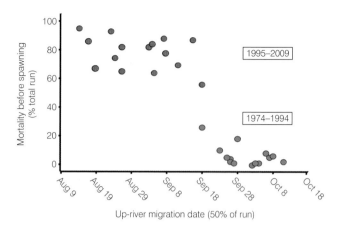

Fig. 14.10

Since 1995, certain populations of sockeye salmon homing to the Fraser River have migrated into freshwater three to six weeks earlier than the historical norm, probably dying as a result.

known about how Pacific salmon, or most fish for that matter, control the timing of their migrations, or how large changes in migration timing leads to extraordinarily poor survival. To fill these gaps in our general knowledge, and to get to the core of this specific problem, field studies began in 2003. Key to these studies was the use both of large-scale observational studies and field experiments that involved the integration of individual-based physiological biopsy of plasma and tissue with positional acoustic and radio telemetry tracking of individually tagged adults (reviewed in Cooke *et al.* 2008). Research in 2003 used a pilot version of the POST acoustic telemetry array and an expanded version in 2006. A companion radio telemetry array was also used in both years (Robichaud & English 2007).

14.3.3 Observational studies: physiological correlates of migration rates and survival

Over those two study years, more than 1,000 adult sockeye were tagged, biopsied, and then tracked from locales 200 or 800 km from the mouth of the Fraser River, using the POST array. Reproductive preparedness was repeatedly identified as a key physiological system driving ocean migration behavior. For example, 200 km from the river, early timed late-run migrants had elevated plasma concentrations of reproductive hormones, including testosterone, 11-ketotestosterone, and estradiol, suggesting that advanced reproductive preparedness may be triggering early migrations (Cooke *et al.* 2006). In a separate study involving both late-run and summer-run sockeye, Crossin *et al.* (2009a) confirmed that plasma testosterone concentrations strongly correlated with migration rates at locales 200 and 800 km distant from the river. The data also suggested that the physiological mechanisms responsible for causing early timed migrations may be triggered in the open ocean (Hinch *et al.* 2006; Crossin *et al.* 2009b).

Marine mortality of tagged fish also showed strong relations to their physiology at time of tagging. Fraser sockeye from several stock groups that perished in the Strait of Georgia before reaching the Fraser mouth were more physiologically stressed, based on plasma ion, glucose, and lactate measures, than those that survived to enter the river (Cooke *et al.* 2006). Late-runs that perished in the ocean were also less physiologically prepared for freshwater entry (that is, higher plasma chloride and total osmolality) than those that survived (Crossin *et al.* 2009a). Reproductive and osmoregulatory preparedness play interacting roles both in river entry timing and subsequent ability to reach spawning areas. Specifically, sockeye that delayed river entry in the Strait of Georgia and subsequently reached spawning areas had initially high somatic energy, low testosterone levels, and low gill Na$^+$,K$^+$-ATPase activity. In contrast, salmon that entered the river directly without delay in the Strait of Georgia and subsequently failed to reach spawning areas had initially lower energy, higher testosterone, and higher gill Na$^+$,K$^+$-ATPase activity (Crossin *et al.* 2009a). The implication is that a successful strategy to reach spawning areas is to spend time in the Strait of Georgia becoming more reproductively mature while restructuring gill physiological systems to enable freshwater entry. Fish that were reproductively mature and entered the river before changing their gill structure were likely to perish during the freshwater migration. Taken together, these results indicate that early migrations and high mortality are related to advanced reproductive schedules and poorly prepared osmoregulatory systems for the transition from salt to freshwater.

14.3.4 Interventional studies: field experiments to test observational hypotheses

14.3.4.1 Maturation enhancement experiment

That advanced reproductive development may cause sockeye to migrate faster was tested by injecting gonadotropin-releasing hormone (GnRH) and/or testosterone, as a means of altering maturation rates, into acoustic tagged Fraser sockeye 800 km from the mouth of the Fraser River (Crossin *et al.* 2009b). No differences in travel rates were detected between hormone injected ($n = 6$) and non-injected ($n = 6$) sockeye to the "first" POST acoustic lines situated at northern Vancouver Island (approximately 460 km from release) (Fig. 14.1). However, pre-injection levels of testo-

sterone did correlate with travel rates, a finding consistent with the original hypothesis (Crossin *et al.* 2009b). These results suggest that maturation rates are set in the open ocean and that coastal hormone interventions may be too late to affect behavioral change. Small sample sizes may also have limited this investigation.

14.3.4.2 Osmoregulation experiment

That ill-prepared osmoregulatory systems may cause sockeye to migrate faster and earlier out of the Strait of Georgia into the Fraser River was tested by capturing late-run sockeye in the ocean near the mouth of the Fraser River, transporting them to a nearby marine laboratory for experimental exposure for about one week under three different salinity conditions (freshwater, 0‰; iso-osmotic, 13‰; saltwater, 28‰) and then acoustic tagging and releasing them to continue their migration. The three salinity treatments produced post-treatment sockeye with different gill Na$^+$,K$^+$-ATPase concentrations (freshwater < iso-osmotic < saltwater) and freshwater-treated sockeye ($n = 7$) entered the Fraser River approximately two days faster than iso-osmotic ($n = 12$) or saltwater ($n = 5$) post-release sockeye. This two-day "estuarine holding" expressed by the saline treatments was a strong effect as it reflected the average Strait of Georgia holding period for late-run sockeye that year (Hinch 2009).

So far, research suggests that the phenomenon of early migration of late-run sockeye may have its roots in the high seas, based on the reproductive hormone findings and recent functional genomics disease assessments (Miller *et al.* 2007), though is likely also driven by coastal environmental variables such as salinity and fish abundance (reviewed in Hinch 2009). Continued experiments using POST arrays are needed to establish how oceanographic conditions trigger or control physiological, and hence behavioral, changes in individual migrating and maturing salmon, and how other environmental factors, both endogenous such as disease states and exogenous such as local abundance, mediate these behavioral changes.

14.4 Summary and New Directions

Even at this early stage, POST has been a surprisingly successful experiment. The varied research applications of the acoustic receiver array have proven that it is possible to monitor both localized and long-distance movements of marine animals on the continental shelf. The high efficiency of marine receiver lines has made it possible to measure survival of juvenile salmon in the ocean, although methodological refinement and testing is needed to improve the accuracy and precision of the measurements. Researchers have discovered very interesting behaviors with implications for conservation and management, and the technology has served as an important component in an extensive, cooperative research effort. Our technical knowledge of large-scale arrays has advanced dramatically since the inception of POST.

Our knowledge of where animals go has increased substantially for green sturgeon, sixgill sharks, and several other species we knew little about. Results have helped to confirm movement patterns of salmonids and have enhanced our understanding of their rates of travel. We are beginning to assemble a valuable library of behaviors that helps us to assess how much variation exists between individuals, populations, and species, and helps account for life-history transitions such as maturation or use of different habitats. The study of sixgill sharks highlights the fact that movement behaviors can seem very different, depending on the temporal and spatial scales on which they are observed, and is a reminder of the importance of year-round monitoring, long-duration studies and large receiver arrays.

One of the consequences of where animals go is whether they survive. Finer-scale estimates of survival in the ocean and the identification of mortality hot spots should help to calibrate life-cycle models and illuminate the practical question of what limits population growth. POST studies have begun to create a baseline understanding of how juvenile salmonid mortality is apportioned between freshwater, residence in estuaries and the transition to saltwater, and the first weeks and months at sea when young salmon typically remain over the continental shelf, and to document inter-annual and inter-population variation in those rates. Researchers are just beginning to estimate survival rates of longer-lived species such as sturgeon using detections of long-lived tags (Lindley *et al.* 2008). As data accumulate, the fitness consequences of various life-history strategies or events may be estimated, and the results can be synthesized into bioenergetic and ecosystem models.

We are beginning to understand the environmental conditions animals experience on their journeys, and we expect rapid progress in this area. Environmental conditions can be assessed by archival tags or by overlaying oceanographic data with animal tracks. In combination with laboratory studies, such information can guide us in understanding the preferences and physiological limits of marine species, which will ultimately help us to quantify and predict the effects of changing ocean conditions on growth and survival.

14.4.1 Unknowns

There are still many unknowns. Many of the results raised new questions. For example, we have observed that stream-type chinook have similar survival rates in the dammed Columbia and the un-dammed Fraser, but we do not know why, or whether those relative rates will hold over time. We know that many juvenile coho and chinook are never detected on the marine lines after they leave river mouths in the Strait of Georgia, but we are not certain whether they die or take up residence in the Strait. We see that neighboring steelhead stocks on Vancouver Island have different survival rates, and our data suggest that migration routes may play a role. We do not understand why Puget Sound steelhead disappear once they leave Puget Sound, nor why juvenile steelhead are not detected on POST's Southeast Alaska line.

We have not tackled one of the original POST questions: where do salmon go in the deep ocean, beyond the shelf? Are they as discriminating about their use of specific areas or features in the deep ocean as they are about their use of freshwater habitats? The inter-annual variability in migration routes that we see on the shelf suggests that salmon are seeking out particular conditions, rather than particular locations. Archival tags may soon begin to yield answers to some of these puzzles.

Interesting new questions include the following. Why are there marked differences among members of the West Coast green sturgeon population in their use of estuaries? What are green sturgeon doing in their "playground" off the northwest end of Vancouver Island, and what attracted an enormous ball of thousands of white sturgeon to the Bonneville dam in 2008? Are sixgill sharks using Puget Sound as a breeding ground? Why did 10 million fewer sockeye than predicted return to the Fraser River in 2009, at the same time that the Columbia River had record sockeye runs?

In most cases, we still are far from understanding the internal mechanisms that drive behavior. These reactions are the key to how species will respond to cyclical annual and decadal changes in ocean conditions (El Niño, Pacific Decadal Oscillation), as well as to long-term directional climate change like ocean acidification, warming, and changes in currents.

Finally, we are barely beginning to use movement data to tell us about species interactions. Understanding species interactions is a lofty goal, but as more species are tagged and tag technology advances, we will be able to ask questions about when and where predators and prey co-occur, an important step to understanding how they impact each other.

14.4.2 Technological developments that could change the game

14.4.2.1 Extending the scope: larger arrays, smaller tags, deeper water

The simplest approach to improving the tracking potential of the POST system is to expand the array. Expansion by in-filling between existing lines will answer questions such as whether chinook and coho that disappear in the Strait of Georgia have died or taken up residence there. Expansion by adding new lines will make new studies possible. A line planned for one of the major straits between the Gulf of Alaska and the Bering Sea will enable study of species interchange between the two seas.

Acoustic tags are getting smaller, and with every decrement in size we can tag earlier life stages and new species. Vemco's newest generation of smaller acoustic tags (V5) will make it possible to tag juveniles of some of the most important salmon stocks, as well as some of the mid-trophic level "forage fish" species that are currently too small to tag. However, smaller tags come with a cost; in this case, a higher-frequency (180 kHz) signal that requires new generation of receivers, and a lower range that requires much closer spacing between receivers. Satellite and archival tags are also getting smaller, and analytical techniques for deriving latitude and longitude from light and temperature data have advanced greatly.

Tests of receiver deployments in deeper water are underway in Alaska to determine the usefulness of acoustic arrays for monitoring valuable groundfish stocks such as halibut and sablefish. The extent of lifelong migratory behavior and seasonal movements by halibut and the question of interchange between seemingly separate stocks of sablefish remain pressing fishery management questions. Deeper receivers may also be useful in the future for linking to data cables from oceanographic arrays.

14.4.2.2 Large-scale monitoring

One of POST's founding Board members founded a worldwide extension of POST, the Ocean Tracking Network (OTN; www.oceantrackingnetwork.org). Collaborators will deploy and maintain their own receiver networks, and OTN will coordinate data management so that a detection of a tagged animal in any country will be discoverable by the researcher who tagged the fish. Many countries are currently working toward implementation of ocean observing systems as part of the Global Ocean Observing System (GOOS). As POST contributes biological observations to regional GOOS organizations such as the Northwest Association of Networked Ocean Observing Systems (NANOOS), opportunities arise to marry acoustic receivers with other sensors on platforms such as oceanographic buoys and sea gliders, and to integrate oceanographic data with tracking data. A richer understanding of biological oceanography will help with efforts to model the movements of very small animals such as larvae (Gawarkiewicz *et al.* 2007), which will remain too small to track for the foreseeable future, and it will sharpen our knowledge of how changes in ocean conditions affect the movement patterns of many species.

14.4.2.3 Using big animals to track smaller animals: business-card tags

An exciting recent technological development has been the manufacture of the "business-card tag", a miniaturized receiver joined with a standard transmitter. The tag not only transmits but also listens. A business-card tag can be put on a large animal such as a seal, sea lion, or shark (Stokesbury 2010). If the animal can be recaptured, the business-card tag will contain a record of every other acoustic tag that swam within the detection radius (approximately 0.2–1 km, depending on the tag). This technology transforms the large animal into a roving reporter. If the animal is double-tagged with a global positioning system (GPS) satellite tag, positioning data can be combined with the receiver log to show exactly where the animal went and what other tagged animals it heard.

14.4.2.4 Integrating archival and acoustic tags: getting to the critical "why" questions

The primary technological goal of the Ocean Tracking Network is the development of joint acoustic/archival tags. These "fully integrated" tags will archive data from sensors that monitor variables such as water temperature and light levels, internal body temperature, stomach pH or heart rate, and will include a miniaturized receiver that listens for other tagged fish. Fully integrated tags will communicate by spread-spectrum acoustic signals to achieve high data rates, so that an animal can download information about where it has been, what it experienced, and what other tags it heard, as it swims past a receiver line. This will require a new generation of receivers that will probably be more expensive and consume much more power. The payoff, however, will be an enormously rich data stream that finally solves the problem of obtaining information about what an animal experienced between the points where it was detected, and it will use large animals to track smaller animals.

14.4.2.5 Synergies from large-scale infrastructure and opportunities for collaborative studies

As arrays grow, the number of species that can be profitably studied increases, the cost per unit of data declines, and opportunities for collaboration increase. The professional data-sharing network that grows with a large array is one of its most valuable products. The study of Fraser River sockeye demonstrates that many tagging studies could be improved by integrating laboratory experiments with manipulative field experiments, observations, other types of tagging, and non-lethal assays, and that telemetry has a powerful role to play in validating and confirming indirect methods of studying movement.

Finally, the POST approach opens doors to the study of forage fish like herring, anchovies, sardines, and eulachon: these are critically important mid-trophic level fish that feed nearly all of the ocean's larger fish, seabirds, and marine mammals, including most of the species that humans exploit. The ability to measure how forage fish react to environmental changes is a key requirement for predicting the impact of climate change on ocean ecosystems.

14.5 Summary

The value of the POST experience has been to show clearly that large-scale acoustic arrays are practical and enormously useful for illuminating the life histories of the host of species that spend most of their lives on the continental shelves. A defining characteristic of results from the POST array so far is that they have been surprising and unpredicted. That realization should be a warning that we may not know enough yet to generalize to new species and situations, and we have barely begun to deal with a key complication, which is that species interact. New technologies like POST bring opportunities to answer old questions, ask new ones, and enrich existing studies with new collaborations. We are confident that the POST infrastructure will continue to be a rich source of new discoveries for many years to come.

Acknowledgments

Core funding for the POST program has been provided by the Alfred P. Sloan Foundation and the Gordon and Betty Moore Foundation, with additional key funding from the US Department of Energy (Bonneville Power Administration), Pacific Salmon Commission, Pacific Salmon Foundation, and the Province of British Columbia (Ministry of Environment). Several people not represented in the author list made major contributions to the early development of POST, including the following: George Boehlert, Bruce Ward, George Jackson, Peggy Tsang, Jayson Semmens, Heather Holden, Jonathan Thar, and former POST board members Ron O'Dor and Paul Kariya. We gratefully acknowledge editorial advice from POST's Management Board, Scientific Management Committee, and staff, and the patient support of Jesse Ausubel, Aileen Lee, Michael Webster, and many other collaborators and colleagues throughout POST's early years.

References

Adams, P.B., Grimes, C., Hightower, J.E., *et al.* (2007) Population status of North American green sturgeon *Acipenser medirostris*. *Environmental Biology of Fishes* 79, 339–356.

Amstrup, S.C., McDonald, T.L. & Manly, B.F.J. (2005) *Handbook of Capture-Recapture Analysis*. Princeton, NJ: Princeton University Press.

Andrews, K.S., Levin, P.S., Katz, S.L., *et al.* (2007) Acoustic monitoring of sixgill shark movements in Puget Sound: evidence for localized movement. *Canadian Journal of Zoology* 85, 1136–1142.

Andrews, K.S., Williams, G.D., Farrer, D., *et al.* (2009) Diel activity patterns of sixgill sharks, *Hexanchus griseus*: the ups and downs of an apex predator. *Animal Behaviour* 78, 525–536.

Araki, H., Cooper, B. & Blouin, M.S. (2007) Genetic effects of captive breeding cause a rapid, cumulative fitness decline in the wild. *Science* 318, 100–103.

Barnett-Johnson, R., Pearson, T.E., Ramos, F.C., *et al.* (2008) Tracking natal origins of salmon using isotopes, otoliths, and landscape geology. *Limnology and Oceanography* 53, 1633–1642.

Beamish, R.J., Noakes, D.J., Macfarlane, G.A., *et al.* (1999) The regime concept and natural trends in the production of Pacific salmon. *Canadian Journal of Fisheries and Aquatic Sciences* 56, 516–526.

Beamish, R.J., Noakes, D.J., Mcfarlane, G.A., *et al.* (2000) Trends in coho marine survival in relation to the regime concept. *Fisheries Oceanography* 9, 114–119.

Beamish, R.J., Sweeting, R.M., Lange, K.L. & Neville, C.M. (2008) Changes in the population ecology of hatchery and wild coho salmon in the Strait of Georgia. *Transactions of the American Fisheries Society* 137, 503–520.

Botsford, L.W., Brumbaugh, D.R., Grimes, C., *et al.* (2009) Connectivity, sustainability, and yield: bridging the gap between conventional fisheries management and marine protected areas. *Reviews in Fish Biology and Fisheries* 19, 65–95.

Bradford, M.J. & Irvine, J.R. (2000) Land use, fishing, climate change, and the decline of Thompson River, British Columbia, coho salmon. *Canadian Journal of Fisheries and Aquatic Sciences* 57, 13–16.

Budy, P. & Schaller, H. (2007) Evaluating tributary restoration potential for Pacific salmon recovery. *Ecological Applications* 17, 1068–1086.

Budy, P., Thiede, G.P., Bouwes, N., *et al.* (2002) Evidence linking delayed mortality of Snake River salmon to their earlier hydrosystem experience. *North American Journal of Fisheries Management* 22, 35–51.

Chittenden, C.M., Sura, S., Butterworth, K.G., *et al.* (2008) Riverine, estuarine and marine migratory behaviour and physiology of wild and hatchery-reared coho salmon (*Oncorhynchus kisutch*) smolts descending the Campbell River, BC. *Journal of Fish Biology* 72, 614–628.

Chittenden, C.M., Butterworth, K.G., Cubitt, K.F., *et al.* (2009a) Maximum tag to body size ratios for an endangered coho salmon (*O. kisutch*) stock based on physiology and performance. *Environmental Biology of Fishes* 84, 129–140.

Chittenden, C.M., Beamish, R.J., Neville, C.M., *et al.* (2009b) The use of acoustic tags to determine the timing and location of the juvenile coho salmon migration out of the Strait of Georgia, Canada. *Transactions of the American Fisheries Society* 138, 1220–1225.

Chittenden, C.M., Melnychuk, M.C., Welch, D.W. & McKinley, R.S. (2010) An investigation into the poor survival of an endangered coho salmon population. *PLoS ONE* 5, e10869. doi:10.1371/journal.pone.0010869.

Cooke, S.J., Hinch, S.G., Crossin, G.T., *et al.* (2006) Physiology of individual late-run Fraser River sockeye salmon (*Oncorhynchus nerka*) sampled in the ocean correlates with fate during spawning migration. *Canadian Journal of Fisheries and Aquatic Sciences* 63, 1469–1480.

Cooke, S.J., Hinch, S.G., Farrell, A.P., *et al.* (2004) Abnormal migration timing and high en route mortality of sockeye salmon in the Fraser River, British Columbia. *Fisheries* 20, 22–33.

Cooke, S.J., Hinch, S.G., Farrell, A.P., *et al.* (2008) Interdisciplinary approaches to the study of the migration biology of telemetered fish. *Fisheries* 33, 321–338.

Coronado, C. & Hilborn, R.M. (1998) Spatial and temporal factors affecting survival in coho salmon (*Oncorhynchus kisutch*) in the Pacific Northwest. *Canadian Journal of Fisheries and Aquatic Sciences* 55, 2067–2077.

Crossin, G.T., Hinch, S.G., Cooke, S.J., *et al.* (2009a) Mechanisms influencing the timing and success of reproductive migration in a capital-breeding, semelparous fish species: the sockeye salmon. *Physiological and Biochemical Zoology.* **82**, 635–652.

Crossin, G.T., Hinch, S.G., Cooke, S.J., *et al.* (2008) Experimental effects of temperature on the behaviour, physiology and survival of river homing sockeye salmon. *Canadian Journal of Zoology* **86**, 127–140.

Crossin, G.T., Hinch, S.G., Welch, D.W., *et al.* (2009b) Physiological profiles of sockeye salmon in the Northeast Pacific Ocean and the effects of exogenous GnRH and testosterone on rates of homeward migration. *Marine Freshwater Behavior and Physiology* **42**, 89–108.

Culum, B. & Kevin, N.L. (2003) Social learning in fishes: a review. *Fish and Fisheries* **4**, 280–288.

Erickson, D.L. & Webb, M.A.H. (2007) Spawning periodicity, spawning migration, and size at maturity of green sturgeon, *Acipenser medirostris*, in the Rogue River, Oregon. *Environmental Biology of Fishes*, pp 255–268.

Fabry, V.J., Seibel, B.A., Feely, R.A. & Orr, J.C. (2008) Impacts of ocean acidification on marine fauna and ecosystem processes. *ICES Journal of Marine Science* **65**, 414–432.

Fortin, D., Beyer, H.L., Boyce, M.S., *et al.* (2005) Wolves influence elk movements: behavior shapes a trophic cascade in Yellowstone National Park. *Ecology* **86**, 1320–1330.

Gawarkiewicz, G., Monismith, S. & Largier, J. (2007) Observing larval transport processes affecting population connectivity: progress and challenges. *Oceanography* **20**, 40–53.

Goetz, F., Jeanes, E. & Morello, C. (2010) Puget Sound Steelhead Telemetry Study: 2006 Study Results. US Army Corps of Engineers, Seattle District, Washington.

Gresh, T., Lichatowich, J. & Schoonmaker, P. (2000) An estimation of historic and current levels of salmon production in the northeast Pacific ecosystem: evidence of a nutrient deficit in the freshwater systems of the Pacific Northwest. *Fisheries* **25**, 15–21.

Grothues, T.M. (2009) A review of acoustic telemetry technology and a perspective on its diversification relative to coastal tracking arrays. In: *Tagging and Tracking of Marine Animals with Electronic Devices* (ed. J.L.L. Nielsen), pp. 77–90. Springer Science+ Business Media B.V.

Habicht, C., Seeb, L.W. & Seeb, J.E. (2007) Genetic and ecological divergence defines population structure of sockeye salmon populations returning to Bristol Bay, Alaska, and provides a tool for admixture analysis. *Transactions of the American Fisheries Society* **136**, 82–94.

Hare, S.R., Mantua, N.J. & Francis, R.C. (1999) Inverse production regimes: Alaska and West Coast Pacific salmon. *Fisheries* **24**, 6–14.

Hinch, S.G. (2009) Overview and synthesis: early migration and premature mortality in Fraser River late-run sockeye salmon. In: *Conference on Early Migration and Premature Mortality in Fraser River Late-run Sockeye Salmon: Proceedings* (eds. S.G. Hinch & J. Gardner), pp. 8–14. Vancouver, BC: Pacific Fisheries Resource Conservation Council. Available at http://www.psc.org/infor_laterunsockeye.htm.

Hinch, S.G., Cooke, S.J., Healey, M.C. & Farrell, A.P. (2006) Behavioural physiology of fish migrations: salmon as a model approach. In: *Behaviour and Physiology of Fish* (eds. K.S. Balshine & R. Wilson), pp. 239–295. New York: Elsevier.

IPCC (2007) Climate Change 2007: synthesis report. Contribution of Working Groups I, II and III to the Fourth Assessment. In: Core Writing Team (eds. R.K. Pachauri & A. Reisinger). Geneva, Switzerland: Intergovernmental Panel on Climate Change. 104 pp.

Irvine, J.R. & Bradford, L.M. (2000) Declines in the abundance of Thompson River coho salmon in the interior of Southern British Columbia, and Canada's Coho Recovery Plan. In: *Proceedings of the Biology and Management of Species and Habitats at Risk Conference* (ed. L.M. Darling), pp. 595–598. Kamloops, BC: BC Ministry of Environment, Lands and Parks, Victoria, BC, and University College of the Cariboo, Kamloops, BC.

Iverson, S.J., Field, C., Bowen, W.D. & Blanchard, W. (2004) Quantitative fatty acid signature analysis: a new method of estimating predator diets. *Ecological Monographs* **74**, 211–235.

Johnson, J.H. (1960) Sonic tracking of adult salmon at Bonneville Dam, 1957. *Fishery Bulletin* **60**, 469–485.

Johnson, L.L., Collier, T.K. & Stein, J.E. (2002) An analysis in support of sediment quality thresholds for polycyclic aromatic hydrocarbons (PAHs) to protect estuarine fish. *Aquatic Conservation: Marine and Freshwater Ecosystems* **12**, 517–538.

Keiper, C.A., Ainley, D.G., Allen, S.G. & Harvey, J.T. (2005) Marine mammal occurrence and ocean climate off central California, 1986 to 1994 and 1997 to 1999. *Marine Ecology Progress Series* **289**, 285–306.

Kritzer, J.P. & Sale, P.F. (2006) *Marine Metapopulations* Amsterdam: Elsevier.

Lacroix, G.L., Knox, D. & McCurdy, P. (2004) Effects of dummy acoustic transmitters on juvenile Atlantic salmon. *Transactions of the American Fisheries Society* **133**, 211–220.

Lindley, S.T., Moser, M.L., Erickson, D.L., *et al.* (2008) Marine migration of North American green sturgeon. *Transactions of the American Fisheries Society* **137**, 182–194.

Mckinnell, S.M., Pella, J.J. & Dahlberg, M.L. (1997) Population-specific aggregations of steelhead trout (*Oncorhynchus mykiss*) in the North Pacific Ocean. *Canadian Journal of Fisheries and Aquatic Sciences* **54**, 2368–2376.

Melnychuk, M.C. (2009) Estimation of survival and detection probabilities for multiple tagged salmon stocks with nested migration routes, using a large-scale telemetry array. *Marine and Freshwater Research* **60**, 1231–1243.

Melnychuk, M.C., Hausch, S., Mccubbing, D.J.F. & Welch, D.W. (2009) Acoustic tracking of hatchery-reared and wild Cheakamus River steelhead smolts to address residualisation and early ocean survival. Vancouver, BC: Canadian National Railway Company, Monitor 2, Project F.

Miller, K., Li, S., Schulze, A., Raap, M., Ginther, N., Kaukinen, K. & Stenhouse, L. (2007) Late-run gene array research. *Final Report To The PSC on Research Conducted in 2006.* Vancouver, BC: Pacific Salmon Commission Southern Boundary Restoration And Enhancement Fund.

Moore, M.E., Berejikian, B.A. & Tezak, E.P. (2010) Early marine survival and behavior of steelhead trout (*Oncorhynchus mykiss*) smolts through Hood Canal and the Strait of Juan de Fuca. *Transactions of the American Fisheries Society* **139**, 49–61.

Moser, M.L., Myers, M.S., West, J., *et al.* (2010) English sole spawning migration and evidence for feeding site fidelity in Puget Sound, U.S.A. with implications for contaminant exposure. *Fisheries Management and Ecology* (in review).

Moyle, P.B. (2002) *Inland Fishes of California.* Berkeley, California: University of California Press.

Muir, W.D., Smith, S.G., Williams, J.G., *et al.* (2001) Survival estimates for migrant yearling chinook salmon and steelhead tagged with passive integrated transponders in the lower Snake and lower Columbia rivers, 1993–1998. *North American Journal of Fisheries Management* **21**, 269–282.

Nelson, T.C., Gazey, W.J. & English, K.K. (2008) Status of white sturgeon in the lower Fraser River: Report on the findings of the Lower Fraser River White Sturgeon Monitoring and Assessment Program 2007. Vancouver, BC: Fraser River Sturgeon Conservation Society.

Pauly, D. & Christensen, V. (1995) Primary production required to sustain global fisheries. *Nature* **374**, 255–257.

Payne, R., Brazier, O., Dorsey, E.M., *et al.* (1983) External features in southern right whales (*Eubalaena australis*) and their use in identifying individuals. In: *Communication and Behavior of Whales* (ed. R. Payne), pp. 371–445. Colorado: Westview Press.

Pearcy, W.G. (1992) *Ocean Ecology of North Pacific Salmonids.* Seattle, WA: University of Washington Press.

Porter, A.D., Welch, D.W., Rechisky, E.R., *et al.* (2009) Pacific Ocean Shelf Tracking Project (Post): Results from the Acoustic Tracking Study on Survival of Columbia River Salmon, 2008. Report to the Bonneville Power Administration by Kintama Research Corporation, Contract No. 2003-114-00, Grant No. 00021107. Available at http://www.efw.bpa.gov/Publications/XXXX.

Rechisky, E.L. & Welch, D.W. (2009) Surgical implantation of acoustic tags: influence of tag loss and tag-induced mortality on free-ranging and hatchery-held spring chinook (*O. tschawytscha*) smolts. In: *Tagging Telemetry and Marking Measures for Monitoring Fish Populations. A Compendium of New and Recent Science for Use in Informing Technique and Decision Modalities* (eds. K.S. Wolf & J.S. O'Neal), pp. 69–94. Duvall, WA: The Pacific Northwest Aquatic Monitoring Partnership and KWA Ecological Sciences.

Robichaud, D. & English, K.K. (2007) River entry timing, survival, and migration behaviour of Fraser River sockeye salmon in 2006. Final Report to the PSC on Research Conducted in 2006. Vancouver, BC: Pacific Salmon Commission Southern Boundary Restoration And Enhancement Fund.

Schaller, H., Wilson, P., Haeseker, S., *et al.* (2007) Comparative survival study (CSS) of PIT-tagged spring/summer chinook and steelhead in the Columbia River Basin: ten-year retrospective analyses report. Comparative Survival Study Oversight Committee and Fish Passage Center.

Schaller, H.A., Petrosky, C.E. & Langness, O.P. (1999) Contrasting patterns of productivity and survival rates for stream-type chinook salmon (*Oncorhynchus tshawytscha*) populations of the Snake and Columbia rivers. *Canadian Journal of Fisheries and Aquatic Sciences* **56**, 1031–1045.

Scott, W.B. & Crossman, E.J. (1973) Freshwater fishes of Canada. *Bulletin of the Fisheries Research Board of Canada* **184**, 1–966.

Seeb, L., Crane, P., Kondzela, C., *et al.* (2004) Migration f Pacific Rim chum salmon on the high seas: insights from genetic data. *Environmental Biology of Fishes* **69**, 21–36.

Stokesbury, M.J.W. (2010) Tracking of diadromous fishes at sea using hybrid acoustic and archival electronic tags. In: *Challenges for Diadromous Fishes in a Dynamic Global Environment* (ed. R. Cunjak). Bethesda, Maryland: American Fisheries Society (in press).

Timia, J.T. (2007) Parasites as biological tags for stock discrimination in marine fish from South American Atlantic waters. *Journal of Helminthology* **81**, 107–111.

Turchin, P. (1991) Translating foraging movements in heterogeneous environments into the spatial distribution of foragers. *Ecology* **72**, 1253–1266.

Waples, R.S., Pess, G.R. & Beechie, T. (2008) Evolutionary history of Pacific salmon in dynamic environments. *Evolutionary Applications* **1**, 189–206.

Weitkamp, L.A., Wainwright, T.C., Bryant, G.J., *et al.* (1995) Status review of coho salmon from Washington, Oregon and California.

Welch, D.W., Batten, S.D. & Ward, B. (2007) Growth, survival, and rates of tag retention for surgically implanted acoustic tags in steelhead trout (*O. mykiss*). *Hydrobiologia* **582**, 289–299.

Welch, D.W., Rechisky, E.L., Melnychuk, M.C., *et al.* (2008) Survival of migrating salmon smolts in large rivers with and without dams. *PLoS Biology* **6**, e265.

Welch, D.W., Ward, B.R. & Batten, S.D. (2004) Early ocean survival and marine movements of hatchery and wild steelhead trout (*Oncorhyncus mykiss*) determined by an acoustic array: Queen Charlotte Strait, British Columbia. *Deep-Sea Research II* **51**, 897–909.

Wetherall, J.A. (1982) Analysis of double-tagging experiments. *Fishery Bulletin* **80**, 687–701.

Williams, J.G. (2008) Mitigating the effects of high-head dams on the Columbia River, USA: experience from the trenches. *Hydrobiologia* **609**, 241–251.

Zale, A.V., Brooke, C. & Fraser, W.C. (2005) Effects of surgically implanted transmitter weights on growth and swimming stamina of small adult westslope cutthroat trout. *Transactions of The American Fisheries Society* **134**, 653–660.

Chapter 15

A View of the Ocean from Pacific Predators

Barbara A. Block[1], Daniel P. Costa[2], Steven J. Bograd[3]

[1]*Hopkins Marine Station, Department of Biology, Stanford University, Pacific Grove, California, USA*
[2]*Long Marine Laboratory, Department of Ecology and Evolutionary Biology, University of California, Santa Cruz, California, USA*
[3]*Environmental Research Division, National Oceanic and Atmospheric Administration, Southwest Fisheries Science Center, Pacific Grove, California, USA*

15.1 Large Pelagic Species in the Marine Ecosystems of the North Pacific

In the oceanic realm, species diversity, population structure, migrations, and gene flow in marine pelagic species are poorly understood. Large marine animals such as whales, pinnipeds, seabirds, turtles, sharks, and tunas have behaviors and life-history strategies that involve high dispersal movements and vagile populations. Most of these animals lack obvious geographic barriers in the marine environment. Obtaining baseline information about seasonal movement patterns, regional habitat use, and location of foraging and breeding grounds has remained elusive. The population status and general oceanic movements of many species and guilds were virtually unknown in the eastern Pacific basin when Tagging of Pacific Predators (TOPP) was initiated as a field program of the Census of Marine Life. Additionally, potential effects of global climate change on species diversity, biomass, and community structure in pelagic environments remain largely unexplored and difficult to assess or predict. These are alarming facts, given that many species face an unprecedented level of exploitation from both directed fisheries and bycatch, and ecosystem management is dependent upon a scientific understanding of habitat use.

The lack of knowledge about large marine species is primarily due to the challenges inherent in studying marine animals. The sheer size of the Pacific Ocean makes it challenging to use traditional techniques such as ship surveys. In an effort to rectify this lack of knowledge the Census initiated two Pacific tagging efforts, TOPP (see www.topp.org), and the Pacific Ocean Shelf Tracking Project (POST) (www.postcoml.org; see Chapter 14). These programs proposed to use electronic tracking technology to study the movements of marine organisms in the North Pacific. In the case of TOPP, biotelemetry or biologging (use of electronic tags carried by the animal to measure its biology) became the major ecological tool of choice for advancing knowledge of multi-species habitat use. We proposed combining electronic tag technology data with oceanographic data acquired through remote sensing and, when possible, with foraging ecology and genetics. We hypothesized that the primary physical forces that influence temporal distribution patterns were temperature and primary production. We expected that by studying multiple species within an ecosystem such as the California Current, we could reveal common locations for feeding, migration highways, potential reproduction regions, and hot spots. The principal factors acting on population dynamics might be revealed, and these would include seasonal variations in thermal regimes and the formation of prey-aggregating physical features, which were assumed to cause variation in food supply, physiological performance, and predation pressures. Understanding behavior of

Life in the World's Oceans, edited by Alasdair D. McIntyre

organisms relative to the physical and biological forces they experience was considered prerequisite to a study of multi-species community structure of large pelagic species in the North Pacific Ocean.

15.2 Tagging of Pacific Predators, 2000–2008

15.2.1 Planning and initiation of TOPP

TOPP scientists proposed a decade-long tagging campaign to elucidate the distribution, movements, behavior, and ecological niche of apex marine predators within the oceanic ecosystems of the North Pacific Ocean. To examine whether it was possible to observe and monitor multiple species in coastal and open-ocean habitats in the North Pacific, a series of planning workshops was initiated. Ninety participants from the USA, Canada, Mexico, France, and Japan met in Monterey, California, USA, for four days in November 2000 to discuss the experimental tagging approaches, selection of organisms, and together developed the foundation deployments of the TOPP program. This workshop led to a formal TOPP plan to use biologging techniques in concert with oceanographic data, feeding studies, and molecular techniques to examine 23 top predators of the North Pacific Ocean (Block *et al.* 2003). After the initial TOPP workshop, ten working groups were formed and met separately to plan the details of the program and experiments, and coordinate the ocean-scale tagging efforts (Box 15.1). The working groups were concentrated around seven organismal teams (tunas, sharks, squid, pinnipeds, seabirds, cetaceans, and turtles), as well as oceanography, tag development, data management, and education and outreach. As with any project of this size and complexity the work reported here could not have been done without the participation and dedication of many individuals to the TOPP program (Box 15.1).

The workshop participants recommended a staged implementation of the TOPP program, beginning with species selection. An emphasis was placed initially on the use of current technologies that had a track record of proven deployment success in large-scale experiments (for example archival/TDR tags), on species that would most likely yield quick data returns (elephant seals, Pacific bluefin tuna). Initiation of pilot studies to explore new and potentially challenging species (for example squid and swordfish), and to test large-scale deployments or new technologies and attachment strategies were also initiated (for example spot tags on shark dorsal fins, archival tags on shearwaters). After the successful completion of this testing and development phase (2003–2005), a second round of field efforts focused on increasing the scale of

the tag deployments throughout the program (2006–2009). Large-scale deployments (approximately 3,000 tags) of existing and new technologies across multiple taxa in a synoptic seasonal pattern occurred. Central to the TOPP plan was selection of approximately two dozen key target species that had distributions within the eastern and central North Pacific Ocean and were tractable for tagging and population studies. The implementation plan built on the known tagging success of key species during the early years, to inform the selection of TOPP target species. The tag data would be combined with *in situ* and remotely sensed oceanographic observations to build a decade-long survey of key species in the North Pacific Ocean. As data began to be collected, TOPP scientists focused on the improvement of electronic tagging technology, implementation of simultaneous large-scale tagging, development of a data acquisition system capable of handling multiple tagging platforms, and integration and visualization of multiple data streams and initiation of four-dimensional display.

Workshop deliberations and early tagging experiments generated a list of key animals that became the research subjects of TOPP (Box 15.2, p. 308). This included a diverse group of species with interesting ecological linkages. One key concept that emerged early in the program was the focus on tagging of guilds of closely related species. The TOPP species guilds include sharks of the family Lamnidae (white, mako, and salmon), fish of the *Thunnus* guild (yellowfin, bluefin, and albacore), albatrosses (black footed and Laysan), pinnipeds (elephant seals, California sea lions, and northern fur seals), rorqual whales (blue, fin, and humpback whales), and sea turtles (loggerheads and leatherbacks). The guild concept turned out to be among the most successful ideas in TOPP for exploration of the oceanic environment harnessing the evolutionary power for related species. By using this classic comparative approach, we were able to examine how organisms of similar phylogeny have diverged in their niche use and foraging ecology and the physiological mechanisms used (Feder 1987; Costa & Sinervo 2004). Additionally, we found that tagging techniques could easily be passed on to multiple species, improving the trophic connections and comparative methodologies (Block 2005). Several additional species that shared similar habitats complemented the original guild species. These species included blue and thresher sharks that overlapped with the lamnid guild, molas that were common in the tuna guild ecosystem, and shearwaters that overlapped with the albatross guild. The guild approach simplified the logistics of the tagging efforts. More than one species within a guild is often found in the same region at the same time of year, making it possible to maximize the success of the tagging effort by using a single platform to tag multiple species. Details of the types of tag used, and how tag technology progressed during TOPP, are given below.

Box 15.1

These individuals represent the members of the various organismal working groups of TOPP. Their work and perseverance in field tagging efforts and dedication to the TOPP program made this ten-year effort possible.

Cetaceans

Bruce Mate[a], Don Croll[b], John Calambokidus[c], Nick Gales[d], Jim Harvey[e], and Susan Chivers[f]

Seabirds

Scott Shaffer[b], Yann Tremblay[b], Bill Henry[b], Henri Weimer-skirch[f], Dave Anderson[g], Jill Awkerman[g], Dan Costa[b], Jim Harvey[e], and Don Croll[b]

Pinnipeds

Dan Costa[b], Dan Crocker[h], Burney Le Boeuf[b], Juan Pablo-Gallo[i], Jim Harvey[e], Mike Weise[b], Mike Fedak[j], Carey Kuhn[b], Sam Simmons[b], Patrick Robinson[b], Jason Hassrick[b], Jeremy Sterling[k], Sharon Melin[k], and Bob DeLong[k]

Sea turtles

Steven Bograd[f], Scott Bensen[f], Peter Dutton[f], Scott Eckert[l], Steve Morreale[m], Frank Paladino[n], Jeff Polovina[f], Laura Sarti[i], James Spotila[o], and George Shillinger[r]

Sharks

Scot Anderson[p], Dave Holts[f], Oscar Sosa-Nishizaki[q], Barbara Block[r], Aaron Carlise[r], Peter Pyle[s], Sal Jorgenson[r], Ken Goldman[t], Peter Klimley[u], Chris Perle[r] John O'Sullivan[v], Kevin Weng[r], Suzanne Kohin[f], Heidi Dewar[f], and Sean Van Sommeran[r]

Tuna

Barbara Block[r], Kurt Schaefer[w], John Childress[x], Heidi Dewar[f], Suzanne Kohin[f], Kevin Weng[r], Steve Teo[r], Chuck Farwell[r], Dan Fuller[w], Chris Perle[r], Dan Madigan[r], Luis Rodriguez[r], Jake Noguiera[r], and R. Schallert[r]

Squid

William Gilly[r]

Institutions

[a]Oregon State University
[b]University of California at Santa Cruz
[c]Cascadia Research Collective
[d]Australian Antarctic Division
[e]Moss Landing Marine Lab
[f]Southwest Fisheries Science Center, NOAA-NMFS
[g]Wake Forest University
[h]Sonoma State University
[i]Centro de Investigación en Alimentación y Desarrollo, Unidad Guaymas
[j]Sea Mammal Research Unit
[k]National Marine Mammal Lab, NOAA-NMFS
[l]Duke University
[m]Cornell University
[n]Indiana University
[o]Drexel University
[p]Point Reyes Bird Observatory
[q]CICESE
[r]Stanford University
[s]Pelagic Shark Foundation
[t]Alaska Fish & Game
[u]University of California at Davis
[v]Monterey Bay Aquarium
[w]Inter-American Tropical Tuna Commission
[x]University of California at Santa Barbara

15.2.2 Developing tag technologies for marine biologging

Biologging technology allows researchers to take measurements from free-swimming marine animals as they move undisturbed through their environment. Importantly, it permits researchers to observe animals beyond the natural reach of humans, and provides extensive data on both the animals' behavior and their physical environment. Biologging can be used for observational studies of animal behavior, controlled experimental studies (translocation experiments), oceanographic observations of the *in situ* environment surrounding the animal, and ecological research such as foraging dynamics.

An early challenge for TOPP was to solve the many technological issues associated with the tag platforms before launching the major deployment phase. For the first three years of TOPP deployments (2001–2004), the priority was to develop efficient deployment strategies of proven technologies, while simultaneously investing in and testing new tags. TOPP worked with four manufacturers of commercial tags and advised significant engineering decisions, which over the course of a decade led to advancements in all major tag types used. Testing included a development phase with new ARGOS transmitters, Fastloc GPS, novel sensors with oceanographic capabilities, increased memory, miniaturization, new track filtering, and data compression algorithms (Teo *et al.* 2004a, 2009; Kuhn & Costa 2006; Tremblay *et al.* 2006, 2007, 2009; Biuw *et al.* 2007; Bailey *et al.* 2008; Kuhn *et al.* 2009; Costa *et al.* 2010a; Simmons *et al.* 2009).

Stable and prototype electronic tags were both available in 2000. They ranged from ARGOS tags to data storage tags (archival) that could be attached or implanted surgically, and pop-up satellite archival tags designed to track large-scale movements and behavior of animals for which the use of real-time ARGOS satellite tags was not possible. The pop-up satellite tags jettisoned from the animal on a pre-programmed date and transmitted data about depth, ambient temperature, pressure, and light to ARGOS satellites (Block *et al.* 1998). Each generation of the tags had technological issues that included the accuracy of sensors, the resilience of tag components, pressure housing problems, and algorithm limitations. In the first phase of TOPP we tested many of the tags on free-ranging animals to ensure that they would be sufficiently robust for the larger-scale deployments in the second phase.

Tag attachment strategies were another early challenge. Strategies were shared among organismal working groups, and the increased communication with multiple taxa specialists helped to facilitate the exchange of ideas and techniques. In addition, the increased memory capacity of the new tags, and the scale of the TOPP deployments, resulted in the need for investment in a novel data management system, improved analytical and visualization tools, and

methods to synthesize a vast array of disparate data streams (Teo *et al.* 2004b; Tremblay *et al.* 2006, 2007, 2009; Bailey *et al.* 2008). Simply put, TOPP is the first large-scale tagging program to implement automation of the ARGOS and geolocation tag management, metadata delivery, and integration of online data display in near-real-time with oceanographic data information. A public display in a browser format (las.pfeg.noaa.gov/TOPP) was developed for the TOPP team members and education and outreach.

Four broad classes of electronic tags were tested, refined and/or developed, and deployed as part of the TOPP program.

15.2.2.1 Archival tags

Archival tags log high-resolution time-series data from sensors that measure pressure, water temperature, body temperature, salinity, and light level (Biuw *et al.* 2007; Simmons *et al.* 2009; Teo *et al.* 2009; Costa *et al.* 2010b). Archival tags are either implanted in the animal (fish), glued to the hair or feathers (birds and mammals), or attached to a leg band (birds). The major limitation of this tag type is that it must be recovered to obtain the data. In fish this tag can only be used on highly exploited species (tunas) where an enhanced reward is offered to recover the tag after capture. Archival tags are also used with species that have a high probability of returning to the colony where they were initially tagged (that is, sea lion, elephant seal, albatross, and shearwater). Archival tags have provided tracks covering up to 1,000 days in northern bluefin tuna (Block *et al.* 2001; Block 2005) and 1,160 days in yellowfin tuna (Schaefer *et al.* 2007). These tags were a mainstay of the TOPP program, with over approximately 1,600 light, temperature, depth (LTD) (Lotek) 2310 tags deployed, primarily on bluefin (Kitagawa *et al.* 2007; Boustany *et al.* 2010), yellowfin (Schaefer *et al.* 2007, 2009) and albacore tunas. This archival tag (D-series) has a pressure accuracy of ±1% of the full-scale reading (up to 2,000 m). The LTD temperature ranged from −5 to 40 °C, with an accuracy of 0.05 °C. The temperature response is less than 2 seconds, so that as an animal dives, it provides a temperature profile with depth (Simmons *et al.* 2009). In the TOPP program, archival tags were used primarily on tunas that were juveniles (3–20 kg) and adolescents (up to 55 kg). The smallest animal tagged was the sooty shearwater (Shaffer *et al.* 2006), weighing approximately 800 g. For this, we used the smallest archival tag the Lotek LTD 2410 that weighs only 5.5 g and is 11 mm in diameter and 35 mm long.

Archival tags often carry sensitive optical detectors that can measure variations in light level quite accurately. The Lotek 2310 tag used by TOPP has a polystyrene light stalk that extends externally from an animal after it has been surgically implanted. The wall of the stalk contains a fluorescent dye that is sensitive to narrow band blue light

(470 nm) passing through the side wall of this optical fiber. When excited, it radiates light in the green wavelengths and focuses the light down the base of the fiber where it is detected by a photodiode. The photodiode is used to convert the flow of photons into a flow of electrons that is measured with an electrometer with about 9 decades of range, allowing detection of sunrise and sunset while a tuna is cruising at depths. On board or post-processing algorithms allow construction of light-level curves and the calculation of local apparent noon to determine longitude. Estimates of time of sunrise and sunset are used to determine day length and latitude using threshold techniques (Hill & Braun 2001; Ekstrom 2004). These locations can be improved by using archival tag observed measurements of sea surface temperature, and TOPP research with double tag datasets provided a robust improvement for these algorithms (Teo et al. 2004a). Archival tags have also been used to estimate in situ chlorophyll concentrations (Teo et al. 2009).

15.2.2.2   ARGOS satellite tags

Satellite tags provide at-sea locations and have the advantage that the data can be recovered in real time and remotely without the need to recover the tag. The satellites are operated by CLS-ARGOS (Toulouse, France, or Landover, Maryland, USA) and the data are acquired from this service over the Internet. Because the antenna on the satellite transmitter must be out of the water to communicate with an orbiting ARGOS satellite, the technology has mainly been used on air-breathing vertebrates that surface regularly (McConnell et al. 1992a, b; Le Boeuf et al. 2000; Polovina et al. 2000; Weimerskirch et al. 2000; Shaffer & Costa 2006). A significant innovation of the TOPP program was the realization that satellite tags could be effectively deployed on sharks (Weng et al. 2005; Jorgensen et al. 2010). For large fish, sharks, or other animals that remain continuously submerged, the ability to transmit to ARGOS at the surface is not possible. For these organisms, a pop-up satellite archival tag (PSAT) was developed (Block et al. 1998, 2001; Lutcavage et al. 1999; Boustany et al. 2002). Pop-up satellite archival tags combine data-storage tags with satellite transmitters. These tags are externally attached and are programmed to release or pop off at a preprogrammed time. The pop-up satellite devices communicate with the ARGOS satellites that serve both to uplink data and calculate an end-point location. Importantly, the tags are fisheries-independent in that they do not require recapture of the fish for data acquisition. The Mk10Pat model, the most used in TOPP, has a pressure sensor with resolution of about 0.5 dBar. The temperature accuracy improved as an external thermistor was tested and used to improve acquisition of oceanographic quality data (temperature accuracy is ±0.1°C with 0.05°C accuracy from about −5 to 40°C) (Simmons et al. 2009). The temperature response

is less than a second over a 70% step in this range. Further advances in the form of data compression have made it possible to get significantly more data through the limitations of the ARGOS system, including detailed oceanographic and behavioral information (Fedak et al. 2001). ARGOS satellite tags are larger than archival tags, with the smallest unit weighing 30 g.

15.2.2.3   GPS tags

With funds from the National Oceanographic Partnership Program (NOPP), TOPP supported the development of a GPS tag for use on marine species (Decker & Reed 2009; Costa et al. 2010b). The advantage of GPS tags is twofold. First, GPS tags provide an increase in the precision of animal movement data to within 10 m compared with the 1–10 km possible with ARGOS satellite tags. Second, GPS tags provide a higher time resolution, with positions acquired every few minutes compared with a maximum of about eight to ten positions a day with ARGOS. However, standard navigational GPS units do not work with diving species, as they require many seconds or even minutes of exposure to GPS satellites to calculate positions and the onboard processing consumes considerable power. At the beginning of the TOPP project, two conceptual solutions to this problem were identified and resources invested in the development and testing of the Fastloc system developed by Wildtrack Telemetry Systems (Leeds, UK). The Fastloc uses a novel intermediate solution that couples brief satellite reception with limited onboard processing to reduce the memory required to store or transmit the location. This system captures the GPS satellite signals and identifies the observed satellites, calculates their pseudo-ranges without the ephemeris or satellite almanac, and produces a location estimate that can be transmitted by ARGOS. Final locations are post-processed from the pseudo-ranges after the data are received using archived GPS constellation orbitography data accessed through the Internet. Although this technology provides a major advance in our ability to monitor the movements and habitat use of marine animals (Fig. 15.1), TOPP researchers have also used these tags to validate the error associated with ARGOS satellite locations (Costa et al. 2010a).

15.2.2.4   Conductivity, temperature, and depth tags

A fundamental component of the TOPP program was a desire to further the development of using animals to collect oceanographic data. This served two functions: the first was to acquire physical oceanographic data that were not otherwise available, and the second was to collect physical environmental data at a scale and resolution that matched the animal's behavior. As with the GPS tag, funds from the NOPP program allowed TOPP to support the development and testing of a reliable and commercially

Fig. 15.1

Basin-scale Pacific map showing all TOPP tracks color-coded by species. **(A)** shows salmon sharks on the top layer, **(B)** shows elephant seals on the top layer.

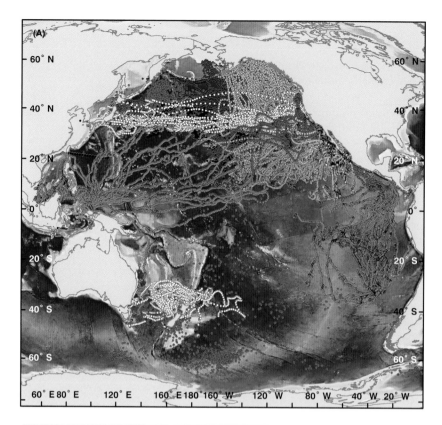

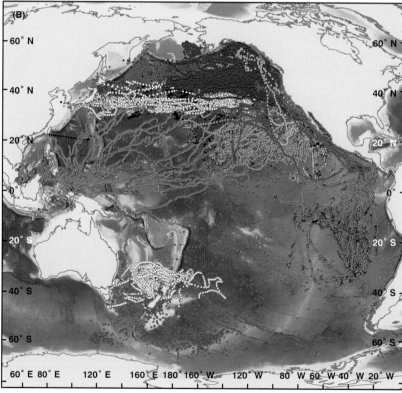

- Humpback whale
- Fin whale
- Sperm whale
- Sooty shearwater
- California sea lion
- Northern fur seal
- Blue whale
- Northern elephant seal

- Thresher shark
- Yellowfin tuna
- Albacore tuna
- Blue shark
- Mako shark
- White shark
- Loggerhead turtle
- *Mola mola*

- Pacific bluefin tuna
- Leatherback turtle
- Salmon shark
- Laysan albatross
- Black-footed albatross
- Humboldt squid

available conductivity, temperature, and depth (CTD) tag for animals, in collaboration with the Sea Mammal Research Unit at St. Andrew's University, UK (SMRU; www.smru.st-andrews.ac.uk). A conductivity–temperature–depth satellite relay data logger (CTD-SRDL) incorporates a Valeport CTD sensor with pressure accuracy ±5 dBar, temperature resolution of ±0.001 °C, with an accuracy of 0.01 °C and an inductive coil for measuring conductivity with resolution of ±0.003 mS cm^{-1}. The tag is optimized to collect oceanographic data at the descent and ascent speeds exhibited by seals (approximately 1 m s^{-1}). In addition to collecting data on an animal's location and diving behavior, it collects CTD profiles (Fig. 15.2). The tag looks for the deepest dive for a 1- or 2-hour interval. Every time a deeper dive is detected for that interval, the tag begins rapidly sampling (2 Hz) temperature, conductivity, and depth from the bottom of the dive to the surface. These high-resolution data are then summarized into a set of 20 depth points with corresponding temperatures and conductivities. These 20 depth points include 10 predefined depths and 10 inflection points chosen by a "broken stick" selection algorithm (Fedak *et al.* 2002). These data are then held in a buffer for transmission by ARGOS. Given the limitations of the ARGOS system, all records cannot be transmitted; therefore a pseudo-random method is used to transmit an unbiased sample of stored records. If the SRDLs are recovered, all data collected for transmission, whether or not it was successfully relayed, can be recovered. The use of these tags has led to insights into both the animal's behavior relative to its environment (Biuw *et al.* 2007) as well as the physical oceanography (Boehme *et al.* 2008a, b; Charrassin *et al.* 2008; Nicholls *et al.* 2008; Costa *et al.* 2010b).

15.2.3 Top predator distributions and the discovery of basin-scale migrations

One of the principal results of the TOPP program was a new understanding of the distribution and migration patterns of a suite of apex marine predators. Over the course of the TOPP program, 4,306 animals, representing 23 species (Box 15.2), were equipped with a variety of sophisticated tags carrying high-resolution sensors (Figs. 15.1A and B). The predators recorded data on their position, the ocean environment, habitat use, and behaviors while traveling remarkable distances underwater. The dataset yielded many surprises and demonstrated for the first time seasonal patterns and fidelity to the eastern Pacific for many species tagged in TOPP and unlimited boundaries when roaming over vast reaches of the Pacific Ocean for others (Figs. 15.3, 15.4, and 15.5). For example, white sharks show a coastal to offshore migration from California nearshore waters to offshore waters of Hawaii and back, resulting in homing behavior (Boustany *et al.* 2002; Weng *et al.* 2007a; Jorgensen *et al.* 2010). Salmon sharks move from the Arctic to the sub-tropical reaches of the North Pacific Ocean and back to the foraging grounds in Prince William Sound (Weng *et al.* 2005), whereas bluefin tuna and loggerhead turtles range across the North Pacific, breeding in the western Pacific but migrating as juveniles and adolescents to the eastern Pacific to take advantage of the highly productive California Current (Peckham *et al.* 2007; Boustany *et al.* 2009). Leatherback turtles tagged on their nesting beaches in Indonesia cross the Pacific basin to feed off central California, whereas sooty shearwaters use the entire

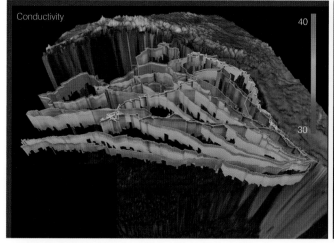 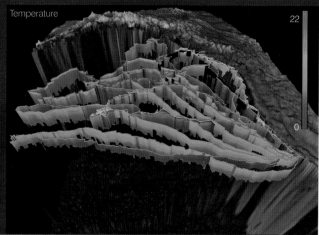

Fig. 15.2

Example of North Pacific conductivity (salinity) and temperature (°C) profiles derived from CTD tags deployed on seven elephant seals. Each seal is represented by a different color line on the top or surface of each track. The curtain effect represents the integrated temperature or conductivity profile.

Pacific Ocean from the Antarctic to the Bering Sea (Shaffer *et al.* 2006). These trans-oceanic journeys require remarkable animal navigation, energetics, and philopatry. The TOPP species that best illustrate these trans-oceanic migrations are the Pacific bluefin tuna, the sooty shearwater, and the leatherback turtle.

15.2.3.1 Pacific bluefin tuna

Pacific bluefin tuna are one of three species of bluefin tuna that inhabit subtropical to subpolar seas throughout the world's oceans. Among *Thunnus*, Pacific bluefin tuna have the largest individual home range, being found throughout the North Pacific Ocean and ranging into the western South Pacific (Collette & Nauen 1983). Pacific bluefin tuna remain in the western Pacific after being spawned (near the Sea of Japan) but a proportion of the juveniles make extensive migrations into the eastern Pacific late in the first or second year (Bayliff 1994; Inagake *et al.* 2001). It has been hypothesized that the trans-Pacific migrations from west to east are linked to local sardine abundances off Japan (Polovina 1996). TOPP researchers deployed over 600

archival tags on juvenile Pacific bluefin in the eastern Pacific. Recovery rates ranged from 50% to 75%, indicative of high mortality on these juveniles. The bluefin tagged displayed a cyclical pattern of movements on a seasonal scale that ranged annually from the southern tip of Baja California to the coast of Oregon (Boustany *et al.* 2010). This seasonal signal was apparent and provided the first clear signal that the California Current has a seasonality that many TOPP species (bluefin, yellowfin, and albacore tunas, blue whales, lamnid sharks, and shearwaters) were following. Approximately 5% of the recovered tagged bluefin tuna migrated back into the western Pacific using the North Pacific Transition Zone (Fig. 15.3). Large adult Pacific bluefin tuna were also tagged in the south Pacific off New Zealand during TOPP (Fig. 15.1). The emerging story is of a Pacific bluefin that travels across the entire North Pacific Ocean as a juvenile and adolescent, and that is capable of post-spawning migrations from the North Pacific to the South Pacific. Taken together, the electronic tag data on adolescents and adults demonstrate this tuna species encompasses one of the largest home ranges on the planet.

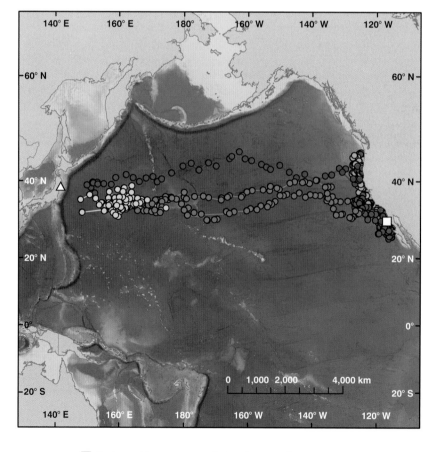

Fig. 15.3
Multiple trans-Pacific migrations of a single 15 kg Pacific bluefin tuna that crossed the North Pacific three times in 600 days. Positions are from a Lotek 2310 tag with threshold geolocation for longitude and sea surface temperature enhanced positions for latitude.

☐ Deploy position   ● Jan.–Feb.   ● Jul.–Aug.
△ Recovery position   ● Mar.–Apr.   ◐ Sep.–Oct.
                      ○ May–Jun.   ● Nov.–Dec.

15.2.3.2 Sooty shearwaters

Equally impressive was the TOPP observation with archival tags that sooty shearwaters that breed off the Southern Islands of New Zealand cross the equator to feed in diverse regions of the North Pacific Ocean, from Japan to Alaska and California (Shaffer *et al.* 2006, 2009). We were able to document this trans-equatorial migration using newly developed miniature archival tags that log data for estimating position, dive depth, and ambient temperature. The Pacific Ocean migration cycle had a figure-eight pattern (Fig. 15.4) while traveling, on average, 64,037 ± 9,779 km round trip over 198 ± 17 days. This is the longest migration of any individual animal ever tracked. Shearwaters foraged in one of three discrete regions off Japan, Alaska, or California before returning to New Zealand through a relatively narrow corridor in the central Pacific Ocean. These migrations allow shearwaters to take advantage of prey resources in both the Northern and Southern hemispheres during the most productive periods; in other words, they exist in an "endless summer."

15.2.3.3 Leatherback turtles

Two distinct populations of leatherback turtles were tagged during TOPP: one that breeds on the Indonesian islands of the western Pacific and another that breeds on the beaches of Costa Rica. A portion of the tagged western Pacific leatherbacks undergo remarkable trans-Pacific migrations, arriving off the coast of California in late summer to forage on dense aggregations of jellyfish. The eastern Pacific popu-lation, in contrast, undergoes a cross-equatorial migration that takes them into the oligotrophic waters of the eastern subtropical South Pacific (Shillinger *et al.* 2008). Nearly all of the tagged eastern Pacific leatherbacks made this journey, traversing a relatively narrow migration corridor through the highly dynamic equatorial Pacific. The apparently different migration and foraging strategies of these two populations of Pacific leatherbacks was another surprise result of TOPP.

15.2.4 Identification of biological hot spots, niche separation, homing, and fidelity

In addition to identifying trans-oceanic migration routes, another fundamental result of the TOPP program was the identification of critical foraging habitat, migration corridors, and regions of high occupancy: that is, biological hot spots. The identification and characterization of biological hot spots is a new focus in marine ecological research, with important implications for understanding ecosystem functions, prioritizing habitat for marine zoning, and managing targeted fisheries populations (Worm *et al.* 2003; Sydeman *et al.* 2006). Although it is well known that top predators, including large predatory fishes, cetaceans, pinnipeds, sea turtles, marine birds, and human fishers, congregate at locations of enhanced prey (Olson *et al.* 1994; Polovina *et al.* 2000), little is known about why particular regions or physical features are hot spots, how the animals locate and

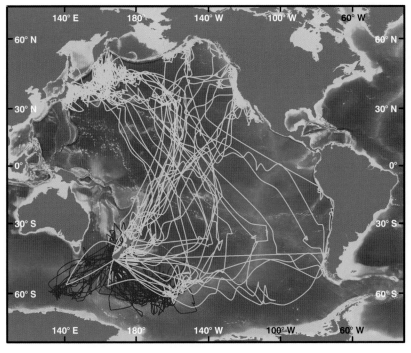

Fig. 15.4
The migrations of sooty shearwaters. From Shaffer *et al.* 2006.

Sooty shearwater ——— Migration tracks ▬▬▬ Breeding tracks

use these features, and what trophic interactions occur within the hot spots. The inaccessibility of open ocean hot spots has precluded systematic field studies that could elucidate the physical characteristics and ecological function of biological hot spots. The TOPP dataset has provided a unique view of North Pacific top predator hot spots, allowing us to address several compelling questions. What conditions occur within hot spots to make these areas of confluence for a diverse set of species? What are the physical characteristics and spatial-temporal persistence of the hot spots? What are the behavioral responses of different species when they enter a hot spot? What trophic interactions occur within hot spots? How can a better understanding of critical habitat result in enhanced conservation and management of these species?

Efforts at studying biological hot spots have relied on two complementary approaches. From a "bottom-up" perspective, standard oceanographic sampling (particularly remote sensing) can be used to describe ocean features of relevance to apex pelagic predators. This approach has several advantages: (1) it can provide global, near-real-time views of the ocean surface; (2) it can identify regions of enhanced biological activity (from ocean color); (3) it can identify features that are persistent or recurrent in space and time, which may be appealing areas for migratory animals; and (4) it can describe the dominant scales of ocean variability, from which hot spots can be classified (that is, based on their spatial-temporal scale, their degree of persistence or recurrence, their forcing mechanisms, and their potential biological impacts) (Palacios et al. 2006).

A more direct "top-down" approach is to let the animals identify and describe hot spots through large-scale tagging studies such as TOPP (Costa 1993; Block et al. 2001, 2003; Block 2005; Costa et al. 2010b). This approach allows (1) a direct detection and observation of the preferred habitats of the animals; (2) a matching of behavioral cues to local oceanographic conditions; (3) a differentiation of behavioral responses in relation to ocean features, allowing a classification of hot spots by their ecological function (for example, aggregation, migration, foraging, diving, and breeding); and (4) a differentiation between species-specific hot spots and hot spots that are biologically diverse and more universally occupied. Understanding the oceanographic, ecological, and physiological factors that are important in attracting apex marine predators to these hot spots, and determining how and why they remain retentive to these species, is a continuing research aim of TOPP and will likely be one of its primary legacies.

Two oceanographic regions that have emerged as key top predator hot spots in the North Pacific Ocean are the California Current and the North Pacific Transition Zone, which are described in detail below.

The California Current ecosystem (CCS) has emerged as a major hot spot for more than a dozen exploited, protected, threatened, or endangered predators that range throughout the eastern Pacific, including leatherback and loggerhead turtles, sooty shearwaters, Laysan and black-footed albatrosses, blue and humpback whales, sea lions, northern fur seals, elephant seals, bluefin, yellowfin and albacore tunas, white, makos, salmon, blue, and thresher sharks. The CCS is a highly productive eastern boundary current system, driven by seasonal coastal upwelling, which maintains numerous economically important marine fisheries. This region is characterized by strong cross-shore gradients in physical and biological fields, is strongly modulated by seasonal wind forcing and Ekman dynamics, and has a complex and energetic current structure (Hickey 1998; Palacios et al. 2006). These dynamics contribute to the aggregation of prey and, hence, top predators. Within the California Current, several areas have emerged as critical multi-species hot spots. These include areas we have designated as the California Marine Sanctuaries Hot Spot (CMHS), the Southern California Bight Hot Spot (SCBHS), and the Baja California Hot Spot (BCHS).

TOPP data have shown that the North Pacific Transition Zone (NPTZ) is equivalent to a major top predator highway across the North Pacific Ocean. Bluefin and albacore tunas, albatrosses, shearwaters, elephant seals, fur seals, and turtles all occur with great frequency in this region. The NPTZ is a complex region encompassing an abrupt north to south transition from subarctic to subtropical water masses, has an abundance of energetic mesoscale features (fronts and eddies), and is dominated by biological patchiness (Roden 1991). Interactions between eddies and the mean flow create transient jets and submesoscale vortices, resulting in up- and downwelling patches a few kilometers across that contribute to this biological patchiness (Woods 1988; Roden 1991). This large basin-scale frontal feature serves as a primary foraging area and migratory corridor for most of the TOPP predators that undergo trans-oceanic movements.

The CCS and NPTZ hot spots support several important ecological functions for North Pacific top predators. TOPP data has also revealed distinct niche separation among TOPP guilds that use these regions (Figs. 15.1, 15.3, and 15.7). At the initiation of the Census we knew that there were differences in the thermal tolerances of marine predators. As a result of the TOPP program, we now know that the physiological tolerance of marine predators, their energetics, and physiological constraints on the cardiac system (Weng et al. 2005) determine the home range of many of the TOPP species. For example, endothermy enables birds, mammals, salmon sharks, and bluefin tuna to range over vast regions of the North Pacific Ocean, including the colder temperate and subarctic oceans. In contrast, species such as yellowfin tuna, mako, and blue sharks are more constrained in their thermal tolerances to the warm temperate to subtropical waters of the North Pacific Ocean. The clear separation in habitat between members of fish or shark guilds occurs primarily by thermal preferences, with sister taxa (for example yellowfin and bluefin) occupying warm to cold environs on a latitudinal scale. Secondly, diet has emerged as a major factor in niche separation. For example, adult white sharks show a marine mammal preference and hence are localized for a portion of the year near pinniped colonies close to shore, whereas mako sharks primarily eat fish and are more frequently associated with the continental shelf and slope areas (Pyle *et al.* 1996; Anderson *et al.* 2008; Goldman & Musick 2008; Jorgensen *et al.* 2010). Advances in electronic tagging in TOPP, combined with genetic research, have made it possible to discern the importance of philopatry for population structure of pelagic marine predators once thought to be panmictic (Jorgensen *et al.* 2010). These results are illustrated below for several of the TOPP guilds.

15.2.4.1 Tuna

TOPP has recorded more than 92,000 days of archival tag data for Pacific, yellowfin, and albacore tunas occupying the California Current (Schaefer *et al.* 2007, 2008; Boustany *et al.* 2010). Pacific bluefin are spawned in the waters off Japan and recruit to the eastern Pacific late in year one or as two-year-olds. TOPP tagging was conducted in the eastern Pacific on individuals from 2 to 5 years of age, and most fish remained in the eastern Pacific for several years (Kitagawa *et al.* 2007; Boustany *et al.* 2010). The principal pan-tuna hot spot was the SCBHS, which has ample bathymetric forcing and seasonal upwelling that may attract the prey these species feed upon. Furthermore, the region is within the thermal preferences of all three species for a large part of the year. Tunas show latitudinal movement patterns that were correlated with peaks in coastal upwelling-induced primary productivity (Boustany *et al.* 2009). Habitat use distributions derived from kernel density analyses of daily geolocations indicate that all three CCS hot spot regions are occupied by one or all three *Thunnus* species in a predictable seasonal pattern. Each species showed distinct thermal preferences that lead to spatial and temporal separation of the species. Bluefin and albacore showed distinct geographic areas of niche overlap, but diel distinctions in their vertical movements. Pacific bluefin fed in the mixed layer, consistent with a forage preference for sardines and anchovy, whereas albacore tended to occupy the deep scattering layer. Interannual variation in the locality of the productivity peaks can be linked to movement patterns of the tunas (Boustany *et al.* 2010).

15.2.4.2 Sharks

Perhaps no marine predators have been more challenging to study than the large predacious sharks of the lamnid guild. This includes closely related species such as the white, mako, and salmon shark. The lamnids, along with blue and thresher sharks, have been heavily studied in TOPP (Weng *et al.* 2005, 2007b, 2008), with more than 50,000 days of

tracking data obtained. Among the lamnid sharks there is a remarkable niche diversification, with salmon sharks occupying the most expansive regions ranging from northern cold-temperate to subpolar habitats and occupying year round temperatures below 6°C. They frequent Prince William Sound, the Alaskan Gyre, the North Pacific Subtropical Gyre, and the CCS. White sharks move from the CCS to the oligotrophic regions of the central Pacific and back. Mako sharks use the CCS throughout the North American continental shelf, overlapping with the blue and thresher sharks. TOPP data have shown that the three lamnid sharks have very little overlap in niche use. In fact, there appears to be diversification of habitat use for most of the year, with each shark group being specialized for foraging on specific prey items. Adult white sharks use marine mammals, salmon sharks eat pollock and herring, and makos and blues appear to specialize on sardines and squid (Pyle *et al.* 1996; Anderson *et al.* 2008; Goldman and Musick 2008; Lopez *et al.* 2009). The tagging has provided an oceanic view of the habitat selection and niche use of closely related groups of sharks. Again, the SCBHS emerges as an important foraging region and nursery ground for blue, mako, thresher, and juvenile white sharks.

White sharks are a cosmopolitan species that occur circumglobally. In TOPP we combined satellite tagging, passive acoustic monitoring, and genetics to reveal how eastern Pacific white sharks adhere to a highly predictable migratory cycle (Fig. 15.5) (Jorgensen *et al.* 2010). Pop-up satellite tagging revealed how individual sharks return to the same network of coastal hot spots following distant oceanic migrations, and comprise a population genetically distinct from previously identified phylogenetic clades. The homing behavior has been hypothesized to have led to the separation of the (TOPP-tagged) northeastern Pacific population after an introduction from Australia/New Zealand migrants during the late Pleistocene. Mitochondrial DNA collected from samples obtained during tagging provided independent evidence for a demographically independent management unit not previously recognized. This fidelity to discrete and predictable migration pathways and locations offers clear population assessment, monitoring, and management options vital for ensuring the white sharks' protection.

Satellite tags directly attached to the dorsal fin of salmon sharks have provided long-term datasets (as long as four consecutive years) on the movements and environmental preferences of female salmon sharks in the eastern Pacific (Weng *et al.* 2005, 2007b). This research has shown that there are repeated annual patterns of migration in the eastern North Pacific of the population being tagged in Alaska, and incredible fidelity to Prince William Sound (the location of deployments). Salmon sharks are believed to use the southern extent of their range as nursery areas; in particular, regions around the NPTZ (Nakano & Nagasawa 1996) and potentially even the CCS (Goldman & Musick

Fig. 15.5
Site fidelity of white sharks tagged along the central California coast during 2000–2007 revealed by PAT records. **(A, B, C, D, E, and F)** Fidelity is demonstrated by six individual tracks (yellow lines; based on five-point moving average of geolocations). Triangles indicate tag deployment locations and red circles indicate satellite tag popup endpoints (ARGOS transmissions) for white sharks returning back to central California shelf waters after offshore migrations. **(G)** Site fidelity of all satellite tagged white sharks ($n = 68$) to three core areas in the Northeast Pacific including the North American continental shelf waters, the waters surrounding the Hawaiian Island Archipelago, and the white shark "Cafe" in the eastern Pacific halfway between North America and Hawaii. Yellow circles represent position estimates from light- and SST-based geolocations (Teo *et al.* 2004), and red circles indicate satellite tag endpoint positions (ARGOS transmissions), respectively. From Jorgensen *et al.* (2010).

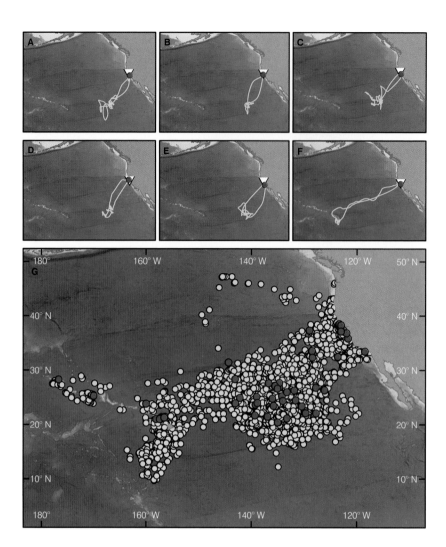

V Deploy location   —— SST Geolocation track
● Popup location   ○ SST Geolocation position

2006) are believed to be parturition areas, suggesting that movements to and within these hot spots may be related to reproductive activities.

15.2.4.3 Seabirds

Albatrosses and shearwaters are highly migratory species, traveling widely across entire ocean basins and passing through many territorial regions. The highly productive CCS is a major destination for seabirds traveling from as far away as New Zealand (Shaffer *et al.* 2006). Previous research has shown that black-footed albatrosses travel from breeding colonies in the Northwest Hawaiian Islands (NWHI) to the west coast of North America to forage within the CCS during breeding (Hyrenbach *et al.* 2002). This was verified with a greater time-series and tracking effort during the TOPP program (Kappes *et al.* 2010) (Figs. 15.4 and 15.6). However, visits to the CCS are typically on the order of days because adults must return to feed

their chicks. We have now tracked adult albatrosses during the post-breeding exodus and their subsequent return to breeding, a period lasting about 160 days. Using archival data loggers, our data show that black-footed albatrosses from NWHI spend several months within the CCS, whereas Laysan albatrosses from the same colony use the habitat of the NPTZ (Shaffer *et al.* 2009; Kappes *et al.* 2010). The oceanographic conditions that black-footed albatrosses experience within the CCS, warmer water temperatures, higher productivity, and lower sea surface height were quite different than those experienced by Laysan albatrosses (Kappes *et al.* 2010), reflecting a difference in preferred foraging habitat.

15.2.4.4 Whales

Blue and humpback whales showed consistent use of the CCS, moving from as far south as the Costa Rica Dome in the eastern tropical Pacific to regions off Oregon

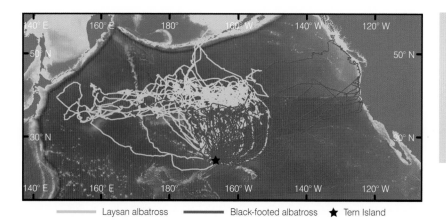

Fig. 15.6

Foraging trips of 37 laysan and 36 black-footed albatrosses breeding at Tern Island, northwest Hawaiian Islands, during the incubation period, as determined by satellite telemetry. Data encompass a period of four breeding seasons, from 2002–03 through 2005–06. ARGOS satellite locations are denoted by marker points along the interpolated tracks, overlaid on ocean bathymetry. From Kappes *et al.* (2010).

——— Laysan albatross ——— Black-footed albatross ★ Tern Island

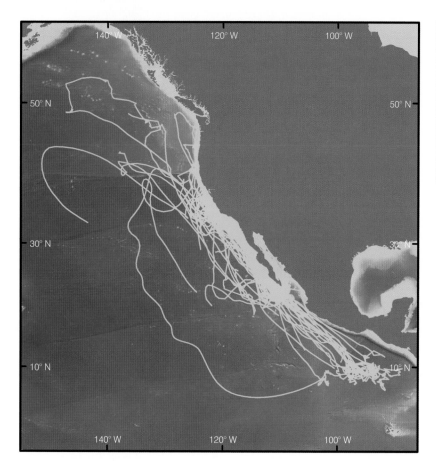

Fig. 15.7

Map of individual SSSM-derived tracks for 92 tags on blue whales, *Balaenoptera musculus*, deployed between 1994 and 2007 that transmitted for more than seven days, color coded by deployment location. In each panel the four tag deployment locations are shown as white circles, the annual climatological position of the CRD as a white contour, and the bathymetry as shades of blue. Map projection: sinusoidal (equal area). From Bailey *et al.* (2010).

and Washington (Fig. 15.7). The occurrence of area restricted search behavior throughout the migration cycle, including the Eastern Tropical Pacific, provides evidence that these animals forage year-round (Bailey *et al.* 2010). The extent of their northward migration from Baja California to Washington during summer/fall varied significantly interannually, likely in response to environmental changes affecting their prey. Most blue whales traveled within 100–200 km of the California coast, although individuals ranged offshore up to 2,000 km. In contrast, humpback whales remained closer to the coast. The blue whales

moved at higher speeds and also appeared to be more transient, with movements up and down the coast, than the humpbacks. One whale that was tracked for more than 16 months repeated much of the previous year's pattern during the fall and winter, with multiple offshore routes during the southward migration. Humpback migration routes extended from California and Oregon to southern breeding areas, including one whale with an extended stay in the Banderas Bay region of Mexico and another venturing farther south off Nicaragua. These tracks provide the first evidence for the actual route, rate of speed, and timing of the southbound migration for California humpback whales to their breeding areas. Although sample sizes are still limited compared with other TOPP species, there are general patterns emerging about the behavior and habitat use of large whales. In general, blue whales move farther from shore into deeper water than the humpback whales, probably reflecting the differences in diet between the two species. Blue whales feed on euphausids, whereas the humpback whales have a more generalized diet, which includes small schooling fish found predominately nearshore (Fiedler *et al.* 1998; Clapham 2009; Sears & Perrin 2009). The long track durations obtained from electronic tagging have provided essential new information about the critical habitats of eastern Pacific whale populations.

15.2.4.5 California sea lions

Just as TOPP has shown that animals move across large ocean basins, TOPP data have also revealed that animal distributions can be significantly affected by climate fluctuations. California sea lions feed throughout the highly productive CCS, with male sea lions moving from the breeding grounds of the Southern California Islands to winter foraging grounds along the coasts of California, Oregon, and Washington. In contrast, female California sea lions remain mostly in and around the Southern California Bight, as they are limited by the need to return to their dependent pups on the breeding islands. However, we have documented changes in the foraging patterns of California sea lions in response to anomalous warming in the CCS. During what could be considered a normal winter of 2003–04, adult male sea lions remained close to the California coast and the durations of their foraging excursions averaged 12 hours, feeding almost exclusively over the continental shelf during trips lasting only 0.8 days. This pattern markedly contrasts with that of the 2004–05 season, where male sea lions traveled 300–500 km offshore for durations of 2.5 days on average, though longer trips also occurred (Weise *et al.* 2006). During spring and summer of 2005, the seasonal upwelling pattern was greatly delayed, resulting in the most spatially extensive and persistent sea surface temperature and primary productivity anomalies in the CCS since the 1997–98 El Niño. Although there was no seasonal variation in the proportion of time male sea lions spent surface swimming during 2003–04, there was a monthly increase during 2004–05 that corresponded to the increasing SST anomaly. Anomalous conditions in 2005 led to large disruptions in the trophic structure of the fish community off California during 2005. These geographic shifts in prey distribution were reflected in the diet of sea lions in central California. Sardine and rockfish were more abundant in the diet during 2005 compared with 2004, whereas market squid and anchovy declined. This redistribution of prey probably forced sea lions to search farther from shore and spend more time swimming at sea in 2005.

15.2.4.6 Elephant seals

Female northern elephant seals forage across the highly productive regions of the NPTZ. However, the use of the NPTZ is seasonal, with the greatest number of females foraging there during their longest at-sea migration over the summer months and the fewest in the winter/spring, when the females are on shore pupping or conducting shorter foraging trips (Fig. 15.8). The TOPP program was able to extend our knowledge of the foraging patterns of elephant seals and track animals from colonies that represent the southern (San Benitos Islands, Baja, Mexico) and northern (Año Nuevo California) limits of their range. These two colonies span the entire known breeding range of northern elephant seals, a distance of 1,200 km. Tracking seals over their extant range has provided a more complete understanding of the foraging ecology of this species throughout the North Pacific Ocean. Interestingly, females from the southernmost regions travel further to reach the same regions of the NPTZ used by females transiting from the northern colonies. Data from repeat trips suggest that female elephant seals rely on the persistence of the NPTZ as a key foraging ground, as they return to the same region year after year. Although these repeat visits have been seen in females over a two- to three-year period, the most impressive repeat track was from a female first tagged in 1995 when she was six years old and then tagged again 11 years later. What is remarkable about this record is the two tracks are almost identical.

15.2.5 Conservation applications

Many TOPP species are harvested by human fishers whereas others are caught indirectly as a byproduct of fishing activities (loggerhead and leatherback turtles, shearwaters, and albatross). Tracking data show that TOPP species (tuna and sharks) do not recognize political boundaries and travel through the exclusive economic zones of many countries, making it clear that these species require multinational protection. For example, Laysan albatrosses tagged at Guadalupe are found within the CCS and within at least three different exclusive economic zones. Pacific bluefin tuna that were spawned in the western Pacific were found to be so overexploited in the eastern Pacific that approximately 65% of the tags were recovered over the course of TOPP archival tagging, and few individuals lived long enough to make trans-Pacific migrations back to the spawning grounds near Japan. Similarly, over 50% of TOPP yellowfin tags were recovered, indicative of high purse seine effort in the Baja California region.

TOPP data have been used by several national and international bodies. For example, TOPP data have been used in listing black-footed albatrosses as an endangered species

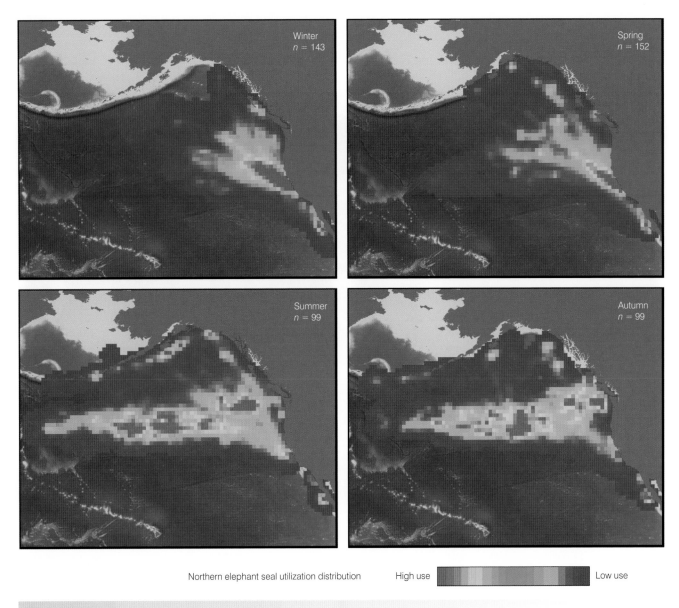

Northern elephant seal utilization distribution   High use   Low use

Fig. 15.8
Kernel density plots showing the seasonal changes in the use of the North Pacific Ocean by female northern elephant seals. The numbers above each panel reflect the number of individual tracks used to create the image.

by the US Fish and Wildlife Service and have been incorporated into BirdLife International and the US Fish and Wildlife Service for deliberations within the international Agreement for the Conservation of Albatrosses and Petrels. The discovery of a persistent and predictable migration corridor for eastern Pacific leatherback turtles (Shillinger *et al.* 2008) also led to an International Union for Conservation of Nature resolution to conserve this endangered species in the open seas. Another example of a successful conservation application in TOPP was the creation of a marine protected area off the coast of Baja California to protect loggerhead turtles (Fig. 15.9) (Peckham *et al.* 2007). The satellite and acoustic tagging of white sharks, combined with a new Bayesian model to provide population estimates, have provided a baseline for future studies focused on monitoring and assessing the white shark population (Jorgensen *et al.* 2010; T. Chapple, unpublished observations). Over the course of the program, the TOPP team has shown that monitoring population trends and designing new tools for the protection of pelagic predators is possible. These efforts can be used as templates for future work or expanded to include multiple species.

Fig. 15.9
Kernel density of loggerhead turtle habitat use in the North Pacific. Inset: positions of tracked loggerheads (yellow) spanned the North Pacific Basin. The 50% use distribution for observed loggerheads consisted of an area of 4,115 km centered 32 km from the Baja California South coast, well within the 55 km range of small-scale fisheries (white line) (Peckham et al. 2007).

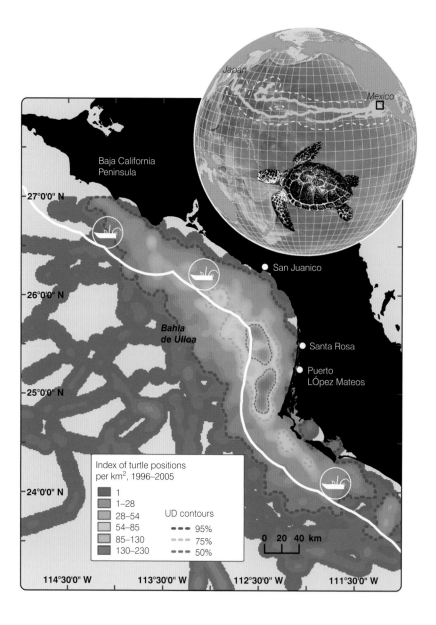

15.3 Limitations to Knowledge of Marine Top Predators

The structure and function of open ocean pelagic ecosystems, including the roles played by top predators, is challenging to understand. Obtaining reliable data on the physiology, behavior, population structure, and ecology of a diverse suite of apex marine predators has been limited by a lack of appropriate observational tools and methods. The mechanisms linking physical forcing, primary and secondary production, prey abundance and distribution to top predator movements are still being elucidated. Furthermore, the influence of climate variations, and potential impacts of climate change, on top predator distributions remain largely unknown. In TOPP, the acquisition of syn-

optic positional data from apex predators simultaneously with environmental data has provided a unique opportunity to synthesize how distinct ecological guilds of top predators use common habitats, as well as a framework for comparative studies between species groups.

Removal of top predators by intense overexploitation by high-seas fisheries is threatening epipelagic ecosystems (Myers & Worm 2003; Sibert et al. 2006) and poses serious, but as yet unquantified, threats to the stability of marine ecosystems (Jackson et al. 2001). The removal of top predators from ecosystems due to the impact of commercial fisheries has been shown to have a cascading effect throughout the food web, resulting in a shift in species composition (Springer et al. 2003; Frank et al. 2005; Pauly et al. 2005). The role of biodiversity in maintaining the structure and integrity of pelagic ecosystems has been challenging to study because of the difficulties associated with monitoring species over such large spatial scales. The resilience of ecosystems in the face of climate variability as well as the top-down elimination of oceanic predators has been difficult to quantify. The uncertainty associated with the inability to appropriately study and visualize the response of pelagic ecosystems to climate effects or human perturbations is hampering our ability to describe the nature of pelagic predator populations.

Knowledge of how predators use the oceanic ecosystem lags far behind terrestrial ecology, partly because of the challenge of following large, highly migratory animals at sea. Limits to our knowledge are due largely to technological limitations, such as our ability to increase battery life, miniaturize electronics, develop inexpensive silicone-based technologies, increase the rate of data transmission to satellites, apply GPS technologies to marine animals, reduce the error associated with geolocations, and develop methods to infer behavior of marine animals from tagging data. TOPP is providing fundamental information that is addressing these unknowns.

15.4 The Future of Marine Biologging

As humans place enormous pressure on marine ecosystems through overexploitation and climate change, the challenges of understanding critical ecosystem processes and learning how to monitor, model, and manage these ecosystems will be of increasing importance. TOPP has provided proof that direct observation of large marine ecosystems is possible. Linking the biotic and abiotic processes is technologically within our reach. The challenge is building the infrastructure and commitment to continue what has been started during the past decade.

Biologging offers new tools for fisheries managers to obtain critical information on exploited and unexploited populations required to better understand ecosystem function. In most cases before TOPP, scientists and managers lacked important distribution, abundance, and ecological information for modeling populations in the North Pacific Ocean. For managing protected areas and building spatially explicit models, appropriate distribution and abundance data on a scale never before obtained is now available for many top predator groups. These data can now support population assessments that are critical for management decisions. By establishing technologies that can be used reliably, remotely, and in combination with satellite data, we can provide inputs of ecological and physiological data that will improve predictive models. By making innovative efforts, TOPP has provided a road map for comprehensive new approaches to animal observation and for improving our understanding of habitat use. This information, in turn, will lay a foundation for improved ecosystem-based management.

Although conventional survey methods provide the potential to fully characterize the biota of a particular ocean region that is accessible (by vessel, submersible, divers, and so forth), the region outside the search area will always remain unknown and in many cases unknowable. Electronic tags report on an animal's whereabouts and the characteristics of its environment wherever the animal may go, completely independently of our ability to reach these areas with conventional survey platforms such as ships. Electronic tags also allow us to identify biodiversity hot spots, areas where the greatest abundance and diversity of tagged animals are aggregated, as well as to understand the oceanic processes responsible for the formation of these biological hot spots. Before TOPP, few studies provided information on top predator distribution and biodiversity on an oceanic scale.

What we as yet cannot do is predict how the ecosystems of the North Pacific Ocean are changing in the long term. We have just begun to understand the foraging responses of a select group of species, and we lack a true understanding of the connectivity and ecological interactions among the full suite of apex marine predators. Until we deploy biologging technologies continuously and on the oceanic scale, it will be difficult to get a more complete picture of how ocean ecosystems work. TOPP has, however, provided the baseline of a decade of study of top predators in the North Pacific from which future monitoring programs can compare or contrast results.

15.5 Conclusions

The maintenance of biodiversity on Earth depends upon our capacity to understand and manage it in our lifetime. TOPP scientists have demonstrated that biologging is a powerful tool for observing and monitoring how animals use the blue ocean ecosystems. In less than a decade, TOPP has shown a remarkable capacity to generate ocean-scale spatial and temporal movement data that have far-reaching impacts for stewardship and management activities in our oceans. The science of biologging is still evolving and incorporation of tagging data into population assessment models is only just beginning.

TOPP has provided a decade-long view of the distribution and movements of 23 top predators in the North Pacific Ocean. The repeatable observation of philopatry in TOPP animals such as white sharks, leatherbacks, elephant seals, and bluefin tuna is indicative of a large-scale biogeographic patterning of the North Pacific. Although some long-lived species remain reproductively active in the western Pacific, their inclination and success when foraging in the eastern Pacific provides the basis for epic trans-oceanic migrations. This biogeographic pattern is only now emerging, and shows the California Current system to be a retentive habitat for some species and an attractive habitat for others. Following up these tracking studies with genetic

Box 15.2

TOPP species. The species in this list were chosen by the organismal working groups as having those biological and ecological attributes meeting the goals of the TOPP program.

Air-breathing vertebrates

Leatherback turtle, *Dermochelys coriacea*
Loggerhead turtle, *Caretta caretta*
Black-footed albatross, *Phoebastria nigripes*
Laysan albatross, *Phoebastria immutabilis*
Elephant seal, *Mirounga angustirostrus*
California sea lion, *Zalophus californianus*
Northern fur seal, *Callorhinus ursinus*
Blue whale, *Balaenoptera musculus*
Fin whale, *Balaenoptera physalus*
Humpback whale, *Megaptera novaengliae*
Sperm whale, *Physeter macrocephalus*
Sooty shearwater, *Puffinus griseus*
Pink shearwater, *Puffinus creatopus*

Fish, shark, and squid

Bluefin tuna, *Thunnus thynnus orientalis*
Yellowfin tuna, *Thunnus albacares*
Albacore tuna, *Thunnus alalunga*
White shark, *Carcharodon carcharias*
Mako shark, *Isurus oxyrinchus*

Salmon shark, *Lamna ditropis*
Blue shark, *Prionace glauca*
Common thresher shark, *Alopias vulpinus*
Humboldt or giant squid, *Dosidicus gigas*
Ocean sunfish, *Mola mola*

(Jorgensen *et al.* 2010) and genomic studies may reveal important information on how the species of the eastern Pacific have come to colonize these highly variable ocean ecosystems, and may reveal critical information on how these species can adapt to the stresses of this unique and highly variable environment.

TOPP has also shown that newly developed sensor capabilities in animal tags are a remarkable new addition to *in situ* ocean observing networks. The high mobility of marine predators, coupled with the remarkable technical advances made in biologging, has enabled the collection of a variety of high-resolution oceanographic data that are important for improving four-dimensional ocean observations. In TOPP, more than 200,000 tagging days and 2,000,000 profiles were collected by our "animal oceanographers". This dataset has high quality ocean data that are now being assimilated into an animal ocean portal that will serve OBIS data (see Chapter 17) to National Aeronautics and Space Administration (NASA) and National Oceanic and Atmospheric Administration (NOAA) portals. Biologging provides the only means to measure *in situ* ocean conditions concomitant with, and at the scale of, the behavioral response of the animals. Most importantly, TOPP has built the capacity for, and demonstrated the efficacy of, an ocean-scale biologging program that is essential for monitoring and sustaining the health of our ocean ecosystems.

Acknowledgments

We thank the TOPP team for inspiring our Pacific-wide effort and for their dedication to long hours at sea deploying instruments. The work would not have been possible without a dedicated data management team throughout the past decade. We are indebted to A. Swithenbank, J. Ganong, M. Castleton, L. deWitt, D. Foley, and R. Weber, who all helped create the outputs and web browser for TOPP. We thank Dr. Randy Kochevar and Don Kohrs of Stanford University for their dedication and outreach for this program. The postdoctoral scholars and students of TOPP have dedicated energy, thought, and enthusiasm to the program. We thank our Steering Committee and the primary funders of this research: Sloan, Packard, Moore, and Monterey Bay Aquarium Foundations, the Office of Naval Research (ONR) and the National Oceanic and Atmospheric Administration. All work was done in accordance with the IACUC requirements of Stanford University and University of California.

References

Anderson, S.D., Becker, B.H. & Allen, S.G. (2008) Observations and prey of white sharks, *Carcharodon carcharias*, at point reyes national seashore: 1982–2004. *California Fish and Game* **94**, 33–43.

Bailey, H., Mate, B.R., Palacios, D.M., *et al.* (2010) Behavioural estimation of blue whale movements in the Northeast Pacific from state space model analysis of satellite tracks. *Endangered Species Research* **10**, 93–106.

Bailey, H.R., Shillinger, G.L., Palacios, D.M., *et al.* (2008) Identifying and comparing phases of movement by leatherback turtles using state-space models. *Journal of Experimental Marine Biology and Ecology* **356**, 128–135.

Bayliff, W.H. (1994) A review of the biology and fisheries for northern bluefin tuna, *Thunnus thynnus*, in the Pacific Ocean. FAO (Food and Agriculture Organization of the United Nations) Fisheries Technical Paper 336, part 2, pp. 244–295.

Biuw, M., Boehme, L., Guinet, C., *et al.* (2007) Variations in behavior and condition of a Southern Ocean top predator in relation to *in situ* oceanographic conditions. *Proceedings of the National Academy of Sciences of the USA* **104**, 13705–13710.

Block, B.A. (2005) Physiological ecology in the 21st century: advancements in biologging science. *Integrative and Comparative Biology* **45**, 305–320.

Block, B.A., Costa, D.P., Boehlert, G.W. & Kochevar, R.E. (2003) Revealing pelagic habitat use: the tagging of Pacific pelagics program. *Oceanologica Acta* **25**, 255–266.

Block, B.A., Dewar, H., Blackwell, S.B., *et al.* (2001) Migratory movements, depth preferences, and thermal biology of Atlantic bluefin tuna. *Science* **293**, 1310–1314.

Block, B.A., Dewar, H., Farwell, C. & Prince, E.D. (1998) A new satellite technology for tracking the movements of Atlantic bluefin tuna. *Proceedings of the National Academy of Sciences of the USA* **95**, 9384–9389.

Boehme, L., Meredith, M.P., Thorpe, S.E., *et al.* (2008a) Antarctic Circumpolar Current frontal system in the South Atlantic: monitoring using merged *Argo* and animal-borne sensor data. *Journal of Geophysics Research* **113**, C08009, doi:10.1029/2007JC004647.

Boehme, L., Thorpe, S.E., Biuw, M., *et al.* (2008b) Monitoring Drake Passage with elephant seals: frontal structures and snapshots of transport. *Limnology and Oceanography* **53**, 2350–2360.

Boustany, A., Matteson, R., Castleton, M.R., *et al.* (2010) Movements of Pacific bluefin tuna (*Thunnus orientalis*) in the Eastern North Pacific revealed with archival tags. *Progress in Oceanography* (in press).

Boustany, A.M., Davis, S.F., Pyle, P., *et al.* (2002) Satellite tagging: Expanded niche for white sharks. *Nature* **415**, 35–36.

Charrassin, J.-B., Hindell, M., Rintoul, S.R., *et al.* (2008) Southern Ocean frontal structure and sea-ice formation rates revealed by elephant seals. *Proceedings of the National Academy of Sciences of the USA* **105**, 11634–11639.

Clapham, P.J. (2009) Humpback whale, *Megaptera novaeangliae*. In: *Encyclopedia of Marine Mammals* (eds. W.F. Perrin, B. Würsig, and J.G.M. Thewissen), pp. 582–585. New York: Elsevier-Academic.

Collette, B.B. & Nauen, C.E. (1983) *Scombrids of The World*. FAO Species Catalog 125.

Costa, D.P. (1993). The secret life of marine mammals: novel tools for studying their behavior and biology at sea. *Oceanography* **6**, 120–128.

Costa, D.P., Robinson, P.E., Arnould, J.P.Y., *et al.* (2010a) Accuracy of ARGOS locations of marine mammals at sea estimated using Fastloc GPS. *PloS ONE* **5**, e8677. doi:10.1371/journal.pone.0008677.

Costa, D.P., Huckstadt L.A., Crocker, D.E., *et al.* (2010b) Approaches to studying climatic change and its role on the habitat selection of Antarctic pinnipeds. *Integrative and Comparative Biology* doi:10.1093/icb/icq054.

Costa, D.P. & Sinervo B. (2004) Field physiology: physiological insights from animals in nature. *Annual Review of Physiology* **66**, 209–238.

Decker, C.J. & Reed, C. (2009) The National Oceanographic Partnership Program: a decade of impacts on oceanography. *Oceanography* **22**, 208–227.

Ekstrom, P.A. (2004) An advance in geolocation by light. *Memoirs of the National Institute of Polar Research* **58**, 210–226

Fedak, M., Lovell, P., McConnell, B. & Hunter, C. (2002) Overcoming the constraints of long range radio telemetry from animals: getting more useful data from smaller packages. *Integrated Comparative Biology* **42**, 3–10.

Fedak, M.A., Lovell, P. & Grant, S.M. (2001) Two approaches to compressing and interpreting time–depth information as collected by time–depth recorders and satellite-linked data recorders. *Marine Mammal Science* **17**, 94–110.

Feder, M.E. (1987) The analysis of physiological diversity: the prospects for pattern documentation and general questions in physiological ecology. In: *New Directions in Physiological Ecology* (eds. M.E. Feder, A.F. Bennett, W. Burggren, & R.B. Huey). pp. 38–75. Cambridge, UK: Cambridge University Press.

Fiedler, P.C., Reilly, S.B., Hewitt, R.P., *et al.* (1998) Blue whale habitat and prey in the California Channel Islands. *Deep-Sea Research II* **45**, 1781–1801.

Frank, K.T., Petrie, B., Choi, J.S. & Leggett, W.C. (2005) Trophic cascades in a formerly cod-dominated ecosystem. *Science* **308**, 1621–1623.

Goldman, K.J. & Musick, J.A. (2006) Growth and maturity of salmon sharks (*Lamna ditropis*) in the eastern and western North Pacific, and comments on back-calculation methods. *Fishery Bulletin* **104**, 278–292.

Goldman, K.J. & Musick, J.A. (2008) The biology and ecology of the salmon shark, *Lamna ditropis*. *Fish and Aquatic Resources Series* **13**, 95–104.

Hickey, B.M. (1998) Coastal oceanography of western North America from the tip of Baja California to Vancouver Island. In: *The Sea, The Global Coastal Ocean, Regional Studies and Syntheses* (eds. A.R. Robinson & K.H. Brink), pp. 345–393. New York: Wiley.

Hill, R.D. & Braun, M.J. (2001) Geolocation by light level – the next step: latitude. In: *Electronic Tagging and Tracking in Marine Fisheries* (eds. J.R. Sibert & J. Nielsen), pp. 315–330. The Netherlands: Kluwer.

Hyrenbach, K.D., Fernandez, P. & Anderson, D.J. (2002) Oceanographic habitats of two sympatric North Pacific albatrosses during the breeding season. *Marine Ecology Progress Series* **233**, 283–301.

Inagake, D.H., Yamada, K., Segawa, M., *et al.* (2001) Migration of young bluefin tuna, *Thunnus orientalis* Temminck et Schlegel, through archival tagging experiments and its relation with oceanographic conditions in the western North Pacific. *Bulletin of the National Research Institute of Far Seas Fisheries* **38**, 53–81.

Jackson, J.B., Kirby, M.X., Berger, K.A., *et al.* (2001) Historical overfishing and the recent collapse of coastal ecosystems. *Science* **293**, 629–637.

Jorgensen, S.J., Reeb, C.A. Chapple, T., *et al.* (2010) Philopatry and migration of Pacific white sharks. *Proceedings of the Royal Society of London B* **277**, 679–688.

Kappes, M.A., Shaffer, S.A., Tremblay, Y., *et al.* (2010) Hawaiian albatrosses track interannual variability of marine habitats in the North Pacific. *Progress in Oceanography* doi:10.1016/j.pocean.2010.04.012.

Kitagawa, T., Boustany, A.M., Farwell, C.J., *et al.* (2007) Horizontal and vertical movements of juvenile bluefin tuna (*Thunnus orientalis*) in relation to seasons and oceanographic conditions in the eastern Pacific Ocean. *Fisheries Oceanography* 16, 409–421.

Kuhn, C.E. & Costa, D.P. (2006) Identifying and quantifying prey consumption using stomach temperature change in pinnipeds. *Journal of Experimental Biology* 209, 4524–4532.

Kuhn, C.E., Crocker, D.E., Tremblay, Y. & Costa, D.P. (2009) Time to eat: measurements of feeding behaviour in a large marine predator, the northern elephant seal *Mirounga angustirostris*. *Journal of Animal Ecology* 78, 513–523.

Le Boeuf, B.J., Crocker, D.E., Costa, D.P., *et al.* (2000) Foraging ecology of northern elephant seals. *Ecological Monographs* 70, 353–382.

Lopez, S., Melendez, R. & Barria, P. (2009) Feeding of the shortfin mako shark *Isurus oxyrinchus* Rafinesque, 1810 (Lamniformes: Lamnidae) in the Southeastern Pacific. *Revista de Biologia Marina y Oceanografia* 44, 439–451.

Lutcavage, M.E., Brill, R.W., Skomal, G.B., *et al.* (1999) Results of pop-up satellite tagging of spawning size class fish in the Gulf of Maine: do North Atlantic bluefin tuna spawn in the mid-Atlantic? *Canadian Journal of Fisheries and Aquatic Sciences* 56: 173–177.

McConnell, B.J., Chambers, C. & Fedak, M.A. (1992a) Foraging ecology of southern elephant seals in relation to the bathymetry and productivity of the Southern Ocean. *Antarctic Science* 4, 393–398.

McConnell, B.J., Chambers, C., Nicholas, K.S. & Fedak, M.A. (1992b) Satellite tracking of grey seals (*Halichoerus grypus*). *Journal of Zoology* 226, 271–282.

Myers, R.A. & Worm, B. (2003) Rapid worldwide depletion of predatory fish communities. *Nature* 423, 280–283.

Nakano, H. & Nagasawa, K. (1996) Distribution of pelagic elasmobranchs caught by salmon research gillnets in the North Pacific. *Fisheries Science* 62, 860–865.

Nicholls, K.W., Boehme, L., Biuw, M. & Fedak, M.A. (2008) Wintertime ocean conditions over the southern Weddell Sea continental shelf, Antarctica. *Geophysical Research Letters* 35, L21605.

Olson, D.B., Hitchcock, G.L., Mariano, A.J., *et al.* (1994) Life on the edge: marine life and fronts. *Oceanography* 7, 52–60.

Palacios, D.M., Bograd, S.J., Foley, D.G. & Schwing, F.B. (2006) Oceanographic characteristics of biological hot spots in the North Pacific: a remote sensing perspective. *Deep Sea Research II* 53, 250–269.

Pauly, D., Watson, R. & Alder, J. (2005) Global trends in world fisheries: impacts on marine ecosystems and food security. *Philosophical Transactions of the Royal Society of London B* 360, 5–12.

Peckham, S.H., Diaz, D.M., Walli, A., Ruiz, G., Crowder, L.B., *et al.* (2007) Small-scale fisheries bycatch jeopardizes endangered pacific loggerhead turtles. *PLoS ONE* 2(10), e1041. doi:10.1371/journal.pone.0001041.

Polovina, J.J. (1996) Decadal variation in the trans-Pacific migration of northern bluefin tuna (*Thunnus thynnus*) coherent with climate-induced change in prey abundance. *Fisheries Oceanography* 5, 114–119.

Polovina, J.J., Kobayashi, D.R., Parker, D.M., *et al.* (2000) Turtles on the edge: Movement of loggerhead turtles (*Caretta caretta*) along oceanic fronts, spanning longline fishing grounds in the central North Pacific, 1997–1998. *Fisheries Oceanography* 9, 71–82.

Pyle, P., Anderson, S.D. & Ainley, D.G. (1996) Trends in white shark predation at the South Farallon Islands, 1968-1993. In: *Great White Sharks* (eds. A. Klimley & D. Ainley), pp. 375–379. Elsevier.

Roden, G.I. (1991) Subarctic–Subtropical Transition Zone of the North Pacific Large-Scale Aspects and Mesoscale Structure. In:

Biology, Oceanography, and Fisheries of the North Pacific Transition Zone and Subarctic Frontal Zone (ed. J. A. Wetherall), pp. 1–38. NOAA Technical Report NMFS, US Department of Commerce.

Schaefer, K.M., Fuller, D.W. & Block, B.A. (2007) Movements, behavior, and habitat utilization of yellowfin tuna (*Thunnus albacares*) in the northeastern Pacific Ocean, ascertained through archival tag data. *Marine Biology* 152, 503–525.

Schaefer, K.M., Fuller, D.W. & Block, B.A. (2008) Comparative vertical movements and habitat utilization of bigeye (*Thunnus obesus*), yellowfin (*Thunnus albacares*) and skipjack (*Katsuwonus pelamis*) tunas in the equatorial eastern Pacific Ocean. In: *Tagging and Tracking of Marine Animals with Electronic Devices* (ed. J.L. Nielsen *et al.*), pp. 121–144 (*Reviews: Methods and Technologies in Fish Biology and Fisheries* 9). Springer Science+Business Media B.V.

Schaefer, K.M., Fuller, D.W. & Block, B.A. (2009) Tagging and tracking of marine animals with electronic devices. In: *Reviews: Methods and Technologies in Fish Biology and Fisheries* (eds. S.J. Fragoso, *et al.*), pp. 121–144 9. Dordrecht and London: Springer eBooks.

Sears, R. & Perrin, W.F. (2009) Blue whale, *Balaenoptera musculus*. In: *Encyclopedia of Marine Mammals* (eds. W.F. Perrin, B. Würsig, & J.G.M. Thewissen), pp. 120–124. New York: Elsevier-Academic.

Shaffer, S.A. & Costa, D.P. (2006) A database for the study of marine mammal behavior: gap analysis, data standardization, and future directions. *Oceanic Engineering* 31, 82–86.

Shaffer, S.A., Tremblay, Y., Weimerskirch, H., *et al.* (2006) Migratory shearwaters integrate oceanic resources across the Pacific Ocean in an endless summer. *Proceedings of the National Academy of Sciences of the USA* 103, 12799–12802.

Shaffer, S.A., Weimerskirch, H., Scott, D., *et al.* (2009) Spatiotemporal habitat use by breeding sooty shearwaters *Puffinus griseus*. *Marine Ecology Progress Series* 391, 209–220.

Shillinger, G.L., Palacios, D.M., Bailey, H. *et al.* (2008) Persistent leatherback turtle migrations present opportunities for conservation. *PLoS Biology* 6, e171.

Sibert, J., Hampton, J., Kleiber, P. & Maunder, M. (2006) Biomass, size, and trophic status of top predators in the Pacific Ocean. *Science* 314, 1773–1776.

Simmons, S.E., Tremblay, Y. & Costa, D.P. (2009) Pinnipeds as ocean-temperature samplers: calibrations, validations, and data quality. *Limnology and Oceanography: Methods* 7, 648–656.

Springer, A.M., Estes, J.A., van Vliet, G.B., *et al.* (2003) Sequential megafaunal collapse in the North Pacific Ocean: an ongoing legacy of industrial whaling? *Proceedings of the National Academy of Science of the USA* 100, 12223–12228.

Sydeman, W.J., Brodeur, R.D., Grimes, C.B., *et al.* (2006) Marine habitat "hotspots" and their use by migratory species and top predators in the North Pacific Ocean: introduction. *Deep Sea Research II* 53, 247–249.

Teo, S.L.H., Boustany, A., Blackwell, S., *et al.* (2004a) Validation of geolocation estimates based on light level and sea surface temperature from electronic tags. *Marine Ecology Progress Series* 283, 81–98.

Teo, S.L.H., Boustany, A., Blackwell, S.B., *et al.* (2004b) Validation of geolocation estimates based on light level and sea surface temperature from electronic tags. *Marine Ecology Progress Series* 283, 81–98.

Teo, S.L.H., Kudela, R.M., Rais, A., *et al.* (2009) Estimating chlorophyll profiles from electronic tags deployed on pelagic animals. *Aquatic Biology* 5, 195–207.

Tremblay, Y.A. Roberts, A.J. & Costa, D.P. (2007) Fractal landscape method: an alternative approach to measuring area-restricted searching behavior. *Journal of Experimental Biology* 210, 935–945.

Tremblay, Y., Robinson, P.W. & Costa, D.P. (2009) A parsimonious approach to modeling animal movement data. *PLoS ONE* **4**(3), doi: 10.1371/journal.pone.0004711.

Tremblay, Y., Shaffer, S.A., Fowler, S.L., *et al.* (2006) Interpolation of animal tracking data in a fluid environment. *Journal of Experimental Biology* **209**, 128–140.

Weimerskirch, H., Guionnet, T., Martin, J., *et al.* (2000) Fast and fuel efficient? Optimal use of wind by flying albatrosses. *Proceedings of the Royal Society of London B* **267**, 1869–1874.

Weise, M.J., Costa, D.P. & Kudela, R.M. (2006) Movement and diving behavior of male California sea lion (*Zalophus californianus*) during anomalous oceanographic conditions of 2005 compared to those of 2004. *Geophysic Research Letters* **33**, L22S10.

Weng, K.C., Boustany, A.M. Pyle, P., *et al.* (2007a) Migration and habitat of white sharks (*Carcharodon carcharias*) in the eastern Pacific Ocean. *Marine Biology* **152**, 877–894.

Weng, K.C., Castilho, P.C., Morrissette, J.M., *et al.* (2005) Satellite tagging and cardiac physiology reveal niche expansion in salmon sharks. *Science* **310**, 104–106.

Weng, K.C., O'Sullivan, J.B., Lowe, C.G., *et al.* (2007b) Movements, behavior and habitat preferences of juvenile white sharks *Carcharodon carcharias* in the eastern Pacific. *Marine Ecology Progress Series* **338**, 211–224.

Weng, K.C., Foley, D.G., Ganong, J.E., *et al.* (2008) Migration of an upper trophic level predator, the salmon shark, *Lamna ditropis*, between distant ecoregions. *Marine Ecological Progress Series* **372**: 253–264.

Woods, J.D. (1988) Scale upwelling and primary production. NATO ASI series. Series C, Mathematical and physical sciences **239**, 7–38.

Worm, B., Lotze, H.K. & Myers, R.A. (2003) Predator diversity hotspots in the blue ocean. *Proceedings of the National Academy of Sciences of the USA* **100**, 9884–9888.

PART V

Oceans Future

Chapter 16

The Future of Marine Animal Populations

Boris Worm, Heike K. Lotze, Ian Jonsen, Catherine Muir

Biology Department, Dalhousie University, Halifax, Nova Scotia, Canada

16.1   Introduction

The Census of Marine Life's overarching goal is to assess and explain the diversity, distribution, and abundance of marine organisms throughout the world's oceans. By stimulating exploration and research in all ocean habitats it has accumulated an unprecedented wealth of new information on the patterns and processes of marine biodiversity on a global scale. Three questions are guiding this research effort. What did live in the oceans? What does live in the oceans? What will live in the oceans? The Future of Marine Animal Populations (FMAP) Project ultimately aims to answer that third question through the analysis and synthesis of available data, and the modeling of patterns and trends in marine biodiversity. This entails all levels of biodiversity, from individuals, to populations, communities, and ecosystems (Box 16.1).

Despite the ultimate focus on future prediction, the synthetic analyses undertaken within the FMAP project inform all three aspects of the Census, past, present, and future. The rationale is that without a solid understanding of past and present trends, it is impossible to make sound future projections. Likewise, our research efforts encompass different levels of organization, from the movements of individual animals through space and time, to broad macroecological patterns of abundance and diversity. Hence, an improved understanding of processes at the level of an individual animal may help inform the interpretation of larger-scale patterns.

Our main analytical tools are meta-analytic models, used to combine and understand species abundance and distribution trends, including both historical and recent data. Models that are effective for synthesis also have potential for prediction, and have been used by others to project potential future effects of fishing and climate change, for example Botkin *et al*. (2007). Moreover, modeling can help define the limits of knowledge: what is known and how firmly, what may be unknown but knowable, and what is likely to remain unknown in the foreseeable future.

FMAP grew out of a workshop held at Dalhousie University in Halifax, Nova Scotia, Canada, in June 2002. Representatives of other Census projects, including the History of Marine Animal Populations (HMAP) project, the field projects, and the Ocean Biogeographic Information System (OBIS), participated and provided guidance in the design of this project. FMAP was originally envisioned and led by Ransom A. Myers, Killam Chair of Ocean Studies at Dalhousie University. His leadership carried the project until his sudden passing in 2007. Two additional FMAP centers were established in 2003 at the University of Iceland with Gunnar Stefansson, and the University of Tokyo with Hiroyuki Matsuda. Since 2007 the project has been co-led by the authors of this chapter.

FMAP's mission has been to describe and synthesize globally changing patterns of species abundance, distribution, and diversity, and to model the effects of fishing, climate change, and other key variables on those patterns. This work has been performed across ocean realms and

Life in the World's Oceans, edited by Alasdair D. McIntyre
© 2010 by Blackwell Publishing Ltd.

Box 16.1

Method and Questions

FMAP engaged primarily in the statistical modeling of ecological patterns derived from empirical data. The emphasis has been on data synthesis, often by means of meta-analysis, which is the statistical integration of multiple datasets to answer a common question (Cooper & Hedges 1994). FMAP researchers have also engaged in field surveys and experimental work, but have mostly focused on analyzing and synthesizing datasets collected by other Census projects and third parties. This approach enabled us to ask broad scientific questions about the status and changes in diversity, abundance, and distribution of marine animals, such as the following:

- What are the global patterns of biodiversity across different taxa?
- Which are the major drivers explaining diversity patterns and changes?
- What is the total number of species in the ocean (known and unknown)?

- How has the abundance of major species groups changed over time?
- What are the ecosystem consequences of fishing and other human impacts?
- How are animal ranges and their distribution in the ocean changing?
- How is the movement of animals determined by behavior and the environment?

The main limits to knowledge have been missing data on species that have not been counted, mapped, or tagged, and in some cases missing access to existing data on species that have been monitored. From a statistical perspective, the main challenge has been to overcome data limitations such as the limited length of most time series, the problem of temporal or spatial autocorrelation, and separating ecologically relevant patterns from environmental noise and measurement error.

---

with an emphasis on understanding past changes and predicting future patterns. The project benefitted throughout from close collaboration with statisticians and mathematical modelers, which enabled the proper processing and analysis of large datasets. FMAP has collaborated with other Census projects to varying degrees, most consistently with HMAP, Tagging of Pacific Predators (TOPP), and OBIS (see Chapters 1, 15, and 17), as well as various deep-sea projects.

This chapter does not intend to provide an exhaustive overview of the research activities within FMAP (see www.fmap.ca for individual projects and publications). Instead, we aim to highlight key areas of interest and discuss major advances that have been made. It is structured along three major research topics, aiming to cover the major research themes of the Census (distribution, abundance, and diversity of marine life): (1) marine biodiversity patterns and their drivers, (2) long-term trends in animal abundance and diversity, (3) distribution and movements of individual animals. In the concluding section we aim to provide some insight into what is unknown, and what is currently unknowable, particularly with respect to predicting the future of marine biodiversity.

16.2 Biodiversity Patterns and their Drivers

16.2.1 Previous work

Before the Census, mapping of the ocean with respect to our knowledge of fundamental patterns of abundance and diversity was limited. The first global study was published in 1999, presenting a pattern of planktonic foraminiferan diversity derived from the analysis of a large sediment core database (Rutherford *et al.* 1999). Another study highlighted global hot spots of endemism and species richness for corals and associated organisms (Roberts *et al.* 2002). Several authors had investigated latitudinal gradients for particular species groups (Hillebrand 2004). Yet compared with our understanding of life on land, synthetic knowledge on marine biodiversity was sparse. It became clear from these early studies, however, that some of the patterns were uniquely different from those seen on land, where biodiversity is generally highest in the tropics (Gaston 2000).

16.2.2 Large marine predators

FMAP studies have mainly focused on large pelagic predators such as tuna and billfish, whales, and sharks, for which global data were available. These species groups were found to peak in diversity in the subtropics, often between 20–30 degrees latitude north or south. Although a similar distribution pattern was first described for Foraminifera (Rutherford *et al.* 1999), we were able to show that this is a more general pattern that applies across very different species groups (Worm *et al.* 2003, 2005). Furthermore, it became clear that this biodiversity pattern is not static, but dynamically changing on both short and long time scales.

Species richness patterns for tuna (Thunnini), billfish (Istiophoridae), and swordfish (Xiphiidae) were derived from a global Japanese longline-fishing dataset (Fig. 16.1). Pelagic longlines are the most widespread fishing gear in the open ocean, and are primarily used to target tuna and billfish. The Japanese data represents the world's largest longline fleet and the only globally consistent data source reporting species composition, catch and effort for all tuna, billfish, and swordfish. Statistical rarefaction techniques were used to standardize for differences in fishing effort and to estimate species richness (the expected number of species standardized per 50 randomly sampled individuals) for each $5° \times 5°$ cell in which the fishery operated.

As seen in Figure 16.1, species richness of tuna and billfish displayed a global pattern with large hot spots of diversity in all oceans in the 1960s. These hot spots faded over time, indicating declining species richness, a pattern most clearly seen in the Atlantic and Indian Oceans. Declining species richness coincided with 5- to 10-fold increases in total fisheries catch of tuna and billfish in all oceans, which may have led to regional depletion of vulnerable species (Worm *et al.* 2005). In the Pacific, however, initial losses of diversity began to reverse in 1977, coinciding with a large-scale climate regime shift, whereas the Pacific Decadal Oscillation changed from a cool to a warm phase. Climatic drivers were also found to be important on an annual scale. Short-term (year-to-year) variation in species richness showed a remarkable synchrony with the El Niño Southern Oscillation (ENSO) index, with increasing temperatures leading to basin-wide increases in species richness (Worm *et al.* 2005). This may be explained by warming of sub-optimal temperature habitats. ENSO-related decreases in diversity were seen in the tropical Eastern Pacific, a region that suffers from greatly reduced productivity and associated mass mortality of marine life during El Niño events. A subsequent study showed that seasonal variation in sea surface temperature is driving the taxonomic richness patterns for deep-water cetaceans (whales and dolphins) as well (Whitehead *et al.* 2008).

For tuna and billfish, as well as cetaceans and Foraminifera, mean sea surface temperature (SST) clearly emerged as the strongest single predictor of diversity, showing a positive correlation over most of the observed temperature range (5–25 °C), but a negative trend above that (Fig. 16.2). This decline of diversity at high temperatures was most pronounced in the western Pacific "warm pool", which has the highest equatorial SST (warmer than 30 °C), and weakest in the tropical Atlantic, which has the lowest equatorial SST (lower than 27 °C). The relation between tuna and billfish diversity and SST could also be independently reconstructed from an analysis of individual species temperature preferences (Boyce *et al.* 2008).

Another factor that explained significant variation in tuna and billfish species richness on a global scale was the steepness of horizontal temperature gradients. Sharp temperature gradients are found around frontal zones and eddies that are typically associated with mesoscale oceanographic variability. Fronts and eddies often attract large numbers of species, likely because they concentrate food supply, enhance local production, and increase habitat heterogeneity (Oschlies & Garçon 1998; Hyrenbach *et al.* 2000). They may also form important landmarks along transoceanic migration routes (Polovina *et al.* 2001). Finally, dissolved oxygen concentrations were positively correlated with diversity. This likely relates to species physiology, as low oxygen levels (less than 2 ml l^{-1}) may limit the cardiac function and depth range of many tuna species (Sund *et al.* 1981). Regions of low oxygen are located west of Central America, Peru, West Africa, and in the Arabian Sea. Despite optimal SST around 25 °C, most of these areas showed conspicuously low diversity.

Knowledge of the relation between SST and diversity for various species groups (Fig. 16.2) allows us to predict how diversity may change as SST changes spatially and temporally with climate variability and climate change. The effects of climate variability, such as ENSO and the Pacific Decadal Oscillation, are discussed above. With respect to long-term climate change, Whitehead *et al.* (2008) combined Intergovernmental Panel on Climate Change (IPCC) scenarios for observed and projected changes in SST between 1980 and 2050 with an empirically derived relation of SST and deep-water cetacean diversity. For the baseline 1980 dataset, diversity was predicted to be highest at latitudes of about 30°, falling towards the equator, and more precipitously towards the poles. With global warming, these bands of maximal diversity were predicted to move pole-wards. The warming tropical oceans were predicted to decline in diversity, while richness was predicted to increase at latitudes of about 50°–70° in both hemispheres (Whitehead *et al.* 2008). These general conclusions were recently corroborated by an analysis of 1,066 exploited fish and invertebrate species (Cheung *et al.* 2009).

16.2.3 Other species groups

Other groups that were investigated with respect to their diversity patterns were deep-water corals and tropical reef

Fig. 16.1

Tuna and billfish species richness over time. Maps depict the number of expected species per 50 individuals as calculated from pelagic longlining catch and effort data using rarefaction techniques. After data from Worm *et al.* (2005).

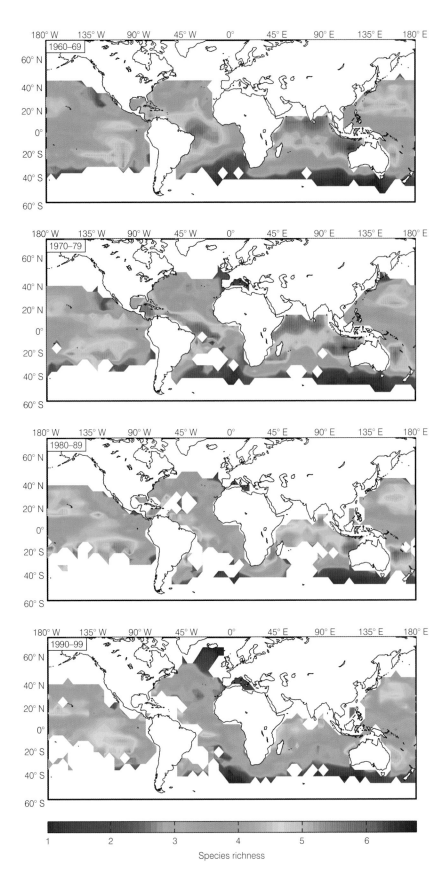

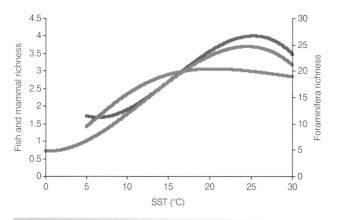

Fig. 16.2
Temperature effects on diversity. Shown are the empirical relationships between sea surface temperature (SST) and species richness for deep-water cetaceans (blue line), planktonic foraminiferans (green line), and tuna and billfish (red line). After data from Worm & Lotze (2009).

fish. The goal was to gain a better understanding of the effects of human impacts such as fishing and ocean acidification on the distribution, abundance, and diversity of different species groups (reviewed by Tittensor *et al.* 2009b).

A study on tropical reef fish at fished and unfished sites in three oceans revealed predictable changes in the species–area relation (SAR). The SAR quantifies the relation between species richness and sampling area and is one of the oldest, most recognized patterns in ecology. Fishing consistently depressed the slope of the SAR, with the magnitude of change being proportional to fishing intensity (Tittensor *et al.* 2007). Changes in species richness, relative abundance, and patch occupancy contributed to this pattern. It was concluded that species-area curves can be sensitive indicators of community-level changes in biodiversity, and may be useful in quantifying the human imprint on reef biodiversity, and potentially elsewhere (Tittensor *et al.* 2007). This study highlighted how human impacts can affect biodiversity through multiple pathways.

Subsequent work focused on cold-water scleractinian corals, an important habitat-forming group of stony corals commonly found on seamounts (Clark *et al.* 2006). Despite their widely accepted ecological importance, records of cold-water corals are patchy and simply not available for most of the global ocean. In an FMAP-CenSeam (Global Census of Marine Life on Seamounts) collaboration (see also Chapter 7), the probable distribution of these corals was derived from habitat suitability models, that incorporated all the available data on cold-water coral distribution in relation to environmental variables such as depth, temperature, and carbonate availability (Tittensor *et al.* 2009a). Highly suitable habitat for seamount stony corals was predicted to occur in the North Atlantic, and in a circumglobal strip in the Southern hemisphere between 20° and 50° S

and at depths shallower than around 1,500 m (Fig. 16.3). Seamount summits in most other regions appeared less likely to provide suitable habitat, except for small near-surface patches. In these models oxygen and carbonate availability played a decisive role in determining large-scale scleractinian coral distributions on seamounts (Tittensor *et al.* 2009a). These results raise concerns about the possible consequences of ocean acidification (Orr *et al.* 2005) and the observed shallowing of oxygen minimum zones in the wake of global climate change (Stramma *et al.* 2008). Both factors would be predicted to limit the distribution of scleractinian corals, and the fauna associated with them.

16.2.4 Total species richness

The number of species is the most basic index used to measure biodiversity and one that plays a fundamental role in the quantification of human-related extinctions and impacts. Unfortunately, the total number of species remains poorly known in the oceans. For example, Grassle & Maciolek (1992) famously suggested that the number of (largely unknown) deep-sea benthic species is more than 1 million, but may even exceed 10 million. The only published estimate of the total number of marine species relied on an inventory of European fauna that was scaled up to the global level (Bouchet 2006). A more analytical approach has recently become possible through the Census' Ocean Biogeographical Information System (OBIS) in combination with newly developed modeling approaches (Mora *et al.* 2008). These modeling methods derive estimates of species richness from "discovery curves" of species sampled over time, and produce confidence limits that allow us to estimate the known and unknown of global species richness. An FMAP pilot project on total marine fish species has estimated that there are approximately 16,000 known species of marine fish, with about another 4,000 awaiting discovery (Mora *et al.* 2008). These methods are currently being used to estimate the known and unknown of total marine species richness.

16.3 Long-term Trends in Abundance

16.3.1 Previous work

Underlying the changing patterns of biodiversity or species richness are changes in the abundance and distribution of individual populations. Most previous work has emphasized variability in population abundance in relation to climate, oceanography, or other factors on yearly to decadal (see, for example, Attrill & Power 2002) or evolutionary time scales (Vermeij 2004; Jackson & Erwin 2006). Changes in marine life over the Anthropocene (the past few

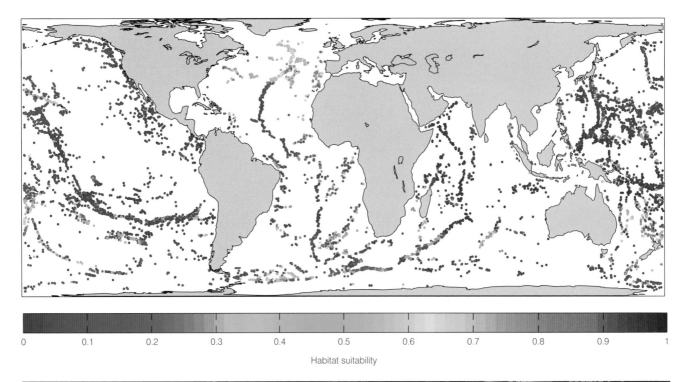

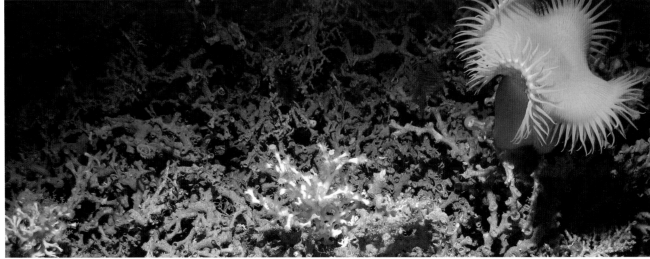

Fig. 16.3
Habitat suitability for cold water corals on seamounts. Colors indicate relative predicted habitat suitability ranging from high (red) to low (blue) as revealed by maximum entropy habitat suitability modeling (after Tittensor *et al.* (2009b)). The photograph depicts *Lophelia pertusa* framework with rich associated invertebrate fauna, Hatton Bank, Northeast Atlantic (UK Department for Business Innovation and Skills (formerly DTI) Strategic Environmental Assessment Programme, c/o Bhavani Narayanaswamy).

hundred years; an epoch dominated by human influences) have only recently received focused attention. This has two reasons: first, the ocean has long been seen as a vast frontier, where human activities would not leave a permanent mark; second, empirical monitoring data are mostly available just for the past 20 to 50 years, which prevented longer-term studies from reaching back beyond the twentieth century.

16.3.2 Synthesizing long-term trends

Over the past decade, the Census at large, and HMAP and FMAP in particular, have partly overcome these limitations. Although HMAP has made enormous progress in unraveling detailed historical, archaeological, and pale-ontological records of past changes in different animal populations and regions (see Chapter 1), FMAP has developed ways of combining and analyzing these data to reveal long-term changes in ocean ecosystems, and uncover their drivers and consequences.

One of FMAP's goals has been to synthesize the long-term trends in the abundance, distribution, and diversity of marine life. This has been pursued for coastal regions over the past centuries and millennia (Lotze & Milewski 2004; Lotze *et al.* 2005, 2006) and continental shelf and open ocean regions over the past 50 years (Myers & Worm 2003, 2005; Worm *et al.* 2005, 2009). These studies have shown that human impacts have resulted in sharply reduced abundance of target and some non-target populations, as well as range contractions and local extinctions that precipitated local and regional losses of species diversity.

To synthesize long-term trends in population abundances of large marine animals, we analyzed 256 records from 95 published studies, many of them from HMAP, FMAP, or other Census projects (Lotze & Worm 2009). Trend estimates for marine mammals, birds, reptiles, and fish were derived from archaeological, historical, fisheries, ecological, and genetic studies and revealed an average decline of 89% (range: 11–100%) from historical abundance levels (Lotze & Worm 2009). Remarkably, the magnitude of depletion was relatively consistent across different species groups (Fig. 16.4A) despite considerable variability in data quality, analytical methods, and time span of the records. Diadromous fish such as sturgeon and salmon, sea turtles, pinnipeds, otters, and sirenia showed the strongest declines with more than 95%. On the other hand, conservation efforts in the twentieth century enabled several whale, pinniped, and coastal bird species to recover from a historical low point in abundance (Fig. 16.4A). These recoveries have reduced the level of depletion across all 256 analyzed species to 84% on average.

Another important dimension of change is the spatial expansion of exploitation, which began in rivers and along the coasts centuries ago and only in the mid-twentieth century moved towards open oceans and the deep sea. Thus, some of the highest population declines can be found in rivers and coastal habitats, with lesser declines found on continental shelves and the open ocean (Fig. 16.4B). Deep-sea habitats differ from this trend, which may be explained by their extreme vulnerability to exploitation (Roberts 2002). Along with this spatial expansion there has been a temporal acceleration in exploitation due to technological advances. Population declines unfolded over hundreds or thousands of years in many rivers and coastal regions, one to two hundred years on the continental shelves, approximately 50 years in the open ocean, and approximately 10–20 years in the deep sea (Lotze & Worm 2009). As a result, the average magnitude of change is almost independent of when exploitation started (Fig. 16.4C). Interestingly though, recoveries are mostly found in species that have been exploited at least 100 years ago and protected in the

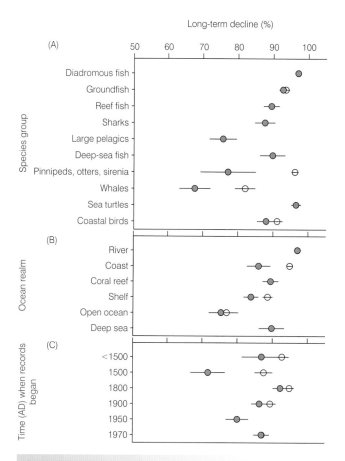

Fig. 16.4
Long-term population declines in large marine animals. Shown are relative changes across **(A)** species groups, **(B)** ocean realms, and **(C)** time period (AD) when exploitation started. There are two measures, the decline to the low point of abundance in the past (open circles) and the decline to today (filled circles), with the difference indicating recovery in population abundance (based on data from Lotze & Worm (2009)).

early to mid-twentieth century, whereas more recently exploited species do not yet show recovery.

Changes in population abundance and distribution have resulted in changes in species diversity. As was discussed previously, there have been remarkable changes in tuna and billfish species richness in the open ocean over the past 50 years (Fig. 16.1). In the coastal ocean, changes in diversity have occurred in two ways: (1) diversity declines have occurred in large marine animals such as mammals, birds, reptiles, and fish due to global, regional, or ecological extinctions; (2) diversity increases have occurred through the invasion of mostly smaller species including invertebrates, plants, unicellular plankton and bacteria, and viruses (Lotze *et al.* 2005, 2006). This shift in diversity from large- to small-bodied animals has resulted in a different species composition, with consequences for ecosystem structure, functioning, and services (see below).

16.3.3 Drivers of long-term change

The underlying drivers of observed long-term changes may include natural and anthropogenic drivers, as well as the cumulative effects of multiple factors. To unravel the relative importance of different drivers on population and ecosystem changes, we have used a variety of methods, including meta-analysis of large datasets, experimental manipulations, and ecosystem models.

For example, an analysis of drivers of long-term population changes in 12 estuaries and coastal seas revealed that exploitation (primary factor) and habitat loss (secondary) were by far the most important causes for the depletion and extinction of marine species over historical time scales (Lotze *et al.* 2006). Pollution, physical disturbance, disease, eutrophication, and introduced predators also contributed to some species declines, although to a lesser extent.

However, the reverse was also true: conservation efforts in the twentieth century, especially reduced exploitation, the protection of habitat, and in some cases pollution control, enabled several species to recover from low abundance. In many cases it was not a single factor, but a combination of exploitation, habitat loss, and other factors that caused a population decline or – in reverse – enabled recovery (Lotze *et al.* 2006). Whereas species invasions and climate change were less dominant drivers of marine biodiversity change in the past, they may increase in importance in the future (Harvell *et al.* 2002; Harley *et al.* 2006; Worm & Lotze 2009).

The cumulative and interactive effects of different drivers have also been explored with multi-factorial laboratory experiments. For example, a three-factorial experiment that used rotifers as a model system showed additive effects between exploitation and habitat fragmentation on population declines and synergistic effects if environmental

Fig. 16.5
Multiple drivers of biodiversity change.
(A) Experimental manipulations of the cumulative effects of harvesting, immigration (as a measure of habitat fragmentation), and environmental warming on population decline in a model organism (after Mora *et al.* 2007, with permission). **(B)** Effects of different socioeconomic and environmental factors on the regional variability of coral reef communities (adapted from Mora (2008)).

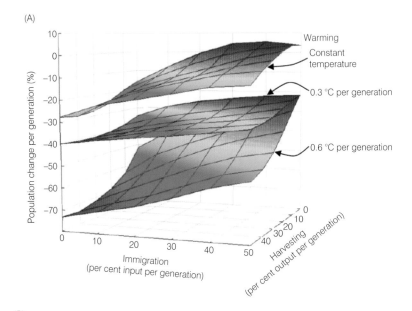

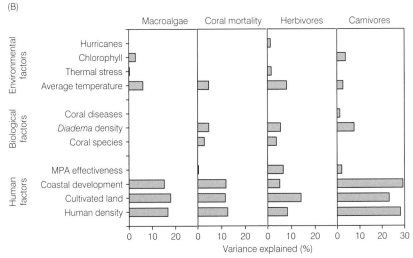

warming was also involved (Mora *et al*. 2007). Although each of these three factors individually caused populations to decline by similar amounts, all factors combined resulted in up to 50 times faster declines of experimental populations (Fig. 16.5A). These results highlight the importance of multiple human drivers for past and future population and biodiversity changes in the ocean.

Many human drivers have their ultimate roots in the social conditions and economic activities of society. Although some individuals, communities, or societies may exercise overexploitation, habitat destruction, and waste dumping, others promote successful stewardship and governance through harvest regulations, pollution controls, and the protection and restoration of species and habitats (Lotze & Glaser 2009; Worm *et al*. 2009). In one study, Mora (2008) analyzed socioeconomic and environmental databases together, to separate the proximate and ultimate drivers of coral reef degradation. Most of the local and regional variability in fishes, corals, and macroalgae was explained by human-related factors such as agricultural land use, coastal development, overfishing, and climate change (Fig. 16.5B). Significant ecological interactions among the different species groups further highlighted the need for a comprehensive management of human influences on coral reef ecosystems (Mora 2008).

16.3.4 Ecosystem consequences

What are the consequences of long-term population changes on the structure and functioning of marine ecosystems today and in the future? And how will changes in ecosystem structure affect the services marine ecosystems provide for human well-being? These questions have been difficult to tackle. First, ecosystem structure, functioning, and services are not easy to quantify and the relevant data are not readily available. This problem was overcome by compiling and analyzing long-term datasets on ecological, environmental, and socioeconomic changes in marine ecosystems. Second, a multitude of species, environmental drivers, and human impacts may interact in ways that are impossible to unravel from simple trend analyses. We therefore used ecosystem modeling approaches to determine overall changes in food web structure, energy flows, and stability. These modeling techniques may also serve to project future scenarios under changing environmental or human conditions, which we are currently exploring.

Long-term changes in population abundance as well as the loss (extinction) and gain (invasion) of species has changed the structure of many coastal ecosystems (Fig. 16.6A). This can have important effects on ecosystem functioning. Many species fulfill important ecological functions, including the provision of spawning, nursery, and foraging habitat by wetlands, underwater vegetation, and reef-building organisms (Fig. 16.6B). Most of these habitats also play an important role in filtering particles, nutrients, and pol-

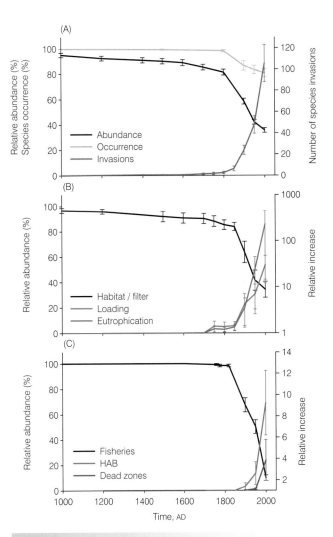

Fig. 16.6

Ecosystem consequences of marine biodiversity change. **(A)** Structural changes in 12 estuaries and coastal seas measured as the relative abundance and occurrence of species (as percent from historical baseline) and number of species invasions. **(B)** Changes in ecosystem functions such as habitat provision and water quality control expressed as nutrient loading and eutrophication response. **(C)** Loss of ecosystem services including the depletion of fisheries, health risks related to harmful algal blooms (HAB), and diminished recreation related to dead zones (data adapted from Lotze *et al.* (2006); Worm *et al.* (2006)).

lutants, thereby maintaining good water quality. If the functioning of coastal ecosystems is compromised, so are the ecosystem services provided for human well-being (Worm *et al*. 2006; Lotze & Glaser 2009). For example, overexploitation, habitat loss, and pollution have depleted many fisheries that previously provided food and employment (Fig. 16.6C). The loss of filter functions together with increasing municipal and industrial discharges have posed health risks to people through harmful algal blooms, contaminants, and disease. Finally, the expansion of oxygen-depleted zones, invasive species, and flooding compromises

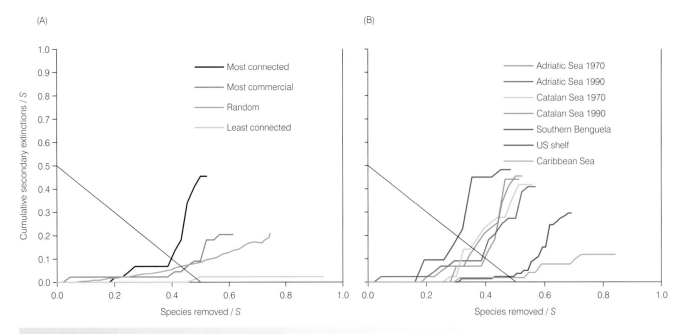

Fig. 16.7
Food web modeling. Robustness of food webs to the simulated extinction of **(A)** species that are most connected, most commercial, random, or least connected in the Adriatic food web in the 1970s, and **(B)** the most connected species in food webs from different regions and time periods. Diagonal black lines indicate where 50% of species are lost through combined removals and secondary extinctions (i.e. robustness) (adapted after Coll *et al.* (2008)).

recreation and shoreline safety (Worm *et al.* 2006). As human impacts spread offshore and expand to other ocean regions (Halpern *et al.* 2008), the observed changes in coastal oceans may forecast potential future changes in other habitats.

To understand better the ecosystem effects of marine biodiversity change, we used two different modeling approaches, stochastic network models and mass-balance food web models. First, basic food and interaction webs are assembled from data on species occurrence, abundance, feeding links, and other ecological information (Coll *et al.* 2008). Next, the two different models enabled us to analyze changes in up to 22 food web properties reflecting (among others) species composition, food-chain length, energy transfer between trophic levels, and the linkage density and complexity of the webs. For example, food webs in the Adriatic and Catalan Seas in the Mediterranean were found to be very similar in terms of their structure and functioning, but were more ecologically degraded compared with food webs from the Caribbean, Benguela, and US continental shelf (Coll *et al.* 2008). Food web properties estimated by both models yielded very similar results, thereby enhancing confidence in our results compared with any single modeling approach.

The network models also allowed us to analyze the robustness of food webs to simulated species loss (Coll *et al.* 2008). It had previously been shown that the removal of highly connected species in the food web results in a much higher rate of secondary extinctions than randomly deleted or less strongly connected species. In our analyses, removing the commercial species caused intermediate rates of secondary extinctions indicating that commercial species are often well connected in the food web (Fig. 16.7A). Finally, we could show that a larger degree of ecological degradation in the Mediterranean food webs resulted in a diminished robustness to species loss and a higher rate of secondary extinctions (Fig. 16.7B). This suggests that the degradation of marine ecosystems may accelerate the rate of biodiversity loss in the future.

16.4 Animal Movements

16.4.1 Previous work

Movement patterns and behavior of individual animals collectively contribute to broader-scale population distribution, species' ranges, and patterns of biodiversity. The development of a statistical toolbox for studying the movements of electronically tagged marine predators and collaboration with animal trackers provides a powerful complement to the global-scale studies of biodiversity and abundance conducted by FMAP. Our aim has been to elucidate the underlying mechanisms that determine animal distributions at the individual scale, and how these contribute to marine predator distribution and biodiversity patterns at broader scales. Much of our previous knowledge on marine animal movement patterns has been inferred

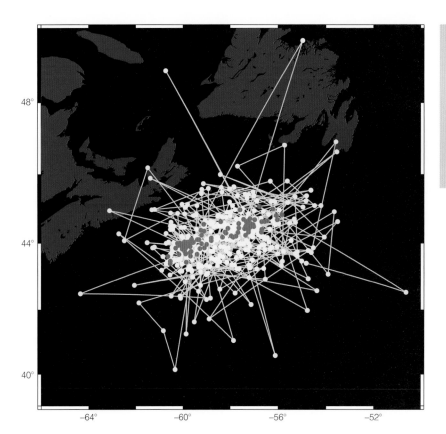

Fig. 16.8

Example of state filtering applied to data for a grey seal tagged on Sable Island, Nova Scotia, with a SMRU (Sea Mammal Research Unit) Argos satellite platform terminal transmitter with a 2-day duty cycle. The white points and lines denote the Argos-observed positions, which contain substantial error (note positions on land and some highly improbable movements). The red points denote the estimated true positions with a 2-day time step. Adapted from Jonsen *et al.* (2005).

from observations of species' departures and arrivals at geographically disparate locations (Carr 1986) or from traditional mark–recapture methods (Hilborn 1990). These studies were necessarily coarse in scale and yielded little insight into the interactions between foraging or migrating animals and their environment. Furthermore, these approaches only revealed movements to locations where observers were present, biasing estimates of movement rates and, more generally, our understanding of marine animal movement patterns.

All this has changed with the introduction of electronic tagging and telemetry technologies that revolutionized our view of animal movement patterns and distribution in the ocean (Block *et al.* 2001; Birdlife International 2004; James *et al.* 2005). Satellite and light-based geolocation tracking technologies now allow us to follow marine animals for protracted periods as they make their living in the ocean (see Chapter 15). The veritable explosion of tracking studies using increasingly refined technologies has revealed, for example, extraordinary 64,000-km round-trip migrations by sooty shearwaters (Shaffer *et al.* 2006), physiological mechanisms underlying niche expansion in salmon sharks (Weng *et al.* 2005), and previously unknown return migrations in white sharks (Bonfil *et al.* 2005). Despite these and many other success stories, our ability to track and document animal movements has far out-stripped our ability to conduct sophisticated analyses of rapidly amassing tracking data. This gap proved to be a fertile area of investigation for FMAP.

16.4.2 Statistical tools and key results

The project has played the leading role in developing a state-of-the-art statistical toolbox for electronic tracking data (Jonsen *et al.* 2003, 2005). This has been a multidisciplinary venture, involving biologists, ecological modelers, and statisticians. Our approach has focused on state-space models (SSMs), which are statistical time-series tools that, in the present context, allow one to estimate true animal positions from error-prone tracking data. State-space models have two components, a process model that describes how animals move from one position to the next and an observation model that relates the true, unobserved positions to the empirical tracking data.

In general, there can be two goals to fitting a SSM to tracking data. The first goal is to estimate the true positions of a tagged animal by accounting for the observation error inherent in the tracking data. This is called state filtering and yields a set of position estimates (and associated uncertainties) that occur over regular time intervals (Fig. 16.8). This kind of filtering is fundamentally different from traditional travel rate filters (McConnell *et al.* 1992) as all observed locations are modeled by a known probability distribution with implausible observations being downweighted rather than discarded. The result is that all information contained in the data is used to estimate the true positions.

Breed *et al.* (2006) used the state filtering approach to gain insight into the sexual segregation of seasonal foraging in adult grey seals breeding on Sable Island, Nova Scotia. Grey seals are an important generalist predator in the Scotian Shelf ecosystem, with a population that has experienced exponential growth over the past 35 years. From October to December and February to March, males used areas along the continental shelf break, whereas females used mid-shelf regions. Breed *et al.* (2006) suggested that this broad-scale segregation may help individuals maximize fitness by reducing intersexual competition during primary foraging periods.

The second goal of fitting a SSM is to estimate biological parameters or behavioral states specified in the process model, thereby allowing inference of unobservable processes that drive the movement and distribution patterns. The ability to construct biologically meaningful models and to estimate their parameters directly from complex, error-prone data is the most compelling facet of the SSM toolbox. For analyses of individual movement datasets, this approach is best achieved when individual datasets are combined meta-analytically (Jonsen *et al.* 2003). Meta-analysis facilitates synthesis of multiple datasets, enabling inference both within and among datasets, and improves parameter estimation from limited datasets.

State-space models allow researchers to think about questions that have no conventional solution. Jonsen *et al.* (2006) highlighted this by showing that endangered leatherback turtles migrating throughout the North Atlantic slow down at night, perhaps to feed on macrozooplankton that migrate toward the ocean surface and/or because their navigation abilities are less precise than during the day. One problem with this is that calculating travel speeds becomes difficult when travel distance is small during a single day or night period (no more than 30 km) compared with the uncertainty in the observed positions (up to 250 km). Conventional analyses of these data are not able to reveal the patterns in day versus night travel rates that the SSM analysis can (Fig. 16.9). Furthermore, Jonsen *et al.* (2006) showed that a Bayesian meta-analytic SSM, a model that estimates day versus night travel rates simultaneously from all datasets, yielded superior estimates for individual turtles and provided the basis for prediction at a population level.

A particularly compelling application of the SSM allows one to infer the (hidden) behavioral state of animals based upon the shifts in movement patterns observed in the tracking data. To derive, for example, the probability of an animal foraging versus migrating, a switching model can be added to the standard SSM (Jonsen *et al.* 2005). Key to this approach are the underlying assumptions that animal movements can be modeled by a small set of correlated random walks, each corresponding to a unique behavioral state, and that animals typically engage in area-restricted type movements (such as slow travel rates with a high frequency of turning) when searching for and consuming prey (Jonsen

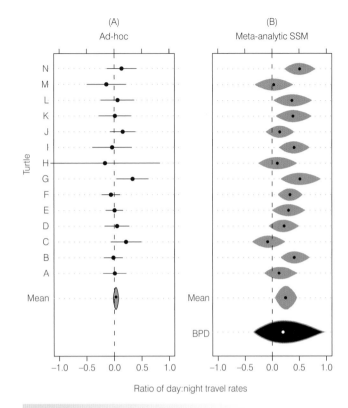

Fig. 16.9

Comparison of the ratio of day-to-night travel rates for a conventional ad-hoc calculation of travel rates **(A)** and a meta-analytic SSM fit to all datasets simultaneously **(B)**. The ratios presented in **(A)** are means with 95% confidence intervals, and those presented in **(B)** are posterior modes with 95% credible interval raindrops. The black raindrop is the Bayesian predictive distribution (BPD), which forms the basis for making predictions at a population or other appropriate level. Adapted from Jonsen *et al.* (2006).

et al. 2007). The switching model is used to estimate the probabilities that an animal is in a particular behavioral state, conditional upon the previous behavioral state. This approach offers a powerful tool to infer animal behaviors from remote tracking data and is being used by FMAP collaborators to analyze migration and foraging patterns in Pacific leatherback turtles (Bailey *et al.* 2008; Shillinger *et al.* 2008). Similar approaches have been adopted for analyzing foraging behaviors of southern bluefin tuna (Patterson *et al.* 2009).

FMAP researchers have used the switching SSM approach to identify potential foraging areas in space and time, a first step in quantifying critical foraging habitat and in developing mechanistic predictions of the potential influence of climate variability and changing distribution of foraging predators. Breed *et al.* (2009) show that adult male and female grey seals forage in different areas of the Scotian Shelf but both sexes tend to focus on relatively small, intensely used foraging sites (Fig. 16.10A). This

(A)

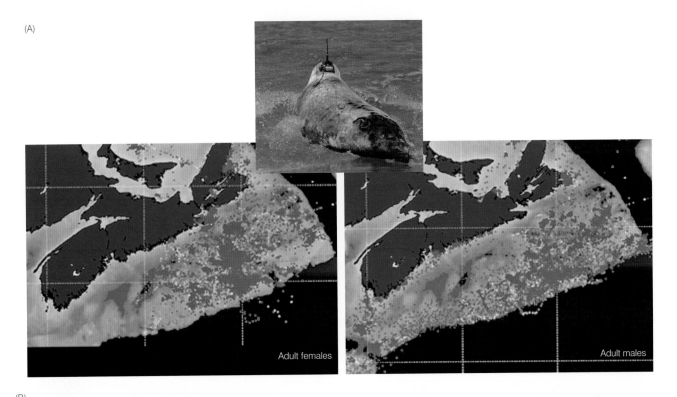

Adult females

Adult males

(B)

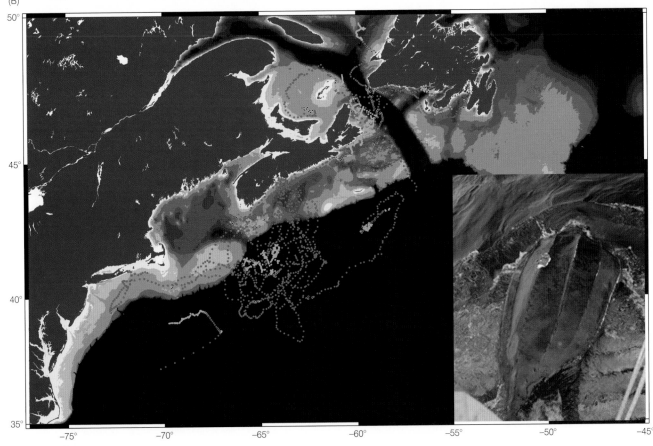

Fig. 16.10

Foraging patterns of adult male and female grey seals **(A)** and adult or sub-adult leatherback turtles **(B)** inferred by a switching SSM. In panel **(A)**, blue points are positions associated with foraging behavior and red points are positions associated with transiting behavior. Color intensity indicates the degree of uncertainty in the behavioral state estimates (less intense is less certain). In panel **(B)**, light blue points are positions associated with foraging behavior, red points are positions associated with transiting behavior, and yellow points are positions with uncertain behavioral state estimates. Image of adult female grey seal leaving Sable Island with an Argos satellite tag and VHF radio tag is courtesy of W.D. Bowen. Image of an Argos satellite-tagged leatherback turtle returning to the water after tagging is courtesy of the Canadian Sea Turtle Network. Adapted from Breed *et al*. (2009) **(A)** and Jonsen *et al*. (unpublished data) **(B)**.

Box 16.2

Key results: Past–Present–Future

- Marine biodiversity has been changing profoundly over the past 100–1,000 years in coastal regions and the past 50 years in open ocean regions.

- A reduction of up to 90% in the abundance of large, commercially exploited megafauna has occurred in these regions, along with a reduction of total animal biomass and the local extinction of particular species.

- These changes in abundance and diversity have negatively impacted ecosystem productivity and resilience, and compromised water quality, fishery yields, and other ecosystem services.

- Conservation and management efforts over the past century have halted or even reversed the decline of biodiversity in some areas.

- In areas where there are few management initiatives in place, the abundance and diversity of marine animal populations continues to decline, mostly in direct relation to multiple human impacts.

- Modeling approaches have allowed us to do the following: (1) think about problems that have no conventional solution; (2) synthesize data of varying quality and type; (3) make quantitative predictions about future trends.

- Spatial patterns of animal behaviors are markedly discrete and predictable both seasonally and inter-annually, implying strong connections to environmental drivers and prey distribution.

- Sea surface temperature is the primary oceanographic driver of marine animal distribution and diversity at global scales. Therefore, in addition to the effects of exploitation, changes in temperature have large effects on marine diversity patterns, as well as the behavior of individual animals.

- The future of marine animal populations may be determined in large part by two key variables: the rate of ocean warming and the rate of exploitation. Where those rates are low, it will increase the chance for adaptation and recovery. Where they continue to rise, the loss of marine biodiversity and associated services will likely be severe.

---

switching SSM analysis builds on previous efforts that suggested grey seals foraged over broad areas of the shelf (Breed *et al.* 2006) by resolving spatial patterns in movement behaviors hidden within the tracking data (Breed *et al.* 2009). Similar patterns of concentrated foraging activity, albeit at a much broader spatial scale, are emerging in an ongoing analysis of leatherback turtle satellite data (Fig. 16.10B). Figure 16.10B presents only a small, but fairly representative, subset of the results so far. Switching SSM analysis of 36 individual tracks, spanning a period from 1999 to 2006, suggests that foraging activity is concentrated along the slope waters of the Scotian Shelf and around Cape Breton in Northeastern Nova Scotia. Both regions are highly productive and likely support large seasonal aggregations of the leatherback's prey, jellyfish.

What is striking about both of these analyses is the distinct patchiness of foraging activity exhibited by both species. Grey seals are benthic foragers and are predominantly tied to shallow banks on the continental shelf, whereas leatherback turtles forage on jellyfish both in the coastal and pelagic realms. Regardless, both species show distinct and annually predictable preferences for relatively discrete regions. Understanding how these behavioral patterns may change in the future as a result of changes in prey distribution and abundance, and as a result of climate variability can provide valuable insight into mechanisms underlying future change in species biodiversity and abundance. Current work is focusing on the biophysical characterization of species' foraging areas and development of switching SSMs that can directly relate environmental gradients to the behavioral patterns hidden within electronic tracking data.

16.5 Concluding Remarks

The FMAP Project has been active from 2002 to 2010 as the modeling component of the Census. Our emphasis has been on data synthesis to reveal broad patterns of diversity, abundance, and distribution of marine animals, with a particular emphasis on large predators. We have developed data analysis and modeling techniques that enabled us to estimate the diversity of known and unknown species in the ocean, to derive long-term trends, short-term dynamics, and spatial patterns of diversity change, and to understand better the distribution and behavior of individual species.

In our communications we have strived to use the knowledge gained to inform society about both the drivers of biodiversity change, as well as current and possible future consequences of biodiversity loss for marine ecosystems and human society. Major results and patterns derived from this project are listed in Box 16.2. In interpreting this body of new knowledge, it is important to realize that these patterns do not play out equally everywhere. A necessary weakness inherent in large-scale data synthesis is that individual differences may be lost in averaging across a large population of animals, different regions, and environmental conditions. It may be these differences, however, existing between one place and another, or one population and another, that hold the key to understanding the ocean's future (Worm *et al.* 2009). There is no doubt that, given our large and growing influence on the marine environment, future societal decisions will drive the trajectory of change in marine animal populations and the ecosystems in which they are embedded. Within FMAP we can highlight past and current trends, and assume different scenarios as to how these may extend into the future. What will actually happen, however, is unknown, as societal choices and technological change will determine future changes. This process will certainly be influenced by the availability of new scientific information, and its perception among the public and decision-makers. It is our hope that the FMAP Project will continue to provide such information into the foreseeable future.

Acknowledgments

This chapter and the research highlighted in it have been supported by the Census of Marine Life, funded by the Sloan Foundation. Additional funding came from the Natural Sciences and Engineering Research Council of Canada, the Lenfest Oceans Program, and the Canadian Department of Fisheries and Oceans. We acknowledge the leadership of the late Ransom A. Myers, former Principal Investigator of FMAP, as well as numerous students and researchers from FMAP and other Census projects that made this work possible. We especially thank FMAP founding members Hiroyki Matsuda and Gunnar Steffansson.

References

Attrill, M.J. & Power, M. (2002) Climatic influence on a marine fish assemblage. *Nature* 41, 275–278.

Bailey, H., Shillinger, G., Palacios, D., *et al.* (2008) Identifying and comparing phases of movement by leatherback turtles using state-space models. *Journal of Experimental Marine Biology and Ecology* 356, 128–135.

BirdLife International (2004) Tracking ocean wanderers: the global distribution of albatrosses and petrels. Results from the Global Procellariiform Tracking Workshop, 1–5 September, 2003, Gordon's Bay, South Africa. Cambridge, UK: BirdLife International.

Block, B.A., Dewar, H., Blackwell, S.B., *et al.* (2001) Migratory movements, depth preferences, and thermal biology of Atlantic bluefin tuna. *Science* 293, 1310–1314.

Bonfil, R., Meyer, M., Scholl, M.C. *et al.* (2005) Transoceanic migration, spatial dynamics, and population linkages of white sharks. *Science* 310: 100–103.

Botkin, D.B., Saxe, H., Araujo, M.B., *et al.* (2007) Forecasting the effects of global warming on biodiversity. *BioScience* 57, 227–236.

Bouchet, P. (2006) The magnitude of marine biodiversity. In: *The Exploration of Marine Biodiversity: Scientific and Technological Challenges* (ed. C. Durate), pp. 31–62. Bilbao, Spain: Fundación BBVA.

Boyce, D., Tittensor, D. & Worm, B. (2008) Effects of temperature on global patterns of tuna and billfish richness. *Marine Ecology Progress Series* 355, 267–276.

Breed, G.A., Bowen, W.D., McMillan, J.I. & Leonard, M.L. (2006) Sexual segregation of seasonal foraging habitats in a non-migratory marine mammal. *Proceedings of the Royal Society of London B* 273, 2319–2326.

Breed, G.A., Jonsen, I.D., Myers, R.A., *et al.* (2009) Sex-specific, seasonal foraging tactics of adult grey seals (*Halichoerus grypus*) revealed by state-space analysis. *Ecology* 90, 3209–3221.

Carr, A. (1986) Rips, FADS, and little loggerheads. *BioScience* 36, 92–100.

Cheung, W.W.L., Lam, V.W.Y., Sarmiento, J.L., *et al.* (2009) Projecting global marine biodiversity impacts under climate change scenarios. *Fish and Fisheries* 10, 235–251.

Clark, M.R., Tittensor, D.P., Rogers, A.D., *et al.* (2006) Seamounts, deep-sea corals and fisheries: vulnerability of deep-sea corals to fishing on seamounts beyond areas of national jurisdiction. Cambridge, UK: United Nations Environment Programme – World Conservation Monitoring Centre (UNEP-WCMC).

Coll, M., Lotze, H.K. & Romanuk, T.N. (2008) Structural degradation in Mediterranean Sea food webs: testing ecological hypotheses using stochastic and mass-balance modelling. *Ecosystems* 11, 939–960.

Cooper, H. & Hedges, L.V. (1994) *The Handbook of Research Synthesis*. New York: Russell Sage Foundation.

Gaston, K.J. (2000) Global patterns in biodiversity. *Nature* 405, 220–227.

Grassle, J.F. & Maciolek, N.J. (1992) Deep-sea species richness: regional and local diversity estimates from quantitative bottom samples. *American Naturalist* 139, 313–341.

Halpern, B.S., Walbridge, S., Selkoe, K.A., *et al.* (2008) A global map of human impact on marine ecosystems. *Science* 319, 948–952.

Harley, C.D.G., Hughes, A.R., Hultgren, K.M., *et al.* (2006) The impacts of climate change in coastal marine systems. *Ecology Letters* 9, 228–241.

Harvell, C.D., Mitchell, C.E., Ward, J.R., *et al.* (2002) Climate warming and disease risks for terrestrial and marine biota. *Science* 296, 2158–2162.

Hilborn, R. (1990) Determination of fish movement patterns from tag recoveries using maximum likelihood estimators. *Canadian Journal of Fisheries and Aquatic Sciences* 47, 635–643.

Hillebrand, H. (2004) Strength, slope and variability of marine latitudinal gradients. *Marine Ecology Progress Series* 273, 251–267.

Hyrenbach, K.D., Forney, K.A. & Dayton, P.K. (2000) Marine protected areas and ocean basin management. *Aquatic Conservation: Marine and Freshwater Ecosystems* 10, 437–458.

Jackson, J.B.C. & Erwin, D.H. (2006) What can we learn about ecology and evolution from the fossil record. *Trends in Ecology and Evolution* 21, 322–328.

James, M.C., Ottensmeyer, C.A. & Myers, R.A. (2005) Identification of high-use habitat and threats to leatherback sea turtles in northern waters: new directions for conservation. *Ecology Letters* 8, 195–201.

Jonsen, I.D., Myers, R.A. & Mills Flemming, J. (2003) Meta-analysis of animal movement using state-space models. *Ecology* 84, 3055–3063.

Jonsen, I.D., Mills Flemming, J. & Myers, R.A. (2005) Robust state-space modelling of animal movement data. *Ecology* 86, 2874–2880.

Jonsen, I.D., Myers, R.A. & James, M.C. (2006) Robust hierarchical state-space models reveal diel variation in travel rates of migrating leatherback turtles. *Journal of Animal Ecology* 75, 1046–1057.

Jonsen, I.D., Myers, R.A. & James, M.C. (2007) Identifying leatherback turtle foraging behaviour from satellite telemetry using a switching state-space model. *Marine Ecology Progress Series* 337, 255–264.

Lotze, H.K. & Glaser, M. (2009) Ecosystem services of semi-enclosed marine systems. In: *Watersheds, Bays and Bounded Seas* (eds. E.R. Urban, Jr. et al.), pp. 227–249. Washington, DC: Island Press.

Lotze, H.K., Lenihan, H.S., Bourque, B.J., et al. (2006) Depletion, degradation, and recovery potential of estuaries and coastal seas. *Science* 312, 1806–1809.

Lotze, H.K. & Milewski, I. (2004) Two centuries of multiple human impacts and successive changes in a North Atlantic food web. *Ecological Applications* 14, 1428–1447.

Lotze, H.K., Reise, K., Worm, B., et al. (2005) Human transformations of the Wadden Sea ecosystem through time: a synthesis. *Helgoland Marine Research* 59, 84–95.

Lotze, H.K. & Worm, B. (2009) Historical baselines for large marine animals. *Trends in Ecology and Evolution* 24, 254–262.

McConnell, B.J., Chambers, C. & Fedak, M.A. (1992) Foraging ecology of southern elephant seals in relation to bathymetry and productivity of the Southern Ocean. *Antarctic Science* 4, 393–398.

Mora, C. (2008) A clear human footprint in the coral reefs of the Caribbean. *Proceedings of the Royal Society of London B* 275, 767–773.

Mora, C., Metzger, R., Rollo, A. & Myers, R.A. (2007) Experimental simulations about the effects of overexploitation and habitat fragmentation on populations facing environmental warming. *Proceedings of the Royal Society of London B* 274, 1023–1028.

Mora, C., Tittensor, D.P. & Myers, R.A. (2008) The completeness of taxonomic inventories for describing the global diversity and distribution of marine fishes. *Proceedings of the Royal Society of London B* 275, 149–155.

Myers, R.A. & Worm, B. (2003) Rapid worldwide depletion of predatory fish communities. *Nature* 423, 280–283.

Myers, R.A. & Worm, B. (2005) Extinction, survival or recovery of large predatory fishes. *Philosophical Transactions of the Royal Society of London B* 360, 13–20.

Orr, J.C., Fabry, V.J., Aumont, O., et al. (2005) Anthropogenic ocean acidification over the twenty first century and its impact on calcifying organisms. *Nature* 437, 681–686.

Oschlies, A. & Garçon, V. (1998) Eddy-induced enhancement of primary production in a model of the North Atlantic Ocean. *Nature* 394, 266–269.

Patterson, T.A., Basson, M., Bravignton, M.V. & Gunn, J.S. (2009) Classifying movement behaviour in relation to environmental conditions using hidden Markov models. *Journal of Animal Ecology* 78, 1113–1123.

Polovina, J.J., Howellb, E., Kobayashia, D.R. & Sekia, M.P. (2001) The transition zone chlorophyll front, a dynamic global feature defining migration and forage habitat for marine resources. *Progress in Oceanography* 49, 469–483.

Roberts, C.M. (2002) Deep impact: the rising toll of fishing in the deep sea. *Trends in Ecology and Evolution* 242, 242–245.

Roberts, C.M., McClean, C.J., Veron, J.E.N., et al. (2002) Marine biodiversity hotspots and conservation priorities for tropical reefs. *Science* 295, 1280–1284.

Rutherford, S., D'Hondt, S. & Prell, W. (1999) Environmental controls on the geographic distribution of zooplankton diversity. *Nature* 400, 749–753.

Shaffer, S.A., Tremblay, Y., Weimerskirch, H., et al. (2006) Migratory shearwaters integrate oceanic resources across the Pacific Ocean in an endless summer. *Proceedings of the National Academy of Sciences of the USA* 103, 12799–12802.

Shillinger, G.L., Palacios, D.M., Bailey, H. et al. (2008) Persistent leatherback turtle migrations present opportunities for conservation. *PLoS Biology* 6, e171.

Stramma, L., Johnson, G.C., Sprintall, J. & Mohrholz, V. (2008) Expanding oxygen-minimum zones in the tropical oceans. *Science* 320, 655–658.

Sund, P.N., Blackburn, M. & Williams, F. (1981) Tunas and their environment in the Pacific Ocean: a review. *Oceanography and Marine Biology* 19, 443–512.

Tittensor, D., Baco, A., Brewin, P., et al. (2009b) Predicting global habitat suitability for stony corals on seamounts. *Journal of Biogeography* 36, 1111–1128.

Tittensor, D., Micheli, F., Nyström, M. & Worm, B. (2007) Human impacts on the species–area relationship in reef fish assemblages. *Ecology Letters* 10, 760–772.

Tittensor, D., Worm, B. & Myers, R.A. (2009a) Macroecological changes in exploited marine systems. In: *Marine macroecology*. (eds. J.D. Witman & K. Roy), pp. 310–317. Chicago: University of Chicago Press.

Vermeij, G.J. (2004) Ecological avalanches and the two kinds of extinction. *Evolutionary Ecology Research* 6, 315–337.

Weng, K.C., Castilho, P.C., Morrissette, J.M., et al. (2005) Satellite tagging and cardiac physiology reveal niche expansion in salmon sharks. *Science* 310, 104–106.

Whitehead, H., McGill, B. & Worm, B. (2008) Diversity of deep-water cetaceans in relation to temperature: implications for ocean warming. *Ecology Letters* 11, 1198–1207.

Worm, B., Barbier, E.B., Beaumont, N., et al. (2006) Impacts of biodiversity loss on ocean ecosystem services. *Science* 314, 787–790.

Worm, B., Lotze, H.K. & Myers, R.A. (2003) Predator diversity hotspots in the blue ocean. *Proceedings of the National Academy of Sciences of the USA* 100, 9884–9888.

Worm, B. & Lotze, H.K. (2009) Changes in marine biodiversity as an indicator of climate change. In: *Climate Change: Observed Impacts on Planet Earth* (ed. T. Letcher), pp. 263–279. Amsterdam: Elsevier.

Worm, B., Sandow, M., Oschlies, A., et al. (2005) Global patterns of predator diversity in the open oceans. *Science* 309, 1365–1369.

Worm, B., Hilborn, R., Baum, J.K., et al. (2009) Rebuilding global fisheries. *Science* 325, 578–585.

PART VI
Using the Data

Chapter 17

Data Integration: The Ocean Biogeographic Information System

Edward Vanden Berghe[1], Karen I. Stocks[2], J. Frederick Grassle[1]

[1]*Institute of Marine and Coastal Sciences, Rutgers University, New Brunswick, New Jersey, USA*
[2]*San Diego Supercomputer Center, University of California San Diego, La Jolla, California, USA*

17.1   Introduction

Informed management of the environment has to be supported by data (Richardson & Poloczanska 2008; Stokstad 2008). Often marine biological data are the result of projects with a limited taxonomic, temporal, and spatial cover. Taken in isolation, datasets resulting from these projects are only of limited use in the interpretation of large-scale phenomena. More specifically, they fail to inform on a scale commensurate with the problems humankind is confronted with: pollution, global change, invasive species, harmful algal blooms, and the loss of biodiversity to name but a few. Individual studies are restricted in the amount of data they can generate but, by combining the results from many studies, massive databases can be created, making possible analyses on a more relevant, much larger scale. It is the ambition of the Ocean Biogeographic Information System (OBIS; www.iobis.org; see Box 17.1) community to provide a sound basis for management decisions by integrating data from many sources, and thus facilitating badly needed regional, ecosystem, and global analyses. OBIS does so by facilitating publication of data, and stimulating open and free access for all potential users. Indeed, OBIS is often mentioned as the organization best suited for this role (see, for example, Poloczanska *et al.* 2008).

OBIS was conceived as the data integration component of the Census of Marine Life (Box 17.1). It is very much a "work in progress": we know that many important datasets are not available through OBIS. However, we do think that the present content is sufficient to start exploring global patterns of biodiversity, taking into account a wide range of life forms; this exercise was not possible before OBIS brought the relevant data together into one consolidated, quality-controlled system.

In the first part of this chapter, we discuss some of the issues we encountered while working on OBIS. In the second part, development of OBIS, in terms of both technology and content, is discussed. In the third part, some of the possible analyses are illustrated, and the content of the database is explored.

17.2   List of Acronyms

CoL:	Catalogue of Life
CPR:	Continuous Plankton Recorder
FAO:	Food and Agriculture Organization
GBIF:	Global Biodiversity Information Facility
GCMD:	Global Change Master Directory
GEO:	Group on Earth Observations
ICES:	International Council for the Exploration of the Sea
iOBIS:	International OBIS (secretariat and portal based at Rutgers University)
IOC:	Intergovernmental Oceanographic Commission

Life in the World's Oceans, edited by Alasdair D. McIntyre
© 2010 by Blackwell Publishing Ltd.

Box 17.1

OBIS "Biography"

The Ocean Biogeographic Information System (OBIS) is an online, user-friendly system for absorbing, integrating, and assessing data about life in the oceans. It is recognized by many as the prime provider of information on the distribution of marine species. OBIS aims to stimulate new research that generates new hypotheses about evolutionary processes and species distributions by providing software tools for data exploration and analysis. All data are freely available over the Internet and interoperable with similar databases. OBIS integrates data from many sources, over a wide range of marine themes, from poles to equator, from microbes to whales. It is the largest provider of information on the distribution of marine species, and one of the largest contributors to Global Biodiversity Information Facility (GBIF). Any organization, consortium, project, or individual may contribute. OBIS was created as the data integration component of the Census of Marine Life; the international portal is hosted by Rutgers University, New Jersey, USA. A global network of 15 Regional and Thematic OBIS Nodes assures the worldwide scientific support needed to fulfill the global mandate.

IODE:	International Oceanographic Data and Information Exchange
LME:	Large Marine Ecosystems
MarBEF:	Marine Biodiversity and Ecosystem Functioning
NOPP:	National Oceanographic Partnership Program (USA)
NSF:	National Science Foundation (USA)
OBIS:	Ocean Biogeographic Information System
OBIS SEAMAP:	OBIS Spatial Ecological Analysis of Megavertebrate Populations
RON:	Regional OBIS Node
SAHFOS:	Sir Alister Hardy Foundation for Ocean Science, www.sahfos.ac.uk/
WOA:	World Ocean Atlas (published by the World Data Center for Oceanography, Silver Spring)
WOD:	World Ocean Database (published by the World Data Center for Oceanography, Silver Spring)
WoRMS:	World Register of Marine Species

17.3 The Data Sharing Challenge

The willingness to share data is a prerequisite to data portals. Advantages of sharing data are clear and numerous, and have prompted many organizations, including the International Council for Science (ICSU) and the Intergovernmental Oceanographic Commission (IOC), to adopt a policy of open access to data. The physical oceanographers have set an example with the World Ocean Database (WOD) and derived products such as World Ocean Atlas (WOA), published by the US National Oceanic and Atmospheric Administration (NOAA) (Boyer *et al.* 2006). Much of our understanding of global patterns is based on these global databases (see, for example, Levitus 1996; Conkright & Levitus 1996). The advantages might be clear, but practice is often lacking. This led the participants at the Ocean Biodiversity Informatics (OBI) conference in Hamburg, 2004, to formulate a public statement summarizing the benefits (Box 17.2) (Vanden Berghe *et al.* 2007a).

Here are a few of the benefits of data sharing.

- Sharing data is a way to avoid data loss related to institutional discontinuities or poor archiving (Froese *et al.* 2003); the very fact of sharing data creates redundancy, and this will assist in recovery of data after accidental destruction of a dataset.

- Sharing data makes the data more visible, and so increases the opportunities to create collaborative ventures with scientists outside the immediate environment.

- It facilitates re-use of the data for purposes that they were not originally collected for; every time a datum is used in some analysis or consulted through a website, society's return on investment in collecting the data increases.

- Not all countries are fortunate enough to have the expertise and/or the resources to set up data management systems of their own; data sharing ventures can be the framework for data repatriation to developing countries, and assist them in fulfilling their

Box 17.2

Public Statement from OBI Conference in Hamburg, 2004

We note that increased availability and sharing of data

- is good scientific practice and necessary for advancement of science
- enables greater understanding through more data being available from different places and times
- improves quality control due to better data organization, and discovery of errors during analysis
- secures data from loss

The advantages of free and open data sharing have been determining factors while developing the data exchange policy of the Intergovernmental Oceanographic Commission of UNESCO.

We call on scientists, politicians, funding agencies and the community to be proactive in recognizing data's

- overall cost/benefit
- importance to science
- long-term benefits to society and the environment
- increased value by being publicly available

We also call upon employers of scientists, academic institutions and funding agencies and editors of scientific journals, to

- promote on-line availability of data used in published papers
- promote comprehensive documentation of data, including metadata and information on the quality of the data
- reward on-line publication of peer reviewed electronic publications and on-line databases in the same way conventional paper publications are rewarded in the hiring and promotion of scientists
- encourage and support scientists to share currently unavailable data by placing it in the public domain in accordance with publicly available standards, or in formats compatible with other users

reporting obligations in the framework of international conventions such as the Convention on Biological Diversity.

- Last but not least, by sharing data it becomes possible to create the large data systems we need to support proper management of our natural resources.

Any initiative relying on the willingness to share data has to take into account the sociology of science: data owners will have to see clearly the advantages of sharing data, and will need incentives to do so. Scientists have to be compensated for the time that they spend making the data available for re-use, and for the loss of exclusive access to the data, and the competitive advantage associated with this. An obvious example of such an incentive is when data are shared between several data providers, with the intent to analyze the pooled dataset and to publish the results jointly. Examples include the North Sea Benthos Project of the International Council for the Exploration of the Sea (ICES) (Rees *et al.* 2007; Vanden Berghe *et al.* 2007b); MacroBen (Somerfield *et al.* 2009; Vanden Berghe *et al.* 2009); and other initiatives of the European Union (EU) Network of Excellence "Marine Biodiversity and Ecosys-

tem Functioning" (MarBEF). The incentive is, in this case, clearly the opportunity to analyze a larger dataset than the one available from a single data provider, and to become a co-author on the resulting papers.

However, the model of co-authorship as incentive for data sharing does not scale: it is not tenable with large databases such as OBIS or the WOD/WOA. There are too many individual data contributors, so papers based on the complete dataset would have to list thousands of authors. Also, even if the number of data contributors were more reasonable, it does not always make sense for people to become co-author; in principle, anyone listed as an author on a paper should have made a direct intellectual contribution to the paper, and share responsibility for the conclusions. A recent trend to include too many colleagues as co-authors is putting pressure on science's credit system (Greene 2007; Sekercioglu 2008). In many cases, citation of the source of the data would be more appropriate. However, this needs a formal system of indexing, just as the citations of "classical" publications are indexed by the Institute for Scientific Information (ISI). And, of course, use or re-use of a dataset should contribute to the career advancement of any person involved in the collection or management of the data. Several initiatives have started to address data citation. There is a working group of the Global Biodiversity Information Facility (GBIF) discussing this issue, organized in response to a discussion at the e-Biosphere conference; another working group, jointly organized by the Scientific Council for Oceanographic Research (SCOR) and the International Oceanographic Data and Information Exchange (IODE), recently published a first report (SCOR & IODE 2008).

When trying to persuade someone to do something, one has the choice of using a carrot or a stick. Data citation and co-authorship are clear examples of the former. However, the stick can also be used creatively and fairly, with everyone having to comply with the same rules. The prime example of appropriate use of the "stick" is the requirement by several major scientific journals to publish gene sequences in GenBank or a similar public and openly accessible repository before the paper is published. The information itself is shared and made public through GenBank, and the papers cite the accession number. At the same time, the GenBank information becomes citable through this accession number, so that it works to the advantage of the scientist depositing the sequence information. It is an excellent example, and a possible model for the biogeographic community. Many journals now have a policy of asking authors to make their data available after publication (see, for example, *Science*; www.sciencemag.org/about/authors/prep/gen_info.dtl#dataavail). However, it seems that these requests are not enforced, and that the GenBank strategy of asking for inclusion of the accession number in the paper is a better guarantee that data will be made public.

Data are often collected using public funding, so many feel that for this reason alone they should be publicly available; sometimes there is a contractual obligation to make data available after publication of results. Funding agencies finance research to further our understanding of the environment; withholding raw data hampers the process by which the results of the funded activities can be used, thus clearly contravening the original intention of the support (Dittert *et al.* 2001). One of the roles of a data portal such as OBIS is to offer a service assisting beneficiaries of public funding in fulfilling their contractual obligations.

Too many datasets are lying dormant, some of them on hard drives, often in difficult-to-access electronic formats; others are only available on paper. The physical oceanographers have set an example with the Global Data Archaeology and Rescue (GODAR) project, through which many datasets, at risk of being lost, were recovered and integrated into the WOD. The cost of "recovering" data is typically only a fraction of the cost of collecting the samples and generating the data. In the case of a Guinean trawling survey, the data recovery cost 0.2% of the initial survey cost (Zeller *et al.* 2005). More important even than these economic arguments is the historic aspect of environmental data: they are irreplaceable, and once lost they cannot be collected again.

Metadata, data about the data, are essential when sharing data. They make it possible for users to judge fitness-for-use (Chapman 2005), so that they are not inadvertently used for purposes for which they are not suited; part of this fitness-for-use statement is a description of quality control and quality assurance methods applied to the data. Metadata facilitate data discovery through their inclusion in metadata repositories such as the Global Change Master Directory (GCMD; gcmd.gsfc.nasa.gov) of the National Aeronautics and Space Administration (NASA). They are essential in creating an audit trail, so that any datum can be traced to its origin. Part of the audit trail is a list of all those involved in collecting, managing, and controlling the quality of the data, which makes it possible to give appropriate credit.

Making data publicly available is a critical step, but it is only the first. To be available for large-scale analysis, data have to be integrated and their quality controlled. Creating these integrated databases is a second step up the ladder from raw data through information and knowledge to wisdom (Fig. 17.1). Data integration requires knowledge about the data being handled, and often is a time-consuming business; it is important to avoid duplication of effort, and to preserve any efforts expended. Without mechanisms to preserve these efforts, any large-scale analysis would have to redo this step of data integration.

An important aspect of the integration of individual datasets is to check for consistency between them, and where inconsistencies are found, to resolve them. Obvious examples here are the spelling of taxonomic names, or detection of outlier distribution points caused by misidentification or errors of georeferencing. This reconciliation process is an extra opportunity for quality control, in addition to what is possible at the level of single datasets; conflicts between datasets are flags for potential problems. Data warehouses such as OBIS can add value by resolving these inconsistencies in consultation with specialists and end users, and with the original data providers. Quality-control procedures have to be documented, so that end users can judge whether data are reliable enough for their purposes.

Fig. 17.1

The Wisdom Pyramid. Reproduced with permission, from a presentation by C. Besancon, UNEP-WCMC.

Neither data managers nor data users should be fooled into thinking that there is such a thing as a database without errors. No matter how much time goes into quality control, there always will be a certain error rate. It is by using the data, sharing it with others to do their analyses, and critically looking at the results that erroneous data can be detected. It is important for any data system to have a mechanism for capturing this information, by making sure that there are mechanisms for user feedback, and by promptly acting on such feedback. In those cases where there are several levels of aggregation (as is the case for many of the OBIS datasets), this can lead to complications: errors detected at a higher level of aggregation (for example, at the level of OBIS or GBIF) have to be communicated to and corrected by the original data provider. Obviously, at any step in this communication things can go wrong, with delays in correcting obvious mistakes, and frustrated end users as a result.

Data integration comes at a price: it is rarely possible to integrate data over many sources without losing detail. Information on sampling devices or sampling effort is difficult to standardize across many data sets. Temporary taxonomic names make sense within one study but not with several studies (Paterson *et al.* 2000). The opportunistic exploitation of available resources will usually result in very unequal sampling in the area of interest, because the sampling effort is governed by external factors that are not under the control of the data manager. Any analysis based on such data collections has to deal with heavy observational bias. However, these drawbacks should be weighed against the larger footprint of the data, and hence stronger signals. For example, combining several datasets to create a consolidated dataset with a much larger latitudinal range will increase any latitudinal gradient, and make this gradient easier to discover. Also, the increased number of observations will result in an increase in statistical power of any analysis done on the combined dataset.

All data published through the iOBIS portal are freely and openly available to anyone who respects the terms of use, as described on the website. In principle, the user is asked to acknowledge the use of the OBIS portal, and to cite datasets downloaded and used in analysis (www.iobis.org/data/policy/citation). An important aspect is also to recognize the limitations of data in OBIS (www.iobis.org/data/policy/disclaimer). Some of the individual datasets have further restrictions, and those are conveyed to the user as part of the metadata of that dataset.

17.4 Development of OBIS

OBIS was created as the data integration component of the Census of Marine Life (Grassle & Stocks 1999; Grassle 2000; Yarincik & O'Dor 2005). From the start it was conceived as a global and distributed system, giving control of data to data providers (Fornwall 2000), with strong ties to existing national and international biodiversity information systems (Fornwall 2000; Grassle 2005). Today, OBIS has evolved into a community of practice, consisting of people and organizations sharing a vision to make marine biogeographic data, from all over the world, freely available over the World Wide Web. OBIS is not limited to data from Census-related projects; any organization, consortium, project, or individual may contribute to OBIS.

From the OBIS portal (the first website page connecting to the data), the user can do the following:

- search where a marine genus and/or species is recorded in the data published through OBIS;
- download data published in OBIS for any species, including location, depth, date and time collected, source datasets, and verified taxonomic name information;
- plot species locations on a range of flat and spherical views of the world, including polar views, using the C-Squares Mapper;
- plot species against background maps of sea temperature, depth, and salinity using the KGS Mapper;
- use environmental data for the locations of these data to predict the species potential range on the KGS Mapper;
- explore relationships between species and environmental data on KGS Mapper to see which parameters best explain a species distribution;
- browse down a taxonomic hierarchy to get lists of all species in OBIS for a phylum, class, order, or other higher taxonomic group;
- plot maps of all data at a higher taxonomic level;
- search for lists of species recorded in OBIS by country (exclusive economic zone), sea or ocean, large marine ecosystems (LMEs), Food and Agriculture Organization (FAO) and ICES fishery areas, Longhurst's pelagic regions, depth, date, and by entering latitude–longitude coordinates;
- connect to other sources of information on the species, including genetic data, published literature, and images.

A workshop critical to the genesis of OBIS was held in Rutgers University Institute of Marine and Coastal Sciences, New Jersey, in October 1997. The framework of the workshop was essentially that different groups were asked which project, to be completed on a scale of five to seven years, would most advance science. The strong consensus of the participants, consisting mainly of benthic ecologists, taxonomists, and statisticians, was to bring together and make publicly available the data that already existed, rather than new sampling campaigns, taking stock of what was known. From this OBIS was defined as "An on-line worldwide marine atlas 'infrastructure' providing scientists with

Table 17.1

Nine original OBIS projects.

- *The Fishnet Distributed Biodiversity Information System.*
  ○ Edward Wiley, Natural History Museum, University of Kansas.

- *Development of a Dynamic Biogeographic Information System: A Pilot Application for the Gulf of Maine.*
  ○ Dale Kiefer, Wrigley Institute of Environmental Studies, University of Southern California.

- *Biogeoinformatics of Hexacorallia (Corals, Sea Anemones, and their Allies): Interfacing Geospatial, Taxonomic, and Environmental Data for a Group of Marine Invertebrates.*
  ○ Daphne Fautin, University of Kansas, and Bob Buddemeier, Kansas Geological Survey.

- *Expansion of CephBase as a Biological Prototype for OBIS.*
  ○ Phillip Lee and James Wood, University of Texas Medical Branch.

- *A Biotic Database of Indo-Pacific Marine Mollusks.*
  ○ Gary Rosenberg, The Academy of Natural Sciences, Philadelphia.

- *ZooGene, a DNA Sequence Database for Calanoid Copepods and Euphausiids: An OBIS Tool for Uniform Standards of Species Identification.*
  ○ Ann Bucklin, University of New Hampshire Durham, New Hampshire; Bruce W. Frost, University of Washington, Seattle, Washington; Peter H. Wiebe, Woods Hole Oceanographic Institution, Woods Hole, Massachusetts; Michael J. Fogarty, NOAA/NMFS Northeast Fisheries Science Center, Woods Hole, Massachusetts.

- *Diel, Seasonal, and Interannual Patterns in Zooplankton and Micronekton Species Composition in the Subtropical Atlantic.*
  ○ Deborah Steinberg, Virginia Institute of Marine Sciences.

- *Census of Marine Fishes (CMF): Definitive List of Species and Online Biodiversity Database.*
  ○ William Eschmeyer, California Academy of Sciences, and Rainer Froese, FishBase Coordinator, Institut für Meereskunde.

- *Seamounts Online*
  ○ Karen Stocks, San Diego Supercomputing Center and Scripps Institute of Oceanography.

the capability of operating in a four-dimensional environment so that analyses, modelling and mapping can be accomplished in response to user demand through accessing and providing relevant data." The key characteristics of the then to-be-developed system were interoperability through common definition of metadata standards and protocols for a distributed, multi-tiered architecture. A website was built to demonstrate the OBIS concept (Stocks *et al.* 2000); this website is being preserved as a reference document, and can still be visited at www.marine.rutgers.edu/OBIS. The first OBIS workshop was held in Washington, DC, in November 1999.

Early growth of OBIS was initiated through the announcement, in May 2000, of eight grants by the US Government Agencies in the National Oceanographic Partnership Program (NOPP) together with the Alfred P. Sloan Foundation. These grants involved researchers in more than 60 institutes in 15 countries, and addressed infrastructural issues as well as taxon-based projects of data acquisition (Grassle 2000; Decker 2001; Zhang & Grassle 2003). A ninth, National Science Foundation (NSF)-funded project (SeamountsOnline; Stocks 2009) was added afterward, and the nine projects formed the core of the early OBIS (Table 17.1). In 2001, an NSF project was awarded to Rutgers University to create an international portal; by February 2002, all NOPP-funded data projects and the NSF-funded SeamountsOnline were made interoperable through the OBIS portal (Zhang & Grassle 2003). At that point, the portal provided access to over 400,000 occurrence records.

Institutionally, OBIS is growing rapidly as a distributed system with an international secretariat and portal (iOBIS) hosted by the Institute of Marine and Coastal Sciences of Rutgers University, and Regional OBIS Nodes (RONs) in all continents (Fig. 17.2 and Table 17.2). RONs were created to serve national or regional needs better and to achieve global coverage. The RON network is still expanding: several RONs were added in 2006 and 2007 (China, Korea, Philippines) and discussions are continuing to create new ones (Arctic, Oman, and possibly Mexico). The RON network has been very active and very successful in connecting datasets. Each RON is self-sustaining and is the geographical backbone for further development of OBIS data content. The institutes hosting the RONs are an asset for OBIS as a network and have proven to be very supportive of OBIS activities and objectives.

In addition to the Regional Nodes, OBIS has thematic nodes for major subsets of marine life. OBIS Spatial Ecological Analysis of Megavertebrate Populations (OBIS SEAMAP), the repository for data on marine birds, turtles,

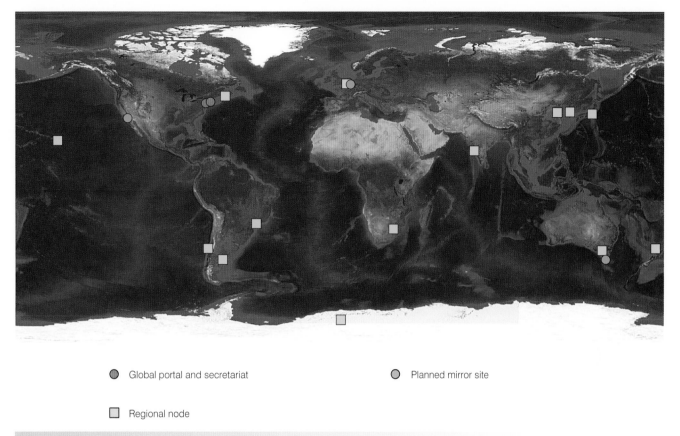

● Global portal and secretariat   ○ Planned mirror site

☐ Regional node

Fig. 17.2
Locations of Regional OBIS Nodes (yellow squares), international secretariat (red circle), and proposed mirror sites (orange circles).

and mammals, is developing new ways to visualize migrations of these animals and to understand their habitats (Halpin *et al.* 2006, 2009). The Biogeoinformatics of Hexacorals website maintains an authoritative, global anemone and coral database (Fautin 2000). FishBase contains comprehensive information on finfishes (Froese & Pauly 2009). The OBIS microorganisms component (MICROBIS) is breaking completely new ground by defining the known world of microorganisms using new molecular approaches to define microbial taxa. The Continuous Plankton Recorder (CPR), managed by the Sir Alister Hardy Foundation for Ocean Science (SAHFOS), provides a unique and very large dataset. One of the strengths of the CPR data is that it has been collected in a standard way for more than half a century (see, for example, Reid *et al.* 1998; Beaugrand *et al.* 2004).

Data generated by the field projects of the Census ultimately will all be available through the OBIS website. This is essential if OBIS is to play its role in integrating Census data, and support the Census Synthesis. All field projects are producing high-quality data. However, as with many projects, data generated by a single project are usually restricted to a single theme defined on the basis of habitat, geographical region, or taxonomic scope. The power of the

OBIS database is the integration of data from all these fields in a single coherent taxonomic framework, presenting a view that is truly global and facilitating analysis across scientific disciplines.

OBIS has strong relationships with several UN organizations. Data are exchanged with the Fisheries Department of the FAO, and links to the species information pages on the FAO site are displayed on the OBIS site. Collaboration with the IOC and its IODE program has centered on data standards and protocols. There have been joint activities on capacity building in Africa, with training workshops on the use of OBIS standards and tools; data logging workshops have been organized, focusing on sponges and on mollusks. Close collaboration between OBIS and IODE has resulted in the formal adoption, in June 2009, of OBIS as an activity of the IOC under its IODE program (see below).

OBIS was one of the earliest Associate Members of GBIF (www.gbif.org) which publishes data on all species. OBIS is a very active participant in GBIF activities, and one of the largest publishers of data to GBIF, reflecting its role as a specialist network for marine species. GBIF recommends that marine data are first published through OBIS, because OBIS can add special value and will manage the subsequent publication of data through GBIF. This also

Table 17.2

List of regional OBIS nodes.

- Antarctica: managed by Bruno Danis, hosted by Belgian Biodiversity Platform, Belgium
  (a) 498,000 records of 2,522 species.

- Argentina: managed by Mirtha Lewis, hosted by Centro Nacional Patagónico (CENPAT) CONICET Argentina
  (a) 171,000 records of 91 species.

- Australia: Tony Rees, National Oceans Office, Commonwealth Scientific and Industrial Research Organisation (CSIRO)
  (a) 827,000 records of 6,000 species.

- Canada: managed by Tana Worcester, hosted by Centre of Marine Biodiversity, Bedford Institute of Oceanography
  (a) 913,000 records of 7,000 species.

- China: managed by Xiaoxia Sun, hosted by Institute of Oceanology, Qingdao
  (a) 57,000 records of 1,200 species.

- Europe: managed by Francisco Hernandez, hosted by Vlaams Instituut voor de Zee (VLIZ), Belgium on behalf of the EU Network of Excellence "Marine Biodiversity and Ecosystem Functioning (MarBEF)".
  (a) 3,543,000 records from 15,000 species.

- Indian Ocean: Managed by Baba Ingole, hosted by National Chemical Laboratory, National Institute of Oceanography India
  (a) 81,000 records of 68,000 species.

- Japan: Katsunori Fujikura, Japan Agency for Marine-Earth Science and Technology
  (a) Currently not active; will come online in the near future

- Korea: Youn-Ho Lee, Korea Ocean Research & Development Institute
  (a) 3,300 records, not yet available through the iOBIS portal.

- South-West Pacific: Don Robertson, National Institute of Water & Atmospheric Research New Zealand
  (a) 430,000 records.

- Sub-Saharan Africa: managed by Marten Grundlingh, hosted by Southern African Data Centre for Oceanography, South Africa
  (a) 3,210,000 records of 23,000 species.

- Tropical and Subtropical Eastern South Pacific: managed by Ruben Escribano, hosted by FONDAP COPAS, Chile
  (a) 28,000 records from 4,000 species.

- Tropical and Subtropical Western South Atlantic: Fábio L. da Silveira and Rubens M. Lopes, University of São Paulo, Brazil
  (a) 43,000 records for 4,000 species.

- USA: managed by Mark Fornwall, hosted by National Biological Information Infrastructure (NBII), Pacific Basin Information Node (PBIN), USA; IT component located in Boulder, Colorado
  (a) 1,698,000 records.

avoids duplication of data being separately published in GBIF and OBIS.

OBIS works closely with other players in the field of biodiversity informatics. OBIS exchanges information and is reciprocally linked with the Barcode of Life (BOL). As the marine component of the latter is being developed OBIS will forge even stronger links. OBIS and its web interface can be used as a geographical window on the BOL information; OBIS distribution records can be used to document occurrence of a species in a region or country, and thus assist in management of property rights to genetic resources. Together with the European Node of OBIS (EurOBIS), OBIS has collaborated on the development of

the World Register of Marine Species (WoRMS, see below). This venture forms the basis of the marine community's contribution to the Catalogue of Life (CoL). For many of the species it contains, the content of WoRMS goes far beyond the pure taxonomic information contained in the CoL. This content is made available to the Encyclopedia of Life (EOL).

17.4.1 Technology

From the outset, OBIS was conceived as a distributed system, leaving control over data publication in the hands of the data custodians (Fornwall 2000). The structure and content of the data exchanged was formatted following the Darwin Core format (Vieglais *et al.* 2000), an extensible markup language (XML)-based standard originally developed at the University of Kansas. Later, the Darwin Core was adopted by the Taxonomic Database Working Group as one of its standards, and further developed. Several "extensions" of the Darwin Core exist: specific user communities have expanded the number of terms defined in the data exchange format to serve the needs of their community better. Also OBIS defined an extension (known as OBIS Schema), to address the specific needs of the oceanographic and marine biology community better. For example, one of the features of the OBIS Schema is that the location of an observation can be ascribed to a set of two points needed to define a transect line instead of a sampling point; this makes it possible to capture accurately the position of data resulting from a trawl. All extensions of the Darwin Core are still compatible with the original standard. It is this compatibility that forms the basis of the compatibility between different content providers and aggregators, and that allows OBIS data to be published through GBIF.

The original protocol defining computer-to-computer communication to exchange the Darwin Core data was the Z39.50 protocol (Vieglais *et al.* 2000); this was soon replaced with the Distributed Generic Information Retrieval (DiGIR; Blum *et al.* 2001). Originally, the OBIS website was built as a pure distributed system, with no data residing in the portal server; exception was only made for datasets from custodians who did not have a provider service set up. All queries to the data provider were performed in real time, as the end user was requesting the data through the OBIS portal (Zhang & Grassle 2003). This proved to be too slow, and too critically dependent on the availability of all providers at all times. For reasons of performance and reliability, a system was developed where all available data (including a link back to the data provider's own website) were stored in a cache, maintained in a database at the OBIS secretariat. This cache also made it possible to build indices on different sets of polygons, and to calculate summary information for the different taxa (Rees & Zhang 2007).

The technology behind the present OBIS system is several years old, and in need of an overhaul. Possible tools and technology for a new incarnation of OBIS have been discussed in the OBIS community and with relevant experts. All developments at iOBIS adhere strictly to the relevant standards wherever they exist. For geographic information system (GIS) and web-based mapping we will work with Open Geospatial Consortium (OGC) compliant tools, and closely collaborate with the people developing GeoServer. Access to OBIS data will no longer be restricted to the iOBIS website, with its canned queries, but will also be possible through standards-compliant web services.

As mentioned above, metadata are an essential element of data warehouses. The DiGIR protocol itself carries some metadata: the data standard is documented, there is room for an abstract to give a verbal description of the original purpose and intent of the data, and contact information, both for technical and for scientific aspects, can be listed; it is also possible to include a universal resource locator (URL) that points back at the website of the data provider. Although these are all the essential elements, many users wanted to include richer metadata: this gives end users the ability to judge the coverage of data in OBIS better, and to assess fitness for use. For this reason, OBIS started collaborating with the GCMD; all OBIS-related metadata are visible as a separate collection on their site (gcmd.gsfc.nasa.gov/KeywordSearch/Home.do?Portal=OBIS&MetadataType=0). One of the great advantages of this system is that users can maintain their own metadata records through the GCMD web interface. OBIS will expand its metadata activities also to accommodate metadata in other widely accepted standards in use by members of the OBIS community.

A taxonomic reference list, including information on classification and synonymy, is an essential tool in the quality control process of taxonomically resolved data. It is needed as a controlled vocabulary, to make sure that data from different datasets are not only compatible at the technical level, but also at the content level. Differently spelled names, or differently interpreted taxonomic names, have to be reconciled before any analysis of the integrated content can be done.

The initial website, launched in February 2002, already included a taxonomy name service, built in partnership with Species 2000 and FishBase. A prototype name service provided common name/scientific name and synonym translation (Zhang & Grassle 2003). Later versions of the portal implemented these taxonomic name services, through integration with the Interim Register of Marine and Non-marine Genera (IRMNG) (Rees & Zhang 2007), developed by Tony Rees of the Australian OBIS Node. One of the objectives was to be able to discriminate between marine and non-marine taxa, and between fossil and extant taxa. Several providers of data to OBIS do not have a simple way of discriminating between these in their databases, so IRMNG was conceived as the basis for this filtering mechanism.

A standard register of taxonomic names of European marine species (European Register of Marine Species, ERMS) was compiled using funding from the European Commission Marine Science and Technology research program (Costello *et al.* 2001). ERMS was made internally consistent, expanded with a consistent classification, and turned into a relational database for use by the European OBIS node with support from the EU Network of Excellence MarBEF. Under the aegis of OBIS, ERMS has developed into WoRMS. WoRMS has nearly 150,000 valid species names, of which 68,700 have at least one record in OBIS. The OBIS website is now using WoRMS as the

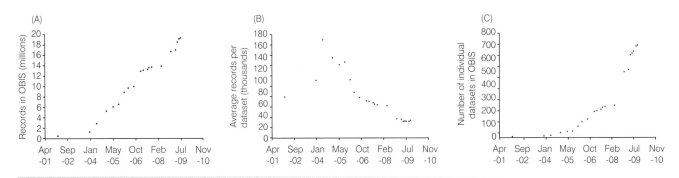

Fig. 17.3
(A) Number of records in the OBIS cache (millions). **(B)** Average number of records per dataset (thousands). **(C)** Number of individual datasets published through OBIS.

Table 17.3

Largest datasets available through OBIS.

Dataset name	Number of records
Marine and Coastal Management – Linefish Dataset (AfrOBIS)	2,744,958
SAHFOS Continuous Plankton Recorder – Zooplankton (The Sir Alister Hardy Foundation)	1,374,170
NODC WOD01 Plankton Database	1,275,382
European Seabirds at Sea (OBIS SEA-MAP)	1,122,884
ICES EcoSystemData (EurOBIS)	735,831
SAHFOS Continuous Plankton Recorder – Phytoplankton (The Sir Alister Hardy Foundation)	721,833
Marine Nature Conservation Review (MNCR) and associated benthic marine data held and managed by JNCC (EurOBIS)	580,008
NMNH Invertebrate Zoology Collections (Smithsonian Institution)	533,822
Fishbase occurrences hosted by GBIF-Sweden (FishBase)	505,852
ECNASAP – East Coast North America Strategic Assessment (OBIS Canada)	466,736
Northeast Fisheries Science Center Bottom Trawl Survey Data (USOBIS)	460,938
North Pacific Groundfish Observer (North Pacific Research Board)	422,150
NIWA Marine Biodata Information System (South Western Pacific OBIS)	377,929
HMAP-History of Marine Animal Populations	255,774
Elephant Seal Sightings, Macquarie Island (Australian Antarctic Data Centre)	221,619
ARGOS Satellite Tracking of animals (Australian Antarctic Data Centre)	213,488
PIROP Northwest Atlantic 1965–1992 (OBIS-SEAMAP)	209,039
Marine and Coastal Management – Demersal Surveys (AfrOBIS)	201,741
Marine benthic dataset (version 1) commissioned by UKOOA (EurOBIS)	175,360
USA Environmental Protection Agency's EMAP Database	173,109

standard source for names of marine species. WoRMS provides correct names for the OBIS community and is recognized as the marine component of CoL.

17.4.2 Content

The number of records in the OBIS databases has grown according to expectations (though there was a setback from November 2007 to May 2008, owing to a change in personnel; Fig. 17.3). The growth in number of records after 2004 is linear after the initial development phase from 2002 to 2004. If the current growth can be sustained, OBIS will publish over 30 million records by October 2010.

An issue worth noting is the size of an average dataset, which has been decreasing steadily (Fig. 17.3). This trend is to be expected, as OBIS has first connected the largest, most important databases. Obviously, this has implications for future planning for OBIS. Smaller average datasets means more work for the same gain. In practice, this will necessitate more data management time in OBIS, either at the level of the secretariat, or at the RONs, or both. In this respect, the linear growth of OBIS content is good news: it means that data acquisition and quality control are becoming more efficient.

Table 17.3 lists the largest datasets available through OBIS. It is gratifying to see two South African datasets in the top 20, a clear example of the strength of the RON network and the global nature of collaboration within OBIS. From the list it is clear that most of the large datasets are monitoring datasets, in many cases fisheries monitoring (for example the South African line fisheries data, several fisheries datasets from the US NOAA, Fisheries and Oceans Canada (DFO), and New Zealand's National Institute of

Water and Atmospheric Research (NIWA)). Several other datasets are in fact aggregations of many individual datasets (for example the WOD01 Plankton database, European Seabirds at Sea, benthic data from the Joint Nature Conservation Committee of the UK, FishBase occurrence records). One of our most valued contributors is the SAHFOS, with the data from the CPR. The Smithsonian Institution's National Museum of Natural History makes the data from its catalogue available, as do many other museums. However, the real value of OBIS is in the 679 datasets that are not listed in this table. The large datasets are often available already online, through the website of the data provider. But many of the smaller datasets would to a large extent be undiscoverable and remain unused, if it were not for OBIS.

The RONs are instrumental in achieving global coverage, and collectively provide about half of the data available through OBIS. The African and the European nodes are the largest with well over 3 million records each. All of the Census projects provide data. The champions here are OBIS SEAMAP with nearly 2.5 million data points, International Census of Marine Microbes (ICoMM) with 1.5 million, and Census of Antarctic Marine Life (CAML) with 900,000. Also, History of Marine Animal Populations (HMAP) contributes a substantial dataset, with 250,000 records and, not surprisingly, extends the time for which data are available (Fig. 17.4).

The map in Figure 17.5 illustrates the very uneven availability of data within OBIS. Most of the data are from coastal waters; the shallow waters of the European Atlantic coast, the Pacific coast of Alaska, and the Atlantic and Gulf of Mexico coasts of the USA are especially well represented. In open waters, the Northern Atlantic is well covered. The large volume of data here is mainly from the CPR. The Northern hemisphere is much better covered than the southern one; exceptions here are South Africa (mainly the west and south coasts), and part of the coast of Argentina. The southern Pacific is particularly poorly represented; the southern Atlantic and Indian Oceans also represent major gaps in coverage. Some of the mega-diverse coastal areas also have a disappointing number of records, such as the coral reefs of eastern Africa and the Red Sea, and the coasts of the Coral Triangle.

The series of maps in Figure 17.6 illustrates that most of the data in OBIS are from surface waters. The top-most map represents essentially the same information as in Figure 17.5, but at a lower resolution. Consecutive maps illustrate the number of records deeper than 100, 500, 1,000 and 2,500 m. respectively. In all five maps, the ocean floor shallower than this depth is drawn in light grey, to illustrate the amount of seafloor at this depth. The bottom map clearly shows that most of the seafloor is completely unexplored. We hope that this part of the oceans will be better represented as the Census deep-sea data become available.

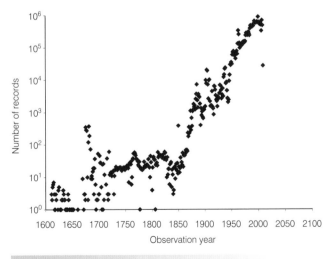

Fig. 17.4

Number of records in OBIS cache, as a function of time. In most cases, this corresponds with the year the observation was made. For historical data, this is the estimated year the organism was alive.

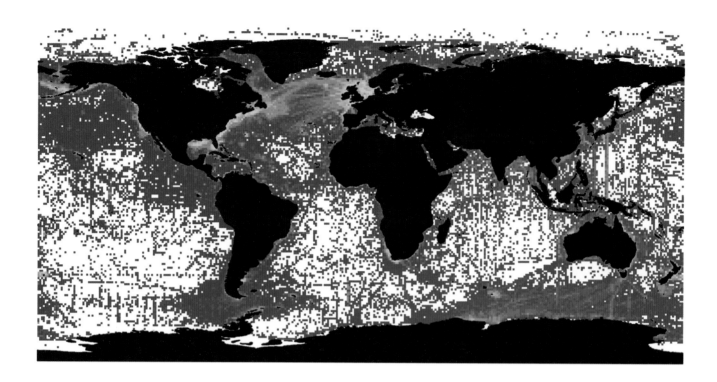

Number of records

Low   High

Fig. 17.5
Number of records in OBIS per 1° × 1° square of latitude and longitude, corrected for differences in surface area of the squares. Red is high numbers, blue low, and white for squares without a single observation.

Not surprisingly, there is also a strong bias in taxonomic coverage. Larger and commercial species are clearly better represented, as is evident from the list in Table 17.4. Of the 50 taxa listed, 37 are vertebrates; of these, 11 are birds and 23 are fish. All fish in this list are species of commercial importance. *Loligo vulgaris reynaudii* d'Orbigny, 1845 is the lone mollusk on the list, very likely so well represented in the database because it is also a commercial species. Apart from the data recorded as phylum Chaetognatha, nearly all other invertebrates are planktonic crustaceans; for both of these groups, this probably accurately reflects their high abundance in the best-sampled waters of the Northern Atlantic. The same is true for the two taxa that are not animals: *Chaetoceros* Ehrenberg, 1844, a genus of diatoms, and *Ceratium fusus* (Ehrenberg, 1834) Dujardin, 1841, a dinoflagellate. Most of the OBIS records are resolved to species (or even subspecies where relevant), but as is apparent from the top 50, there are exceptions. In the case of groups that are difficult to identify such as Euphausiacea, Decapoda, or Chaetognatha, this is not completely unexpected.

Table 17.5 further illustrates the bias towards larger and commercially important species, and reflects the completeness of our knowledge. The percentage completeness and degree of cover is calculated for WoRMS. Because WoRMS is not complete, the estimates of the total number of marine species compiled by Bouchet (2006) are also listed. Fish and other vertebrates are virtually complete, and well covered, with a high number of records per species. For other groups, such as the mollusks, the percentage completeness, even measured against WoRMS, is very low; also WoRMS is quite incomplete for this very species-rich group. Within the mollusks, the cephalopods are well covered, with two-thirds of the species having at least one record in OBIS, and an average of nearly 500 records per species in WoRMS. Bryozoa are poorly represented in both OBIS and WoRMS; there are records for only 690 species, where Bouchet estimates that there are 5,700 in total.

As has been noted before, OBIS is a work in progress. There are clear gaps in geographical and taxonomic coverage. Some of these gaps no doubt are the result of the

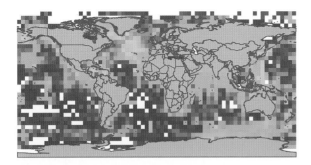

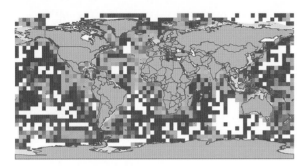

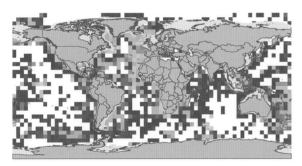

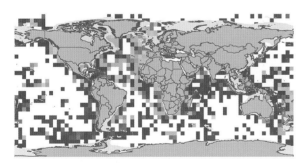

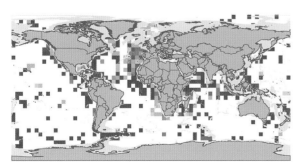

uneven distribution of scientific work: some places such as open oceans and polar seas are difficult and costly to sample; some groups of organisms are more difficult and less "interesting" to study. In these cases, sparse data coverage reflects our uneven knowledge of nature, and could provide interesting guidelines to set priorities for future work. In other cases, data exist but are not available through OBIS. One of the highest priorities is to identify such datasets; this inventory will assist in defining priorities for data assimilation.

Missing data are a problem, wrong data are an even greater worry. Yet, no data system is without mistakes, and that is definitely the case for OBIS. For example, a 2008 study of OBIS content found wrong records for over a third of the species present in OBIS (Robertson 2008). Responsibility for the accuracy of the data in a multi-level aggregation system such as OBIS is not a simple issue. One argument could be that OBIS is only the publisher of the data, and just as it cannot take credit as "owner" of the data, it cannot take responsibility for the mistakes in it, just like Google cannot be held responsible for the information that shows up on its pages (R. Froese, personal communication). However, Google does not claim to have expertise in the subject matter of all the sites it indexes. We like to think that OBIS has a certain degree of competence in biogeography. This makes it possible to implement at least a minimum level of quality control, which is applied to all incoming data and gradually to all data retrospectively. OBIS works with its data providers to improve the quality not only at the level of the international portal, but also at the level of the data provider. Of course, no system is perfect, and Robertson's advice of "caveat emptor" should be kept in mind. The best way of detecting errors in a database is to work with the data. We hope that any user finding errors will not be discouraged from using OBIS data, but work together with OBIS staff at the secretariat and its data providers to improve the content.

17.5 Using OBIS

OBIS is both a secure repository for data and a ready source of data for a growing user community of scientists and educators throughout the marine sciences community. Education and outreach are being achieved by developing modules for use in schools and broadening our end-user

Fig. 17.6
Number of observations in OBIS deeper than a given depth, per 5° × 5° degree square. Depths are 0, 100, 500, 1,000 and 2,500 m, respectively. Ocean floor deeper than this depth is shaded light grey. Color coding same as in Figure 17.5.

Table 17.4
Most-recorded taxa

Taxon	Count	Taxon	Count
Thyrsites atun (Euphrasen, 1791)	385,547	*Sula capensis* (Lichtenstein, 1823)	88,187
Fulmarus glacialis (Linnaeus, 1761)	378,807	*Chaetoceros* Ehrenberg, 1844	86,796
Limanda limanda (Linnaeus, 1758)	367,976	*Paracalanus* Boeck, 1865	85,410
Loligo vulgaris reynaudii d'Orbigny, 1845	319,209	*Larus argentatus* Pontoppidan, 1763	83,910
Mirounga leonina (Linnaeus, 1758)	240,640	*Seriola lalandi* Valenciennes, 1833	80,819
Calanus Leach, 1816	232,264	*Pterogymnus laniarius* (Valenciennes, 1830)	75,338
Argyrosomus De la Pylaie, 1835	227,182	*Euphausia superba* Dana 1850	74,223
Uria aalge (Pontoppidan, 1763)	224,180	*Thunnus alalunga* (Bonnaterre, 1788)	73,832
Gadus morhua Linnaeus, 1758	207,556	*Oithona* Baird, 1843	71,917
Pachymetopon blochii (Valenciennes, 1830)	167,612	*Epinephelus* Bloch, 1793	70,147
Rissa tridactyla (Linnaeus, 1758)	164,598	*Ceratium fusus* (Ehrenberg, 1834) Dujardin, 1841	65,854
Copepoda	155,452	*Bucephala albeola* Linnaeus, 1758	62,202
Morus bassanus (Linnaeus, 1758)	141,322	Decapoda Latreille, 1803	62,158
Argyrozona argyrozona (Valenciennes, 1830)	140,257	*Hippoglossoides platessoides* (Fabricius, 1780)	61,955
Merluccius Rafinesque, 1810	137,734	*Caretta caretta* (Linnaeus, 1758)	59,770
Euphausiacea Dana, 1852	135,396	*Sebastes capensis* (Gmelin, 1789)	58,748
Merlangius merlangus (Linnaeus, 1758)	131,017	*Merluccius bilinearis* (Mitchill, 1814)	57,666
Laridae	128,793	*Rhabdosargus globiceps* (Valenciennes, 1830)	56,889
Calanus finmarchicus (Gunner, 1765)	127,848	*Acartia* Dana, 1846	55,834
Chrysoblephus laticeps (Valenciennes, 1830)	112,514	*Squalus acanthias* Linnaeus, 1758	54,570
Chaetognatha	98,821	*Balaenoptera physalus* (Linnaeus, 1758)	52,209
Atractoscion aequidens (Cuvier, 1830)	98,354	*Fratercula arctica* (Linnaeus, 1758)	51,379
Chrysoblephus puniceus (Gilchrist & Thompson, 1908)	92,439	*Larus marinus* Linnaeus, 1758	50,877
Pygoscelis adeliae (Hombron & Jacquinot, 1841)	91,682	*Oncorhynchus kisutch* (Walbaum, 1792)	50,268
Selachii	90,963		
Cheimerius nufar (Valenciennes, 1830)	89,892		

community. There is hardly any downtime and the number of visitors and records downloaded from the OBIS website increases steadily and now averages 80,000 per day (Fig. 17.7). The OBIS website will continue to host species-level links to most other species-referenced marine databases. Through the OBIS website, active (and in most cases reciprocal) links at the species level are made with CoL, Integrated Taxonomic Information System (ITIS), Barcode of Life, FishBase, FAO, and GenBank among others.

The content of the OBIS database is growing and maturing; it is now possible to use the OBIS database to answer

Table 17.5

Completeness of OBIS, per phylum or class. "Records" is number of records in OBIS for taxa belonging to this higher taxon. "Species" is the number of species with distribution records in OBIS. "WoRMS" is the number of species in the World Register of Marine Species. "%C" is the percentage completeness, namely the percentage of species in WoRMS for which there are distribution records in OBIS. "r/s" is the number of records per species in WoRMS. "Bouchet" is the estimate in Bouchet (2006) of the number of species in this taxon.

Taxon name	Records	Species	WoRMS	%C	r/s	Bouchet 2006
Nemertina	10,788	253	1375	18.40	7.85	1,180–1,230
Arthropoda	2,631,678	13023	38,827	33.54	67.78	
Crustacea	2,593,885	12106	35,559	34.04	72.95	44,950
Chelicerata	13,510	694	2627	26.42	5.14	2,267
Ctenophora	4,440	21	174	12.07	25.52	166
Cnidaria	398,217	5648	11,195	50.45	35.57	9,795
Sipuncula	15,801	92	164	56.10	96.35	144
Echiura	629	81	203	39.90	3.10	176
Entoprocta	236	13	168	7.74	1.40	165–170
Tardigrada	90	36	171	21.05	0.53	212
Rhombozoa	6	5	95	5.26	0.06	82
Orthonectida	1	1	25	4.00	0.04	24
Rotifera	970	65	194	33.51	5.00	50
Gnathostomulida	18	16	99	16.16	0.18	97
Bryozoa	69,771	690	1,678	41.12	41.58	5,700
Phoronida	5,597	9	11	81.82	508.82	10
Brachiopoda	4,578	179	419	42.72	10.93	550
Echinodermata	267,894	2906	5,992	48.50	44.71	7,000
Hemichordata	6,029	29	108	26.85	55.82	106
Vertebrata	1,0829,052	15233	18,942	80.42	571.70	
Pisces	7,388,961	14546	17,670	82.32	418.16	16,475
Aves	2,708,767	475	915	51.91	2960.40	
Mammalia	584,646	108	162	66.67	3608.93	110
Reptilia	93,099	35	105	33.33	886.66	
Agnatha	7,378	69	90	76.67	81.98	
Tunicata	153,714	1136	3,141	36.17	48.94	4,900

Taxon name	Records	Species	WoRMS	%C	r/s	Bouchet 2006
Cephalochordata	5,549	13	33	39.39	168.15	32
Acanthocephala	42	27	142	19.01	0.30	600
Gastrotricha	281	173	527	32.83	0.53	390–400
Chaetognatha	75,429	59	207	28.50	364.39	121
Cycliophora	5	1	2	50.00	2.50	1
Kinorhyncha	577	49	162	30.25	3.56	130
Loricifera	23	8	23	34.78	1.00	18
Nematomorpha	5	1	5	20.00	1.00	5
Priapulida	1,741	9	20	45.00	87.05	8
Mollusca	1,101,758	5383	18,371	29.30	59.97	52,525
Porifera	63,380	1325	8,256	16.05	7.68	5,500
Platyhelminthes	14,272	315	3,902	8.07	3.66	1,500
Nematoda	97,218	2500	5,729	43.64	16.97	12,000
Annelida	812,515	4522	12,839	35.22	63.28	12,000
Plantae	274,631	2687	8,473	31.71	32.41	
Rhodophyta	212,527	1854	6,289	29.48	33.79	6,200
Chlorophyta	51,890	759	1,811	41.91	28.65	2,500
Phaeophyceae	108,696	629	1,996	31.51	54.46	1,600
Fungi	10,205	103	571	18.04	17.87	500
Protoctista	586,752	1545	6,069	25.46	96.68	
Ciliophora	39,851	173	1,074	16.11	37.11	
Rhizopoda	401	30	189	15.87	2.12	
Foraminifera	88,812	387	1,919	20.17	46.28	1,0000
Dinomastigota	380,923	613	1,925	31.84	197.88	4,000
Radiolaria	2,044	25	194	12.89	10.54	550
Bacillariophyta	628,780	917	2,342	39.15	268.48	5,000
Monera	47,033	225	651	34.56	72.25	4800
Cyanobacteria	12,329	225	405	55.56	30.44	1,000

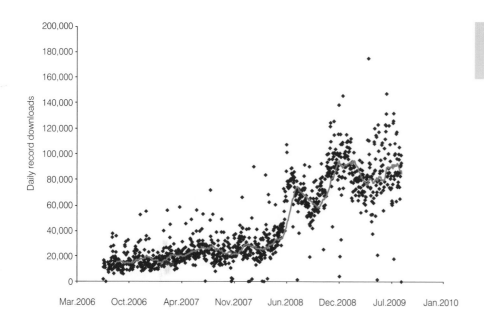

Fig. 17.7
Number of records downloaded, per day, from the OBIS website.

scientific questions and to investigate broad patterns of distribution of biodiversity. A first series of maps was created and distributed through newsletters and conferences (for example the LME conference in Qingdao, September 2007; Group on Earth Observations (GEO) IV meeting in Cape Town, November 2007; Ocean Sciences conference in Orlando, March 2008; the October 2007 newsletter of Global Ocean Ecosystem Dynamics (GLOBEC)). As an illustration, the global map of Hurlbert's Index (es(50), the expected number of distinct species in a random sample of 50 distribution records from the database; Hurlbert 1971) is reproduced here (Fig. 17.8). This is actually the first map of biodiversity of all taxa, on a global scale; previous studies were restricted either in taxonomic or in geographical scope; the OBIS integration of datasets across its many data providers makes it possible to present this comprehensive picture. A similar analysis formed the basis of maps published in National Geographic's *Ocean: An Illustrated Atlas* (Grassle & Vanden Berghe 2009). A second application of Hurlbert's Index is shown in Figure 17.9, illustrating the latitudinal gradient in species richness.

Figure 17.10 illustrates another use of OBIS data. Yellow dots are actual observations of lionfish (*Pterois volitans* (Linnaeus, 1758)), an invasive species with its home range in the Red Sea. Through environmental envelope modeling, the range to which this invader could spread can be calculated. The red area in Figure 17.10 displays the region with similar environmental conditions to that in which the species was found, and so might be expected to spread. Environmental envelope modeling is a good demonstration of the power of data sharing and integration. It can combine data from different sources of biogeography, and overlay these with physical and chemical oceanography data, allowing multi-disciplinary analysis. Other potential applications using this and other modeling techniques are the study of shifts in species distribution in response to global change.

Publicly available fisheries data in OBIS have already played an important role in documenting examples of overfishing in the ocean (Worm & Myers 2004; Baum & Worm 2009). Other examples of use of the database are a study of the completeness of our knowledge of fish communities (Mora *et al.* 2007), global distribution patterns of Myxinidae (Cavalcanti & Gallo 2008) and of Cephalopoda (Rosa *et al.* 2008a, b). It is expected that the number of papers based on data obtained from the OBIS database will grow rapidly, now that the content of OBIS has sufficiently matured and grown.

17.6 Future of OBIS

Participation in OBIS is open to any interested individual, country, or organization committed to the long-term maintenance of an accessible, relevant, biogeographic database. Present members of the federation include the NOPP-funded OBIS programs, the Census projects, the RONs, and many independent data custodians interested in developing ties with the OBIS international system of databases. The international OBIS secretariat, through the international portal, is responsible for making the entire system interoperable, maintaining standards for data exchange, and coordinating data acquisition. Each member of the Federation will, in addition to maintaining their own database systems, be committing to provide data through the OBIS portal. One of the priorities for OBIS at this point is to fill some of the gaps in the available data by forging relationships with more organizations, and to expand the federation.

OBIS is one of the main outputs from the Census – a four-dimensional atlas of marine life, accessible online and

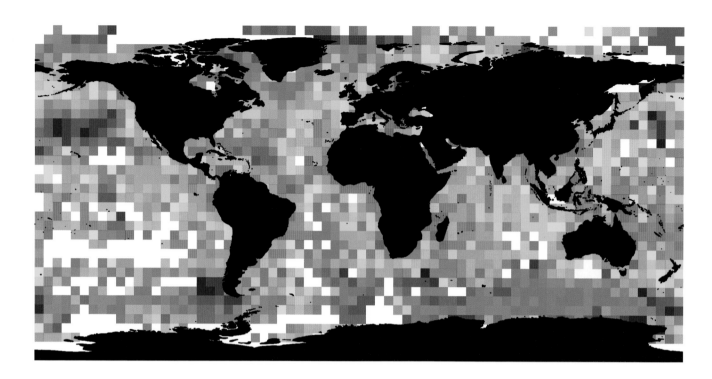

Hurlbert's index: es(50)

Low   High

Fig. 17.8
Map of Hurlbert's index, es(50) – the expected number of distinct species in a random sample of 50 distribution records, calculated per squares of 5° × 5°. Red indicates high species richness, blue low. White areas are where there are fewer than 50 distribution records in a square.

Fig. 17.9
Latitudinal gradient in species richness, as measured by Hurlbert's index, es(50).

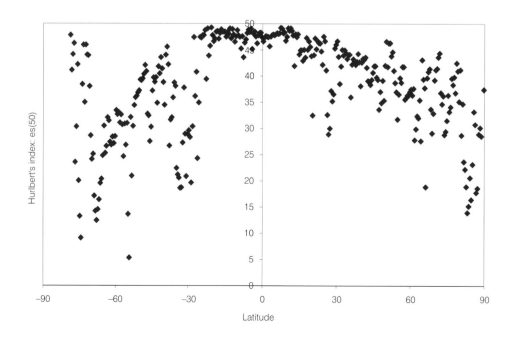

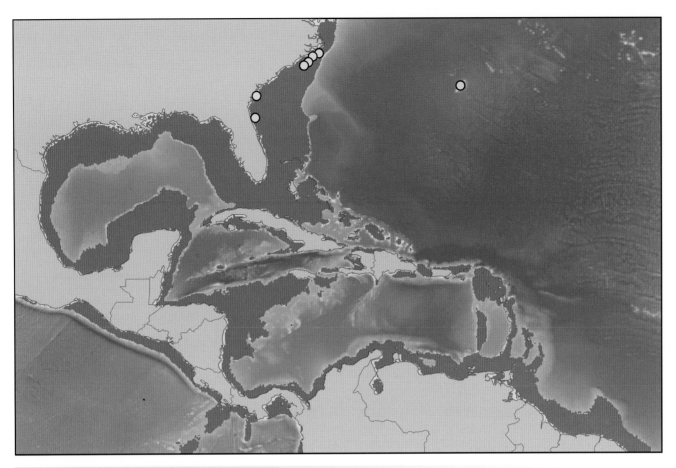

Fig. 17.10
Predicted potential range for *Pterois volitans* (Linnaeus, 1758) (lionfish), an invader from the Red Sea. Yellow dots are actual observed occurrences. Red area represents areas with similar oceanographic conditions to the one where the observations were made, and so where conditions might favor the spread of this species.

analyzable to test hypotheses and make predictions about diversity, distribution, and abundance of marine life. This data system will be used in ocean management, including fisheries, conservation planning, and risk assessment of invasive species. Although the Census culminates in 2010, OBIS will live on as a major legacy of Census and a community of practice, maintaining an informatics infrastructure for managing, researching, and educating about living marine resources. OBIS is establishing itself as an integral part of the international scientific infrastructure. Its regional development, as exemplified by the establishment of RONs, will ensure that it can serve these needs both locally and globally.

In June 2009, OBIS was adopted by the IOC of the United Nations Educational, Scientific and Cultural Organization (UNESCO) as one of the activities of its IODE program. This is a clear recognition by the IOC member states that OBIS is part of the international scientific infrastructure, and gives a formal intergovernmental status to OBIS activities. This will be important in soliciting resources

to fund further activities, to attract more data, and to achieve wide acceptance of OBIS data in the process of environmental decision making.

The future data needs of ocean science and ocean resource management will require a seamless coupling of biological data with physical oceanographic processes. This biophysical data framework will be built through the active integration of data from a large and diverse number of sources, including physical, chemical, and biological oceanography. GEO and its Global Earth Observation System of Systems (GEOSS) is an international federation bringing together relevant players in this field. The Global Ocean Observing System (GOOS), the marine component of GEOSS, is hosted by the IOC. OBIS is poised to play a significant and expanding role in GOOS, and to take on the responsibility for marine biogeographic information, through involvement in GEO's Biodiversity Observing Network (GEO BON). Its position within IOC will assist in achieving this ambition.

OBIS data have been used for scientific purposes, and it is expected that this use will grow. Another objective of OBIS is to inform management of the marine environment; for example, OBIS data have been used in the preparation of scientific background documents for the Convention on Biological Diversity through an International Union for the Conservation of Nature (IUCN) project to identify areas of special ecological or biological significance. If OBIS is to reach its full potential, it needs to be made interoperable with data systems on socio-economic data, including use data. Although there are mature systems that can easily serve as sources for global physical oceanography, there seems to be no equivalent for socio-economic data.

OBIS is now at the stage where it is an essential international source of data and Web-based tools for defining habitats, communities, and biogeographical units in the marine environment. However, it still is far from a comprehensive source for all biogeographic data that have been collected; and there are large gaps in the coverage. The OBIS portal expects continued growth, and counts on input from the international community of OBIS users, including the Census National and Regional Implementation Committees (NRIC) and the Regional OBIS Nodes, to help this happen.

Acknowledgments

We are grateful for the generous support and guidance OBIS received from the Alfred P. Sloan Foundation and its staff. Parts of the development of OBIS were funded through NSF grants to Fred Grassle and Yunqing (Phoebe) Zhang. Phoebe was instrumental in building the IT infrastructure for OBIS. OBIS would not exist without the input from others, including the numerous data providers, node managers, and the members of the International Committee and Governing Board. We are also grateful for the trust of this OBIS community, and for the opportunity to develop OBIS.

References

Baum, J. & Worm, B. (2009) Cascading top-down effects of changing oceanic predator abundances. *Journal of Animal Ecology* **78**, 699–714.

Beaugrand, G., Edwards, M., John, A. & Lindley, A. (2004) Continuous Plankton Records: Plankton Atlas of the North Atlantic Ocean 1958–1999. *Marine Ecology Progress Series* (Suppl.): 1–75.

Blum, S., Vieglais, D. & Schwartz, P.J. (2001) DiGIR – distributed generic information retrieval. Available at http://digir.sourceforge.net/events/20011106/DiGIR.ppt.

Bouchet, P. (2006) The magnitude of marine biodiversity. In: *The Exploration of Marine Biodiversity: Scientific and Technological Challenges* (ed. C. Duarte), chapter 2. Spain: Fundacion BBVA.

Boyer, T.P., Antonov, J.I., Garcia, H.E., *et al.* (2006) *World Ocean Database 2005* (ed. S. Levitus). NOAA Atlas NEDIS 60. Washington, DC: US Government Printing Office. DVD, 190 pp.

Cavalcanti, M.J. & Gallo, V. (2008) Panbiogeographical analysis of distribution patterns in hagfishes (Craniata: Myxinidae). *Journal of Biogeography* **35**, 1258–1268.

Chapman, A.D. (2005) *Principles of Data Quality, version 1.0*. Report for the Global Biodiversity Information Facility, Copenhagen.

Conkright, M.E. & Levitus, S. (1996) Objective analysis of surface chlorophyll data in the northern hemisphere. In: *Proceedings of the International Workshop on Oceanographic Biological and Chemical Data Management*. NOAA Technical Report NESDIS 87, 33–43.

Costello, M.J., Emblow, C. & White, R. (eds) (2001) European Register of Marine Species. A check-list of the marine species in Europe and a bibliography of guides to their identification. *Patrimoines Naturels* **50**, 463 pp.

Decker, C. (2001) The Census of Marine Life: an update on activities. In: *Proceedings of the PICES.COML.IPRC Workshop on Impact of Climate Variability on Observation and Prediction of Ecosystem and Biodiversity Changes in the North Pacific* (eds. V. Alexander, A.S. Bychkov, P. Livingston & S. M. McKinnell,), pp. 5–9. PICES Scientific Report 18. Sidney, Canada: North Pacific Marine Science Organisation (PICES). V + 205 pp.

Dittert, N., Diepenbroek, M. & Grobe, H. (2001) Scientific data must be made available to all. *Nature* **412**, 393.

Fornwall, M. (2000) Planning for OBIS: examining relationships with existing national and international biodiversity information systems. *Oceanography* **13**(3), 31–38.

Fautin, D. (2000) Electronic Atlas of Sea anemones: an OBIS pilot project. *Oceanography* **13**, 66–69.

Froese, R., Lloris, D. & Opitz, S. (2003) The need to make scientific data publicly available – concerns and possible solutions. In: *Fish Biodiversity: Local Studies as Basis for Global Inferences* (eds. M.L.D. Palomares, B. Samb, T. Diouf, *et al.*), pp 267–271. Brussels. 281 pp.

Froese, R. & Pauly, D. (eds.) (2009) FishBase. World Wide Web electronic publication. www.fishbase.org, version 09/2009.

Greene, M. (2007) The demise of the lone author. *Nature* **450**, 1165.

Grassle, J.F. (2000) The Ocean Biogeographic Information System (OBIS): an on-line, worldwide atlas for accessing, modelling and mapping marine biological data in a multidimensional geographic context. *Oceanography* **13**(3), 5–7.

Grassle, J.F. (2005) Data management and communications plan for research and operational integrated ocean observing systems, 1. Interoperatable Data Discovery, Access and Archive, Part III. Appendices. Appendix 7, pp 285–292. Biological Data Considerations, Ocean.US, Clarendon Boulevard, Suite 1350, Arlington, VA 22201-3667, USA.

Grassle, J.F. & Stocks, K.I. (1999) A Global Ocean Biogeographic Information System (OBIS) for the Census of Marine Life. *Oceanography* **12**(3), 12–14.

Grassle, J.F. & Vanden Berghe, E. (2009) Census of Marine Life. In: *Ocean: An Illustrated Atlas* (eds. S.A. Earle. & L.K. Glover) Washington, DC: National Geographic. 352 pp.

Halpin, P.N., Read, A.J., Best, B.D., *et al.* (2009) OBIS-SEAMAP 2.0: developing a research data commons for the ecological studies of marine mammals, seabirds and seaturtles. *Oceanography* **22**(2), 104–115.

Halpin P.N., Read A.J., Best B.D., *et al.* (2006) OBIS-SEAMAP: developing a biogeographic research data commons for the ecological studies of marine mammals, seabirds, and sea turtles. *Marine Ecology Progress Series* **316**, 239–246.

Hurlbert, S.H. (1971) The nonconcept of species diversity: a critique and alternative parameters. *Ecology* **52**, 577–586.

Levitus, S. (1996) Interannual-to-decadal variability of the temperature–salinity structure of the world ocean. In *'Proceedings of the international workshop on oceanographic biological and chemical data management.* NOAA Technical Report NESDIS 87, 51–54.

Mora, C., Tittensor, D.P. & Myers, R.A. (2007) The completeness of taxonomic inventories for describing the global diversity and distribution of marine fishes. *Proceedings of the Royal Society B* 275, 149–155.

Paterson, G., Boxshall, G., Thomson, N. & Hussey, C. (2000) Where are all the data? *Oceanography* 13(3), 21–24.

Poloczanska, E., Hobday, A.J. & Richardson, A.J. (2008) Global database is needed to support adaptation science. *Nature* 453, 720.

Rees, H.L., Eggleton, J.D., Rachor, E. & Vanden Berghe, E. (2007) Structure and dynamics of the North Sea Benthos. ICES Cooperative Research Report 288. Copenhagen. 259 pp.

Rees, T. & Zhang, Y. (2007) Evolving concepts in the architecture and functionality of OBIS, the Ocean Biogeographic Information System. In: *Proceedings of Ocean Biodiversity Informatics: An International Conference on Marine Biodiversity Data Management Hamburg, Germany, 29 November – 1 December, 2004* (eds. E. Vanden Berghe, *et al.*), pp. 167–176. IOC Workshop Report, 202, VLIZ Special Publication 37.

Reid, P.C., Edwards, M., Hunt, H.G. & Warner, A.J. (1998) Phytoplankton change in the North Atlantic. *Nature* 391, 546.

Richardson, A.J. & Poloczanska, E. (2008) Under-resourced, under threat. *Science* 320, 1294.

Robertson, D.R. (2008) Global biogeographical data bases on marine fishes: caveat emptor. *Diversity and Distributions* 14, 891–892.

Rosa, R., Dierssen, H.M., Gonzalez, L. & Seibel, B.A. (2008a) Ecological biogeography of cephalopod molluscs in the Atlantic Ocean: historical and contemporary causes of coastal diversity patterns. *Global Ecology and Biogeography* 17, 600–610.

Rosa, R., Dierssen, H.M., Gonzalez, L. & Seibel, B.A. (2008b) Large-scale diversity patterns of cephalopods in the Atlantic open ocean and deep sea. *Ecology* 89, 3449–3461.

SCOR & IODE (2008) SCOR/IODE Workshop on Data Publishing, Oostende, Belgium, 17–19 June 2008. IOC Workshop Report No. 207. Paris: UNESCO. 23 pp.

Sekercioglu, C.H. (2008) Quantifying coauthor contributions. *Science* 322, 371.

Somerfield, P.J., Arvanitidis, C., Vanden Berghe, E., *et al.* (2009) MarBEF, databases and the legacy of John Gray. *Marine Ecology Progress Series* 382, 221–224.

Stocks, K (2009) SeamountsOnline: an online information system for seamount biology. Version 2009-1. Available at http://seamounts.sdsc.edu.

Stocks, K., Zhang, Y., Flanders, C. & Grassle, J.F. (2000) OBIS: Ocean Biogeographic Information System. The Institute of Marine and Coastal Science, Rutgers University. Available at http://marine/rutgers.edu/OBIS.

Stokstad, E. (2008) Proposed rule would limit fish catch but faces data gaps. *Science* 320, 1706–1707.

Vanden Berghe, E., Appeltans, W., Costello, M.J. & Pissierssens, P. (eds.) (2007a) *Proceedings of "Ocean Biodiversity Informatics": An International Conference on Marine Biodiversity Data Management Hamburg, Germany, 29 November – 1 December, 2004.* Paris, UNESCO/IOC, VLIZ, BSH, 2007. vi + 192 pp.

Vanden Berghe, E., Claus, C., Appeltans, W., *et al.* (2009) MacroBen integrated database on benthic invertebrates of European continental shelves: a tool for large-scale analysis across Europe. *Marine Ecology Progress Series* 382, 225–238.

Vanden Berghe, E., Rees, H.L. & Eggleton, J.D. (2007b) NSBP 2000 data management. In: *Structure and dynamics of the North Sea Benthos* (eds. H.L. Rees, J.D. Eggleton, E. Rachor, & E. Vanden Berghe), pp 7–20. Copenhagen: ICES Cooperative Research Report 288. 259 pp.

Vieglais, D., Wiley, E.O., Robins, C.R. & Peterson, A.T. (2000) Harnessing museum resources for the Census of Marine Life: the FISHNET project. *Oceanography* 13(3), 10–13.

Worm, B. & Myers, R.A. (2004) Managing fisheries in a changing climate. *Nature* 429, 15.

Yarincik, K. & O'Dor, R. (2005) The Census of Marine Life: goals, scope and strategy. *Science Marine* 69 (Suppl. 1), 201–208.

Zeller, D., Froese, R. & Pauly, D. (2005) On losing and recovering fisheries and marine science data. *Marine Policy* 29, 69–73.

Zhang, Y. & Grassle, J.F. (2003) A portal for the Ocean Biogeographic Information System. *Oceanologica Acta* 25, 193–197.

Index